Accession no.
36188332

KU-722-323

# Biochemistry

**Raymond S. Ochs**
St. John's University

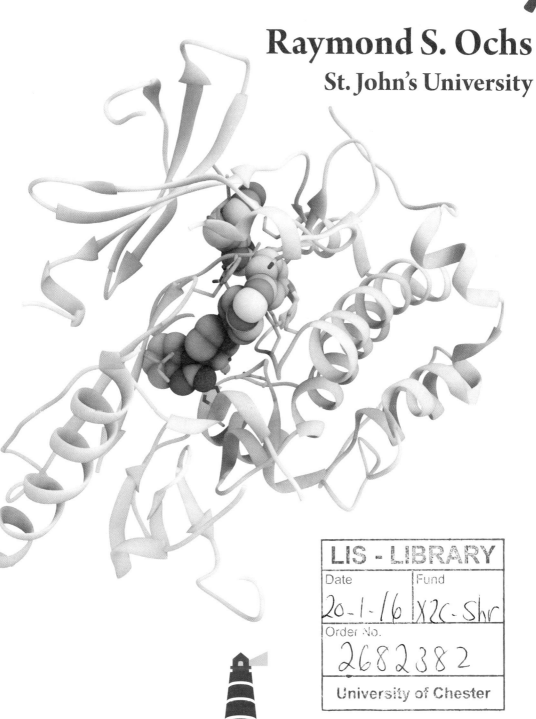

**LIS - LIBRARY**

| Date | Fund |
|------|------|
| 20-1-16 | X2c-Shr |

Order No.
2682382

University of Chester

JONES & BARTLETT
LEARNING

*World Headquarters*
Jones & Bartlett Learning
5 Wall Street
Burlington, MA 01803
978-443-5000
info@jblearning.com
www.jblearning.com

Jones & Bartlett Learning books and products are available through most bookstores and online booksellers. To contact Jones & Bartlett Learning directly, call 800-832-0034, fax 978-443-8000, or visit our website, www.jblearning.com.

Substantial discounts on bulk quantities of Jones & Bartlett Learning publications are available to corporations, professional associations, and other qualified organizations. For details and specific discount information, contact the special sales department at Jones & Bartlett Learning via the above contact information or send an email to specialsales@jblearning.com.

Copyright © 2014 by Jones & Bartlett Learning, LLC, an Ascend Learning Company

All rights reserved. No part of the material protected by this copyright may be reproduced or utilized in any form, electronic or mechanical, including photocopying, recording, or by any information storage and retrieval system, without written permission from the copyright owner.

*Biochemistry* is an independent publication and has not been authorized, sponsored, or otherwise approved by the owners of the trademarks or service marks referenced in this product.

**Production Credits**
Chief Executive Officer: Ty Field
President: James Homer
SVP, Editor-in-Chief: Michael Johnson
SVP, Chief Marketing Officer: Alison M. Pendergast
Publisher: Cathleen Sether
Senior Acquisitions Editor: Erin O'Connor
Editorial Assistant: Rachel Isaacs
Editorial Assistant: Michelle Bradbury
Production Manager: Louis C. Bruno, Jr.
Senior Marketing Manager: Andrea DeFronzo
V.P., Manufacturing and Inventory Control: Therese Connell
Manufacturing and Inventory Control Supervisor: Amy Bacus
Composition: Circle Graphics, Inc.
Cover Design: Scott Moden
Rights & Photo Research Associate: Lauren Miller
Cover and Title Page Image: © Ramon Andrade/Photo Researchers, Inc.
Printing and Binding: Courier Companies
Cover Printing: Courier Companies

**About the Cover:** The human enzyme akt1, shown in a computer ribbon drawing, acts on a peptide substrate, shown as a space-filling representation. Akt1 is a protein kinase; it incorporates phosphate groups into protein substrates. The enzyme is part of the insulin signaling pathway and is involved in other signaling pathways such as those of cellular growth factors.

To order this product, use ISBN: 978-1-4496-6137-3

**Library of Congress Cataloging-in-Publication Data**
Ochs, Raymond S.
  Biochemistry / Ray Ochs. — 1st ed.
    p. ; cm.
  Includes bibliographical references and index.
  ISBN 978-0-7637-5736-6 (alk. paper)
  I. Title.
  [DNLM: 1. Biochemistry. 2. Biochemical Phenomena. QU 4]
  572—dc23
                                    2012008575

6048

Printed in the United States of America
16 15 14 13   10 9 8 7 6 5 4 3 2

*To my wife, Jessica, for her immeasurable support.*

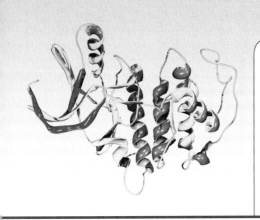

# Brief Contents

Image © Dr. Mark J. Winter/Photo Researchers, Inc.

# Contents

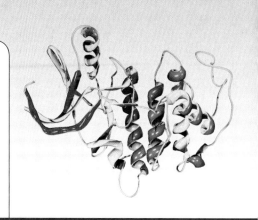

Image © Dr. Mark J. Winter/Photo Researchers, Inc.

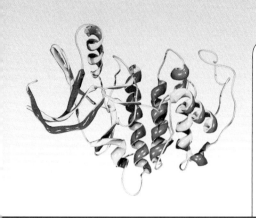

# Preface

## Considering Metabolism

Metabolic diseases such as diabetes and others associated with the current obesity crisis have thrust metabolism into the forefront of popular thinking. In many ways, metabolism is the central science of biochemistry. This is the view I have adopted as the core concept of this textbook.

Having a research background in metabolism, a long-standing interest in the fundamental topics of biochemistry—including kinetics and thermodynamics—and having taught a one-semester biochemistry course for over 25 years, I have long wanted to write a book that reflects the whole of the subject with the unifying theme of metabolism.

## Brevity

A key consideration when writing this text was to keep the book short enough and approachable enough that a student can read it in one semester. The alternative approach—having a book far too long for continuous reading and having the instructor suggest which portions to omit—is already well represented. In my experience, for students who are not biochemistry majors, the fragmentation resulting from parsing longer texts leads to less reading and, therefore, less understanding. The areas of special interest to the instructor can be readily augmented with primary sources, while the textbook provides continuity and context.

I have emphasized recurring ideas such as commonalities in chemical reaction mechanisms and pathway construction as much as possible. The decision of whether the book is for teaching or reference is decidedly in favor of teaching; for example, only a few protein domains are presented. The wealth of information available on the Web (notably ncbi.nlm.nih.gov) can substitute for a more extensive collection. The present text will provide students with an introduction to the foundations of biochemistry.

## Organization

The presentation of biochemistry topics is mostly a classical one, with a slight divergence: the introduction to lipids directly follows the chemistry of water. This is in keeping with presenting molecular classes in the order of increasing chemical complexity: water, lipids, carbohydrates, nitrogen compounds. It also has the virtue of contrasting water solubility with water insolubility in the case of lipids.

Image © Dr. Mark J. Winter/Photo Researchers, Inc.

When using this book for medically oriented courses, the chapter on photosynthesis may be omitted. In that case, however, the photosynthetic "dark reactions" should at least be referenced, as they are similar to those in the pentose cycle; this analogy is made explicit in the chapter on carbohydrate pathways.

## Key Features

- **Dual diagrams of enzymatic reactions.** A unique method of presenting electron flow for reaction mechanisms was devised for this book. For each mechanism, the substrate, intermediates, and product are presented on the top line. Below a separator (dotted line), the molecules are redrawn with their electron flows, using the traditional curved arrows. This separation allows the student to visualize the result of the electron flow, which is commonly obscured by the need to show the electron arrows for the next transformation.

- **Word Origins feature box.** Included in most chapters is a box that provides a short history of certain words that are rich in meaning, without which it is often more difficult to understand the underlying concept. Providing an explanation of their origins helps students become familiar with these important terms and achieve a better understanding of what is being described.

- **Thermodynamics treatment.** The development of thermodynamics for metabolic purposes leads to the distinction between two classes of enzymes: near-equilibrium and metabolically irreversible. This allows a simplification, as near-equilibrium reactions are not sites of cellular regulation. The roles of standard, actual, and near-equilibrium states for free energy are distinct and consistently presented to aid student understanding.

- **Chemical mechanisms.** A study of biochemistry should impart a viewpoint enriched by understanding the underlying chemistry of events in living systems. This extended view of biology is best achieved by understanding how enzymatic reactions function. All of the background chemistry needed for this text should be covered by prerequisite courses of chemistry. Some further information is presented in the appendix; a review of organic chemistry reaction mechanisms (most critically, nucleophilic reactions) may be necessary for those who are less comfortable with the material.

- **Enzyme kinetics treatment.** The text emphasizes direct plots of substrate concentration against initial velocity, introducing double-reciprocals only after a complete development of the subject. While in widespread use, the double-reciprocal form is difficult to visualize and leads to the false impression that memorizing patterns of lines for enzyme inhibition provides insight into how inhibitors work. Instead, direct plots, with an emphasis on the behavior of reaction velocity at different substrate concentrations, deliver the message clearly. Coupled with everyday descriptions of the different types of inhibition, these critical ideas are easily grasped. A further distinct notion is the use of the kinetic term $V_{max}/K_m$ rather than $K_m$ in developing kinetics. Too often, "$K_m$" becomes a focal point and is treated as an equilibrium constant rather than a steady-state constant. This common misuse of $K_m$ versus $V_{max}/K_m$ is part of the reason that many students have difficulty understanding enzyme kinetics, and it is a problem this text carefully avoids.

- **Minimalist molecular biology treatment.** The essence of molecular biology is included in the final two chapters. The emphasis is on providing an overview with a chemical perspective. For example, stacking interactions in DNA, the basis for forming the double helix, are explained in simple, chemical terms.

- **The pathway view.** The distinction between a *reaction* view and a *pathway* view is clearly emphasized to facilitate student comprehension. For example, the distinction between a bound cofactor like $FADH_2$ from a free cofactor like NADH becomes obvious: only NADH can transfer electrons between metabolic reactions.
- **Appendix.** The appendix of this text contains both entries for students needing a little extra help and more advanced material that, while perhaps not appropriate for the level of this particular text, is nonetheless important to the field of biochemistry as a whole. For remediation, basic ideas of mathematics and chemistry with which students traditionally have difficulty can be found in the appendix, as can certain extended pathways like amino acid and cholesterol routes. The advanced material includes the mechanism of aconitase and the role of fluoroacetate, which represent milestone achievements in biochemistry. These may be of interest to the more inquisitive student wishing to go beyond the fundamentals.

## Resources

### For Instructors

An *Instructor's Media CD*, compatible with Windows® and Macintosh® platforms, provides instructors with the following resources:

- The PowerPoint® ImageBank contains all of the illustrations, photographs, and tables (to which Jones & Bartlett Learning holds the copyright or has permission to reproduce electronically). These images are inserted into PowerPoint slides. Instructors can quickly and easily copy individual images into existing lecture slides.
- The PowerPoint Lecture Outline presentation package provides lecture notes and images for each chapter of *Biochemistry*. Instructors with the Microsoft® PowerPoint software can customize the outlines, art, and order of the presentation.

To receive a copy of the *Instructor's Media CD*, please contact your sales representative.

Also available for qualified instructors to download from the Jones & Bartlett Learning website, www.jblearning.com, are the text files of the *Testbank*.

### For Students

To further enhance the learning experience, Jones & Bartlett Learning offers the following ancillary materials:

The *Student Companion Website* (http://science.jbpub.com/biochemistry) provides content exclusively designed to accompany *Biochemistry*. The site hosts an array of study tools including chapter outlines, study quizzes, an interactive glossary, animated flashcards, crossword puzzles, and web links for further exploration of the topics discussed in this book.

## Acknowledgments

This book has come a long way from my original, draft manuscript. It has gone through several rounds of reviews and edits, and it benefitted from the feedback of many individuals to whom I owe great thanks.

Sergio Abreu, Fordham University
Lois Bartsch, Graceland University
Sajid Bashir, Texas A&M University—Kingsville
Mrinal Bhattacharjee, Long Island University
Debra Boyd-Kimball, University of Mount Union
Barbara Bowman, University of California—Berkeley
Jeanne Buccigross, College of Mount St. Joseph
Mickael Cariveau, Mount Olive College
James Cheetham, Carleton University
Zhe-Sheng Chen, St. John's University
David Eldridge, Baylor University
Susan Evans, Ohio University
John Fain, University of Tennessee Health Science Center
Sue Ford, St. John's University
Matthew Gage, Northern Arizona University
Eric Gauthier, Laurentian University
Neil Haave, University of Alberta, Augustana Campus
David Hilmey, St. Bonaventure University
Blaine Legaree, Keyano College
Lisa Lindert, California State University—Sacramento
Meagan Mann, Austin Peay State University
Nick Menhart, Illinois Institute of Technology
Abdel Omri, Laurentian University
Gordon Rule, Carnegie Mellon University
Mary Railing, Wheeling Jesuit University
Gerald Reeck, Kansas State University
Abbey Rosen, Marian College of Fond du Lac
Frank Schmidt, University of Missouri
Michael Sehorn, Clemson University
Kavita Shah, Purdue University
Andrew Shiemke, West Virginia University
Amruthesh Shivachar, Texas Southern University
Todd Silverstein, Willamette University
Madhavan Soundararajan, University of Nebraska—Lincoln
Salvatore Sparace, Clemson University
Vicky Valancius-Mangel, Governors State University
David Watt, University of Kentucky
Wu Xu, University of Louisiana—Lafayette

I would also like to thank the entire team at Jones & Bartlett Learning: Shoshanna Goldberg, former editor Molly Steinbach, Erin O'Connor, Cathleen Sether, Rachel Isaacs, Lauren Miller, Scott Moden, and Louis Bruno. Without all of your support and guidance, this book would never have come to fruition.

Ray Ochs
St. John's University
ochsr@stjohns.edu

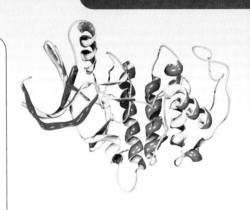

# Foundations

**CHAPTER OUTLINE**

Image © Dr. Mark J. Winter/Photo Researchers, Inc.

**B**iochemistry is the study of the chemical nature of biology. Biochemists use chemical ideas and tools to understand living systems. We begin with a review of the chemical concepts of equilibrium, kinetics, steady state, and energy. Additional chemical principles are discussed in the appendix. Some foundational ideas of biology complete the chapter, namely, the cell theory, evolution, and species hierarchies.

## 1.1 Origins of Biochemistry

Compared to its component sciences, biochemistry is a young discipline. The word *biochimie* was introduced by the German scientist Hoppe-Seyler in 1877. He also edited the first biochemistry journal, *Biological Chemistry*, still in existence today. It was during this period, the later part of the 19th century, in which two erroneous ideas— **spontaneous generation** and **vitalism**—were dispelled, clearing the way for a discipline that required new thinking.

According to the hypothesis of spontaneous generation, living organisms arise from nothingness, such as bacteria appearing in a nutrient-rich broth. In the 1860s, however, Louis Pasteur demonstrated that bacteria exist in the air; no bacteria appear in the broth if the container is isolated from the atmosphere. This led to the **cell theory**, which states that cells are the fundamental unit of living systems, arising from other cells. Despite this advance, Pasteur was himself a vitalist; that is, he believed that living systems do not obey the same chemical principles as inert materials.

Two challenges to vitalism bracketed the work of Pasteur. In 1828, Wohler discovered that urea (formed in living systems) could be synthesized in the laboratory from ammonia and bicarbonate. This *in vitro* synthesis of an organic compound therefore did not need the "aid of a kidney" as Wohler put it (**BOX 1.1**). In 1897, the German chemists (and brothers) Eduard and Hans Buchner showed that fermentation could exist in an extract from ruptured cells, thus dispelling the notion that cellular organization is required for processes that occur in living systems.

In the 20th century, biochemistry was dominated first by organic chemistry, as metabolic pathways were discovered, then by enzymology, then bioenergetics, and later by molecular biology as the role of DNA and information molecules emerged. As biochemistry plays an increasingly significant role in both the physical and chemical sciences today, all of these disciplines overlap to some extent. What remains, however, is the distinctive viewpoint of perceiving biology from a chemical perspective. While both the chemical and the biological understanding of many areas of study have greatly expanded in recent times, principles of biochemistry have emerged that are unique to this discipline and have been invaluable in elucidating a large number of biological phenomena. These underpinnings are presented in subsequent chapters. Presently, though, we consider some fundamental concepts of chemistry and biology.

## BOX 1.1. WORD ORIGINS

### Organic

Among its rich meanings, the word *organic* implies a sense of the whole, retained in the word *organism* or *organ*. Historically, however, it was used by vitalists to identify substances that could be produced only by a living organism. After vitalism was disproved, *organic* was redefined, rather than cast aside. Today, *organic chemistry* simply identifies a branch of chemistry that specifically deals with compounds of carbon. Strictly speaking, *inorganic chemistry* refers to the study of all other types of molecules. Yet another meaning for *organic* has emerged that is closer to its historical roots. That is, it describes farming methods that do not use synthesized chemicals (e.g., fertilizers and pesticides for plants and hormones for animals). Thus, growing plants and raising animals in this way is said to produce *organic foods*. Whether this is a definite health benefit is debatable; for example, the absence of pesticides can lead to a greater bacterial content in food. Moreover, this use of *organic* is less strict and may vary from one producer to another. Despite the three distinct definitions of organic, we are concerned in this book only with the middle one, the chemistry of carbon compounds.

# 1.2 Some Chemical Ideas

In order to determine whether you need a basic review, or can instead skip to the next section, take the following self-test:

- Without specifying a value, what is the meaning of Avogadro's number?
- Distinguish between atoms, electrons, molecules, and moles.
- When is it appropriate to use mol units as opposed to grams?
- Why is the equilibrium constant for a reaction equal to the products multiplied together, divided by the substrates multiplied together?
- How are equilibrium and kinetics related?

Mastering the ideas of the mole, Avogadro's number, atoms, and molecules is the first step in understanding more complex information in biochemistry. We consider here the notions of kinetics and thermodynamics with the assumption that the more elementary ideas are well in hand.

## Reactions and Their Kinetic Description

If substances A and B react to form substances C and D, then the generic reaction for this transformation can be written as follows:

$$A + B \rightleftarrows C + D \tag{1.1}$$

A and B are called *substrates*, whereas C and D are called *products*. Each is a molecule, but each can also be called a compound or a metabolite. In order to visualize what is happening, let us relax our molecular thinking and represent the molecules schematically as shown below.

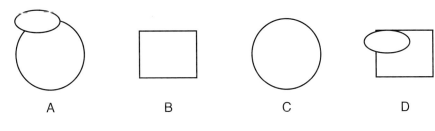

Based on this representation, the reaction involves removing a piece of molecule A and placing it onto molecule B, thus creating molecules C and D. This is a "mechanistic" view. To more fully characterize the reaction, we need to know the rates in both the forward and reverse directions. Consider first the *forward* direction, which proceeds from left to right as the reaction is written:

$$A + B \rightarrow C + D \tag{1.2}$$

Suppose there are three A and four B molecules, as shown in FIGURE 1.1. Each A can combine with any of 4 B's, so there are 12 possible collisions (3 × 4) between the three A molecules and the four B molecules.

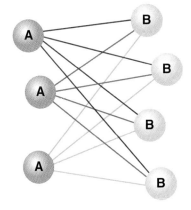

FIGURE 1.1 **Collision Theory of Reaction Rates.** Three A molecules react with four B molecules and each possible interaction of an A with a B is indicated. There are a total of 3 × 4 = 12 possible collisions. In general, the rate is proportional to the number of A molecules times the number of B molecules in a fixed volume; that is, rate is proportional to [A] × [B].

According to **collision theory**, the rate of a reaction is proportional to the number of collisions. In this example, the rate is proportional to $3 \times 4$. In general, the rate is proportional to $[A] \times [B]$, where $[A]$ and $[B]$ are the concentrations of A and B, respectively.

The rate under consideration, the forward rate, can be designated as $rate_f$. This rate is equal to a constant multiplied by the product of $[A]$ and $[B]$:

$$rate_f = k_f[A][B], \tag{1.3}$$

where $k_f$ is the constant of proportionality between the rate and the concentrations and is called the **rate constant**.

For the reverse reaction, $C + D \rightarrow A + B$, we can follow the same derivation and arrive at a similar expression:

$$rate_r = k_r[C][D] \tag{1.4}$$

where $rate_r$ represents the reverse reaction rate. The expressions for reaction rate shown in Equations 1.3 and 1.4 can be generalized to any reaction. The rate constant, while independent of changes in concentration, can vary with conditions such as temperature or ionic strength. Experimentally, reaction rates are measured by determining the decrease in substrate concentration or the increase in product concentration with time. Equations 1.3 and 1.4 summarize the results of those experimental studies. Thus, the reaction rate has units of concentration per time.

A rate, by definition, has only one direction, yet chemical reactions are commonly written with double arrows between substrates and products. This notation, which indicates that both forward and reverse reactions occur, is commonly used when a reaction attains a position of *equilibrium*.

## Equilibrium

Perhaps the most important fundamental idea in chemistry is that of **equilibrium**. A state of equilibrium exists when the forward and reverse rates of a reaction are equal. Once a reaction achieves equilibrium, no further observable change occurs in the substrate or product concentrations, unless there is a change in external conditions. Reactions not yet at equilibrium have a driving force toward the equilibrium state, just as a rolling ball eventually comes to rest as the forces acting on it balance. To become familiar with the state of equilibrium, we examine the origins of the equilibrium expression and the equilibrium constant.

According to the definition of equilibrium, $rate_f = rate_r$, so we can equate Equations 1.3 and 1.4:

$$k_f[A][B] = k_r[C][D] \tag{1.5}$$

Sometimes to emphasize the fact that we are describing an equilibrium, an "eq" subscript is added to the concentration terms (i.e., $[A]_{eq}$,

$[B]_{eq}$, etc.). In Equation 1.5, though, we will just assume the concentrations are those at equilibrium under our conditions. Rearranging,

$$k_f / k_r = [C][D]/[A][B] = K_{eq}, \tag{1.6}$$

where $K_{eq}$ is the equilibrium constant. Thus, the rate constants ($k_f$ and $k_r$) and the equilibrium constant ($K_{eq}$) are related.

Equation 1.6 relates kinetics (a rate process intrinsically tied to a time element) to equilibrium (a process independent of a time element). A single reaction may exist in equilibrium, but so, too, may multiple reactions:

$$A \rightleftarrows B \rightleftarrows C \rightleftarrows D \rightleftarrows E \tag{1.7}$$

Under biological conditions, though, multiply connected reactions are better represented using the *steady state*.

## The Steady State

One problem with applying the equilibrium model to living cells is that living cells are never actually at equilibrium. Thus, a more elaborate system must be used to model metabolism: the **steady state**. The steady state applies when the intermediates of a process are constant with time, yet the overall process is changing with time. To understand this concept, consider just two components, a substrate (S) and a product (P), which are the absolute minimum requirements for an equilibrium:

$$S \rightleftarrows P \tag{1.8}$$

The equilibrium constant for this reaction is:

$$K_{eq} = [P]/[S] \tag{1.9}$$

To describe a steady state, we need a minimum of three components: a substrate (S), an intermediate (I), and a product (P):

$$S \rightarrow I \rightarrow P \tag{1.10}$$

Equation 1.10 is really two reactions, $S \rightarrow I$ and $I \rightarrow P$, much in the same way that an equilibrium reaction is really two reactions (i.e., a forward reaction and a reverse reaction). In the steady state, there is a net flow of material from S to P in Equation 1.10, so the system overall is *not* constant with time. Commonly, multiple intermediate species are present in a steady state rather than just one. In Equation 1.11, for example,

$$S \rightarrow I \rightarrow J \rightarrow K \rightarrow L \rightarrow P \tag{1.11}$$

all of the individual rates (e.g., $S \rightarrow I$, $I \rightarrow J$, etc.) are equal. Not only that, but the overall rate, $S \rightarrow P$, is the same. The result is that the intermediate concentrations (i.e., [I], [J], [K], and [L]) do not vary with time. Instead, the concentrations of the intermediates are constant because each one is formed at the same rate that it is removed.

The concentrations of S and P need not be constant, however. Suppose, for example, that S is **saturating** for the first reaction. Saturation with a substrate means that its concentration is very high, effectively unchanging during the reaction, thus fixing the rate of the reaction S → I. According to Equation 1.11, the product P cannot revert back to L. As a result, P can accumulate and its concentration has no effect on the rate of its formation from L.

Metabolic pathways that are linear in structure, as in Equation 1.11, are readily modeled by a steady state. In order to clearly picture the situation, consider an everyday example of a line of people forced to move single-file through a gate. Suppose, as diagrammed in FIGURE 1.2, that people are admitted at a rate that enables them to flow without backup through an entry gate to a building, and then allowed to exit through a gate that is also controlled to allow no buildup at the exit. The number of people entering is analogous to [S], whereas the number of those leaving is analogous to [P]. We can view the steady-state intermediates (analogous to I, J, K, and L in Equation 1.11) by imagining a window in the building that lets us look at the line. At any given time there are eight people visible through the window. Each time we look, there will be a different eight people in view, but the number is the same every time. This is the essence of the steady state.

While the steady state is distinct from an equilibrium, equating the two is a common scientific error. Often, an equilibrium is used for all situations where there is an element of balance, not recognizing that this is inappropriate when a series of chemical reactions produces a net flow. In that situation, the steady-state model is more appropriate.

Finally, it is important to stress that both equilibrium and steady state are models. Thus, while they often are appropriate to the situation, they are always approximations. In some instances they are wildly inaccurate ones. For example, prior to an enzymatic reaction or series of reactions reaching constant intermediate concentrations,

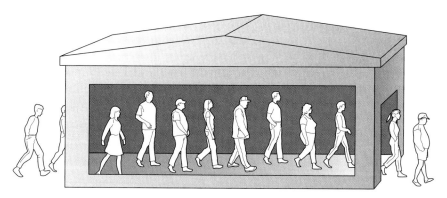

FIGURE 1.2 **The Steady-State Window.** People are moving through a building at a constant rate. As long as the rate of entry and exit is the same, the number of people visible through the window (the intermediate state) will be the same (8). At any instant in time, though, the eight people in this intermediate state will be different.

**CHAPTER 1** FOUNDATIONS

neither model applies. This is called a *pre-steady state condition*, which requires a different model entirely.

# 1.3 Energy

**Energy** is a concept that is at once simple and complex. It is simple because it is part of everyday parlance: a child knows about "having a lot of energy," the relative difficulty of walking uphill, and the existence of friction. It is complex because all forms of energy, such as electrical, gravitational, pressure–volume work, and heat flow, can be interconverted. The energy of a waterfall, a galloping horse, a chemical reaction, or within a single molecule can be understood through a study of thermodynamics, the science of energy changes.

Thermodynamics has two faces: one that is classical and one that is statistical. The *classical* approach uses a few postulates and definitions but makes no assumptions about the exact nature of the systems under investigation; even the existence of molecules is unnecessary. The *statistical* approach considers the behavior of large collections of molecules and allows more mechanistic conclusions, but it is more narrowly applicable. Both approaches lead to a consistent set of equations and together they provide a reasonable understanding of the nature of energy. To start, we need strict definitions of three entities: **internal energy**, **enthalpy**, and **entropy**.

Internal energy is the energy of the system under study, such as a chemical reaction or a pot of boiling water. It is the sum of the **work** and **heat** of the system under study. Work is an energy of motion, the product of force and distance.

Enthalpy (from the Greek *enthalpos*, meaning "putting heat in") is closely related to internal energy. Enthalpy is the heat released or absorbed by a reaction at constant pressure, a common condition in the laboratory as well as in the chemistry of living systems. As a result, values for reaction enthalpies rather than internal energies are listed in tables of thermodynamic data.

The statistical approach to thermodynamics reveals entropy as the number of ways that energy can be distributed as a result of a process. For example, if we suddenly apply brakes to a speeding car, the energy of motion will be redistributed to particles of tire rubber on the street, as well as into frictional heat. Entropy *increases* with an increase in energy dispersion.

When discussing energy in a biochemical context, we are invariably referring to a combination of changes in enthalpy and entropy, called **free energy**. The free energy change for a system can be used to determine whether a reaction can proceed in the direction written. Issues of thermodynamics, free energy, and the relationship between free energy and equilibrium are discussed elsewhere in this book. Here, we will just assume that chemical energy is equivalent to free energy.

We commonly speak of both molecules and portions of molecules as having *high energy*. Two **high-energy molecules** of particular importance in biochemistry are the reduced form of nicotinamide adenine

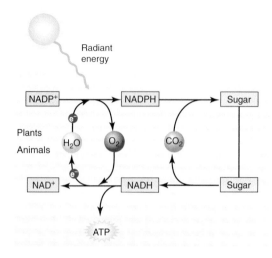

FIGURE 1.3 **The Global CO₂ Cycle.** Radiation from the Sun is used to energize electrons in the chloroplasts of plants, thus producing high-energy electrons in the form of NADPH. These electrons can be used to drive the photosynthesis of sugar from atmospheric CO₂ in plants. Subsequently, the plants are eaten by animals, and the sugars (and other molecules) are broken down to CO₂, which is released to the atmosphere, completing the cycle. Energy is trapped as electrons in NADH and is used to produce ATP for cell processes.

dinucleotide (NADH) and adenosine triphosphate (ATP). Both NADH and ATP are mobile cofactors that allow communication between hundreds of different reactions within the cell. NADH transfers electrons; consider it a donor of **high-energy electrons**. ATP, on the other hand, transfers a terminal phosphoryl group; think of it as a **high-energy phosphate** compound.

The connection between electron flow and ATP formation is a profound one in biochemistry. In the global energy cycle (**FIGURE 1.3**), radiation from the sun induces the formation of high-energy electrons by driving the formation of the reduced form of nicotinamide adenine dinucleotide phosphate (NADPH), a close analog of NADH. NADPH then donates electrons to convert atmospheric carbon dioxide ($CO_2$) to sugars and other molecules. Subsequently, animals consume those sugars, converting their electrons to NADH, which is used to form ATP. In the process, sugars and other molecules form $CO_2$, which is once again utilized by plants.

# 1.4 Cell Theory

The *cell*—the smallest unit of life—is a key organizing principle in biology. Single-celled organisms, such as bacteria, yeasts, and protozoans (e.g., the paramecium), comprise the largest number of species. Our major focus, however, is on multicellular organisms, primarily human. The features of a typical mammalian cell are illustrated in **FIGURE 1.4**, which shows internal organelles and their arrangement within

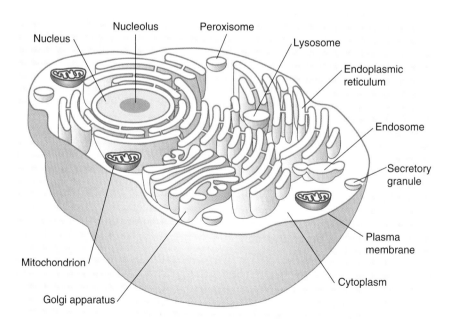

FIGURE 1.4 **The Cell.** A representative drawing of the cell shows the compartments that are created by membranes. The cell itself is separated from its exterior by a plasma membrane. Several interior membranes, such as those for the mitochondria, the lysosome, and the endoplasmic reticulum, define separate reaction spaces within the cell.

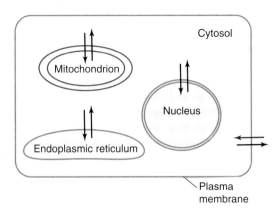

FIGURE 1.5 **The Cell as a Concept.** The cell can be represented as separate reaction spaces, set apart by semipermeable membranes. Specific exchanges across the relatively few membrane-delimited spaces are indicated by the arrows drawn across the membranes in both directions.

the cell. A much simplified functional representation is depicted in FIG-URE 1.5, where the properties relevant to biochemical analysis can be readily seen. Consider, for example, the outer *plasma membrane* that encloses the cell, serving as its boundary with the outside world. The plasma membrane, like all cell membranes, is **semipermeable**. Some molecules readily cross the membrane, such as $O_2$, $H_2O$, and $CO_2$, whereas many others can traverse the membrane only if there is a carrier protein that selectively allows them to cross. From a chemical viewpoint, the membranes create separate water spaces to isolate chemical reactions. Only those molecules that can communicate across these spaces—those which can diffuse, or those which have a specific transporter—can participate in the reactions. The major water space within the cell is called the **cytosol**. Other membrane-delimited organelles shown in Figure 1.5 that define separate water spaces for reaction are the **mitochondria**, the **endoplasmic reticulum**, and the **nuclear membrane**. In addition to defining reactive spaces, membranes are the sites of lipid biosynthesis.

## 1.5 The Species Hierarchy and Evolution

*Evolution* is a fundamental principle of biology. It is also commonly misunderstood, perhaps due to the popularization of the phrase "survival of the fittest." It is ironic that this phrase is so strongly embedded in the popular lexicon, because it does not exist in *On the Origin of the Species,* Darwin's landmark work on evolution.

Evolution was first suggested from the extremes of plant and animal species that exist on isolated islands, notably the Galápagos of present day Ecuador. The idea was that species arose from other species, and those that could adapt best to their environment were able to survive because they lived to reproduce. Eventually, adaptive characteristics emerged. It is now commonplace to apply evolutionary principles to biochemistry, mapping the formation of numerous enzymes and DNA sequences from early origins to divergent present forms.

In order to catalog the complexity of the immense number of known species, currently believed to number in the tens of millions, scientists have traditionally turned to the classification hierarchy. The classical organization is kingdom, phylum, class, order, family, genus, and species. While there are currently considered to be five kingdoms, we will focus on just three: animals, plants, and prokaryotes. Our emphasis is on mammalian species, although plant, bacterial, and occasionally yeast examples will be included. Most discoveries in biochemistry have come from only a very small sampling of the biological universe, largely for practical purposes, but there are many indications that a unity of pathways and mechanisms exists between different organisms. Some differences between organisms will be considered to get a sense of how biological variation is expressed.

## 1.6 Biological Systems

There is another hierarchy that identifies the viewpoint of the investigator toward biological systems. FIGURE 1.6 shows a ranking with organisms at the top level and subatomic particles at the bottom. There are even higher levels of organization than organisms, such as populations, which have medical as well as scientific importance (as in studies of the transmission of infections between organisms). However, levels beyond that of organism are usually of interest in other fields, such as the social sciences.

The rankings in Figure 1.6 also can be used to exemplify two extremes in the analysis of biochemistry. A view close to the top, taking in as wide a swath as possible, is called **holistic**. Those who favor the holistic view argue that it leads to answers that are physiologically relevant. A view close to the bottom, examining specific chemicals or molecular interactions, is called **reductionist**. Those who favor the reductionist view argue that it alone allows firm conclusions to be drawn because the number of variables is much smaller than in the holistic approach. Both are essential, and we will move between them in the text. We begin with a reductionist approach, examining the chemical properties of water, the molecule most essential for the existence of life.

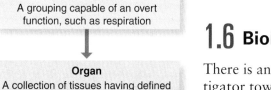

**Views**

Holistic

Reductionist

**Organism**
Free living unit capable of reproduction

**Organ system**
A grouping capable of an overt function, such as respiration

**Organ**
A collection of tissues having defined structure and function, such as the liver

**Tissue**
A group of cells with a single function, such as connective tissue

**Cell**
The smallest unit of life; some cells are also organisms (unicellular)

**Subcellular**
Organelles, extracts, complexes

**Molecules**
Large (e.g., proteins), and small (e.g., most metabolites)

**Atoms**
Molecular building blocks; the roughly 100 elements

**Subatomic particles**
Electrons, neutrons, and protons

FIGURE 1.6 **Hierarchies of Biological Study.** Different levels of investigation are possible in the experimental study of biology. Physiologists typically study the higher levels, such as organisms, organ systems, and organs, whereas physicists typically study subatomic particles. Biochemists usually examine the intermediate levels. If the study is closer to the top of the hierarchy (i.e., if the approach is holistic), it is considered to be *physiologically relevant,* but there is greater uncertainty about the findings. If the study is closer to the bottom (i.e., if the approach is reductionist), it is considered to be more exact, but less relevant. Both approaches, while in conflict, require each other.

# Key Terms

cell theory

collision theory

cytosol

endoplasmic reticulum

energy

enthalpy

entropy

equilibrium

free energy

heat

high-energy electrons

high-energy molecules

high-energy phosphate

holistic

internal energy

mitochondria

nuclear membrane

rate constant

reductionist

saturation

semipermeable

species hierarchies

spontaneous generation

steady state

vitalism

vitalist

work

# References

1. Ord, M. G.; Stocken, L. A. *Foundations of Modern Biochemistry.* In *Early Adventures in Biochemistry,* M. G. Ord, L. A. Stocken, Eds.; JAI Press: Greenwich, CT, 1995; Vol. 1.

   Part of a multivolume history of biochemistry; the information in the present chapter is found mostly in the first two chapters of *Foundations of Modern Biochemistry.*

2. Morowitz, H. J. *Entropy for Biologists. An Introduction to Thermodynamics;* Academic Press: New York, 1970.

   A gentle introduction to thermodynamics with fewer equations than the usual introduction. Biological examples are emphasized.

3. Margulis, L.; Schwartz, K. V. *Five Kingdoms. An Illustrated Guide to the Phyla of Life on Earth,* 2nd ed.; W. H. Freeman and Co.: New York, 1988.

   In addition to a catalog displaying drawings and photographs of examples of the major life forms, there is an introduction to the science of taxonomy, the hierarchies of life.

4. *Cells,* 2nd ed.; Cassimeris, L., Lingappa, V. R., Plopper, G., Eds.; Jones and Bartlett Publishers: Boston, 2011.

   A recent cell biology textbook that provides broad introductions to life at the cellular level.

5. Piontkivska, H.; Hughes, A. L. Evolution of vertebrate voltage-gated ion channel alpha chains by sequential gene duplication. *J. Mol. Evol.* 56:277–285, 2003.

   A study of the evolutionary relationship of a specific calcium ion channel.

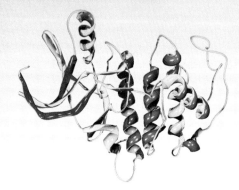

# 2

# Water

Image © Dr. Mark J. Winter/Photo Researchers, Inc.

Humans are comprised mostly of water. Our daily weight fluctuates largely because of a gain or loss of water. At a more philosophical level, most feel an almost mystical connection to water. This is why Ishmael's claim in *Moby-Dick* that we all yearn for the sea still stirs a sentiment in readers 150 years after it was written.

In most cells, the number of water molecules relative to others is even more impressive than its preponderance by weight. This is a consequence of the low molecular mass of water (18 Daltons). In fact, for every single protein molecule in cells, there are 75 lipid molecules, 100 $Na^+$ ions, and 20,000 water molecules.

Although we generally refer to the molecule itself as water, each phase has its own name: *ice* for the solid, *water* for the liquid, and *steam* or *vapor* for the gas. While commonplace, water is curiously distinct from other substances, especially in the liquid phase. One unique property of water is its ability to dissolve so many other substances. It falls short of being a "universal solvent," but the number and type of substances that dissolve in water is an important biological consideration. Water also can retain an enormous amount of heat compared to other liquids. Moreover, acid–base reactions in water are assisted by the ionization of the water molecule itself. Finally, water has unusual phase transitions, such as a very high boiling point, and unusual behavior at low temperatures, such as the fact that the solid is less dense than the liquid. All of these properties stem from the molecular structure of the water molecule, which we consider next.

## 2.1 Structure of Water

Water vapor, like other substances in the gas phase, can be thought of as molecules in isolation. Distinct ways of viewing phases of substances are illustrated in FIGURE 2.1. The gas and liquid are called **fluid phases**; liquid and solid alternatively can be grouped together as **condensed phases**. A study of each phase and the transitions between them provide insight into both the behavior of water and its interaction with other molecules.

### Gas Phase Water

The essential feature of the gas phase is that there is effectively no interaction with other molecules. Many properties of gases (regardless of their chemical identity) are essentially the same, as long as they are studied under conditions we would consider "average." One way of determining such conditions and ensuring that they are the same for all observers is to establish them by committee. This has been done by the International Union of Pure and Applied Chemists (IUPAC): a pressure 1 atm, and a temperature of 0°C is called standard temperature and pressure (STP).

The properties of greatest biochemical consequence are those that arise from the arrangement of electrons and nuclei within the water molecule. It is evident from the $H_2O$ formula that two atoms of

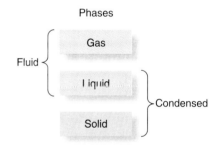

FIGURE 2.1 **Liquid water is a condensed fluid.** The three phases of matter can be alternatively characterized as *fluid* or *condensed* (incompressible). The liquid state of water (as with other substances) shares the properties of being both fluid and condensed.

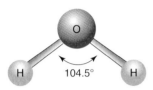

**FIGURE 2.2 Two-dimensional structure of water.** The three atoms of water can be drawn in a plane; the two bonded hydrogens make an obtuse angle, the value of which can only be explained by a consideration of its three-dimensional structure.

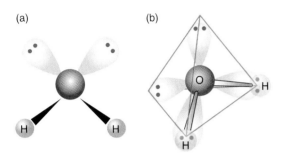

**FIGURE 2.3 Three-dimensional structure of gas-phase water.** The four orbitals of the water molecule are shown; water has a tetrahedral structure. (a) The space-filling diagram shows bonded hydrogen orbitals coming out of the page and nonbonded electron pairs behind the plane of the page. (b) The tetrahedral orientation depicts the oxygen in the center of the tetrahedron and the nonbonded orbitals and bonded hydrogen atoms at the vertices.

|  | Period | | | | | | | |
|---|---|---|---|---|---|---|---|---|
|  | I | II | III | IV | V | VI | VII | VIII |
| 1 | H |  |  |  |  |  |  |  |
| 2 | Li | Be | B | C | N | O | F |  |
| 3 | Na | Mg | Al | Si | P | S | Cl |  |
| 4 | K | Ca |  |  |  |  |  |  |
|  |  |  |  |  |  |  |  |  |

**FIGURE 2.4 An abbreviated periodic table.** The periods IA through VIIIA are shown without the traditional intervening transition elements. Those elements towards the right and towards the top have the greatest electronegativity. Only the elements N, O, and F (shaded) participate in hydrogen bonding. Elements of biological importance are highlighted in blue.

hydrogen are attached to a single atom of oxygen, and the molecule is neutral overall. When the atoms are drawn on paper as they are known to be arranged geometrically, the two-dimensional structure in FIGURE 2.2 is the result. The H–O–H bond angle is about 104°. The underlying reason water adopts this structure is only apparent when we consider the three-dimensional representation in FIGURE 2.3. An oxygen atom contains eight electrons: two inner electrons (which are not involved in bonding) and six valence electrons. The number of valence electrons corresponds to the number of the column oxygen occupies in the periodic table (FIGURE 2.4).

In the water molecule, the six valence electrons on oxygen mix with the two from the hydrogen atoms, recombining into four molecular orbitals that are oriented roughly as a tetrahedron, as in Figure 2.3. This tetrahedral orientation results from the mutual repulsion of the electrons in these orbitals. Because they are all attached to a central oxygen atom, and thus in forced proximity, the orbitals assume a geometrical arrangement in which they can be as far apart from one another as possible. In water, the two nonbonding orbitals contain a "lone pair" of electrons, and occupy a somewhat greater volume than the orbitals with bonded electrons. As a result, the H–O–H angle is somewhat less (i.e., 104°) than it would be in a perfect tetrahedron (i.e., 109.5°), such as occurs in methane ($CH_4$).

## Partial Charges and Electronegativity

A covalent bond is by definition a sharing of electrons between the nuclei of two atoms, but the sharing is usually unequal. In fact, the sharing is only truly equal in covalent bonds between identical atoms, as in $H_2$, $N_2$, or $O_2$. In the case of the single bond between O and H in water, the electrons are located closer to the oxygen atom. This uneven sharing leads to *partial charges* that can be assigned to each atom involved in the bond: the O is *partially negative* and the H *partially positive*.

In general, atoms differ in their inherent ability to attract the electrons that are involved in bonding. The ability of an atomic nucleus to attract electrons in a bond to itself is called **electronegativity**. Values of electronegativity can be directly assigned to each atom, based upon measurements of bond energies involving different pairs of atoms. A normalized scale runs from 0 (no electronegativity) to 4 (greatest electronegativity). A clear trend in electronegativity values is present in the arrangement of the elements in the periodic table. Those elements in the top right (e.g., N, O, and F) are the most electronegative, whereas those in the bottom left (e.g., Fr, Ra, and Cs) are the least electronegative. These periodic trends indicate the different powers atoms have in attracting bonding electrons to them. Elements on the right side are more stable when they attract electrons, in part because these electrons complete their outer shells. Elements on the top of the table have fewer electrons to shield the charge of the nucleus. Thus, the nuclei of these atoms have greater power of attracting bonding electrons. The electronegativity values for the most important elements in biochemistry are as follows: oxygen (3.5), nitrogen (3.0), carbon (2.5), sulfur (2.5), hydrogen (2.1), and phosphorus (2.1); the element with the highest electronegativity overall is fluorine (4.0).

We are now in a position to examine the electronegativity values for the O–H bond in water. Because hydrogen has a value of 2.1, and oxygen 3.5, the electronegativity difference is 1.4. It has been estimated that this makes the bond about 39% ionic. This merely quantifies the previous point that the electrons in this bond are more strongly attracted to the O than to the H. Thus, while the bond is in one view covalent, it is in fact a **polar bond**, or a **partially ionic bond**.

A symbolism used to indicate the polar bond is a vector (i.e., a directed line segment) drawn with the origin at the H and the arrowhead at the O. The vector is further embellished with a little "plus sign" at the origin to denote the relatively positive (electron deficient) part of the distribution. Using the vector notation for the water molecule in three dimensions, it is possible to take a vector sum of all of the individual polar bonds (**FIGURE 2.5**). Because the molecule has a nonzero net sum, it has a net polarity. The two orbitals that are not bonded (i.e., those containing electrons only) do not contribute to the overall polarity of the molecule. Not all molecules with polar bonds have a *net* polarity; for example, $CO_2$ has polar bonds, but the vectors cancel out, as shown in **FIGURE 2.6**.

Because water has a net polarity due to the overall vector sum of its bonds, it is a **polar molecule**. We can assign partial charges to the atoms within the molecule. By convention, polarity is indicated by the Greek letter sigma, superscripted with a negative or positive sign to indicate the partial charges as in **FIGURE 2.7**. These features of the water molecule in the gas phase give rise to the unusual interactions that occur between water molecules in the condensed phases, and provide an explanation for the distinct properties of water.

## Condensed Phase Water: Hydrogen Bonding

Just as people change their behaviors in social settings, molecules in the condensed phases (i.e., the liquid and solid phases) display new properties. The molecules are in such close proximity in these states that they are also known as incompressible states, which means that external pressure cannot cause them to move perceptively closer together. Most of our interest will be in the liquid state. Still, some consideration of solid phase water as well as the transitions between states (phase changes) will enhance our understanding of water.

The manner in which neighboring water molecules are held together is illustrated for two water molecules in **FIGURE 2.8**. The hydrogen atom between the two O nuclei in the figure has one bond indicated as a solid line and the other as a dotted line. The solid line represents a covalent bond, whereas the dotted line represents a **hydrogen bond**, one of the most important interactions in biochemistry. Because hydrogen has only one electron to share in the covalent bond, the hydrogen bond can be thought of as an electrostatic attraction to one of the lone pairs of the other water molecule. See **BOX 2.1** for a discussion on covalent and ionic bonds.

Covalent bonds are much stronger than hydrogen bonds. The average bond energy for a covalent O–H bond is 467 kJ/mol whereas a hydrogen bond is on the order of 10 kJ/mol. Despite the relative

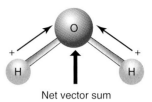

**FIGURE 2.5** **The net polarity of water.** The vectors resulting from electronegativity differences are indicated with plus annotations at their positive ends. The sum of these vectors is the overall polarity of the molecule.

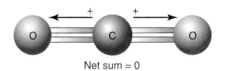

**FIGURE 2.6** **The lack of polarity in carbon dioxide.** Despite having electronegativity differences within the molecule, the sum of the vectors in carbon monoxide is zero, so it is a nonpolar molecule.

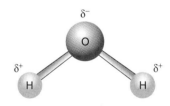

**FIGURE 2.7** **Partial charges in gas-phase water.** The differences in electronegativity are indicated by partial positive and partial negative charges within water. As a result, the bonding is only partly covalent.

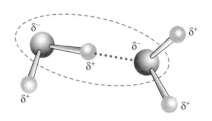

**FIGURE 2.8** **The hydrogen bond.** As a result of the charge interactions, two water molecules are shown interacting through a hydrogen bond. The bond consists of one hydrogen atom attracted to two oxygen nuclei at once. Colloquially, the dotted portion of the bond is often identified as the hydrogen bond, while in fact the entire structure contained in the loop in the figure is the hydrogen bond.

## BOX 2.1: Covalent and Ionic: Discreet or Continuous Distinction?

In your first encounters with chemistry, a bond is defined as being either ionic or covalent. The division is important in distinguishing chemical behavior. For example, an ionic solid has a much higher melting point than a solid of a covalently bonded compound, because the forces that must be overcome in ionic solids involve numerous species in a crystal. Yet, ionic compounds dissolve readily in water and separate completely into ions, as opposed to covalently bonded compounds, which may or may not dissolve readily in water, but retain their molecular identity when they do. In order to understand deeper problems in chemistry—beginning, say, with why some organic solvents dissolve in water and others don't—we need to explore whether the bonding qualities of these types of compounds are entirely discreet or whether there is a continuum that extends from strictly ionic to strictly covalent. The notion that the continuous situation *must* be the answer is not true of all physical principles. For example, the fundamental changes of quantum physics are discreet steps. However, bonding characteristics are a macroscopic quality that appears as a continuous process, involving *partial ionic* bonds. We could consider water, in fact, from the opposite extreme as it is normally viewed: namely, as a compound of $O^{2-}$ associated with two $H^+$ ions that are attached on one side, so that the molecule has a preponderance of negative charge on one side and a preponderance of positive charge on the other. Because of the incomplete dissociation of water in its liquid state, we would conclude that water is not actually an ionic compound but rather a *partially covalent* one.

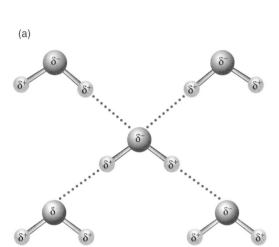

(a)

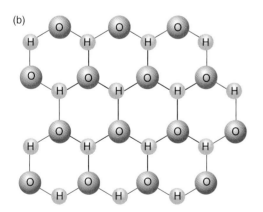

(b)

FIGURE 2.9 **Condensed phase water.** Two views of condensed phase water are shown. (a) The oxygen atoms are shown to bond to four positions throughout the structure. (b) A side view of condensed phase water showing a hexagonal structure that results from a plane of water molecules in a regular array. This structure accurately portrays ice. Liquid water approximates this structure, but continuously breaks and reforms bonds so that the structure that results is both irregular and changing.

weakness of individual hydrogen bonds, in the aggregate they become a dominant force and largely explain the peculiar properties of water. The three-dimensional view depicted in FIGURE 2.9 applies to both liquid and solid water. The latter is the regular crystal from which the figure is taken. The liquid state is sometimes considered to be a *flickering crystal*; the distinguishing fluid property, conforming to the shape of any vessel, is due to the greater motion of the molecules in liquid water and their ability to exchange positions in the liquid crystal-like structure. Another way to view the liquid state is to consider that the structure, while largely resembling the crystal, has an occasional defect that allows the water molecules to rearrange themselves.

Molecules in the solid state are usually more dense than they are in the liquid state. For example, liquid ethanol has a density of 790 kg/m³, whereas the solid form is 950 kg/m³. This is because molecules are usually at their closest approach in a crystalline solid, and an increase in temperature (which is needed to convert a solid to a liquid) tends to cause the molecules to move further apart (which decreases their density). Liquid water, however, is already highly ordered, and the small amount of motion at temperatures close to the freezing point causes water molecules to move closer to one another on the average than in ice. As ice melts to water, its density increases, reaching a maximum at 4°C. At higher temperatures, the increased motion does reduce the density of liquid water (FIGURE 2.10). Thus, water has a maximum density at 4°C, which explains why ice floats on water.

## 2.2 Properties of Water Follow from the Hydrogen Bonded Structure

One of the anomalous properties of water compared to other common liquids is its relatively high boiling point (Table 2.1).

The extensive hydrogen bonding in water produces a cohesiveness that can resist the tendency of molecules to escape into the gas phase as the temperature is increased. Thus, a higher temperature (and a correspondingly greater energy) is needed to boil water.

Another way of thinking about water is to view it as a hydride of oxygen. Other hydrides of closely related elements are HF (hydrogen fluoride), $H_2S$ (hydrogen sulfide), and $NH_3$ (ammonia). HF does not occur biologically; $H_2S$ is a product of bacterial respiration; and ammonia, an intermediate in mammals, is excreted as a final waste product of nitrogen metabolism in fish. The boiling points of these substances are shown in Table 2.2.

Two factors—electronegativity differences and molecular geometry—can be used to explain the striking differences between the boiling points of water and these other polar hydrides.

The electronegativity of F (4.0) is greater than the electronegativity of O (3.5). The fact that HF has a considerably lower boiling point than water is due to a difference in spatial arrangement. Three-dimensional views of these molecules are shown in FIGURE 2.11.

We have seen how water interacts with other water molecules, creating an extensive network of hydrogen bonds. It is not possible for either HF or $NH_3$ to form such extensive hydrogen bonded networks; they can form only limited intermolecular hydrogen bonds. Oxygen and sulfur are in the same column of the periodic table, so you might expect hydrogen sulfide to have properties that are very similar to those of water. The S–H bond is insufficiently polar, however, because the electronegativities of S (2.5) and H (2.1) are so similar. As a result, hydrogen sulfide (indeed, any SH-containing molecule) does *not* form hydrogen bonds. Accordingly, $H_2S$ has an extraordinarily low boiling point compared to that of $H_2O$. Indeed, in our everyday experience both $H_2S$ and $NH_3$ occur only as a gas.

Thus, the uniqueness of water stems from its combination of electronegativity differences and the geometry of its hydrogen-bonded interactions. This leads to a well-oriented and extensive lattice that

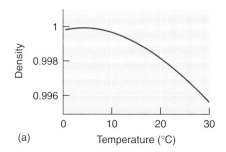

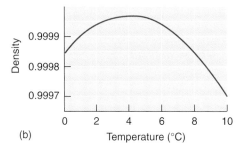

FIGURE 2.10 **How the density of water varies with temperature.** The density of water has a maximum at 4°C. The variation of density with temperature is shown over a range of (a) zero to 30°C and (b) zero to 10°C. The change is modest but significant. There are two opposing effects as energy (temperature) is increased: closer approach of more active molecules and increased separation as the temperature is increased further.

### TABLE 2.1 Boiling Points of Common Liquids

| Fluid | Boiling Point (°C) |
| --- | --- |
| Water | 100 |
| Benzene | 80 |
| Carbon tetrachloride | 77 |
| Chloroform | 61 |
| Ethanol | 79 |

### TABLE 2.2 Boiling Points of Hydrides

| Hydride | Boiling Point (°C) |
| --- | --- |
| $H_2O$ | 100 |
| HF | 20 |
| $NH_3$ | −33 |
| $H_2S$ | −61 |

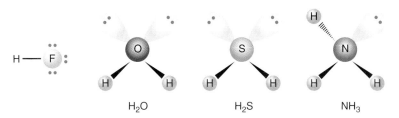

FIGURE 2.11 **Comparing water to similar structures.** Three-dimensional structures of the hydrides of the atoms F, O, and S indicate that $NH_3$ and $H_2S$ are very similar to water, yet they do not exhibit water's extensive hydrogen bonding due to geometric differences ($NH_3$) and lack of sufficient electronegativity differences ($H_2S$).

is not shared even by other very similar molecules. This combination of properties is the key to understanding how water interacts with other substances. At one extreme are molecules with virtually no interactions with water. This is the basis of the *hydrophobic effect*, a concept discussed in the next section. At the other extreme are molecules that strongly interact with water, in which case they tend to dissolve in water.

## 2.3 The Hydrophobic Effect

The aphorism that "oil and water don't mix" can be readily understood as a direct consequence of the condensed state of water. Consider the molecule decane:

$$CH_3-CH_2-CH_2-CH_2-CH_2-CH_2-CH_2-CH_2-CH_2-CH_3$$

It is composed entirely of C–C and C–H bonds, which are nonpolar (C–C; the electronegativity difference is $2.5 - 2.5 = 0$) or barely polar (C–H; the electronegativity difference is $2.5 - 2.1 = 0.4$). Even this small polarity is offset by the geometry of the structure, as the polarity vectors (pointing to the carbon chain) offset each other (FIGURE 2.12). Hence, decane and compounds with similar chemical compositions are *nonpolar* and *cannot* form hydrogen bonds with water. Another nonpolar substance is the oil in vinegar-and-oil dressing. It is common experience that the dressing settle into separate layers, even after they have been vigorously mixed.

The finding that nonpolar molecules separate from water is called the **hydrophobic effect** (BOX 2.2). This term should be taken literally: an effect rather than an accurate description of the underlying chemistry. Although *hydrophobic* means "water hating," alkanes such as decane do not hate water at all. In fact, decane has a stronger affinity for water molecules than it has for other decane molecules! The water–alkane attraction results from water inducing a dipole in the electrons of the alkane. The induced dipole attraction, however, is a minor factor. Focus should be trained instead on the water itself, because water prefers to interact with other water molecules far more than it does with lipids. This is sometimes described as water squeezing out nonpolar molecules so that the latter form their own phase. For a more thorough consideration of the forces underlying the hydrophobic effect, see BOX 2.3.

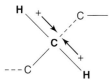

FIGURE 2.12 **The lack of polarity in methylene (–CH₂–) residues.** Using the method of vector sums, the small polarities due to C–H bonds cancel out, explaining the nonpolar nature of methylene groups in lipids.

## BOX 2.2: WORD ORIGINS

### Hydrophobic

The word *hydrophobic* was in use as a medical term for hundreds of years as a synonym for the disease *rabies.* A consequence of late-stage rabies infection is an inability to swallow, so those afflicted have a "fear of water," or hydrophobia. The use of "hydrophobic" for the exclusion of water from nonpolar substances is more recent, stemming from the early part of the 20th century. Despite the fact that the scientific usage is well established and that there is no debate about what it means, its strict translation does imply that water and lipid molecules display repulsive forces, which is not the case at all. The term only describes the phenomenon that water and nonpolar substances form separate phases *as if* they had some antipathy for each other. Ironically, as an archaic medical term for rabies, it is far removed from the underlying pathology. As an embedded term for water–nonpolar interactions, it is similarly a description of the phenomenon, far removed from the underlying chemistry.

## 2.4 Molecules Soluble in Water

The hydrogen bonding between water molecules in pure solutions of water is extensive. A large number of other molecules nonetheless can dissolve in water if a stronger interaction can result from the solute–water interaction than from the hydrogen bonds in pure water.

## BOX 2.3: A Thermodynamic Look at the Hydrophobic Effect

In our discussion of the hydrophobic effect, we considered the formation of two phases, colloquially an oil and water phase, and suggested that the principal forces were due to the interactions of water with itself, rather than with the oil. While this is true, the underlying reason for it is somewhat subtle and requires that we apply the ideas of thermodynamics.

In particular, the **free energy** of a process is determined by two factors: the **enthalpy** and the **entropy**. To apply these notions, consider the process of moving from one state to the next. Suppose our process moves individual lipid molecules (substrate) into coalesced lipid molecules (product). Studies of nonpolar molecules in water have shown that the hydrogen bonding of water molecules still occurs when lipid molecules are forced into water; a representative drawing of two such molecules in a water solution is shown in FIGURE B2.3a. The bonding energy (the enthalpy) of water arranged around lipid molecules is similar to the bonding energy of water molecules in the pure water solution itself. Thus, enthalpy is not the driving force for the hydrophobic effect. The "product," which represents the change that we know occurs, is modeled in FIGURE B2.3b, where the two lipid molecules have coalesced and a cage of hydrogen-bonded waters surrounds both of them.

The placement of water molecules around the lipids is believed to be far more *ordered* than in pure water; alternatively we can say that water in this cage has less freedom of movement. In thermodynamic language, we would say that we have increased the **entropy** of the system depicted in Figure B2.3b. Our picture of entropy requires a statistical view: an increase in entropy means molecules are more disperse in both space and in energy distribution.

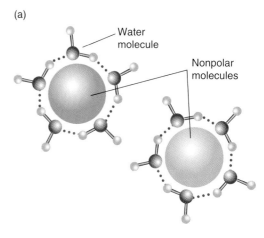
(a)

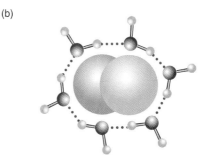
(b)

**FIGURE B2.3 Hydrophobic effect.**
(a) Two nonpolar molecules, introduced into water, become individually surrounded by hydrogen-bonded water molecules. The enthalpy of these water molecules is similar to that of the bulk water (not shown, but assumed as background in the graphic). However, the entropy is *less* as these waters are more ordered than bulk water. (b) The same molecules are now shown but associated together. The amount of water surrounding the two of them is much less than that in (a), so that the overall entropy of this situation is greater than in (a) and is thus favored.

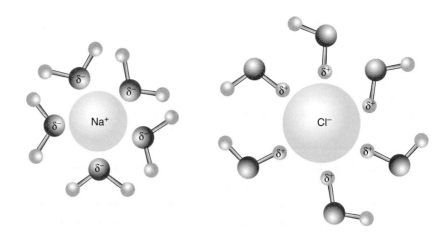

FIGURE 2.13 **Solubilizing NaCl.** Entry into solution involves water molecules giving up their own hydrogen-bonded interactions in favor of an ionic interaction between the partial charge of the oxygen atom of water and the sodium ion and between the partial charge of the hydrogen atom of water and the chloride ion. Thus, solubilized NaCl has a different and literally separate identity from the solid.

Imagine you are trying to dissolve table salt (NaCl) in water. In the solid form, NaCl is a regular array of interspersed $Na^+$ and $Cl^-$ ions. There is a considerable difference in the electronegativities of Na (0.9) and Cl (3.0), so electrons are *not* shared between these ions. Instead, $Na^+$ bears a full charge of +1, $Cl^-$ bears a full charge of −1, and the attraction is purely electrostatic. Once added to water, the $Na^+$ ion becomes surrounded by the negative ends of the water molecules and the $Cl^-$ ion becomes surrounded by the positive ends of the water molecules, as illustrated in FIGURE 2.13.

This can be represented as a reaction, in which solid NaCl is converted to isolated $Na^+$ and $Cl^-$ ions that are surrounded by water. The process of surrounding ions with water molecules is called *hydration* and the individual ions that are surrounded by water molecules are said to be **hydrated**. It is customary to write the hydration of NaCl as follows:

$$NaCl \rightarrow Na^+ + Cl^- \tag{2.1}$$

Water is omitted from this chemical equation because it is not strictly a reactant or product in the *formal* sense. Still, the reaction is impossible without it. What is happening is that the energy of hydration—the strength of the interaction of the ions for water—overcomes both the strength of the interaction of water for other water molecules, as well as the ionic bonding of the NaCl lattice, so the ions are brought into solution.

In a similar way, many other compounds can dissolve in water through the interactions of charges with the partial charges of water. Alternatively, uncharged polar compounds such as alcohols (ROH, where R = $CH_3-$ or $CH_3CH_2-$, for instance) can form hydrogen bonds to water with their −OH groups, again displacing the interactions between water molecules.

## 2.5 High Heat Retention: The Unusual Specific Heat of Liquid Water

Another feature of liquid water that results from hydrogen bonding is its exceptionally high specific heat. Specific heat is the amount of energy needed to increase the temperature of 1 g of a substance by 1°C. Effectively, the high specific heat of water means that water acts as a "temperature buffer"; that is, heat loss or gain is smaller in water compared to other substances. This stems from the strength of the hydrogen bonds. As the temperature is increased, the bonds bend but do not break (i.e., the water molecules adhere to each other very tightly). Heat goes into *vibrational* energy of the bonds, but the lattice structure remains; that is, it resists **translational** changes. Hence, the temperature rise is minimal in water. In the solid form (ice), the bending modes are unavailable, so ice does not share the high specific heat of water. The specific heat of ice is more similar to that of an alcohol such as ethanol ($CH_3CH_2OH$).

## 2.6 Ionization of Water

Pure water dissociates to a very small extent into ions. The reaction can be understood as a reaction between two water molecules, in which a proton migrates from one to the other as in FIGURE 2.14. In solution, the protonated water molecule —$H_3O^+$— exists, but there is very little of the $H^+$ form. However, it is customary to represent protonated water as just $H^+$, so the dissociation equation is written as:

$$H_2O \rightleftharpoons H^+ + OH^- \tag{2.2}$$

Although the **ionization of water** takes place to only a very slight extent, it is nonetheless an extremely important aspect of biochemistry. With pure water, the dissociation produces an equal amount of $H^+$ and $OH^-$. However, many substances dissolve in water and alter this balance, which leads us to the subject of acids, bases, and pH.

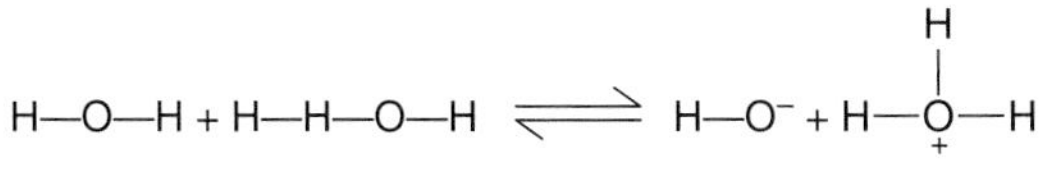

FIGURE 2.14 **Water equilibrium.** A different view of water dissociation in which ions are produced from pure water by removing a hydrogen ion from one water and attaching it to the other. The hydrogen ion moves without its electrons, leaving behind a negatively charged hydroxide ion and creating a positively charged hydronium ion with three hydrogens. This picture is abbreviated in the text by using the simpler $H^+$ rather than the more elaborate hydronium ion (hydrated proton; $H_3O^+$).

## 2.7 Some Definitions for the Study of Acids and Bases

While several definitions for acids and bases exist, the following simple set suffices for our study of water solutions:

- Acids are species that *increase* the hydrogen ion concentration when added to water.
- Bases are species that *decrease* the hydrogen ion concentration when added to water.

Based on these definitions, we can define neutral, acidic, and basic solutions as follows:

- In a neutral solution, $[H^+] = [OH^-]$.
- In an acidic solution, $[H^+] > [OH^-]$.
- In a basic solution, $[H^+] < [OH^-]$.

Species such as NaCl dissociate completely in water, but do not affect the concentration of $H^+$ or $OH^-$. Hence, NaCl dissolves in water to form a neutral solution.

Hydrochloric acid (HCl, the acid in your stomach that helps digest food) and acetic acid ($CH_3COOH$, the principal component of vinegar) dissolve in water according to the following chemical equations:

$$HCl \rightarrow H^+ + Cl^- \tag{2.3}$$

$$CH_3COOH \rightleftarrows CH_3COO^- + H^+ \tag{2.4}$$

Equation 2.2 for HCl is similar to the dissociation of NaCl in that both species dissociate completely into ions in water. Unlike NaCl, the dissociation of HCl leads to an increase in $[H^+]$, thus making HCl an acid. The reaction in Equation 2.2 is essentially irreversible, so HCl is said to be a **strong acid**.

The dissociation of $CH_3COOH$ in Equation 2.3 also meets our definition of an acid; that is, $[H^+]$ increases. Because $CH_3COOH$ is only partially dissociated, which is indicated by the double arrows in Equation 2.3, $CH_3COOH$ is said to be a **weak acid**.

Sodium hydroxide (NaOH) and ammonia ($NH_3$) are bases:

$$NaOH \rightarrow Na^+ + OH^- \tag{2.5}$$

$$NH_3 + H_2O \rightarrow NH_4^+ + OH^- \tag{2.6}$$

NaOH dissociates completely in water, making it a **strong base**, whereas $NH_3$ reacts with water to a limited extent, making it a **weak base**. The $OH^-$ produced in both Equations 2.4 and 2.5 will react with any $H^+$ present to form $H_2O$. Thus, by displacing the water dissociation equation (Equation 2.1) in the direction of $H_2O$ formation, bases such as NaOH and $NH_3$ lower the concentration of $H^+$.

# 2.8 The pH Scale

The pH scale is used to represent the very small numbers that occur for the values of [H+]. By definition,

$$pH = -\log[H^+] \tag{2.7}$$

The log term in Equation 2.6 is the common (base-10) log. Some typical pH values and the corresponding hydrogen ion concentrations, [H+] are shown in Table 2.3.

One useful feature of pH is obvious: it is a compact way of describing the concentration of hydrogen ions. However, there are some clear pitfalls. The pH is a double transformation of hydrogen ion concentration. First, it is a negative value, so that increasing pH means less hydrogen ion. Second, it is a logarithm, so a change of one unit of pH is equivalent to a 10-fold change in [H+]. Thus, a solution with a pH of 2 ([H+] = 0.01 mol/L) has 10 times more H+ than a solution with a pH of 3 ([H+] = 0.001 mol/L).

To become more familiar with pH, consider the range of the pH function shown in FIGURE 2.15. Scale A shows pH values typically encountered in the human body fluids. The values range from about 2 to about 8. Scale B is the readout of the pH meter. This is the most familiar representation, and many have come to regard this as the actual range of pH itself. It is important to understand that it is simply another representation of the pH concept, this time based on how we measure it. In fact, scale B is somewhat generous, because the electrode of this instrument is poorly selective above a pH of about 12. Scale C is the actual range of the pH function, from negative infinity to positive infinity. This corresponds to the true range of [H+], which varies from zero to infinity. It is important to recognize the artificiality of the middle scale as merely the markings on the pH meter. For example, the pH of a 2 M solution of HCl has a negative value (pH = −log 2 = −0.30).

| TABLE 2.3 pH and [H+] Values | | |
|---|---|---|
| pH | [H+] (mol/L) | [H+] in Scientific Notation (mol/L) |
| 5 | 0.00001 | $10^{-5}$ |
| 7 | 0.0000001 | $10^{-7}$ |
| 9 | 0.000000001 | $10^{-9}$ |

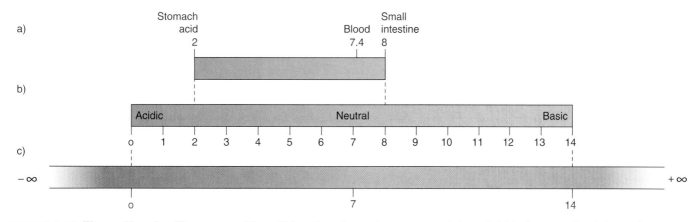

FIGURE 2.15 **Three pH scales**. The range of the pH function depends upon your interest. (a) In human physiology, the range is about 2 to 8, with some of the key values indicated. (b) Measurements of pH using the pH electrode and meter are typically marked from zero to 14, giving the popular conception that this is *the* range of pH. (c) The actual range of the pH function is from negative infinity to positive infinity. The value of 7 appears to be in the center only because pure water at STP has this value; otherwise, even that entry is arbitrary.

## 2.9 The Henderson–Hasselbalch Equation

Most biological acids and bases are weak rather than strong. As an example, the equilibrium of acetic acid:

$$CH_3COOH \rightleftarrows CH_3COO^- + H^+ \qquad (2.8)$$

We may represent $CH_3COOH$ as HAc and $CH_3COO^-$ as Ac$^-$. Equation 2.7 can be characterized by the following equilibrium constant expression:

$$K_a = [Ac^-][H^+]/[HAc] \qquad (2.9)$$

Taking the log of both sides of Equation 2.8 yields Equation 2.9:

$$\log K_a = \log([Ac^-][H^+]/HAc) \qquad (2.10)$$

Using properties of logarithms, Equation 2.9 becomes:

$$\log K_a = \log[H^+] + \log[Ac^-]/[HAc] \qquad (2.11)$$

In the previous section (see Eq. 2.6), pH was defined as,

$$pH = -\log[H^+] \qquad (2.12)$$

In an analogous way, we can define the negative log of the equilibrium constant as,

$$pK_a = -\log K_a \qquad (2.13)$$

which has advantages for representing the equilibria of weak acids in a similar way to those for hydrogen ion. For example, consider the two equations for the acetic acid dissociation:

$$K_a = 1.8 * 10^{-5}$$
$$pK_a = 4.7$$

Using the definitions for pH (Eq. 2.6) and $pK_a$ (Eq. 2.12), rearranging Eq. 2.10 produces the Henderson–Hasselbalch equation:

$$pH = pK_a + \log[Ac^-]/[HAc] \qquad (2.14)$$

As a sample problem that makes use of this equation, suppose that acetic acid ($pK_a = 4.7$) was present at a pH of 7.4 (the value for normal blood). Using the Henderson–Hasselbalch equation, we can immediately determine that most of the acetic acid would be in the salt form, that is, as Ac$^-$. Inserting the values for the pH and $pK_a$ into Equation 2.13, we get:

$$7.4 = 4.7 + \log([Ac^-]/[HAc]) \qquad (2.15)$$

Thus,

$$\log[Ac^-]/[HAc] = 2.7 \tag{2.16}$$

Raising each side of Equation 2.17 to the power of 10 (i.e., taking anti-logarithms), we get:

$$[Ac^-]/[HAc] = 10^{2.7} = 500 \tag{2.17}$$

It is evident that despite the less than intuitive nature of a negative logarithm transformation, as discussed in the prior section, using this equation simplifies calculations. We need only to perform arithmetic to subtract $pK_a$ from pH. From the last equation, even if we didn't know the exact value of $10^{2.7}$, we know $10^2$ is 100, and this must be much greater than 100 (but less than 1000), so we can immediately say the species in the anionic form, $Ac^-$, greatly outnumber those in the acid form, HAc. In fact, there are 500 ions of $Ac^-$ for every 1 molecule of HAc. To a very close approximation, $Ac^-$ prevails; thus the overall molecular charge is $-1$.

The Henderson–Hasselbalch equation applies to weak bases, too. A weak base, such as ammonia, can be written as a dissociation in the same form as acetic acid:

$$NH_4^+ \rightleftarrows NH_3 + H^+ \tag{2.18}$$

Written in this way, the protonated form—the ammonium ion, $NH_4^+$—is an acid. It is a weak acid, as it is only partially dissociated, and can be characterized by the following equilibrium constant expression:

$$K_a = [NH_3][H^+]/[NH_4^+] \tag{2.19}$$

The corresponding form of the Henderson–Hasselbalch equation:

$$pH = pK_a + \log[NH_3]/[NH_4^+] \tag{2.20}$$

A more general form of the Henderson–Hasselbalch equation, which applies to both weak acids and weak bases, is:

$$pH = pK_a + \log[X]/[XH] \tag{2.21}$$

where X is the base (i.e., least protonated) form of the species and XH is the acid (i.e., more protonated) form. Equation 2.20 is tremendously useful in the study of the numerous weak acids and bases that we encounter throughout our study of biochemistry. We can calculate the ratio of the two forms of the species by merely subtracting the values for the pH and $pK_a$.

## 2.10 Titration and Buffering

The Henderson-Hasselbalch equation provides specific pH values, given the ratio of the basic and acidic forms of a weak acid and the constant $pK_a$. A different view of pH as a function of the same ratio is provided

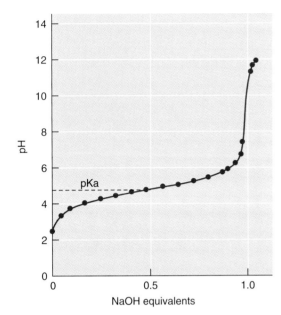

**FIGURE 2.16 Acetic acid titration curve.** Acetic acid is placed in solution and NaOH added until its amount in solution reaches the same number of equivalents, which is moles per charge. As both species have a unit charge, equivalents and moles are the same in this case. The middle, relatively flat region shows buffering, that is, a minimal response of pH to added acid or base. The center of the buffering region, at 0.5 equivalent volumes, corresponds to the p$K_a$ on the ordinate.

by a **titration curve**. FIGURE 2.16 shows a titration of acetic acid, starting with the fully protonated form ($CH_3COOH$) and ending with the completely anionic form (acetate ion, $CH_3COO^-$). In this titration, the strong base NaOH is added in small increments, and the pH is recorded continuously. Note the relatively flat **buffering region** centered in the midpoint of the titration, where pH changes very little as NaOH is added. The center of the buffer region—where pH changes the least—corresponds to the p$K_a$ value. All weak acids and weak bases are buffers, and thus effectively blunt the response to added acid or base. The reason for the behavior can be explained mathematically as follows. When pH = p$K_a$, the Henderson–Hasselbalch equation reduces to

$$0 = \log [Ac^-]/[HAc] \tag{2.22}$$

Taking the antilog of both sides yields

$$[Ac^-]/[HAc] = 1 \tag{2.23}$$

A small amount of added NaOH will only slightly raise this ratio, because substantial and nearly equal amounts of $Ac^-$ and $HAc$ are present. However, once the ratio becomes a factor of 10 larger or smaller, the same addition of NaOH will more profoundly change the ratio, and hence the pH.

Perhaps the most significant physiological buffer is the blood bicarbonate buffer that maintains the extracellular pH in mammals. This system has two equilibria:

$$CO_2 + H_2O \rightleftharpoons H_2CO_3 \rightleftharpoons H^+ + HCO_3^- \tag{2.24}$$

and is maintained by having a relatively high concentration of $CO_2$. The actual concentration of $CO_2$ can be adjusted by changes in respiration. The concentration of the bicarbonate ion, $HCO_3^-$, can be adjusted by excretion in the kidney. Note that our definition of acids as species that increase the hydrogen ion concentration in solution also applies to $CO_2$, which itself has no protons, because it causes an increase in [$H^+$] in solution when it dissolves in water, thus making it an acid.

## Summary

Water is the most abundant molecule in biological systems and a ubiquitous part of everyday life. However, compared to other liquids, water has unusual properties, including an exceptionally high boiling point indicative of intermolecular cohesiveness, and an ability to dissolve a wide variety of polar molecules.

The unusual properties of water are due to the hydrogen bond, which is a consequence of the electronegativity difference between H and O, and the geometry of the water molecule. An extensive hydrogen-bonded network provides the bonding strength that explains the high boiling point. Due to the fixing of molecules in a three-dimensional lattice, heat energy is stored in the bond vibrations of water rather than in translational energy, accounting for its high specific heat.

The ready dissociation of water into protons and hydroxide ions, and the ability to solvate ions accounts for the ability of water to support the dissociation of acids and bases. Weak acids, which are partially dissociated, are particularly important biologically. These can be profitably analyzed using the Henderson–Hasselbalch equation, a logarithmic transform of the equilibrium expression for acid dissociation. This allows quick calculations of pH or the ratio of dissociated to undissociated forms of weak acids, given the $pK_a$. Finally, a full titration of weak acids reveals a buffering region, in which changes in acidity have only small effects on solution pH.

## Key Terms

| | | |
|---|---|---|
| buffering region | hydrated | strong acid |
| condensed phases | hydrogen bond | strong base |
| electronegativity | hydrophobic effect | titration curve |
| enthalpy | ionization of water | translational |
| entropy | partially ionic bond | weak acid |
| fluid phases | polar bond | weak base |
| free energy | polar molecule | |

## Review Questions

1. Gas-phase water (i.e., water vapor) has properties similar to most other gases, yet condensed water (i.e., liquid and solid water) is unusual.
   a. Why are the properties the same in the gas phase?
   b. What are the unusual phase transitions of water?
   c. What accounts for the unusual properties of water in general?
2. In a solution of pure water, it is possible to calculate the concentration of water itself. What is its value?
3. Electronegativity values are assigned to individual elements, and yet the concept of electronegativity applies only to bonds. Explain this disparity.
4. Oxygen and sulfur are both nucleophiles, they have the same valence, and both form dihydrides. Yet, $H_2S$ does not form hydrogen bonds to other molecules of $H_2S$; indeed, sulfur does not participate in hydrogen bonding at all. Why is that?
5. Oxygen and nitrogen appear to be very different. Whereas oxygen appears in acidic compounds and tends to assume a negative charge, nitrogen appears in basic compounds and tends to assume a positive charge. Both elements, however, participate in hydrogen bonding. Explain these facts.
6. Why is it that a molecule like $CO_2$, which contains polar bonds but is not polar overall, cannot attract polar molecules?
7. Ammonia is a polar molecule by virtue of its lone pair. When protonated to ammonium, however, it is symmetrical and would seem, by vector addition, to have no polarity. Yet, the ammonium ion readily dissolves in water. Why?
8. The hydrophobic effect is due to attractive forces only and not repulsive ones. Rationalize this with the term *hydrophobic*.
9. A rule of thumb in chemistry is "like dissolves like." Explain how this applies differently to polar and nonpolar substances.
10. When enough NaCl is added to water, the salt no longer dissolves and solid NaCl falls out of solution. At this point, the solution is said to be saturated. Explain this phenomenon.

11. Water is a *temperature buffer,* which means that it resists changes in temperature more than other liquids. Explain this based on the structure of water.

12. Water ionizes to a very small extent, although this can be increased with increasing temperature. Provide a reasonable explanation.

13. Protons produced by the self-ionization of water have an unusually rapid mobility in solution; they move faster than can be accounted for by diffusion. Provide a possible mechanism to account for this *proton jumping.*

14. The definition of an acid as an entity that increases the proton concentration in water (known as the Arrhenius definition) has the advantage that we can call $CO_2$ an acid. How could we handle this situation if instead we used the Brönsted (proton donor) definition?

15. While useful, the pH scale can also be confusing because it is a double transform: both the negative of the number and the log of the number are taken. How does this affect your intuition about a change in proton concentration?

# References

1. Johnson, J. L. H.; Yalkowsky, S. H. A three-dimensional model for water. *J. Chem. Ed.* 79:1088–1091, 2002.

   A model-building exercise that leads to the construction of a three-dimensional water molecule that illustrates key properties of the condensed phase, such as the "flickering crystal" quality of the liquid and the density change in going from liquid to solid.

2. Pauling, L. *The Nature of the Chemical Bond and the Structure of Molecules and Crystals; an Introduction to Modern Structural Chemistry;* Cornell University Press: Ithaca, NY, 1960, p. 644.

   Pauling's classical treatise clearly describes the topics of electronegativity and hydrogen bonding, among many others.

3. Tanford, C. *The Hydrophobic Effect: Formation of Micelles and Biological Membranes;* John Wiley & Sons: New York, 1980.

   Another landmark publication is this monograph on Tanford's expertise, the hydrophobic effect. The text includes detailed descriptions of the net forces in the water–lipid interactions.

4. Tanford, C. How protein chemists learned about the hydrophobic factor. *Protein Sci.* 6:1358–1366, 1997.

   A historical introduction to the notion of hydrophobicity, which largely became a controversy among protein chemists, but later was used to clearly explain lipid–water interactions, which are much simpler.

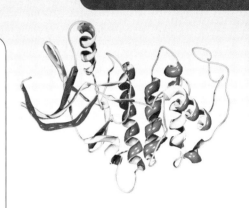

# 3

# Lipids

Image © Dr. Mark J. Winter/Photo Researchers, Inc.

A defining characteristic of **lipids** is their relative insolubility in water. Chemically, lipids are relatively simple—consisting mostly of unreactive hydrocarbons—making this topic an appropriate next step in our study. Lipids play a role in virtually every aspect of the cellular biology and in disease states of the body. The myriad of forms that serve this purpose arise from relatively modest variations in structure. Here, we consider the major types of lipid molecules and their principal roles.

## 3.1 Significance

The two major functions of lipids are *energy storage* and *membrane formation*. Beyond this, lipids can form micelles and serve as signal molecules.

Distinct structures correspond to each of the two major functions. For storing energy, the exclusion of lipid from the water phase is absolute. The major chemical species involved are triacylglycerols, which are classified as **nonpolar lipids**. For the formation of membranes, a portion of the molecule interacts with water. These are **polar lipids**, the most prominent class of which is the phospholipids. We will begin by considering **fatty acids**, the building blocks of lipids.

## 3.2 Fatty Acids

Fatty acids are molecules with two distinct parts: a long hydrocarbon segment, called a *tail,* and a smaller region that typically consists of a carboxyl group, called the *head* (FIGURE 3.1). The hydrocarbon tail consists mostly of chemically unreactive methylene groups with a variable number of double bonds. Table 3.1 lists some examples and nomenclature of the fatty acids.

Table 3.1 succinctly describes the fatty acids, each of which varies in the number of carbon atoms and the number of double bonds. Table 3.1 also introduces the delta notation, which designates the position of the first carbon involved in the double bond, with the carboxyl group taken as carbon number 1. Some of the common names for fatty acids reflect their biological origins, such as *butyric*

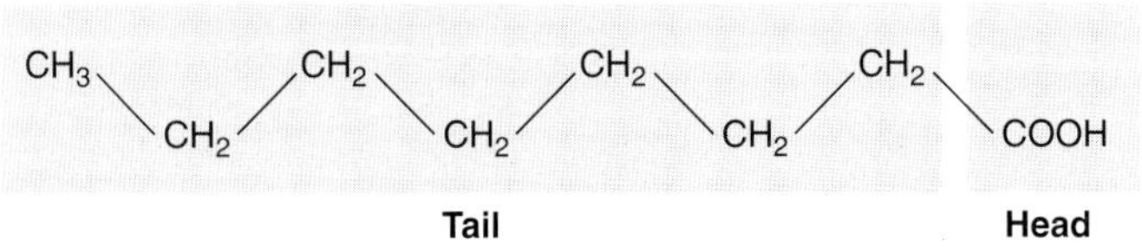

**FIGURE 3.1 Parts of a Fatty Acid.** The hydrophobic portion is the tail; the hydrophilic portion is the head. Numbering systems usually consider the carboxyl group as carbon 1.

## TABLE 3.1 Fatty Acids

| Common Name | Structure | Carbons: Double Bonds | Delta Designation | Source |
|---|---|---|---|---|
| Butyric acid | $CH_3CH_2CH_2COOH$ | 4:0 | – | butter |
| Palmitic acid | $CH_3(CH_2)_{14}COOH$ | 16:0 | – | palm oil |
| Oleic acid | $CH_3(CH_2)_7CH=CH(CH_2)_7COOH$ | 18:1 | $18\Delta^9$ | olive oil |
| Linoleic acid | $CH_3(CH_2)_7CH=CHCH_2CH=CH(CH_2)_7COOH$ | 18:2 | $18\Delta^{9,12}$ | plants |
| Arachidonic acid | $CH_3(CH_2)_7 (CH=CHCH_2)_4CH_2COOH$ | 20:4 | $20\Delta^{5,8,11,14}$ | animal membranes |

*acid* (found in butter), *oleic acid* (from olive oil), and *palmitic acid* (from palm trees).

Fatty acids containing double bonds are said to be **unsaturated**, whereas fatty acids containing no double bonds are said to be **saturated**. The terminology refers to the maximum number of hydrogen atoms that are attached to the carbon atoms in the fatty acid chains. In the absence of carbon–carbon double bonds, hydrogen atoms "saturate" the carbon atoms. When more than one double bond exists, the fatty acid is said to be **polyunsaturated** (**BOX 3.1**)

## BOX 3.1: On Steak and Fish: Polyunsaturated Fats, *Trans* Fats, and Health Risks

A high-quality (prime) steak is said to be *marbled,* which means that it has a large amount of fat dispersed through the meat. The steak is essentially a cross section of muscle tissue that is interspersed with adipocytes (visceral fat). When cooked, the fat melts and mixes with the meat, solubilizing its flavors into one that most omnivores like. Animal fat has a higher content of saturated fat than fish. Diets rich in animal fat are associated with an increased risk of cardiovascular disease, for reasons that remain elusive.

Certain fatty acids derived from fish are especially effective in promoting cardio-vascular health: the *omega fatty acids.* These polyunsaturated fatty acids may act by substituting for some **eicosanoids**, a group of endogenous lipid signaling molecules. The *omega* in the name refers to a numbering system distinct from that in Table 3.1. For example, linoleic has a double bond six carbons away from the terminal $CH_3$ group, so it would be called an *omega-6* fatty acid.

*Trans* fatty acids, because of their structural similarity to saturated fatty acids, pose a similar risk but are known to be worse because they are metabolized more slowly. *Trans* fatty acids are not found in nature but are produced synthetically. For example, margarine is produced by the chemical reduction of a mixture of triglycerides (with a reducing agent such as sodium borohydride). This is followed by chemical oxidation to introduce a controlled amount of unsaturation, allowing any desired physical consistency, such as "tub margarine." Unfortunately, this produces a mixture of *cis* and *trans* isomers.

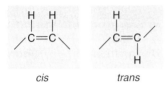

**FIGURE 3.2 Stereochemistry of the Double Bond.** The two possibilities for the stereochemistry of double bonds are *cis,* in which both hydrogens are on the same side of the chain, and *trans,* where the hydrogens are on opposite sides.

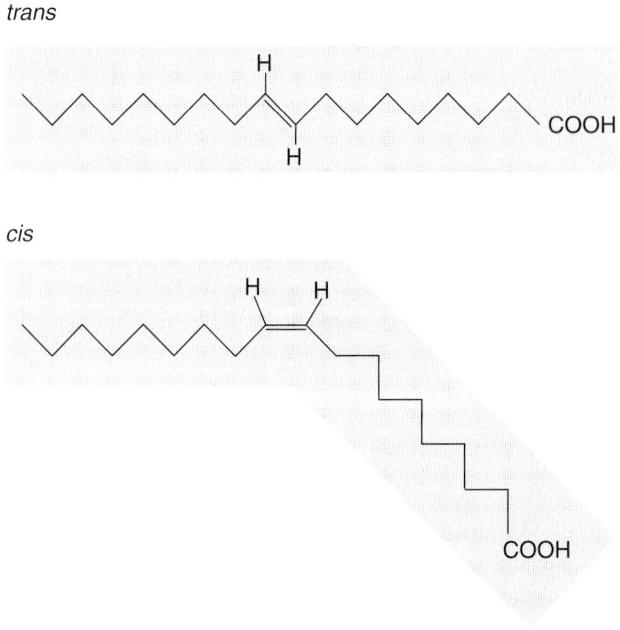

*trans*

*cis*

**FIGURE 3.3 *Cis* and *Trans* Fatty Acids.** The *trans* fatty acid molecules pack in regular arrays. However, the *cis*-oriented double bond produces a kink in the chain, so that packing is interrupted. This causes the molecules to stack less readily and leads to lower melting points.

Isolated double bonds          Conjugated double bonds

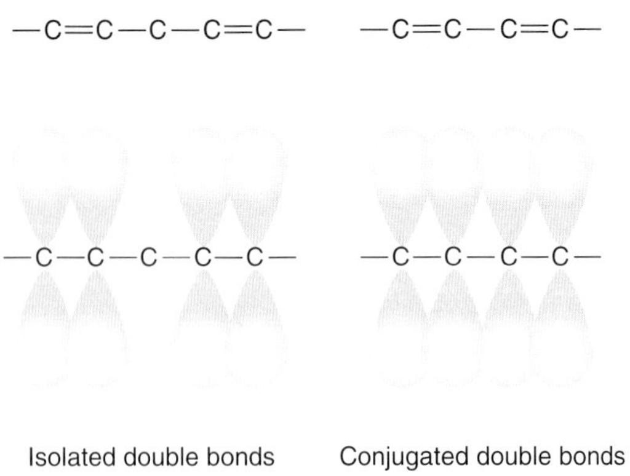

**FIGURE 3.4 Isolated and Conjugated Double Bonds.** Polyunsaturated fatty acids contain more than one double bond, and they are always isolated, as shown in the left panel. By contrast, in the conjugated double bond as shown in the right panel, every other carbon is involved in a double bond, and the electrons in the pi orbitals overlap.

The hydrogens attached to the carbon atoms that form double bonds can be stereochemically arranged in two distinct orientations: on the same side of the chain (*cis*) or on opposite sides of the chain (*trans*), as illustrated in FIGURE 3.2. Carbon atoms connected by single bonds exhibit free rotation around the bond axis. However, two carbons engaged in double bonds are geometrically constrained.

The structural consequence of this stereochemical distinction is illustrated in FIGURE 3.3. Only the *cis* fatty acid produces a bend in the chain. The *trans* and saturated fatty acids, on the other hand, appear "straight" rather than bent. As a result, *trans* fatty acids and saturated fatty acids tend to pack in an orderly way, like crystals (or shipping boxes), facilitating the formation of the solid phase. By contrast, fatty acids with *cis* double bonds do not stack together well, so these molecules tend to be liquids, not solids. Alternatively, we could say that *cis* fatty acids have lower melting points than *trans* or saturated fatty acids.

When more than one double bond occurs in a fatty acid, they are typically separated by two single bonds, as shown for linoleic and arachidonic acid in Table 3.1. These double bonds are said to be **isolated,** in contrast to the **overlapping** double bonds as shown in FIGURE 3.4. The electrons in the pi bonds of isolated double bonds do not overlap with the other pi electrons in polyunsaturated fatty acids. As a result, they are more reactive than the overlapping or **conjugated** double bonds, where the electrons are delocalized. The latter case, where electrons can spread out over more than two nuclei, is a condition called **resonance**

(FIGURE 3.5). The yellowish color of most oils results from the reactivity of isolated double bonds with oxygen and light in the fatty acid chains. Cells contain antioxidant systems to minimize the same reactions of isolated double bonds in their lipids.

The carboxylic acid head group (FIGURE 3.6) has the usual dissociation properties of a weak acid. Its $pK_a$ is 4.8, so it is almost entirely in the anionic form (i.e., –COO⁻) at pH 7, which explains why this portion of the molecule is so soluble in water. Thus, fatty acids have both polar and nonpolar parts within a single molecule, a quality that makes fatty acids **amphipathic**. Most fatty acids, however, exist in humans as covalently linked forms, either as triglycerides or as phospholipids, which we consider in Sections 3.3 and 3.4, respectively.

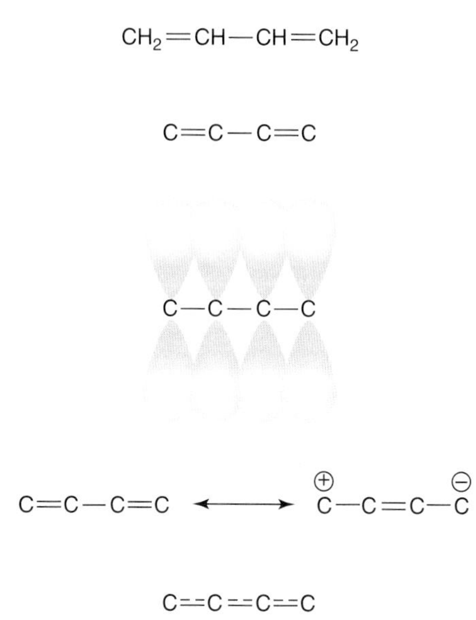

# 3.3 Triacylglycerols

Most of the fatty acids found in mammals are in the form of *triacylglcyerols* (also known as **triglycerides**), which is an energy storage form. Structurally, triacylglycerols consist of three fatty acids esterified to a glycerol backbone. The ester bond, formed between an alcohol and an acid, with release of water, is shown in FIGURE 3.7. While fatty acids are amphiphilic, triacylglycerols are hydrophobic, because the ester bond does not form hydrogen bonds with water. This is a benefit for storage, because excluding water reduces the overall weight of fat cells, which are essentially vessels containing a large fat droplet.

Whereas animals primarily use lipids (as triacylglycerol) for energy, plants use carbohydrates instead. Carbohydrates, having substantial amounts of oxygen, are correspondingly heavier than lipids. This is not a problem for plants, however, because they are immobile ("planted"). The seed is the only mobile phase of a plant's life cycle. Accordingly, lipid storage in plants is largely confined to the seed stage. All of our familiar plant lipids are derived from seeds, such as corn oil, peanut oil, or the generic "vegetable oil" (mostly composed of soybean oil).

The compositions of triglycerides vary in the same way the fatty acids vary themselves, namely, the tail portions have different lengths and degrees of saturation. Saturated side chains lead to more solid lipids, typical of triacylglycerol deposits near organs, called **visceral fat**. Less saturated chains are found in the deposits just under the skin, called **subcutaneous fat**. Even more unsaturated triacylglycerides are found in the seed lipids; these are referred to as **oils**. The great diversity of natural fatty acids within triglycerides can be appreciated by examining those found in pizza (**BOX 3.2**). In common practice, lipids are considered *oils* if they are in liquid form and *fats* if they are in solid form, at room temperature. This is clearly an arbitrary definition, however. For example, in the standard state (i.e., at 0°C), most oils would be solids!

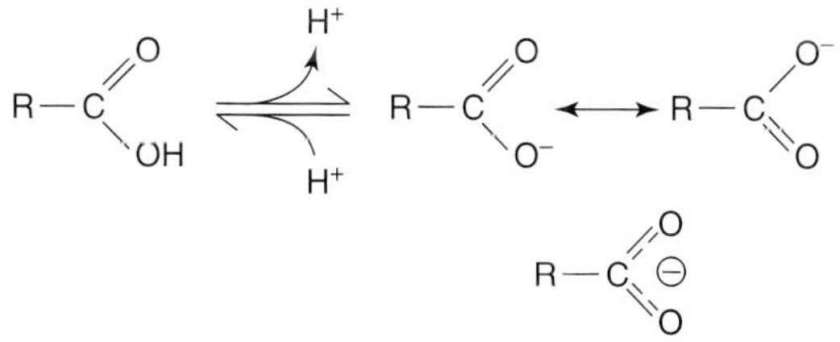

FIGURE 3.5 **Resonance in Butadiene.** The representations of butadiene show first the molecule with hydrogens explicitly represented, then with just the carbon backbone. Next, the electron clouds for the pi bonds are shown to overlap each other, spread out above and below the carbon chain. The last two represent the resonance (delocalized) structures. The two structures separated by a double-headed arrow show a dual representation and *not* a movement of electrons. The last representation shows each carbon in a "bond-and-a-half" representation. Delocalized electrons are less reactive (more stable) than localized electrons.

FIGURE 3.6 **Carboxyl Group Dissociation and Resonance.** Once the carboxyl group dissociates to the anion, it has a resonance form, represented by the double-arrow method as well as by the "bond-and-a-half" method. Unlike butadiene, however, both resonance forms have the same energy, an indicator of great resonance stability.

**Ester formation**

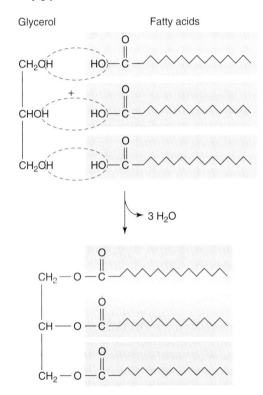

**Triacylglyceride formation**

FIGURE 3.7 **Esters and the Triester of Triacylglycerol.** Ester formation is formally the removal of water between an alcohol and an acid, as shown in the top panel. This occurs three times between glycerol and three fatty acid molecules, shown in the bottom panel, to form triacylglycerol.

## BOX 3.2: Lipid Composition of Pizza

The lipids found in pizza are almost entirely triacylglycerols, which vary in their fatty acid composition. These are shown as pie charts in FIGURE B3.2. The more abundant fatty acids (those representing at least 1% of the total) are displayed as Figure B3.2a. Only four fatty acids account for the bulk: myristic acid (14:0), palmitic acid (16:0), stearic acid (18:0), oleic acid (18:1 or, more specifically, $18\Delta^9$), and a mixture of 18:2 fatty acids, including the common linolenic acid ($18\Delta^{9,12}$). About half of the fatty acids are unsaturated and half saturated. Moreover, virtually none of the fatty acids has an odd number of carbons. The reason for this is that they are synthesized two carbon atoms at a time.

A fuller display of fatty acids shows the wide array of less abundant fatty acids (Figure B3.2b), which includes those with three and four double bonds. In particular, arachidonic acid ($20\Delta^{9,12,15,18}$), while present in small amounts, is important as a precursor to a large number of signal lipid molecules called the *eicosanoids*. There are two listings for 18:3, which denote distinct positions of the double bonds. One is characterized as n-3 in the omega numbering system (see Box 3.1).

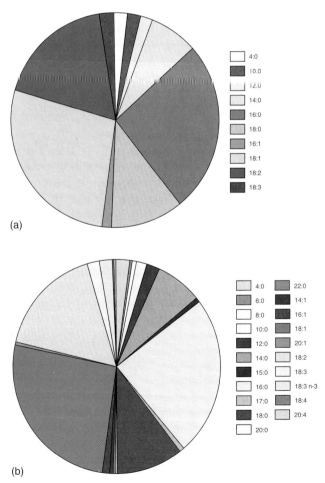

FIGURE B3.2 **Pie Chart of Pizza Lipids.** (a) Major fatty acids. (b) More complete fatty acid composition.

# 3.4 Phospholipids

The key polar lipids are the phospholipids, the principal components of cell membranes. Phospholipids are derivatives of phosphatidic acid (FIGURE 3.8). They resemble triacylglycerols, except that the third position of the glycerol is esterified to phosphate.

An isolated phosphate group is called **inorganic phosphate** and, like a carboxylic acid, is a weak acid and can combine with an alcohol to form an ester. However, unlike the carboxylic acids, phosphates are polyprotic acids, with multiple equilibria (FIGURE 3.9). Of the three $pK_a$ values, only $pK_2$ (6.6) is close to physiological pH; the others are near 2 and 12. Inorganic phosphate is a significant contributor to intracellular buffering, achieving a pH of about 7.1. Aside from its multiple dissociation equilibria, phosphate can react with alcohols to form multiple esters (FIGURE 3.10). In both the acid dissociation and ester formation functions, the phosphate group behaves as if the P=O portion and each of the –OH portions can act separately as an acid (Figures 3.9 and 3.10).

In the commonly occurring phospholipid phosphatidylcholine (FIGURE 3.11a and BOX 3.3), the phosphate group is part of a **diester**, in which one ester is formed with glycerol and the other ester with the alcohol choline. Phospholipids have considerable heterogeneity that arises from variations in the attached fatty acids at the first two positions. There are also different alcohols esterified to the phosphate; two examples of such phospholipids are shown in Figure 31.1b. Membranes also contain less frequently occurring, but nonetheless physiologically important, amphiphiles related to phospholipids, such as amides (sphingolipids) and ethers (plasmalogens).

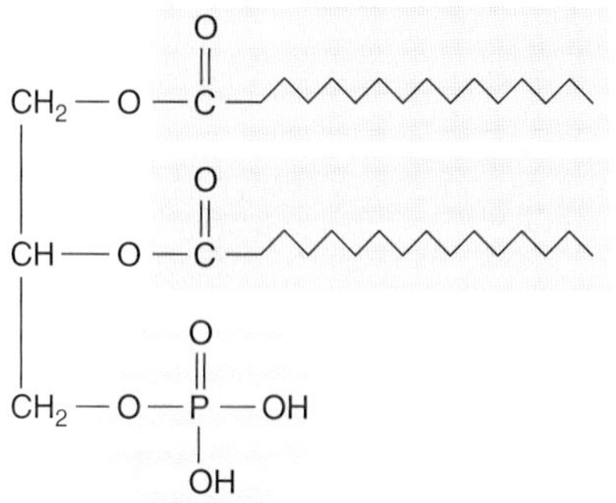

FIGURE 3.8 **Phosphatidic Acid.** This is the simplest phospholipid; the third position (called *sn*-3) of phosphatidic acid is esterified to a phosphate group. The *sn* refers to a naming system, an abbreviation for "stereochemical numbering."

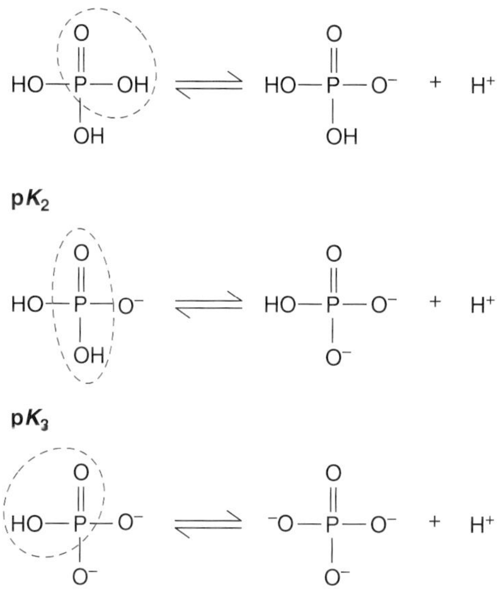

FIGURE 3.9 **Inorganic Phosphate: Acid Behavior.** Inorganic phosphate ($P_i$) has three $pK_a$ values: $pK_1 = 2$, $pK_2 = 6.6$, and $pK_3 = 12$. These three values correspond to the three dissociation equilibria shown.

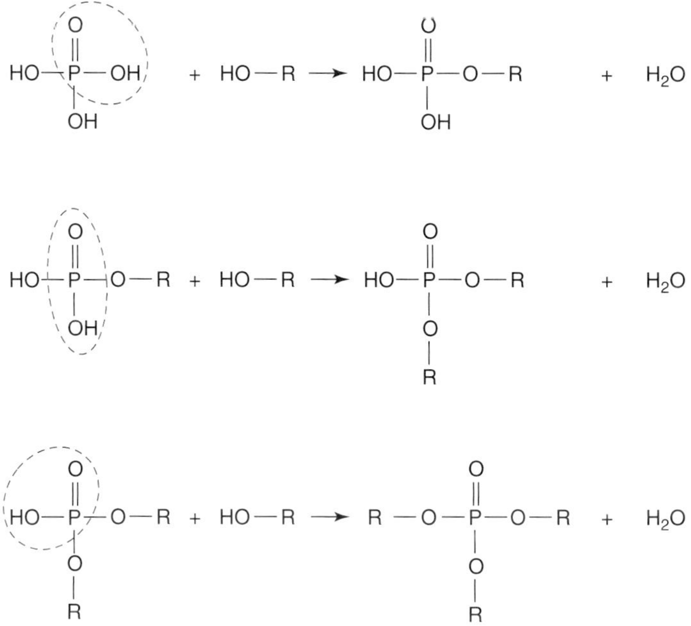

FIGURE 3.10 **Inorganic Phosphate: Ester Behavior.** Inorganic phosphate ($P_i$) can form the three esters indicated: a monoester, a diester, or a triester. In biology, the most common forms are the monoester and diester.

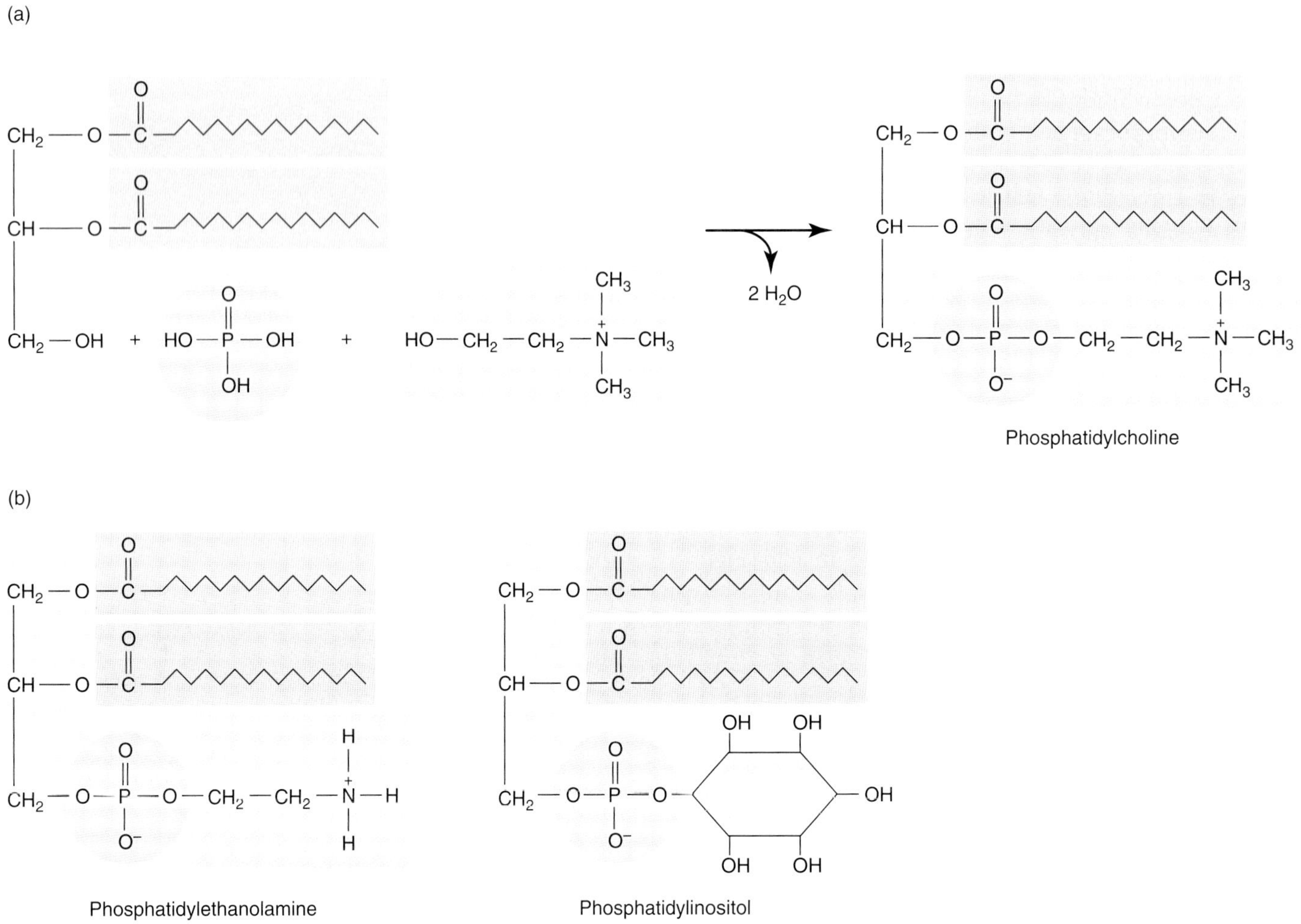

**FIGURE 3.11 Phospholipid Molecules.** (a) The most abundant phospholipid in mammalian membranes is phosphatidylcholine. It formally arises as a diester between phosphate and two alcohols: one from the third position of glycerol, the other from choline. (b) Two other common phospholipids of membranes: phosphatidylethanolamine and phosphatidylinositol.

## BOX 3.3: Lecithin and Emulsification

Lecithin (Greek word for egg yolk) is commonly used by biochemists as synonymous for phosphatidylcholine. In the food and pharmaceutical industries, a mixture of phospholipids extracted from animals or plants—largely containing phosphatidylcholine—is known as **lecithin.** Two common materials are "egg yolk lecithin" (a mildly redundant term considering the origins) and "soybean lecithin." While phosphatidylcholine is the major component of these materials, they contain other phospholipids (e.g., phosphatidylethanolamine) as well as lysophospholipids (i.e., phospholipids with just one fatty acid ester). The preparations are essentially organic solvent extracts of the biological material and are often used for the purposes of emulsification.

**Emulsification** refers to the dispersion of insoluble lipids, such as triglycerides, into water solutions. Lecithins are commonly used for this purpose, for example, in drug delivery. Because the hydrophobic portion of the phospholipid can interact with the triglyceride and the hydrophilic portion can interact with the solution, this has the effect of spreading the oil throughout the material. This is the principle behind the creation of mayonnaise as well as the formation of dispersed lipids in the first portion of the intestine in the digestive process, thus allowing the breakdown of dietary lipids.

## 3.5 Cholesterol

**Cholesterol** is the best known example of the **steroids**, a class of lipids having fused, flat rings. While popularly viewed as a dangerous component when present in high concentrations in blood, it is an essential component of animal cell membranes.

Structurally, cholesterol is a mostly hydrophobic molecule, having four rings fused together and a hydroxyl group bonded to one of them (FIGURE 3.12). Many other steroid molecules are derivatives of cholesterol, such as the sex steroids testosterone and estradiol, and the salt-retention hormone aldosterone, also shown in Figure 3.12. Cholesterol and the steroid derivatives, along with fatty acids and phospholipids, are all amphipathic molecules, because they share the quality of having both hydrophilic and hydrophobic regions within the same molecule. In Section 3.6, we explore how these compounds interact with water.

## 3.6 Lipid–Water Interactions of Amphipathic Molecules

Amphipathic molecules must satisfy two conflicting natures simultaneously: water solubility and lipid solubility. This is achieved by forming either of two distinct aggregates that are a compromise between these mutually exclusive constraints: **micelles** or **liposomes** (FIGURE 3.13).

In micelles, the polar portions face the aqueous exterior and the nonpolar portion forms an interior core excluded from water. Fatty acids form micelles in water (FIGURE 3.14). As fatty acids are added to a water solution, their concentration increases up to a point that

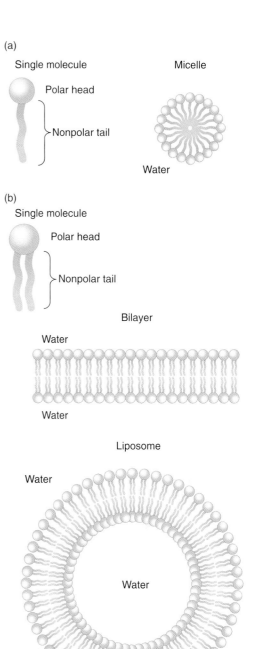

**FIGURE 3.13 Micelles and Liposomes.** (a) The micelle structure allows the polar portion of the amphiphile to interact with water on the outside of the structure while packing the nonpolar tails together in the interior. (b) In the liposome, two water phases are separated by a nonpolar interior phase, forming a sphere. A portion of that sphere can be viewed as a bilayer, an important conceptual division when considering transport and membrane studies.

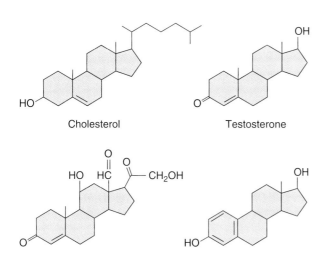

**FIGURE 3.12 Cholesterol and Other Steroids.** Steroids have four flat, fused rings and various substituents. Both cholesterol and estradiol are alcohols (sterols), whereas testosterone and aldosterone are ketones.

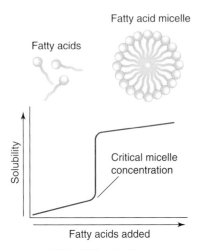

**FIGURE 3.14 Critical Micelle Concentration.**
Micelle-forming molecules, such as the fatty acids, have a break in their solution curves as they form the micelle structure. The concentration at which fatty acids form micelles is called the critical micelle concentration.

represents the limit of their solubility in water, called the **critical micelle concentration (CMC)**. As more fatty acids are added beyond this point, they form new micelles, but the concentration of free fatty acids in the water solution stays constant. Table 3.2 lists the CMC values for two distinct fatty acids and cholesterol. These vary over several orders of magnitude, indicating that a molecule like octanoate is very soluble in water compared to the others. However, unlike physical constants such as the melting point, the CMC depends strongly on experimental conditions, such as salt concentrations. As a result, the number of molecules in each micelle can vary. Thus, while useful, the CMC is only an empirical characteristic of the molecular aggregate.

An alternative to micelle formation is the liposome shown in Figure 3.13b. This aggregate consists of three phases: an exterior water phase, an interior water phase, and a nonpolar phase sandwiched between them. For example, phosphatidylcholine forms liposomes in water, and the resulting structure is the essence of a biological membrane. Figure 3.13b also shows an expansion of a portion of the liposome structure, indicating that an alternative description of this structure is the bilayer.

The two different types of aggregates each resolve the conflicting solubility demands of amphipathic molecules, raising the question of why a lipid structure would form one or the other. To address this question, consider the relative volumes of the hydrophobic and hydrophilic portions of the molecules (FIGURE 3.15). If these are unequal, then the molecules tend to pack like slices of a pie, forming a micelle. If they are roughly equal, then they will pack like bricks, forming a bilayer that, when wrapped around itself to form a sphere, becomes a liposome. While this notion of "packing" may seem simplistic, it is similar to the formation of crystals. For example, fatty acids have an unequal lipid and polar portion, indicating that they pack as wedges and produce micelles. Other micelle-forming lipids may not have the same "packing unit" shape, but nonetheless have hydrophobic and hydrophilic regions with unequal volumes.

The hydrophobic and hydrophilic regions of phosphatidylcholine have roughly equal volumes, so this amphiphile packs as a bilayer, and forms a liposome in water. Physiologically, we are usually most interested in bilayers as these aggregates are the essence of cellular membranes. One important example of micelles is the formation of **bile salts** in the small intestine. These contain polar cholesterol derivatives that form mixed micelles with dietary lipids, facilitating their hydrolysis to fatty acids, which can then enter the cells lining the intestine.

| TABLE 3.2 Critical Micelle Concentration (CMC) Values for Two Fatty Acids and Cholesterol | |
| --- | --- |
| Molecule | CMC (mM) |
| Palmitate (16:0) | 0.002 |
| Octanoate (9:0) | 400 |
| Cholesterol | 0.040 |

## 3.7 Water Permeability of Membranes and Osmosis

The liposome vesicle serves as a model for cellular membranes. One consequence of two water phases separating a membrane phase is that two solutions with distinct compositions can exist on either side of the membrane. Unlike most hydrophilic substances, water itself can readily cross the membrane. This occurs despite the limited solubility of water in the hydrophobic core, due to the extremely large concentration of water. In cellular membranes, water flow can be accelerated further or regulated by the presence of **aquaporins**, which are protein channels selective for water transport.

Most of the channels and transport proteins in cell membranes exist for the purpose of selectively transporting solutes (e.g., $Na^+$ or glucose) across the membrane. Selective transport is another way of saying that physiological bilayers are **semipermeable**. The process can be modeled by a classic experiment using an apparatus called a *U-tube*, illustrated in FIGURE 3.16. A glass tube is bent into a U shape and a semipermeable membrane is placed in its center. In Figure 3.16, small dots represent permeable water molecules, whereas larger dots represent an impermeable solute dissolved in water.

The first frame (Figure 3.16a) shows the two solutions the instant after they are placed in the apparatus. At this point, the left-hand compartment has a lower water concentration. Accordingly, water will move from right to left. Any movement of molecules due to spatial differences in concentration is called **diffusion**. The water movement under consideration here is a type of diffusion more specifically known as **osmosis** (BOX 3.4) The process is complete when water is equalized across the membrane, as in the middle frame (Figure 3.16b). Note that the solution has been elevated to a greater height on the left side. If an external pressure is applied through a plunger until the heights of the sides are equal, we arrive at the situation in the third panel (Figure 3.16c). The applied pressure needed to equalize the fluid levels in the U-tube is the **osmotic pressure**; its value can be read from the gauge as indicated. In this process, water has been forced to move from a region of low concentration to a region of high concentration. This converts the solution on the left-hand side into more pure water

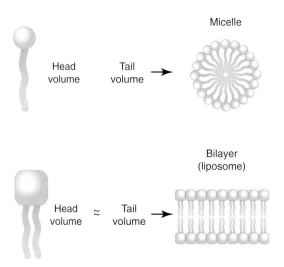

**FIGURE 3.15 Rationale for Micelle or Liposome/ Bilayer Formation.** Micelles form when the amphiphiles have different volumes for the polar head and nonpolar tail structure. When the volumes occupied by each is approximately equal, the individual molecules can stack like bricks and form bilayers, which can wrap to form liposomes.

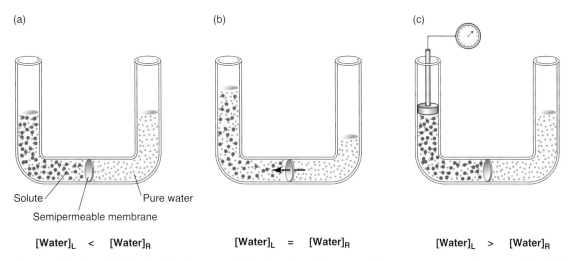

**FIGURE 3.16 Osmosis in the U-tube.** Solute (large dots) is dissolved in one solution, and pure water (small dots) is present in the other. Initially, these are placed in the U-tube at the same volume (a). Due to diffusion, water moves from right to left, equilibrating its concentration across the semipermeable membrane in (b). In (c), a piston is introduced into the left side to push the solution down so that both sides become equal again. The pressure reading on the gauge (pressure = force × area) indicates the osmotic pressure of the solution.

## BOX 3.4: WORD ORIGINS

### Osmosis

Osmosis originated as a technical term to describe water movement in cells. Originally separated as *endoosmosis* and *exoosmosis,* the simplification to osmosis reflects the Greek origins of the word, meaning a push or impulse. Osmosis is a simple driving force based on an even simpler one: diffusion. Molecules move by diffusion simply because there are more of them in one place than another, and statistically, their locations tend to even out. When this occurs across a membrane, and when this membrane is selectively permeable, the phenomenon of osmosis emerges, as described in the text. Long recognized for the movement of water within plant cells and across the membranes of all cells, the non-scientific use of the word suggests that this process is mysterious, typically as the ability to obtain new knowledge by its flow into the brain from the nether regions.

on the right-hand side. The technique is the principle behind *reverse osmosis,* used to desalinate water.

## 3.8 Effect of Lipids on Membrane Composition

While all biological membranes are essentially phospholipid bilayers (also containing other lipid components and some proteins), there is extensive variation in the phospholipid molecules themselves. Part of this variation relates to the phospholipid head group. Another factor that contributes to membrane diversity is the different fatty acyl chains, which leads to alterations in **membrane fluidity**. This concept is analogous to the solid–liquid transitions in other molecules, but in membranes refers the relative ability of the individual molecules to move within the bilayer. The presence of double bonds in the fatty acyl chains favors an increase in membrane fluidity. A further increase in fluidity can be obtained with more than one double bond (i.e., a polyunsaturated tail), although the extent of this increase is known to be modest.

The presence of cholesterol also contributes to the fluidity of membranes. At high temperatures, cholesterol tends to decrease the fluidity of the membrane, whereas at low temperatures, cholesterol decreases membrane fluidity. Thus, cholesterol, which is most abundant in the outer leaflets of the plasma membranes of animal cells, tends to dampen temperature-induced changes in membrane fluidity.

There are a variety of lipids beyond triacylglycerols, phospholipids, and sterols that have specialized biological purposes. Many polar lipids are conjugates with other types of molecules. For example, lipopolysaccharides are molecules formed by the joining lipids with sugar. To understand such species, we must first examine the sugar molecules themselves.

## Summary

Lipids are used primarily for energy storage and as membrane components. Their dominant quality is insolubility in water. The building block of most lipids is the fatty acid, which consists of a hydrophobic hydrocarbon tail and a hydrophilic carboxylic acid head. When the tail contains double bonds, the fatty acids is said to be unsaturated. The double bond in naturally occurring fatty acids is in the *cis* configuration, with both hydrogens on the same side of the chain. The more unsaturated and the shorter the chain, the more soluble the molecule is in water. When the carboxylic acid head group of a fatty acid forms ester linkages with the alcohol glycerol, the result is a triacylglycerol. Triacylglycerols are the major storage molecules of lipids and are extremely hydrophobic. Phospholipids are similar to triacylglycerols, except that the third position on the glycerol is esterified to phosphate. In the common phospholipid molecule phosphatidylcholine, the phosphate is also esterified to the alcohol choline. Both phospholipids and the steroid molecule cholesterol are found in cell membranes, and both are amphiphiles, having both hydrophilic and hydrophobic portions in the same molecule. Fatty acids are also amphiphiles. Micelles are formed when the volumes of the hydrophilic and hydrophobic portions of amphiphiles are different, such as in the case of fatty acids. Micelles have a hydrophobic core and a hydrophilic exterior. Liposomes form when the hydrophilic and hydrophobic portions of the amphiphile have similar volumes, such as the case of phosphatidylcholine. When a liposome or a membrane forms, water can pass through but most other water-soluble molecules cannot. The barriers, thus, is said to be semipermeable. The flow of water across a membrane in the direction that balances the concentrations on both sides is called osmosis.

## Key Terms

amphipathic
aquaporins
bile salts
cholesterol
conjugated (double bonds)
critical micelle
   concentration (CMC)
diester
diffusion
emulsification
fatty acids
inorganic phosphate

isolated (double bonds)
lecithin
lipids
liposomes
membrane fluidity
micelles
nonpolar lipids
oils
osmosis
osmotic pressure
overlapping
   (double bonds)

polar lipids
polyunsaturated
resonance
saturated
semipermeable
subcutaneous fat
steroids
triglycerides
unsaturated
visceral fat

## Review Questions

1. Bacteria that flourish in extremes of temperature have several chemical strategies that allow them to adapt. What sort of fatty acyl chains would you expect to find in extreme thermophiles (those living near the boiling point of water)? What about plants that must adapt to lower temperatures?
2. If the p$K_a$ of an average fatty acid is about 5, what effect would there be on solubility at pH 2, 5, and 7?
3. What features make phospholipids suitable for a membrane but not an energy storage molecule? What features make triacylglycerols suitable for energy storage but not membrane function?
4. Oils from plant seeds are typically yellow in color, but fats are typically almost colorless. Explain the reason for this difference.
5. Red blood cells are commonly used to illustrate the dramatic actions of osmosis. A solution that is lower in osmotic strength than the red cell cytosol is called *hypotonic;* one that is higher is called *hypertonic,* and one that is the same is called *isotonic.* Explain what each type of solution would do to a red blood cell.
6. Soap is made by the hydrolysis of triglycerides, using lye (NaOH). The resulting solution contains sodium ions, fatty acids at alkaline pH, and glycerol. The fatty acids form a micelle when in solution. Explain the use of a micelle in washing your hands with soap.

## References

1. Doyle, E. Trans fatty acids. *J. Chem. Ed.* 74:1030–1032, 1997.
   The chemistry of formation and the health consequences of *trans* fatty acids are detailed. The author concludes with the intriguing observation that a *Pseudomonas*

bacteria has an enzyme that converts *cis* fatty acyl chains in the membrane to *trans*, thus decreasing membrane fluidity when the organism enters a stationary phase.

2. Nawar, W. W. Chemical changes in lipids produced by thermal processing. *J. Chem. Ed.* 61:299–302, 1984.

   The author describes changes in lipids induced by heat as well as how light can alter fatty acids.

3. Jain, M. K. *Introduction to Biological Membranes;* John Wiley: New York, 1988, p. 423.

   This book thoroughly describes membranes, including the distinction between bilayer and micelle formation. Many details of different membranes and different possible phases of lipid materials are presented also.

4. Cantor, C. R.; Schimmel, P. R. *Biophysical Chemistry;* W. H. Freeman: San Francisco, 1980.

   This text provides a detailed description of physical biochemistry, including a good explanation of both diffusion and osmotic pressure.

5. Van Holde, K. E.; Johnson, W. C.; Ho, P. S. *Principles of Physical Biochemistry;* Pearson/Prentice Hall: Upper Saddle River, NJ, 2005, p. 710.

   This is a more succinct physical biochemistry text, with appropriately more brief treatments of the essentials of diffusion and osmosis.

**CHAPTER 3** LIPIDS

# 4

# Carbohydrates

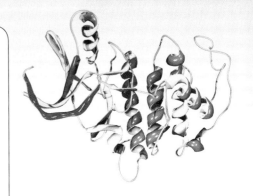

## CHAPTER OUTLINE

Image © Dr. Mark J. Winter/Photo Researchers, Inc.

**C**arbohydrates are most commonly encountered as the sweetener sucrose and as starch, the nutrient of bread and pasta. The carbohydrates are literally *hydrates of carbon*, having the empirical formula $CH_2O$. Because they posses both hydroxyl and carbonyl groups, they are a step up in complexity from lipids. Unlike lipids, carbohydrates are generally very soluble in water. The direct translation of sugar into Latin is *saccharide*, which remains in common use in the term *polysaccharides*.

Sugars are key nutrients in both animals and plants. Animals use sugars for rapid (i.e., minute-to-minute) energy production and lipids for long-term storage. Plants use sugars for energy exclusively, except at the seed stage. Because sugars are more dense than lipids (having more oxygen atoms), and often associate with water, their greater weight impairs mobility. Hence, lipids are the long-term energy source used for mobile lifestyles (i.e., animals and plant seeds), and carbohydrates are the immediate energy source for animals.

Aside from providing energy, sugars have a wide variety of roles once they are chemically modified, such as cell–cell recognition and signaling. Sugars attached to lipids comprise specific blood groups, and sugars are covalently linked to proteins secreted from cells. The basis for understanding sugars rests on the study of the simplest ones: the monosaccharides.

# 4.1 Monosaccharides

Monosaccharides have a single carbonyl group, which occurs at the first or second carbon atom. All remaining carbons have an attached hydroxyl group. There are two classes of monosaccharides, called aldoses and ketoses. **Aldoses** (aldehydes) have the carbonyl at carbon one, whereas **ketoses** (ketones) have the carbonyl at carbon two.

The "ose" suffix (as in ald*ose* and ket*ose*) is commonly used to define a compound as a sugar. The general category name for sugars is shown in Table 4.1, starting with the smallest, **triose**. Exceptions to this naming rule are found in the simplest monosaccharides: dihydroxyacetone and glyceraldehyde (FIGURE 4.1).

Dihydroxyacetone is a symmetrical molecule, because a plane of symmetry can be drawn through its central carbon atom. Glyceraldehyde, however, has an asymmetric central carbon, called **chiral** (Latin for *handedness*). Any carbon to which four distinct groups are attached is a chiral center. As drawn in two dimensions, two different forms of glyceraldehyde can be represented in a **Fischer projection**, shown in FIGURE 4.2. The molecule on the left is called L, whereas the one on the right is called D, an arbitrary but agreed-upon convention. The letters themselves stand for Levo and Dexter, from the Latin for "left" and "right." In three dimensions, the four groups attached to the chiral carbon are spatially as far apart from one another as possible, because the bonding electrons repel each other. Each group attached to the chiral carbon is at the corner of a virtual

**TABLE 4.1 General Nomenclature of Sugars**

| Carbons | Sugar Category Name |
|---------|---------------------|
| 3 | Triose |
| 4 | Tetrose |
| 5 | Pentose |
| 6 | Hexose |
| 7 | Heptose |
| 8 | Octose |

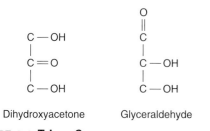

Dihydroxyacetone        Glyceraldehyde

FIGURE 4.1 **Triose Sugars.**

tetrahedron, as shown in FIGURE 4.3. Note that carbon number 1 is assigned, by chemical convention, to the most oxidized carbon in the molecule. In the case of a ketose, the carbonyl becomes carbon number 2.

The two forms of glyceraldehyde are called L-glyceraldehyde and D-glyceraldehyde. They are mirror reflections or **enantiomers** (indicated in Figure 4.2) of each other. As a result, the two are not superimposable on each other, just as your left and right hands are not superimposable on each other.

Enantiomers share many chemical properties, such as identical melting and boiling points. They were first identified through their distinct interaction with polarized light. Polarized light is produced by special filters that eliminate all light rays other than those oriented in one plane. Experimentally, the interaction of a compound with polarized light is measured as degrees of rotation from a reference point, and a separate naming system is used (**BOX 4.1**). From a biochemical perspective, the greatest significance of enantiomers is their selective interaction with enzymes. For example, most living systems can metabolize D-sugars, but not L-sugars.

Larger sugar molecules (e.g., those with four or more carbon atoms) have multiple chiral carbons. FIGURE 4.4 shows the Fischer projection of a four-carbon sugar (a tetrose). It is an aldose because it has an aldehyde group at the first carbon. The next two carbons are both chiral, as indicated with asterisks. The convention for naming such a sugar unambiguously is to first differentiate D-sugars from L-sugars. The carbon furthest from the carbonyl determines if the entire molecule is D or L. In this example, it is carbon 3. By comparing this molecule to glyceraldehyde, as indicated, the molecule is a D-sugar. All that remains is to resolve the ambiguity of the remaining chiral center. This is done by simply assigning a name to the molecule: this one is called erythrose. Our focus will be on D-sugars because they are far more common than L-sugars. The other four-carbon D-sugars are D-threose and D-erythrulose, shown in FIGURE 4.5. Unless we are making an explicit stereochemical point, we will usually omit the D indicator.

Sugars that differ at chiral centers other than the D and L positions are called epimers. Thus, threose (Figure 4.5) is the epimer of erythrose (Figure 4.4). Larger sugars have correspondingly greater numbers of isomers, but the naming conventions hold. After assigning the D and L forms, each molecular arrangement of the remaining chiral centers has a unique name. Examples of commonly occurring D-sugars are shown in FIGURE 4.6: ribose, glucose, galactose, and fructose.

## 4.2 Ring Formation in Sugars

Sugars having five or more carbons exist in solution primarily in a ring form. The reaction that leads to ring formation in sugars is an *intramolecular* version of the more general one that forms **hemiacetals**

**Fischer projection**

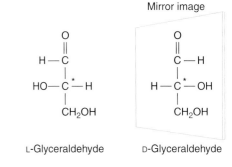

L-Glyceraldehyde          D-Glyceraldehyde

**3D view**

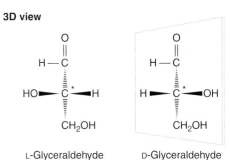

L-Glyceraldehyde          D-Glyceraldehyde

**FIGURE 4.2 D- and L-Glyceraldehyde.** For each form, the Fischer projection is shown above a representation of the three-dimensional view. Chiral carbons are indicated by asterisks.

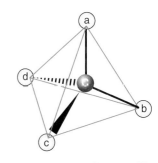

**FIGURE 4.3 Tetrahedral Carbon.** The carbon with four substituents, each as far apart from one another as possible, forms a tetrahedron in space.

LIBRARY. UNIVERSITY O CHESTER

## BOX 4.1: Stereochemical Conventions: Little *d/l*

In addition to the D/L naming convention, a separate system uses the lower case letters *d* and *l* (alternatively, + and −) to refer to the direction of rotation of plane-polarized light: *d* or + for clockwise and *l* or − for counterclockwise. To better understand the use of the *d/l* system, consider the following practical experiment. Suppose you take a pair of polarized sunglasses, pop out the lenses, and place them at either end of a tube filled with water (FIGURE B4.1). This device you have just made is a **polarimeter**. If you look through the tube, rotating one sunglass lens will entirely eliminate the light that gets through at a particular point. If you continue to rotate it, light will appear again until, with continued rotation, it darkens once again. Mark the darkest point. This is where the polarized planes in the lenses are at right angles to one another. This follows from the operation of a Polaroid filter, which allows light beams to travel only in one direction. If you now replace the water in the tube with a solution of an optically active (chiral) molecule, and start with the marked point, one lens will have to be rotated to some extent to restore the dark point. That displacement is measured in degrees, right or left (i.e., *d* or *l*). This is caused by the interaction of light with the molecules, and reveals the presence of asymmetrical interactions in the solution.

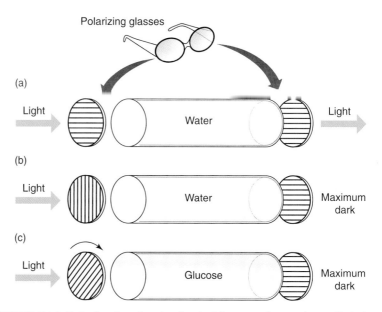

**FIGURE B4.1 Polarimetry.** A pair of polarizing sunglasses (or untinted polarizing glasses) can be used as polarized filters. Removing the lenses and placing each on opposite sides of a sealed tube filled with solution is the essence of a polarimeter. Its operation is shown in three situations. (a) The tube is filled with water. Both of the filters are oriented so that maximal light passes through. The polarization is indicated by lines on the filters, showing that they are in parallel. (b) Again, the tube is filled with water. The filters are oriented at 90° to one another, in which case no light passes through. (c) The tube is filled with a glucose solution. In this case, the point of maximum darkness is achieved by rotating one filter clockwise; this indicates that glucose is a *d*-sugar or, equivalently, a + sugar. Because this is independent of its D/L status, we would call this D-(+)-glucose. The same experiment performed with fructose would show that the filter would have to be rotated counterclockwise; hence, fructose is D-(−)-fructose.

or **hemiketals** (FIGURE 4.7). As described in the figure, the reaction joins a carbonyl group with a hydroxyl group (a nucleophilic substitution). Note that the carbonyl oxygen becomes a hydroxyl group in the product.

The carbonyl carbon of the straight chain form becomes the **anomeric carbon** in the ring form. Because both the carbonyl group and the alcohol group come from the same molecule, the hemiacetal or hemiketal reaction is more favorable than if the reactants were from different molecules. The other consequence of this intramolecular reaction is that the product becomes a ring. Of the different OH groups that *could* join with the carbonyl, only five- and six-membered rings are stable. Open and ring forms for the sugars fructose and glucose are shown in FIGURE 4.8.

The representation of ring forms as joined, straight-chain diagrams in Figure 4.8 is unrealistic. Most commonly, sugar rings are drawn as **Haworth projections**, as shown in the middle structure of FIGURE 4.9. When constructing this diagram, you need only place the groups appearing on the right side of the straight-chain Fischer projection below the ring in the Haworth projection to achieve the correct orientation. Note that the Haworth is a simplified representation of the more realistic chair form, also shown in Figure 4.9.

We now turn our attention to the hydroxyl group that is connected to the anomeric carbon, colored blue in these diagrams. Because the anomeric carbon has become a new chiral center in the ring form, this hydroxyl group can have two distinct stereochemical forms (FIGURE 4.10). If the hydroxyl group is on the opposite side of the ring from the position determining the D-form of the sugar, it is called the **alpha** form. If the hydroxyl group is on the same side of the ring as the position determining the D-form of the sugar, it is called the **beta** form. As drawn in the Haworth projection shown, the alpha-hydroxyl group is below the ring, and the beta-hydroxyl group is above the ring. The alpha and beta forms are in equilibrium with each other as well as with the open-chain form, so that, metabolically, any of them can be utilized in reactions. Figure 4.10 provides the full name of the sugar, with both the designation D (to identify its stereochemical similarity to D-glyceraldehyde) as well as the α and β forms resulting from ring formation.

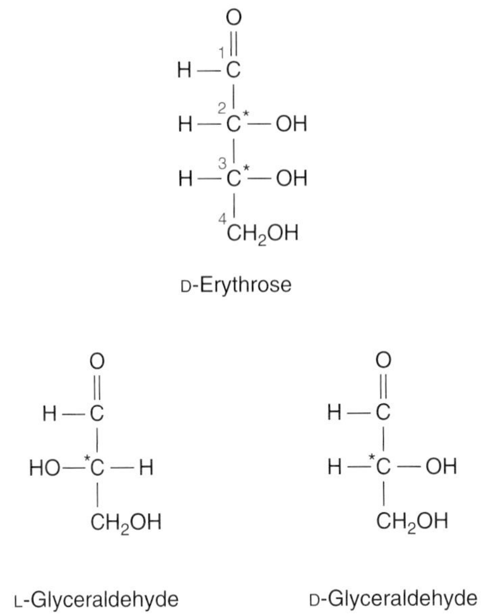

FIGURE 4.4 **D-Erythrose.** The configuration at carbon 2 defines this structure as erythrose; the configuration at carbon 3, the one furthest from the carbonyl, defines this as the D form. The reference compounds, L- and D-glyceraldehyde, are shown for comparison. Chiral carbons are indicated by asterisks.

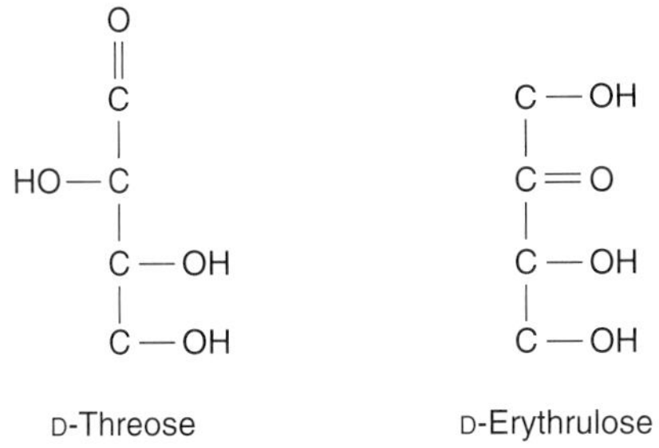

FIGURE 4.5 **D-Threose and D-Erythrulose.**

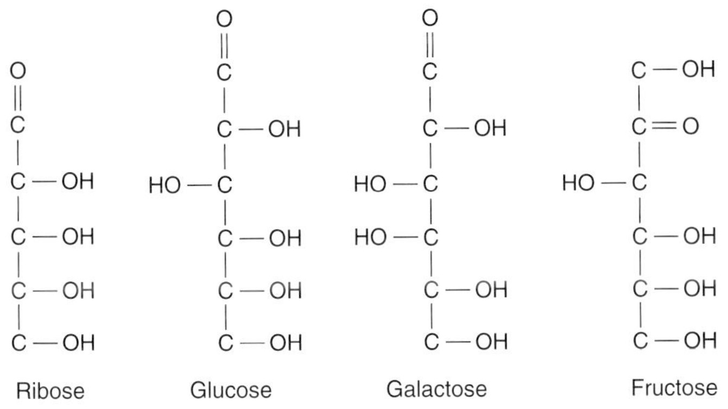

FIGURE 4.6 **Common Monosaccharides.** Only the D forms of these common sugars are shown, because they are the prevailing biological isomers.

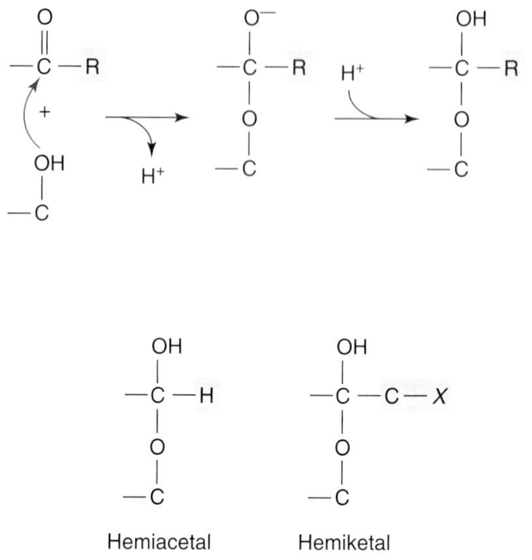

FIGURE 4.7 **Hemiacetal and Hemiketal Formation**. The reaction of an alcohol group with a carbonyl is one of simple nucleophilic substitution.

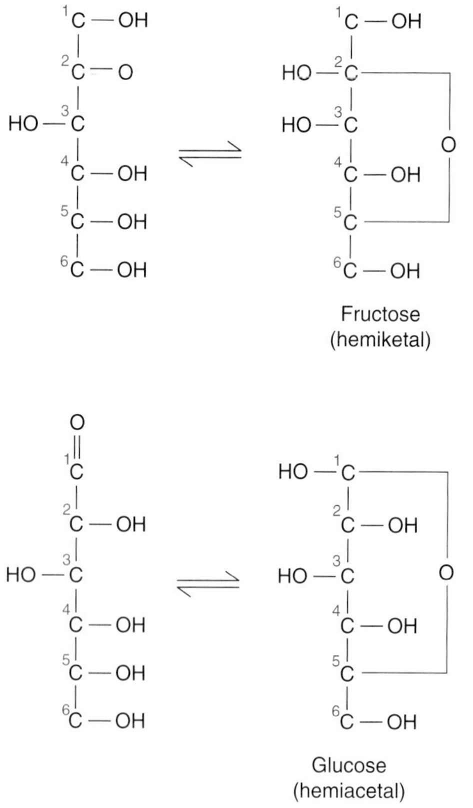

FIGURE 4.8 **Ring Formation in Sugars**. The reaction shown in Figure 4.7 takes place intramolecularly in sugars having at least five carbons. Because the aldehyde and alcohol are part of the same molecule, the product is a ring.

To summarize, we have discussed three types of stereochemistry that allow us to unambiguously identify a sugar:

1. D or L, which refers to the configuration of the chiral carbon furthest from the carbonyl group
2. Different names for other epimers (e.g., mannose or galactose)
3. Alpha or beta, which refers to the configuration of the anomeric carbon

The ring forms of ribose and fructose, shown as FIGURE 4.11, are also in equilibrium with their open-chain forms. The fact that common sugars like these form rings is important not merely because this is the predominant solution form, but also because the hydroxyl group attached to the anomeric carbon is used in bonding sugars together. The first example of this kind of linkage is in the disaccharides.

## 4.3 Disaccharides

Two sugars bonded together are called **disaccharides**; some examples are shown in FIGURE 4.12. Of these, the best known is sucrose; it is what we popularly call "sugar" itself. Sucrose is widely found in plants but is heavily concentrated in just a few, such as sugar cane and sugar beets. Maltose is a product of the partial digestion of starch and is common in the production of beer. The partial digest itself is called a "malt," hence the name "malt liquor." Lactose is milk sugar, which is produced by mammals for feeding their newborn. Trehalose is uncommon in plants but found in bacteria, fungi, and invertebrates and plays a role analogous to sucrose in those organisms.

In each of these disaccharides, one of the sugar units is glucose; the other is glucose, galactose, or fructose. The linkage between the two sugars is called a **glycosidic bond**. The glycosidic bond of maltose,

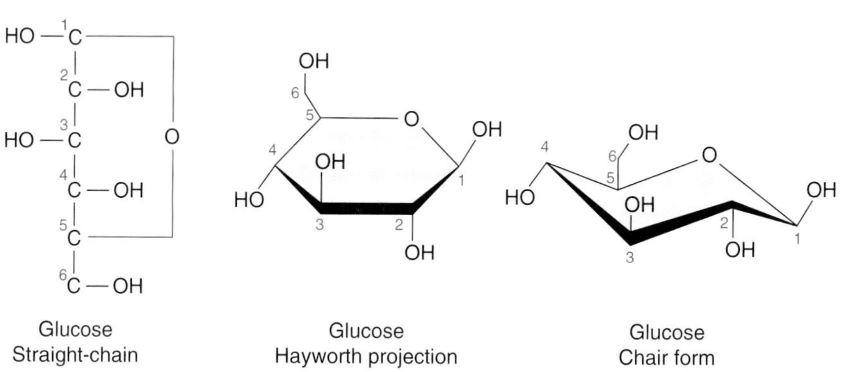

FIGURE 4.9 **Views of Sugar Rings**. The straight-chain form (Fischer projection) can be recast as the Hayworth projection by taking groups positioned to the right in the Fischer projection above the ring, and those to the left below it. The chair form is a solution structure that more closely represents the three-dimensional structure of the sugar molecule.

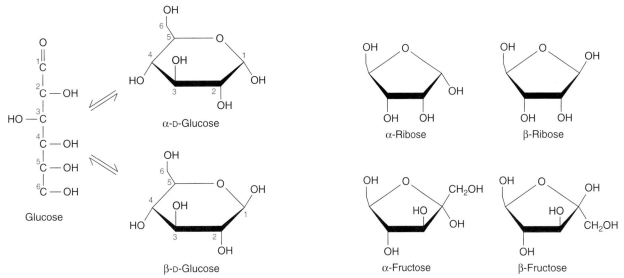

**FIGURE 4.10 Multiple Ring Forms of Glucose.** Two separate ring forms of glucose are possible, depending on which face of the carbonyl is attacked by the hydroxyl group attached to carbon 5.

**FIGURE 4.11 Ring Forms of Ribose and Fructose.**

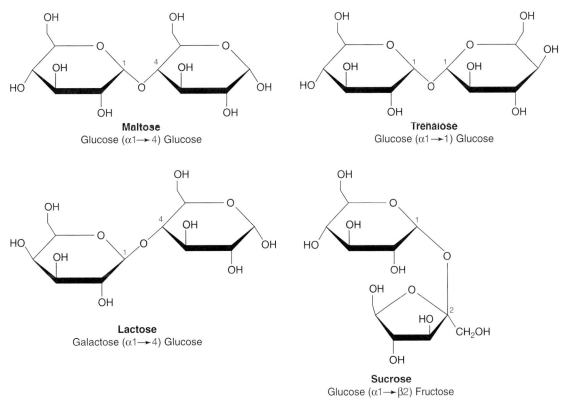

**FIGURE 4.12 Disaccharides.** The two sugars in the disaccharide are linked by a glycosidic bond. This locks the anomeric hydroxyl in either the alpha or the beta configuration. Either one (maltose and lactose) or both (sucrose and trehalose) sugar units have their anomeric carbons in a glycosidic bond.

for example, is *formally* a removal of water from two glucose molecules, as indicated in FIGURE 4.13. The metabolic route to glycoside formation involves energy input and intermediate steps.

Once a disaccharide has formed, the stereochemistry of the anomeric carbon involved in the glycosidic bond is fixed: it is either α or β, as indicated in Figure 4.12. The notation for the glycosidic bond between the sugars provides the number of the anomeric carbon (with its α or β oxygen position indicated) separated by an arrow from the number of the connected carbon on the other sugar. Only one of the anomeric carbons is involved in the glycosidic bonds of maltose and lactose, whereas both anomeric carbons are involved in the glycosidic bonds in sucrose and trehalose.

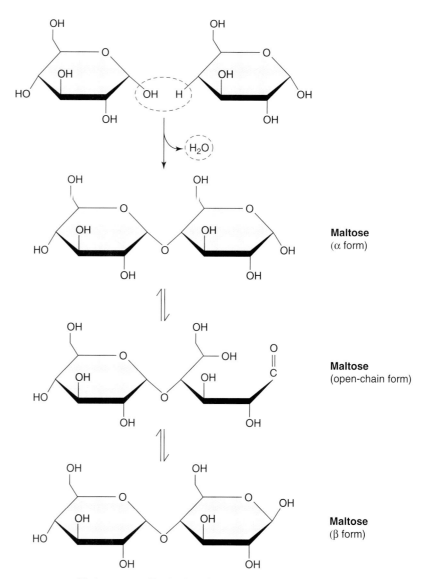

FIGURE 4.13 **Maltose as a Reducing Sugar**. Maltose consists of a glucose bearing an alpha anomeric carbon locked to a second glucose molecule. The second glucose molecule is in equilibrium with an open-chain form the alpha form, and the beta form. The presence of the carbonyl makes the entire maltose molecule a reducing sugar.

When just one anomeric carbon is involved in glycoside formation, as in the maltose and lactose, a free anomeric carbon remains in equilibrium with an open-chain form (see Figure 4.13 for maltose). Because at least a small amount of this open-chain form exists in solution, its carbonyl group can reduce certain metal ions in diagnostic tests. Any saccharide that has at least one free anomeric carbon (i.e., an anomeric carbon *not* involved in a glycosidic bond) is therefore called a **reducing sugar**. Maltose and lactose are reducing sugars. By contrast, sucrose and trehalose are **nonreducing sugars** because they have no free anomeric carbon atoms. To better understand the chemical meaning of the word *reducing*, see **BOX 4.2**.

# 4.4 Polysaccharides

When more than two sugars are linked via glycosidic bonds, they are said to be either **oligosaccharides** (meaning "a few") or **polysaccharides** (meaning "many"). The distinction is inexact; some consider chains of up to a dozen or so linked sugars to be oligosaccharides. Commonly, polysaccharides have thousands of monosaccharide residues bound together. Such molecules have extremely large molecular weights, which leads to properties distinct from smaller molecules. In general, large molecules that are constructed from small, repeating units are called **polymers**. The polysaccharides are our first example of biological polymers; the other two major classes are proteins and nucleic acids. While lipids can form large aggregates with new properties, they are not biological polymers, because they consist of noncovalently associated small molecules. Nonbiological polymers, such as nylon, were discovered and studied at the same time as biological polymers. Nonbiological and biological polymers share some properties, including the methods used for their analysis.

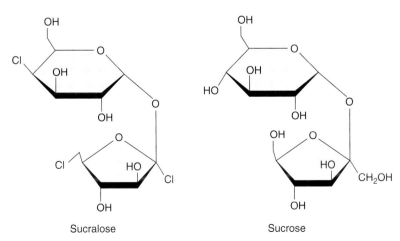

Sucralose

Sucrose

FIGURE B4.2 **Sucralose and Sucrose.**

## BOX 4.2: WORD ORIGINS

### Reducing

When a word with rich meanings such as *reducing* is applied to sugar chemistry, its interpretation may be obscured. There are distinct sociological, mathematical, biological, and chemical definitions for this word. In the everyday sense, the word originally meant to bring back, or to restore, an original condition. A later use was related to conquest, and subsequently, a general sociological meaning of *lowering*.

Mathematically, the term refers to the factor-label method of manipulating units algebraically in a separate way from the numbers they label. In biology, meiosis involves a reductive division, in which the number of chromosomes is halved. In chemistry, reduction is the decrease in oxidation number for an atom. A compound is reduced if it gains electrons, so the term *reduction* is somewhat misleading as a description of the chemical process. Reducing sugars transfer electrons from the carbonyl group to copper or silver ions. In most cases, a further electron transfer is present to produce a color reaction, confirming the result visually.

None of these definitions describes the dieter's dream molecule: a truly reducing sugar. Still, there is a long history of artificial sweeteners. Sucralose, for example, is a modern artificial sweetener made by chlorinating sucrose (**FIGURE B4.2**). As a result, sucralose can attach itself to the taste receptors (the sweet receptors in cells of the tongue), but is metabolically inert. Thus, sucralose provides the sweetness of sucrose, but none of the calories.

Sucrose itself is the standard of sweeteners for both artificial and natural sugars. The measurement scale is less objective than others, because it requires human tasters. This is especially troublesome for the discovery of new sweeteners that, unlike sucralose, may bear no structural similarity at all to sugars, making it impossible to predict in advance whether a compound is a potential candidate. Typically, a sweetener is discovered by accident; someone working with a chemical product has actually tasted it (and survived!).

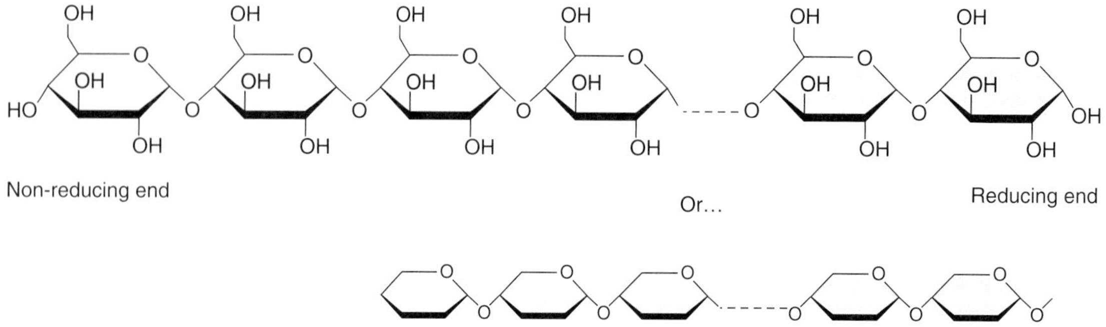

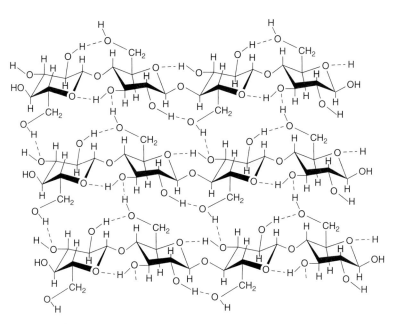

**FIGURE 4.14 Amylose Has Distinct Ends.** The straight-chain glucose polymer amylose has α1→4 bonds. At the left-end residue, the anomeric carbon is engaged in a glycosidic bond, and thus is nonreducing. At the right end residue, the anomeric carbon is free and, thus is reducing. All polysaccharides have these distinctive ends.

## Linear Polysaccharides

Amylose is a biological polysaccharide composed of glucose molecules linked via α1→4 glycosidic bonds (FIGURE 4.14). Like all linear polymers, the amylose chains have distinctive ends, called the **reducing end** and the **nonreducing end**. This property of dissimilar ends also applies to disaccharides. Maltose also has a reducing end and a nonreducing end, albeit a chain of just two monosaccharide units. All unbranched polysaccharides have exactly one reducing end and one nonreducing end. In solution, amylose forms hydrogen bonds between water molecules and the free hydroxyl groups at positions 2, 3, and 6. This makes it an extended polymer with considerable flexibility.

Cellulose, another linear polymer of glucose, has β1→4 linkages, and assumes the very regular three-dimensional solution structure shown in FIGURE 4.15. The specific arrangement results from the

**FIGURE 4.15 Cellulose.** Cellulose does not interact with water because all of its hydroxyl groups are engaged in intracellular hydrogen bonding. As a result, cellulose is completely insoluble in water.

**CHAPTER 4** CARBOHYDRATES

glycosidic bond being placed above the plane of the glucosyl residue. All of the hydroxyl groups in the chain interior are engaged in hydrogen bonds *with each other*. The resulting intramolecular hydrogen bonding gives the molecule enormous internal strength, but it also renders cellulose unable to bind to water. Cellulose, therefore, is entirely *insoluble* in water, despite the existence of multiple hydroxyl groups (**BOX 4.3**). Cellulose is the most abundant molecule in the world, principally found in plant cell walls; it is the major component of wood.

An important reason for using polysaccharides as energy storage molecules in animals has to do with osmosis. Osmolarity depends only on the number, rather than the size (or other qualities) of molecules. Thus, having multiple glucose molecules covalently linked greatly reduces the osmolarity within the cell compared to individual molecules, thereby minimizing osmotic pressure while maintaining a virtual glucose reservoir within the cell. This is critically important for animal cells, which must maintain the same osmolarity on either side of the plasma membrane. By contrast, plants simply have a cell wall—for which cellulose is a key structural component—allowing plant cells to maintain very high internal osmotic pressures without bursting.

## Branched Polysaccharides

Amylopectin is a branched polysaccharide that occurs in plants. In addition to $\alpha1\rightarrow4$ linkages, amylopectin also has $\alpha1\rightarrow6$ linkages commonly called **branch points**. FIGURE 4.16 shows how a single $\alpha1\rightarrow6$ glycosidic bond creates a branch in the overall structure. Branched polysaccharides have a more compact three-dimensional structure than linear chains.

## BOX 4.3: Irony of Intramolecular Hydrogen Bonds

Hydrogen bonding is a key feature that makes certain compounds soluble in water. However, when the groups responsible for hydrogen bonding are linked to a group other than water, they lose their solubility. This commonly occurs when hydrogen bonding exists within the same molecule; that is, the hydrogen bonding is intramolecular. Cellulose, a polymer of glucose found in plants, is a perfect example of a compound that is insoluble in water due to intramolecular hydrogen bonds. These kinds of interactions exist in all biological polymers, however, including nucleic acids and proteins.

A practical application of this property is the development of synthetic molecules called **methylcelluloses**. These are formed in the laboratory by creating methyl esters with some of the hydroxyl groups of cellulose. Partial methylation prevents the remaining free hydroxyl groups from forming internal hydrogen bonds, creating new polymers whose partial solubility in water can be controlled. These polymers can be used as solvents for lipid-soluble drugs. The glucose molecules in cellulose are connected by beta linkages, for which humans have no enzymes to break down, so methylcellulose is biologically inert.

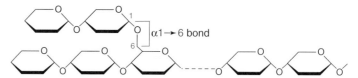

FIGURE 4.16 **A Polysaccharide Branch Point**. The α1→6 bond, found in glycogen and amylopectin, creates a branch point in the polysaccharide structure. With one branch point, there are now *two* nonreducing ends and still one reducing end.

Starch is a mixture of amylose and amylopectin that occurs in plants. Different plants have various proportions of the two polysaccharides, with different polymer lengths and different degrees of branching accounting for the distinctive qualities of corn starch, potato starch, and wheat starch (flour).

A purely branched polysaccharide is **glycogen**, the major form of carbohydrate storage in animal cells. Molecules of glycogen are more branched than amylopectin and are much larger, having hundreds of thousands of glucose residues. While glycogen exists in a variety of animal cells, it is found in large amounts in just two: liver and muscle. Liver stores glycogen in the fed state and releases glucose units from it to the blood during times of fasting. By contrast, muscle takes up blood glucose and uses the glucose residues of glycogen for its own metabolism during active contraction. Thus, the liver's role in glucose balance is *altruistic*, providing for the entire body, whereas the muscle's role is *selfish*, being used only for muscle metabolism. Most of the glycogen molecules in the human body are found in muscle, which has about 1% of its weight as glycogen. Liver can have up to 10% of its weight as glycogen (during the fed state), but represents less of the total body glycogen due to its relatively small total mass.

# 4.5 Carbohydrate Derivatives

A large number of molecules have many of the qualities of carbohydrates, because they are metabolically derived from them. We will consider two categories of such derivatives: simple modifications and substituted carbohydrates. The simple modifications consist of only slightly modified monosaccharides and polysaccharides. In the more heavily substituted carbohydrates, the sugar portion is no longer the dominant chemical property.

## Simple Modifications

The modification of hydroxyl groups by a phosphate ester occurs in many of the metabolic intermediates we will encounter in later chapters. For example, glucose-6-P (the phosphate group is abbreviated as P) is glucose with a single phosphate ester and fructose-1,6-*bis*-$P_2$ has

**CHAPTER 4** CARBOHYDRATES

two phosphate esters (FIGURE 4.17). Another simple modification is to change the oxidation state of one of the carbons. Reduction of the carbonyl carbon results in a sugar alcohol, or polyalcohol, such as glycerol (FIGURE 4.18). Glycerol is a product of fat breakdown and a major additive in prepared foods and drugs. Conversely, oxidation of the carbonyl group produces a **sugar acid**, such as glyceric acid. Oxidation can also occur at other carbons, leaving the carbonyl group intact, as in glucuronic acid. Glucuronic acid is compared to the carbonyl-oxidized gluconic acid in FIGURE 4.19. Note that gluconic acid cannot exist in a ring form.

Replacing an oxygen atom with a nitrogen atom forms another derivative, exemplified by glucosamine (FIGURE 4.20), which is found as a subunit of secreted proteins and in connective tissue. Derivatives in which one of the hydroxyl groups is replaced with a hydrogen atom, as in deoxyribose (Figure 4.20) are called deoxy sugars.

Polysaccharides can also be modified to produce derivatives with altered properties. For example, the molecule **agarose** (found in certain seaweeds) has a repeating D-galactosyl unit linked with a $\beta1\rightarrow4$ glycosidic bond to an L-galactosyl residue that is modified by having an additional internal bridge oxygen between carbons 3 and 6 and a sulfate group esterified to carbon 2 (FIGURE 4.21). Agarose is a linear molecule of this repeating unit, although not all of its L-galactosyl groups are sulfated. A similar polysaccharide is **agaropectin**, which has branch points. Together, the mixture of agarose and agaropectin is known as **agar**.

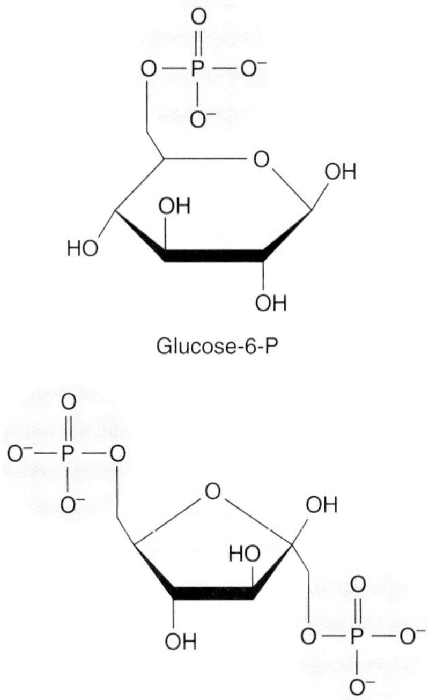

## Substituted Carbohydrates

Nucleotides, lipopolysaccharides, and proteoglycans all contain sugar molecules, but they are so heavily modified by other groups that their chemical features greatly diverge from less substituted sugars.

**Nucleotides** are modified sugar molecules that contain a nitrogenous base and at least one phosphate ester. The bases are either pyrimidines or purines, as shown in FIGURE 4.22. Nucleosides have bases joined to either a ribose or a deoxyribose sugar in an N-glycosidic link, as shown in FIGURE 4.23. Note that the nitrogen attached to the sugar is the 1-position of the pyrimidines but the 9-position of the purines. The sugar portion is further substituted in nucleotides: a phosphate is esterified to one of the available hydroxyl groups. These can be monophosphates or chains of phosphates. FIGURE 4.24 shows the three common nucleotides known as AMP, ADP, and ATP.

Nucleotides are key energy transfer molecules that are widely used in metabolic reactions in cells. Additionally, nucleotides are joined together to form the polymers DNA and RNA, where the sugar component of DNA is deoxyribose and the sugar component of RNA is ribose. DNA exists as two strands noncovalently linked by hydrogen bonds between the bases (FIGURE 4.25). The sugar component of DNA (i.e., deoxyribose) has no remaining free hydroxyl groups. The $2'$ position is substituted by a hydrogen atom and all others are involved

Glucose-6-P

Fructose-1,6-bis-P$_2$

FIGURE 4.17 **Phosphorylated Sugars.**

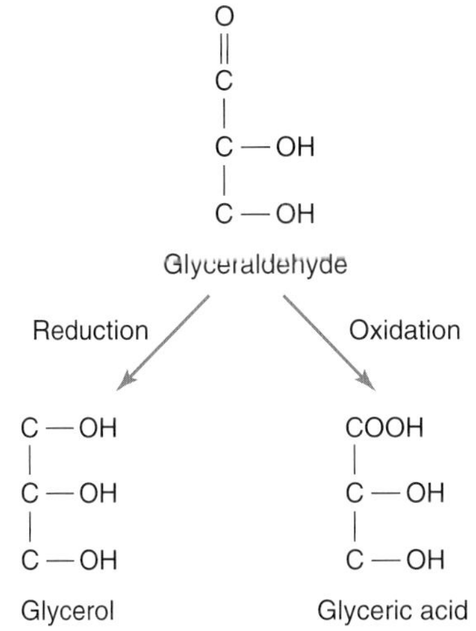

Glyceraldehyde

Reduction / Oxidation

Glycerol          Glyceric acid

FIGURE 4.18 **Redox Derivatives of Sugar.**

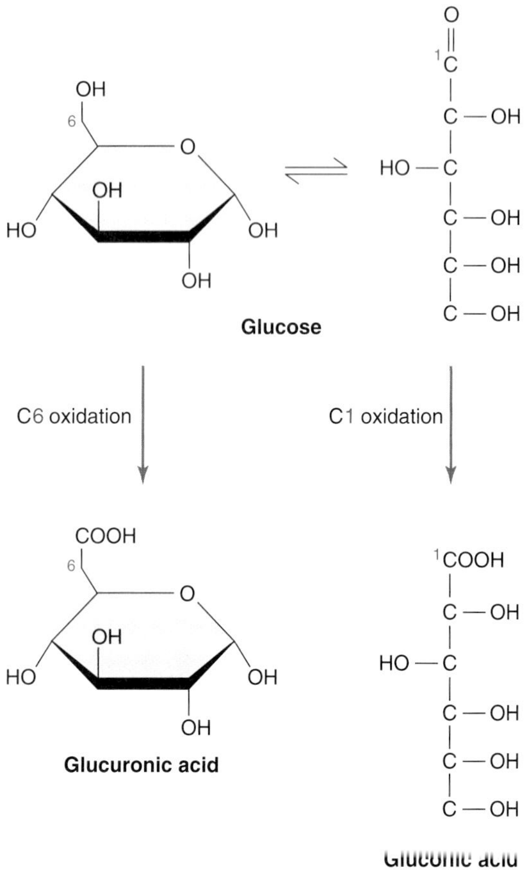

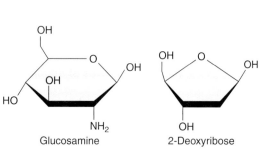

**Glucose**

C6 oxidation  →  C1 oxidation

**Glucuronic acid**

¹COOH
|
C — OH
|
HO — C
|
C — OH
|
C — OH
|
C — OH

Gluconic acid

**FIGURE 4.19 Different Oxidized Glucose Molecules**. Oxidizing glucose at the anomeric carbon forms gluconic acid, which can no longer form a ring or become polymerized. If any other carbon in glucose is oxidized, yielding a compound such as glucuronic acid, it can still form a glycosidic bond and thus be part of a polysaccharide.

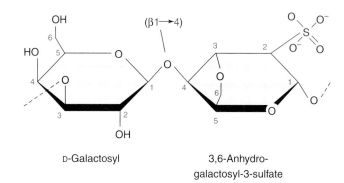

D-Galactosyl        3,6-Anhydro-
galactosyl-3-sulfate

**FIGURE 4.21 Agarose.** The repeating unit of the polysaccharide agarose, a linear modified polysaccharide. The two residues shown are a galactosyl unit and a modified galactose in a β1→4 linkage.

in the linkages that bind the nucleotides together. RNA, unlike DNA, is usually a single-stranded polymer (**FIGURE 4.26**).

DNA provides a template for two cellular processes. First, the molecule can duplicate itself in preparation for cell division, an activity known as **replication**. Second, small regions of DNA can serve as a template for messenger RNA (mRNA) formation, an activity known as **transcription**. In turn, mRNA provides a template for protein synthesis (**translation**). These processes are critical to the formation of new cells or of new expression for cells.

Sugars form conjugates with other biological molecules. For example, some phospholipids are sugar-lipid derivatives such as phosphatidylglycerol and phosphatidylinositol (**FIGURE 4.27**). Sugar–protein conjugates (called *glycoproteins*) are usually divided into two classes: O-linked and N-linked (**FIGURE 4.28**a and b). In the O-linked example, a glycosidic bond is formed between the sugar (the modified sugar N-acetylgalactosamine is shown) and a hydroxyl group from a

Glucosamine        2-Deoxyribose

**FIGURE 4.20 Sugar Derivatives Replacing Hydroxyl Groups.**

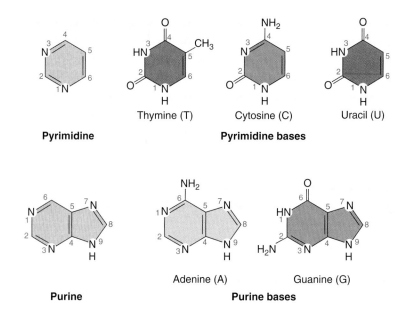

Pyrimidine        Thymine (T)        Cytosine (C)        Uracil (U)

**Pyrimidine bases**

Purine        Adenine (A)        Guanine (G)

**Purine bases**

**FIGURE 4.22 Purines and Pyrimidines.**

protein, here contributed from the side group of an amino acid called serine. The N-linked example in Figure 4.28b shows an N-glycosidic bond, similar to those found in nucleotides. Here, the nitrogen is contributed by the side group of an amino acid called asparagine. An additional protein–sugar conjugate arises when an amine from a lysine side chain in hemoglobin reacts with glucose, followed by rearrangement to the stable ketone shown (Figure 4.28c). This relatively slow reaction produces a form of hemoglobin abbreviated as $HbA_{1C}$. Measuring $HbA_{1C}$ levels provides a measure of the average blood glucose over a period of weeks. $HbA_{1C}$, therefore, is of considerable importance in monitoring diabetics.

## Summary

Sugars have an empirical formula of $CH_2O$, with multiple hydroxyl groups and one carbonyl group in the simple sugars. The smallest monosaccharides are the trioses glyceraldehyde and dihydroxyacetone.

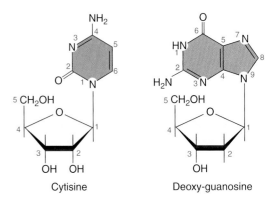

FIGURE 4.23 **Nucleosides.** A base from those shown in Figure 4.22 is attached to a ribose sugar or a deoxyribose sugar. A large number of other combinations are possible, with two exceptions: uracil is not found attached to deoxyribose, and thymine is not found attached to ribose.

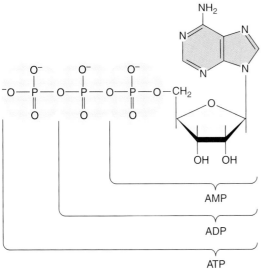

FIGURE 4.24 **AMP, ADP, and ATP.** A nucleotide is a phosphorylated nucleoside. The three examples shown here are the most common nucleotide forms of adenine, used in energy transfer reactions.

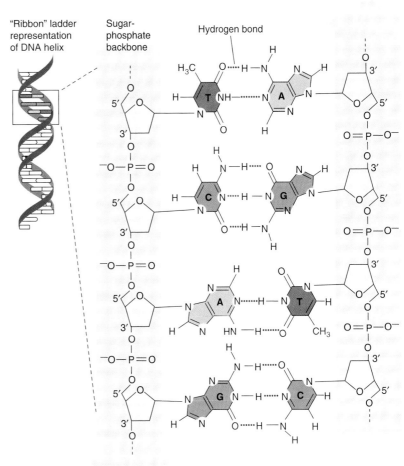

FIGURE 4.25 **DNA.** The two strands of the double-stranded DNA molecule are linked by hydrogen bonding. Within each strand, the sugars are covalently linked via phosphate diester bonds.

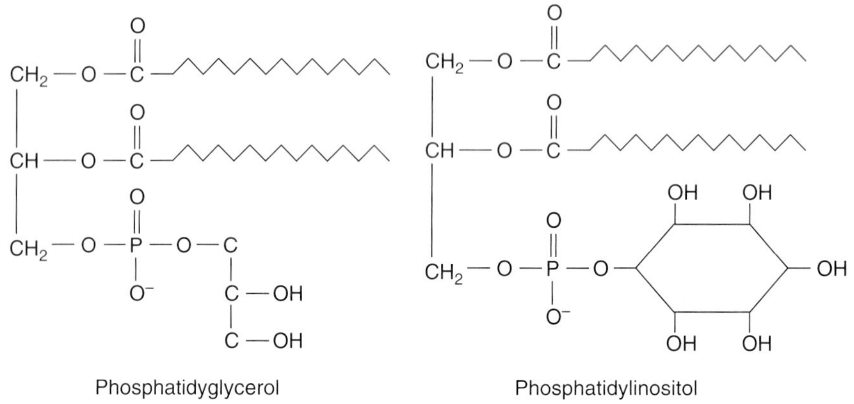

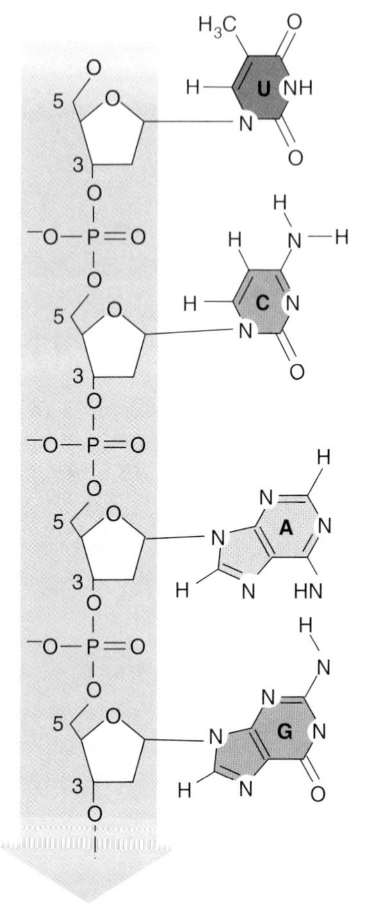

FIGURE 4.26 **RNA.** Like DNA, the sugar portion of RNA binds the chain together, using phosphate diester bonds as shown. The purine and pyrimidine bases also are attached to the sugars of RNA like those of DNA. Unlike DNA, the RNA molecule is typically a single chain.

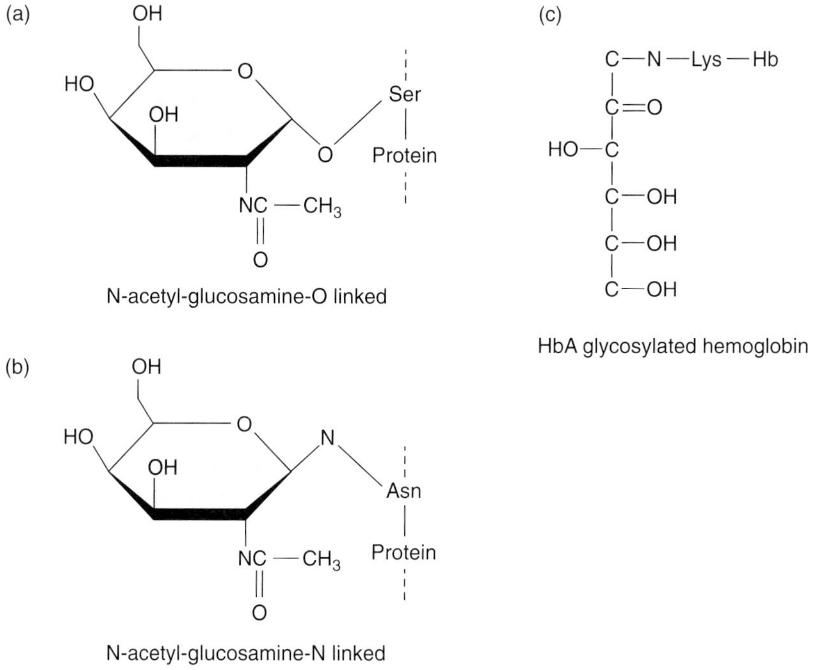

FIGURE 4.27 **Phosphatidylglycerol and Phosphatidylinositol.**

There are two stereochemical forms of glyceraldehyde because it has a chiral center. These are denoted D and L, but biological sugars are overwhelmingly in the D form. Sugars containing more carbon atoms have correspondingly larger numbers of stereoisomers, which are identified by assigning unique names. For example, one six-carbon sugar is glucose; another, differing in configuration at the 4-carbon, is galactose, equivalently expressed as the 4-epimer of glucose. All sugars having at least five carbon atoms form rings in a reaction between a carbonyl group and a hydroxyl group. The formation of rings produces a new chiral center, because a new hydroxyl group emerges from the

FIGURE 4.28 **Sugar-Modified Proteins.** The most common modifications are (a) O-linked and (b) N-linked polysaccharides; these are glycosidic and N-glycosidic bonds, respectively. (c) A nonenzymatic reaction of glucose with hemoglobin produces a specially modified glycosylated hemoglobin.

## BOX 4.4: Medical Connections: Sugars and Digestion

Sugars are normally absorbed by the small intestine. If appreciable amounts are not removed by the time they reach the large intestine, they are metabolized there by the resident bacteria. The bacterial end products include gases which cause distension of the colon, bloating, and pain. Two of the most common causes of this bloating are **lactose intolerance** and **raffinose.** Lactose intolerance is a common genetic deficiency of lactase, an enzyme that breaks down the sugar lactose. Lactase is present in abundance in infants and less so in adults. Because lactose is only present in milk and milk products, and milk is a normal component of the diet only in infancy in all animals but humans, lactose intolerance is strictly a human problem.

A common intestinal disturbance is caused by the presence of raffinose, a sugar found in foods such as cabbage and beans. Raffinose is poorly digested due to the presence of the galactose-$\alpha 1 \rightarrow 6$-glucose bond in the first two sugar moieties of the trisaccharide. As a result, it is also metabolized in the large intestine, with attendant gas production. However, due to the far smaller amounts of raffinose in those foods, the clinical symptoms of raffinose digestion are less severe than those in lactose intolerance.

carbonyl group. The carbon at this locus is called the anomeric carbon and is always involved in joining sugar moieties together, either as disaccharides like sucrose and maltose, or in polysaccharides, such as glycogen and cellulose. The link between sugars is called a glycosidic bond, and it locks the position of the anomeric hydroxyl into two separate stereochemical forms ($\alpha$ and $\beta$). The three-dimensional structure that results from this distinctive orientation is the difference between glycogen, the water-soluble storage carbohydrate in animals, and cellulose, the water-insoluble storage carbohydrate that is abundant in plants. Any sugar that contains an anomeric carbon that is not connected via a glycosidic bond can equilibrate with the open-chain form. Because it can react with a test solution containing an ion that can be reduced to produce a color reaction, it is called a reducing sugar. This describes all monosaccharides and polysaccharides and some disaccharides such as maltose. Both of the anomeric carbons of the bound sugars forming sucrose are engaged in a glycosidic bond, making this disaccharide a nonreducing sugar. Polysaccharides have at least one end—more if there are branches—in which the anomeric carbon is engaged in a glycosidic bond. This is called a nonreducing end. A myriad of carbohydrate derivatives exist in biology, both of the simple monosaccharides (such as phosphorylated sugars and deoxy sugars) and the polysaccharides (such as the sulfated polymer agarose). More elaborate substitutions produce molecules where the properties of the sugar itself are lost, and the sugar merely becomes a connecting molecule, such as the role deoxyribose plays in the structure of DNA. Finally, sugars form conjugates with other biological molecules, as in lipopolysaccharides and proteoglycans.

## Key Terms

agar
agaropectin
agarose
aldoses
alpha (form)
anomeric carbon
beta (form)
branch points
cellulose
chiral
disaccharides
enantiomers
epimers

galactose
glycogen
glycosidic bond
Fischer projection
Haworth projections
hemiacetals
hemiketals
ketoses
lactose intolerance
methylcelluloses
nonreducing end
nonreducing sugars
nucleotides

oligosaccharides
polymers
polysaccharides
polarimeter
raffinose
reducing end
reducing sugar
replication
sugar acid
transcription
translation
trehalose
triose

## Review Questions

1. Raffinose is a sugar that, upon hydrolysis of its glycosidic bonds, yields galactose, glucose, and fructose. The galactose–glucose bond is an $\alpha1 \rightarrow 6$ linkage, and the trisaccharide is a nonreducing sugar. Draw the structure of raffinose.

2. In the DNA structure, the sugar is completely modified. List each modification and identify an analogous, simpler substitution that exists in another sugar molecule. What sugar properties remain in the DNA molecule?

3. Glucose in solution has two ring forms and one open-chain form. How many structures exist in equilibrium for a polysaccharide of glucose that has multiple $\alpha1 \rightarrow 4$ bonds and three $\alpha1 \rightarrow 6$ bonds?

4. Reducing sugars are a mixture of alpha and beta forms. Why does one predominate over another in different sugars?

5. Related to the previous question, suppose you have equal concentrations of glucose and fructose at room temperature. In both cases, beta-ring forms predominate, but with different percentages. In each case, however, when a reducing sugar measurement is made, the same amount of indicator ion is reduced in each case. Moreover, the amount of directly reacting species for the indicator ion is far less than 1% of the total forms. Explain these findings.

6. One indication of the distinction in physical properties between simple sugars (like monosaccharides and disaccharides) and polysaccharides is their observed behavior in solution. For example, only starch can form a paste in water. Speculate on how that can be explained from the structural differences between these types of molecules.

## References

1. DeMan, J. M. *Principles of Food Chemistry*; Springer Publishing: New York, 1999, Chapter 4, pp. 163–208.
   A treatise on the chemistry of food, the chapter on carbohydrates includes many of the biochemical properties of sugars in more extensive detail than is described in the present text.

2. Semenza, G.; Auricchio, S. Small-Intestinal Disaccharidases. In *The Metabolic Basis of Inherited Disease*; Scriver, C. R.; Beaudet, A. L.; Sly, W. S., Eds.; McGraw-Hill: New York, 1995, Chapter 22, pp. 4451–4480.
   This is an authoritative reference work on inherited diseases that affect metabolic pathways. In this chapter, the inborn errors associated with lactose and other disaccharides are described in detail.

3. Sinnott, M. L. *Carbohydrate Chemistry and Biochemistry: Structure and Mechanism*; Royal Society of Chemistry: Cambridge, UK, 2007, p. 748.
   This is a more chemically oriented description of carbohydrates.

**CHAPTER 4** CARBOHYDRATES

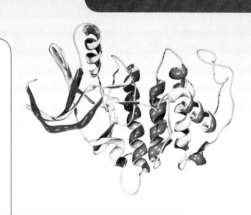

# 5

# Amino Acids and Proteins

## CHAPTER OUTLINE

Image © Dr. Mark J. Winter/Photo Researchers, Inc.

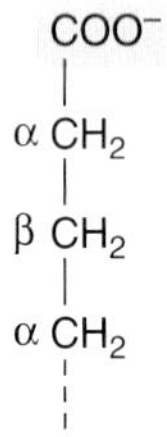

FIGURE 5.1 **Organic Carboxyl Bearing Chain Numbering**. This system is usually used only for the first few carbons adjacent to a carboxyl group.

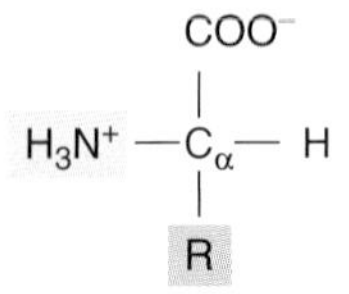

FIGURE 5.2 **General Amino Acid.**

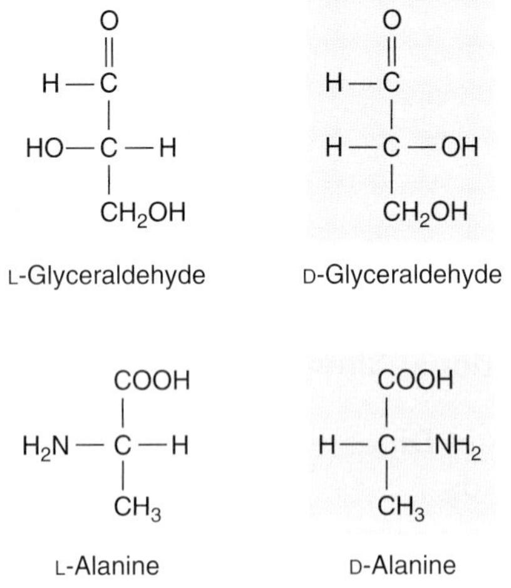

FIGURE 5.3 **Amino Acid Stereochemistry.** Glyceraldehyde is the reference compound. Most biological amino acids have the L-configuration.

The word **protein** is of Greek origin, meaning *first place* or *primary*. Berzelius used this label in 1835 to identify a substance found in plant fibers that was of prime importance to animal nutrition. This was well before the discovery of proteins themselves. In fact, proteins are the most diverse biomolecules. They include structural proteins (such as collagen), binding proteins (such as albumin), and enzymes, the biological catalysts (such as lactase). In this chapter, we examine protein structure and some elements of their binding behavior.

# 5.1 Identity and Roles of Amino Acids

All **amino acids** contain an amine group (most commonly a primary amine), and a carboxyl group (the acid portion), attached to the same carbon. The latter is called the **α-carbon**, from an organic chemistry nomenclature system that assigns Greek letters to carbon atoms adjacent to carboxyl groups: α, β, etc. (FIGURE 5.1). Note that this is a distinctive use of Greek lettering from the carbohydrates. Also bonded to the α-carbon are a hydrogen atom (the **α-hydrogen**) and a variable group designated as R (FIGURE 5.2). Twenty different R groups comprise the *common amino acids*, meaning those incorporated into most proteins. Except for glycine (in which R is a hydrogen atom), the α-carbon is chiral and designated as either L or D by comparison to the reference molecule glyceraldehyde. This is illustrated in FIGURE 5.3 for the case of alanine (in which R = $CH_3$). The carboxyl group of the amino acid is most similar to the carbonyl of glyceraldehyde; the amine group of the amino acid is most similar the OH group of glyceraldehyde. In nature, virtually all amino acids are present in the L form.

# 5.2 Amino Acid Individuality: The R Groups

The R groups chemically distinguish amino acids from one another. Table 5.1 lists the names, abbreviations, and $pK_a$ values of the 20 common amino acids. We categorize them by polarity and by functional group. These different views provide insight into amino acid behavior.

## Polarities

The overall polarity of amino acids depends on their R groups. FIGURE 5.4 shows the polar amino acids subdivided into neutral, acidic, and basic classes. FIGURE 5.5 shows the nonpolar amino acids, which are further subdivided into alkyl chains, branched chains, aromatics, and a pair of unique nonpolar structures.

## Functional Groups

FIGURE 5.6 presents a chemical categorization of some of the amino acids. The acids include not only aspartate and glutamate, but also

## TABLE 5.1  Properties of Amino Acids

| Name | Abbreviations | | p$K_a$ COOH | p$K_a$ NH$_3$ | p$K_a$ R group |
|---|---|---|---|---|---|
| Alanine | Ala | A | 2.35 | 9.87 | |
| Arginine | Arg | R | 2.18 | 9.09 | 13.2 |
| Asparagine | Asn | N | 2.18 | 9.09 | 13.2 |
| Aspartic Acid | Asp | D | 1.88 | 9.6 | 3.65 |
| Cysteine | Cys | C | 1.71 | 10.78 | 8.33 |
| Glutamic Acid | Glu | E | 2.19 | 9.67 | 4.25 |
| Glutamine | Gln | Q | 2.17 | 9.13 | |
| Glycine | Gly | G | 2.34 | 9.6 | |
| Histidine | His | H | 1.78 | 8.97 | 5.97 |
| Isoleucine | Ile | I | 2.32 | 9.76 | |
| Leucine | Leu | L | 2.36 | 9.6 | |
| Lysine | Lys | K | 2.2 | 8.9 | 10.28 |
| Methionine | Met | M | 2.28 | 9.21 | |
| Phenylalanine | Phe | F | 2.58 | 9.24 | |
| Proline | Pro | P | 1.99 | 10.6 | |
| Serine | Ser | S | 2.21 | 9.15 | |
| Threonine | Thr | T | 2.15 | 9.12 | |
| Tryptophan | Trp | W | 2.38 | 9.39 | |
| Tyrosine | Tyr | Y | 2.2 | 9.11 | 10.07 |
| Valine | Val | V | 2.29 | 9.74 | |

cysteine, histidine, and tyrosine, as these can act as acids under some biological conditions. All of the p$K_a$ values of these acids are given in Table 5.1. The bases include lysine, arginine, and histidine. Thus, histidine can act as *both* an acid and a base, befitting its p$K_a$ value of about 6.0.

The three amino acids bearing hydroxyl groups in Figure 5.6 are distinct. Serine has a primary hydroxyl, whereas threonine has a secondary hydroxyl and tyrosine has an aromatic hydroxyl. As a result, these three amino acids have different reactivities. For example, while all of the hydroxyl-containing amino acids can become phosphorylated when they are part of a protein chain, their chemical rates of dephosphorylation increase in going from ser to thr to tyr. Sulfur appears in two amino acids: as a sulfhydryl in cysteine and as a thioether in methionine. Unlike oxygen, sulfur is a **soft nucleophile**, because its valence electrons are more distant and more shielded from the nucleus. As a result, it more readily reverses its nucleophile additions than oxygen. In proteins found in the cell exterior, two separate cysteine –SH groups can form a bridge with each other, –S–S–, called a disulfide link. Proteins inside of cells are maintained largely in the sulfhydryl form.

For all amino acids, alterations in pH affect the charge of the molecules due to their dissociable acidic and basic groups. This is the feature of the amino acids we consider next.

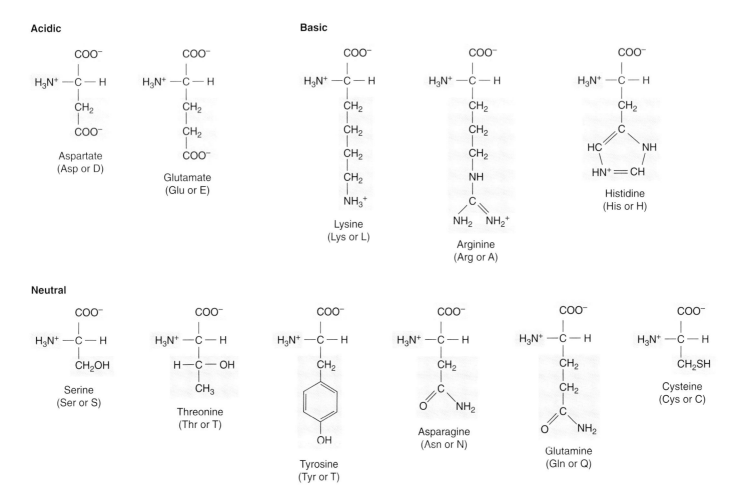

**Acidic**

Aspartate
(Asp or D)

Glutamate
(Glu or E)

**Basic**

Lysine
(Lys or L)

Arginine
(Arg or A)

Histidine
(His or H)

**Neutral**

Serine
(Ser or S)

Threonine
(Thr or T)

Tyrosine
(Tyr or T)

Asparagine
(Asn or N)

Glutamine
(Gln or Q)

Cysteine
(Cys or C)

FIGURE 5.4 **Polar Amino Acids.** The three subdivisions shown are acidic, basic, and neutral.

# 5.3 Acid–Base Properties and Charge

### Titration and Net Charge

The fundamental relationship between acid–base dissociation and charge is defined in the Henderson–Hasselbalch equation, which expresses the ratio of dissociable groups in terms of the $pK_a$ and pH. In this section, we explore the consequences of having more than one dissociable group in a single molecule, which is always the case for amino acids.

FIGURE 5.7 shows a titration curve of glycine, in which we start with a fully protonated molecule and add increasing amounts of a base, such as NaOH. Notice that the shape of this titration curve resembles that of a single dissociable group, but repeated so that two rather than one equivalent of NaOH is required to remove both protons from the amino acid. There are two **buffering regions**, corresponding to the carboxyl group and the amine. These flat regions represent a minimal change in pH with added acid or base, over a span of about $\pm$ 1 pH unit. The center point of a flat region, extrapolated to the pH axis, corresponds to the $pK_a$ value for a dissociable group. All of the $pK_a$ values for the amino acids are listed in Table 5.1.

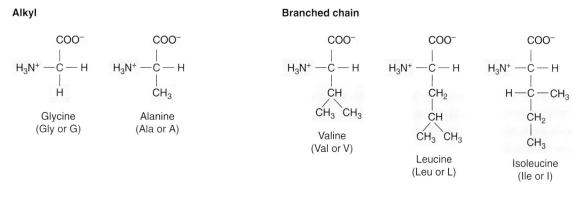

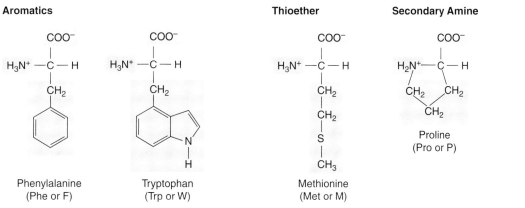

**FIGURE 5.5 Nonpolar Amino Acids.** The three major subdivisions are alkyl, branched chain, and aromatic. Methionine and proline are the only examples of a thioether and secondary amine, respectively.

Five regions of the curve are marked A through E; these structures are displayed in FIGURE 5.8, which shows the relationship between molecular charge, pH, and p$K_a$. Structure A exists at a pH value well below p$K_1$. In this region, both groups in A are protonated, and the molecule has an overall charge of +1. Structure E exists at pH values well above p$K_2$; both groups are deprotonated and the overall charge is −1. Structure C prevails when the pH is above p$K_1$ but below p$K_2$, so

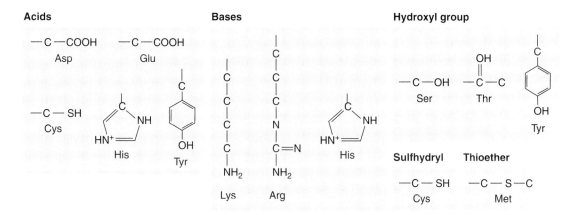

**FIGURE 5.6 Functional Groups in Amino Acids.** A cross-categorization of amino acids groups together all acids, bases, and hydroxyl groups. Cysteine and methionine are the only examples of a sulfhydryl group and a thioether, respectively.

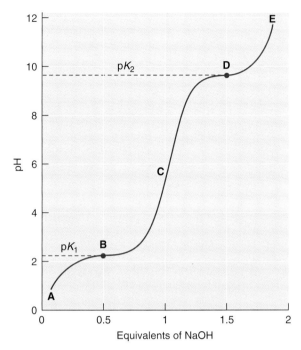

FIGURE 5.7 **Titration Curve for Glycine.** The molecule has two titratable groups; the titration curve appears as if two separate dissociable groups were titrated sequentially. The flat regions centered at B and D are also called *buffer regions* because altering the acidity or basicity has little effect on the pH in those regions. The center of those regions, corresponding to the midpoint of the equivalent of base added, extrapolates to the p$K_a$ value on the pH scale.

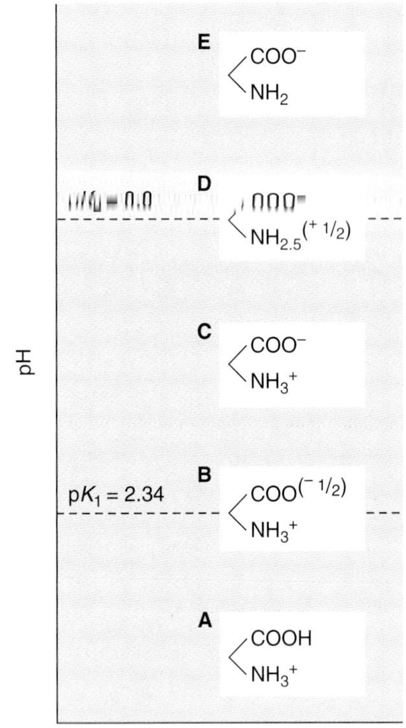

this molecule has a net charge of zero: −1 for the carboxyl group and +1 for the amine. Structure B, which corresponds to the point where pH = p$K_1$, the carboxyl group has a charge of −½. This is because p$K_1$ corresponds to the COOH group, and half of the molecules have a zero charge and half have a −1 charge. Because the amine proton has a charge of +1, the overall molecule has a charge of −½ + 1 = ½. In a similar way, the net charge of structure D is −½ (**BOX 5.1**).

## BOX 5.1: Fractional Charges

For a chemical species, we expect the value of the charge to be a whole number, such as +1 for the sodium ion. For weak acids, however, the charge depends on the pH. We have seen that this value may still be closely approximated by a whole number at certain extremes of pH. But at pH values close to the p$K_a$, a significant amount of both charged and uncharged species exist in solution. In this case, we must calculate the *net charge* of the molecules in solution. Fractional charges thus do not apply to any one molecule, but to an average. For a single dissociable group, this fraction lies between 0 and −1, or between 0 and +1, depending upon whether the group is acidic or basic. We have considered only the case where pH = p$K_a$. In this instance, exactly half of the molecules are charged and half of them are neutral. If this group is acidic, the *net charge* will be −½, whereas if this group is basic, the *net charge* will be +½).

FIGURE 5.8 **Charged Forms of Glycine Over Its Titration Range.** Different points taken from Figure 5.7 are drawn for glycine in this "conceptual titration" figure. At very low pH, the molecule is entirely protonated, having a charge of 1.0 (A). At very high pH, the molecule has a charge of −1.0 (E). Form (C) has exactly zero charge in the middle point of pH between p$K_1$ and p$K_2$. The other forms have half-charges because half of the groups at the exact titration midpoint—that is, at the p$K_a$— have a charge of zero and half have a charge of −1 (for B) or +1 (for D).

## Zwitterions

Species C in Figure 5.8 has two charged groups, and yet the molecule has zero charge overall. This form is known as a **zwitterion**, from the German word *zwitter,* meaning a hybrid of two forms. At most physiological pH values, amino acids without additional acidic or basic groups are in this form. The zwitterion is favored over a wide range of pH values because of the electrostatic interaction between the carboxyl group and the amine. Note in Table 5.1 that most values for $pK_1$ are near 2.0. An isolated carboxyl group, however, such as the one in acetic acid, has a $pK_1$ of 4.8. Typical values of $pK_2$ are just above 10.0. The $pK_a$ of an isolated amine such as methylamine is somewhat less, about 9.8. Because both carboxyl and amine groups are attached to the same carbon, they interact (a **through-space** effect). A proton tends to leave the carboxyl group more readily and attach to the amine due to the attraction of opposite charges. This explains the divergence of the $pK_a$ values for the groups compared to the isolated groups used for comparison.

The pH at which the zwitterion has *exactly* a zero charge is called the **isoelectric point**. Symbolized as pI, its value is calculated by averaging the two $pK_a$ values:

$$pI = (pK_1 + pK_2)/2 \qquad (5.1)$$

While seemingly a straightforward calculation, this average *requires* the use of $pK_a$ values rather than $K$ (dissociation equilibrium constant) values (see Appendix: Geometric Mean).

## Multiple Dissociable Groups

Several amino acids have dissociable groups attached to the α-carbon in addition to the carboxyl and amine groups. For example, acidic amino acids have extra carboxyl group and basic amino acids have an extra amine. Any other functional group that can dissociate can be treated similarly, such as the –SH group of cysteine.

Lysine has three dissociable groups. Its titration curve is very similar to Figure 5.7, except for a third segment, with yet another flat region surrounding the third $pK_a$. The major species and regions are shown in FIGURE 5.9. There are now three $pK_a$ values and seven structures represented. Again, the net charge of the molecule is simply the sum of the charges of each functional group. In order to determine the pI, we first find the region in which the species with zero charge exists. From Figure 5.9, this is structure (E), which is flanked by $pK_2$ and $pK_3$. Hence, for lysine, pI = $(pK_2 + pK_3)/2$. Thus, determination of pI requires that we first find the zero-charged species, and then average the two $pK_a$ values that surround it.

The ideas discussed here can be extended to more than three charges, although when $pK_a$ values are relatively close together, the individual titration curves tend to run together, rendering the buffering regions indistinct. Moreover, when brought together in a protein structure, through-space interactions may influence the $pK_a$ values.

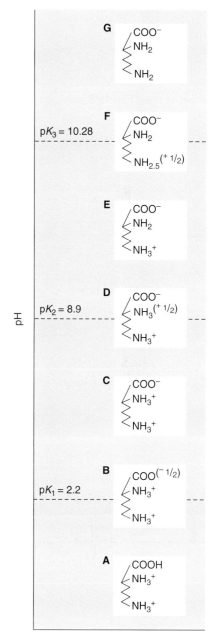

FIGURE 5.9 **Charged Forms of Lysine Over Its Titration Range.** Similar to the forms of glycine in Figure 5.8, lysine has an extra dissociable group, three $pK_a$ values, and thus seven regions, with charge ranging from +2 (acidic) to −1 (basic).

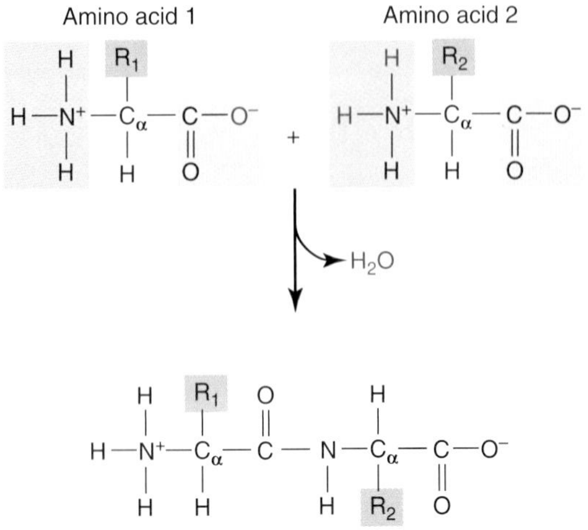

Amino acid 1                    Amino acid 2

Formation of the Peptide Bond.

(a)

(b)

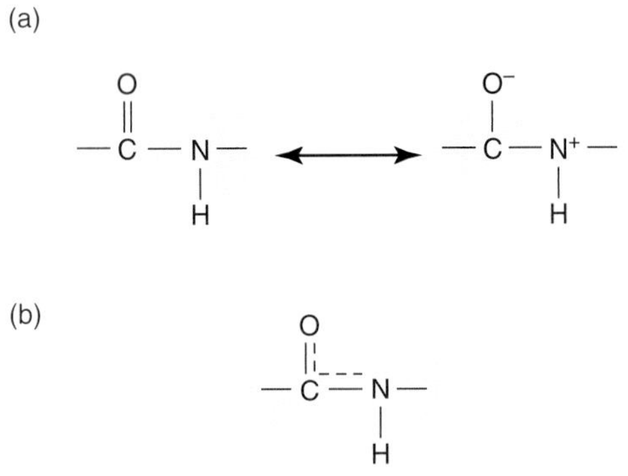

FIGURE 5.10 **Resonance in the Peptide Bond.**

How amino acids are modified once they occur in proteins largely depends upon how they are joined. The joining itself is a covalent linkage called the *peptide bond*.

# 5.4 The Peptide Bond

Amino acids are joined together by covalently linking the α-amino group of one amino acid with the α-carboxyl group of another. Formally, the reaction is a dehydration (i.e., water if formed) and it can be written as in the chemical formulae below. The actual mechanism is more complex than this chemical equation suggests. During protein synthesis, this reaction requires several proteins and a nucleic acid scaffold. Here, we are concerned only with the bond that is formed and its properties.

The bond is an **amide**; when it occurs between amino acids, it is commonly called a **peptide bond**. The peptide bond has resonance properties, as shown in FIGURE 5.10a. The electrons between the O, C, and N, are delocalized, accounting for the increased stability of this bond. The resonance can also be represented by drawing partial double bonds between the three atoms as in Figure 5.10b. Because of this resonance, the peptide bond is chemically more stable than the corresponding ester bond. The C–N bond of the amide, moreover, has partial double bond character that prevents free rotation about this bond. This in turn restricts the possible three-dimensional arrangements for structures containing peptide bonds.

Once the amine and carboxyl groups react to form a peptide bond, they no longer have acid–base characteristics. Only the carboxyl group on one end of the amino acid chain and the amine group on the other end retain dissociable groups, along with the side chains (i.e., the R groups) that extend from the chain. The ends of the chain are therefore distinct: there is one **amino end** (called the **N** end) and one **carboxyl end** (called the **C** end). Beyond the loss of the numerous dissociations, the polymer itself has distinct properties. These depend in part on how many amino acids are joined via peptide bonds.

# 5.5 Peptides and Proteins

Short chains of amino acids—usually less than 25 or so amino acids—are called **peptides**; longer chains are called **proteins**. While the size cut-off is somewhat arbitrary, there is an operational difference. Peptides behave more like *small molecules*, whereas proteins have distinctive polymer behavior. This is analogous to the distinction between small molecular weight sugars and polysaccharides; the polymers are generically known as **macromolecules**.

Two examples of small peptides, both of which control blood pressure in humans, are presented in FIGURE 5.11. Angiotensin II is a linear peptide, whereas vasopressin is cyclic, the result of a bridge formed

**CHAPTER 5** AMINO ACIDS AND PROTEINS

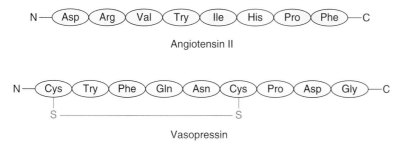

**FIGURE 5.11 Angiotensin II and Vasopressin: Representative Peptides.**

between the sulfhydryl groups of the cysteine residues. Hence, vasopressin has a more constrained, compact structure than angiotensin II.

Some peptides are formed from proteins by breakdown processes, such as the one that occurs in the stomach as part of the mammalian digestive process (**BOX 5.2**). The more elaborate structures of proteins are traditionally placed into four *levels* of structure.

# 5.6 Levels of Protein Structure

The *levels* of protein structure form a hierarchy: primary, secondary, tertiary, and quaternary. **Primary structure** is the sequence of amino acids in a linear protein chain. **Secondary structure** refers to a regular, repeating structural element that is formed when a segment of the chain folds in three dimensions. **Tertiary structure** refers to the entire three-dimensional arrangement, or conformation, of a protein chain in three dimensions. Finally, **quaternary structure** is the three-dimensional arrangement of multi-subunit protein chains that associate noncovalently. A separate level of protein structure, called the **domain**, has been inserted into this hierarchy between the secondary and tertiary structures. A domain is a combination of secondary structural elements that has a specific function, such as a binding pocket in an enzyme. Domains exist within many different proteins.

## BOX 5.2: Protein Digestion: Proteins, Peptides, and Amino Acids

Proteins are broken down into amino acids in a multistep process in humans. First, proteins are precipitated by the strongly acidic environment of the stomach, and then they are exposed to a **protease**, an enzyme that hydrolyzes peptide bonds. After a few hours in the stomach, most of the proteins are converted to peptides, which next enter the intestine. In the intestine, they are subjected to further proteolysis by a large number of distinct proteases that produce amino acids, dipeptides, and tripeptides. Dipeptides and tripeptides are further hydrolyzed by enzymes attached to intestinal epithelia (i.e., the lining cells of the intestine, called *enterocytes*) to free amino acids that are then transported into the epithelia, and ultimately into the bloodstream. In short, proteins enter the stomach, peptides leave it, and amino acids enter the bloodstream.

## Primary Structure

The amino acid sequence—the primary structure—is the fundamental physical description of a protein. The primary structure is a complete description for the peptides. Proteins, however, have more fixed three-dimensional structures. The primary structure of proteins is usually determined experimentally from the coding sequence of the messenger RNA for the protein. In the past, the primary structure was established by degrading proteins with proteases. Certain proteases are selective for amino acids on the N or C side of peptide bonds. Analysis of the resulting protein fragments provides clues the amino acid sequence of the protein. This procedure is still used for the small peptides that result from proteolysis.

The primary structure of a protein determines its complete spatial arrangement, assuming that the chain folds into its lowest-energy conformation. The theoretical prediction of that three-dimensional structure, however, is far from simple. Despite decades of study and increasingly sophisticated computer power, it is still not possible in most cases to predict the three-dimensional structure computationally.

## Secondary Structure

Secondary structures are single elements of regular three-dimensional structure that are recognizable in most proteins. The two main secondary structures are the α-helix and the β-sheet (FIGURE 5.12). A third type of secondary structure is called either a **loop** or a **random coil**. Random coils are indeterminate in structure and serve to link the other secondary structures together.

### α-Helix
Within the spiraling chain of the α-helix (FIGURE 5.13), hydrogen bonds form between the N and O atoms of aligned peptide bonds. This produces

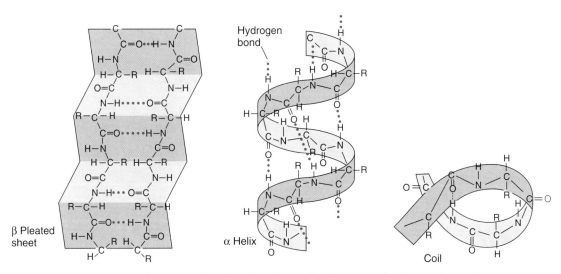

FIGURE 5.12 **Secondary Structures in a Protein**. Shown in this generalized protein are the secondary structures: the α helix, the β–sheet, and the random coil.

**CHAPTER 5** AMINO ACIDS AND PROTEINS

an inner core that is entirely hydrophobic, for the same reason as cellulose: All of the hydrogen bonds are satisfied intramolecularly. This leaves none to interact with water. The water solubility of the entire segment is determined by the nature of the R groups. If the latter are all hydrophobic, then the entire helix is hydrophobic. Commonly, membrane-embedded proteins have at least one segment that passes through the membrane; this exists as an α-helix composed entirely of hydrophobic amino acids. Two examples are illustrated in FIGURE 5.14. In one, the protein makes one pass through the membrane. In the other, the protein makes multiple passes, a common arrangement for receptor proteins and ion channels.

Certain amino acids favor the formation of a helix, whereas others hinder it. For example, hydrophobic amino acids or isolated charged side chains generally permit helix formation. On the other hand, electrical repulsion due to neighboring charges (say from multiple aspartate residues) prevents the regular spiral from forming; these amino acids thus act as **helix breakers**.

### β-Sheet

The β-sheet is also known as the β-pleated sheet because it resembles a corrugated plate (FIGURE 5.15). The β-sheet is stabilized by hydrogen bonds between backbone peptide bonds, as in the α-helix. Note that forming parallel strands requires a turn to exist in the structure as the chain must fold back on itself. The amino acids involved in the turn do not participate in hydrogen bonding. The alignment of the chains can occur in two different ways. If the strands run in the same direction, then the sheet is said to be **parallel**; if they run in opposite directions, then it is said to be **antiparallel**. This is schematically represented in FIGURE 5.16 by drawing an arrow to represent the direction of the strand.

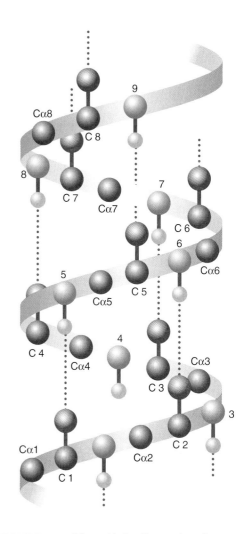

FIGURE 5.13 **The α Helix**. Several amino acids are numbered 1 through 8. The repeated backbone of $C_\alpha C'N\ C_\alpha C'N\ C_\alpha C'N \ldots$, where C represents the carbonyl carbon, makes up the spiral of the helix. The intrachain hydrogen bonds between the carbonyl oxygen (red) and the NH (blue) are indicated as dotted redlines. Note that all possible hydrogen bonds are engaged, making the entire internal segment hydrophobic.

### Domains

Certain groupings of secondary structure that occur in many different proteins are called **domains**. Domains are considered to exist

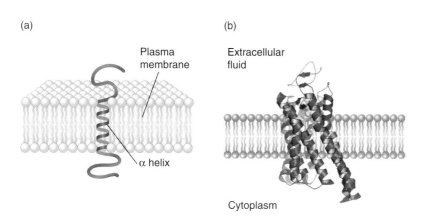

FIGURE 5.14 **Membrane-Spanning α Helices**. (a) The α helix spans the membrane a single time; certain hormone receptors have this structure. (b) The protein spans the membrane several times, so that multiple α helices are present within the membrane. (Courtesy of Heidi E. Hamm and Will Oldham, Vanderbilt University Medical Center.)

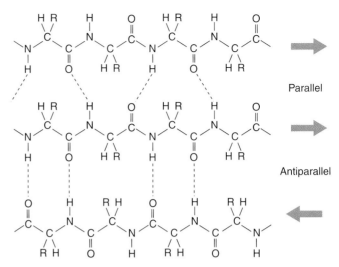

**FIGURE 5.15 The β Sheet.** Hydrogen bonding in the β sheet stabilizes this structure into a pleated arrangement.

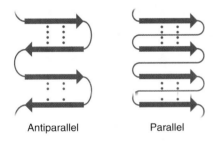

**FIGURE 5.16 Antiparallel and Parallel β-Sheets.** The schematic arrangement of both types of β-sheets are indicated, with large arrows representing the β-sheet portion and curved lines representing the random coil sections that join them.

between the secondary and tertiary levels of structure. The reason for the interest in domains is that they often correspond to a definite function. The interchangeable terms **motif** and **supersecondary structure** are synonyms for the domain structure (**BOX 5.3**).

The **TIM domain** FIGURE 5.17, is donut-shaped, with an inner portion composed of a β-sheet and an outer portion of α-helices. The inner portion also exists on its own in some proteins, where it is known as a β-barrel. The TIM domain was named for the enzyme **triosephosphate isomerase**, in which this domain is a portion of the active site. The TIM domain is the one most commonly found in proteins.

The **SH2** domain and the **PTB** domain, illustrated in FIGURE 5.18, have similar functions. Both of these domains bind phosphorylated tyrosine residues in proteins, serving as part of signal transduction systems. The SH2 domain was first found in the **src** or sarcoma virus protein, from which the name for this domain—Src Homology—is formed. PTB stands for phosphotyrosine binding domain.

FIGURE 5.19 illustrates the **NAD binding domain**, which serves to bind an important redox transfer cofactor used by a large number

## BOX 5.3: Domain Names

We have used the term *domain* to indicate a structural level that exists between secondary and tertiary structures. Others have used terms such as *motif* and *supersecondary structure,* which also refer to a structural level between secondary and tertiary. While it is possible to distinguish between motifs and domains, some chemists use them interchangeably. Because there are no formal distinctions between these terms, we treat them as synonyms. Note that the identities of the various protein domains themselves have no systematic classification as yet, despite the fact that domains have been known for 30 years.

**CHAPTER 5** AMINO ACIDS AND PROTEINS

of enzymes. The **EF-hand** domain (FIGURE 5.20) selectively binds $Ca^{2+}$ ions in both enzymes and dedicated binding proteins such as **calmodulin**. This relatively small domain (about 40 amino acid residues) is a helix-loop-helix design that serves to selectively and tightly bind $Ca^{2+}$. Calmodulin is itself bound to certain proteins, and confers $Ca^{2+}$ sensitivity to its attached protein through a conformational change. This is one way in which an elevation in cellular $Ca^{2+}$ can be sensed by a protein that does not itself bind $Ca^{2+}$: the ion binds the EF hand domain, which alters the structure of calmodulin and subsequently a target protein.

## Tertiary Structure

The three-dimensional structure of an entire protein chain is called the *tertiary structure*. This includes all aspects of lower level structures. We may also view the tertiary structure as a collection of domains joined by loops. For the many proteins that consist of a single chain of amino acids, this is the final level of structure.

## Quaternary Structure

Proteins that consist of more than one chain have quaternary structure. The individual subunits typically are bound together noncovalently, often through charge interactions or hydrophobic interactions at the interface of the subunits. The individual protein chains (subunits) are called **monomers**. The entire protein may contain two, three, four, or occasionally more monomers. Some proteins contain identical monomers, which occurs in many enzymes. Other proteins are composed of different subunits, which may be structurally similar, as in the case of hemoglobin, or extremely different, as is the case in most ion channels.

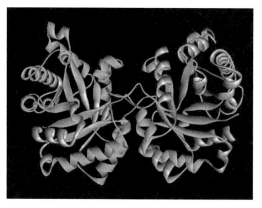

**FIGURE 5.17 The TIM Domain.** This domain has a central region of β sheet and an outer ring of α helices. The most common of the domain structures, it is named for triose phosphate isomerase, the enzyme in which it was first discovered. Protein Data Bank 3TA6. Connor, S.E., Capodagli, C.G., Deaton, M.K., and Pegan, S.D. 2011. *Acta Crystallogr.* 67:1017–1021.

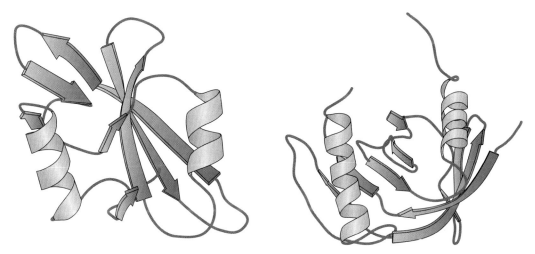

**FIGURE 5.18 Phosphotyrosine-Residue-Binding Domains.** Both the SH2 (for src-homology) and PTB (phosphotyrosine binding) domains recognize the same structure: a phosphorylated tyrosine residue of a protein.

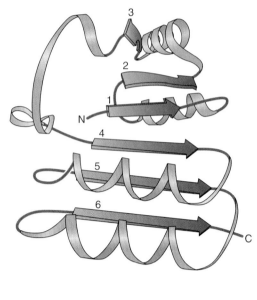

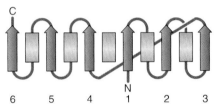

FIGURE 5.19 **The NAD-Binding Domain.** This domain is present in many enzymes that bind a cellular redox cofactor called NAD⁺. Due to the widespread occurrence of NAD⁺, there is a correspondingly widespread occurrence of this binding domain.

In proteins with quaternary structure, an alteration in the conformation of one subunit may affect a neighboring subunit. This allows for more extensive regulatory interactions. However, many proteins possess quaternary structure and yet have no known regulatory features.

With this final level of protein structure, we have a complete description of a protein's spatial arrangement. We have not yet considered the basis for the assumption that the primary structure determines all higher levels of structure, however, and address this topic next.

# 5.7 Protein Folding

The enzyme ribonuclease (RNase) is a digestive enzyme secreted from the pancreas. The enzyme is commonly used to study enzyme properties as it is relatively easy to purify and is exceptionally stable. An experiment conducted in 1959 by Christian Anfinsen using RNase is the basis of the assertion that the primary structure determines the higher levels of protein structure. The essence of this experiment is illustrated in FIGURE 5.21. First, the enzyme was inactivated with heat and then allowed to cool. On cooling, the preparation demonstrated enzyme activity once again; that is, it was able to catalyze the hydrolysis of phosphodiester bonds in RNA. This experiment suggests that proteins have an active conformation that can be disrupted (unfolded) and that it can subsequently refold into an active configuration. Because this was a purified preparation, no other components were necessary for the renaturation. Evidently, only the amino acid sequence is necessary for a protein to achieve its active, three-dimensional structure.

The discovery of the bacterial protein GroEL (FIGURE 5.22) in the 1970s greatly enhanced our understanding of protein folding. During the synthesis of cellular proteins, GroEL helps them fold correctly. Hence, while proteins can refold *in vitro*, the rate of this assembly is generally inadequate within cells. GroEL and similar proteins, called **chaperone proteins**, serve to catalyze the folding process within the cell. Figure 5.22 shows a portion of the active assembly; several of these rings stack and form a cage within which correct protein assembly is accelerated.

# 5.8 Oxygen Binding in Myoglobin and Hemoglobin

Myoglobin and hemoglobin are two oxygen-binding proteins in mammals. They have very similar protein structures, except that myoglobin is a single chain, whereas hemoglobin has four monomer units. In myoglobin, the protein chain has a molecular weight of about 17,000, and it is tightly bound to a molecule of **heme.** In hemoglobin, each protein chain is also about 17,000 in molecular weight (there are two slightly different chains; 2α and 2β subunits); a total of four heme molecules bind each hemoglobin tetramer. The function of heme in both cases is to reversibly bind oxygen. Hence, these proteins only *indirectly* bind to oxygen, relying on oxide formation by the heme iron. In general, a small molecule like heme, which is bound to a protein to assist its function, is called a **prosthetic group**.

The structure of heme (FIGURE 5.23) shows four linked **pyrrole** rings (a nitrogen heterocycle); each has a nitrogen atom chelated to a central $Fe^{2+}$. The $Fe^{2+}$ has a **coordination number** of six, which means that it forms six directed bonds to electron-rich atoms. Beyond the four already noted, there are two more not shown in Figure 5.23. One of these is bound to a histidine residue of myoglobin itself. The sixth and final coordination site is occupied by oxygen (in oxygen-bound myoglobin) or is unoccupied (in free myoglobin). Abbreviating myoglobin as Mb, the binding to oxygen can be written as an equilibrium dissociation:

$$MbO_2 \rightleftarrows Mb + O_2 \tag{5.2}$$

The equilibrium constant for this reaction is:

$$K_{diss} = [Mb][O_2]/[MbO_2] \tag{5.3}$$

In order to render this expression experimentally useful, we modify it in two ways. First, oxygen is a gas, so we must replace the concentration term with the *partial pressure* of $O_2$, $P_{O_2}$. Secondly, rather than using the equilibrium constant $K_{diss}$, we use a related measure called the **fractional saturation** of Mb with oxygen, symbolized by Y. Y is defined as:

$$Y = [MbO_2]/([Mb] + [MbO_2]) \tag{5.4}$$

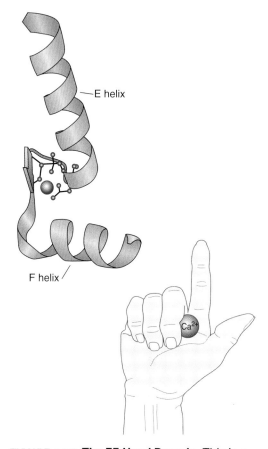

FIGURE 5.20 **The EF-Hand Domain.** This is a $Ca^{2+}$-binding domain that occurs in proteins that selectively bind this ion. The flexible joints between the helices allow movement when $Ca^{2+}$ is bound, altering the shape and triggering a response to the presence of $Ca^{2+}$ in the cell.

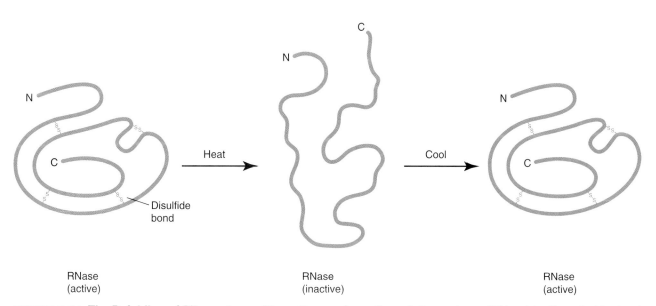

FIGURE 5.21 **The Refolding of Ribonuclease.** The active conformation of ribonuclease (RNase) is disrupted by heat and restored by subsequent cooling. Activity is determined by measuring its ability to catalyze the hydrolysis of the phosphate ester bonds in an RNA chain.

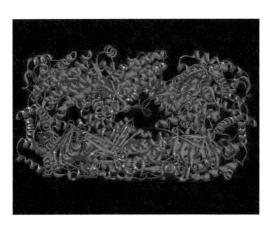
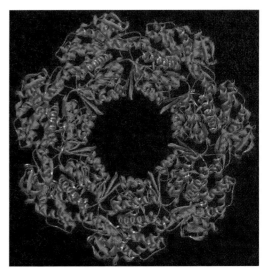

FIGURE 5.22 **The GroEL Segment.** Side (left) and top (right) views of a segment of the GroEL protein, a portion of a chaperone that assists proteins in folding as they are synthesized. Seven of these are stacked one on top of the other to make half of the chamber; two of these are a complete GroEL. Once a newly synthesized protein is within this assembly, a smaller protein (GroES) forms a cap, which is also essential for the acceleration of protein folding. Protein Data Bank 1SS8. Chaudry, C., Horwich, H.L., Brunger, A.T., and Adams, P.D. 2004. *J. Mol. Biol.* 342:229–245.

FIGURE 5.23 **Heme.** Heme has a central iron ion, represented as Fe in the Figure, but which is actually in the +2 oxidation state (i.e., $Fe^{2+}$). Four of its six ligand positions are shown, namely, to the N atoms of linked pyrrole rings. The other two are indicated as dotted lines: one connects to a histidine of the protein that binds the heme, and the other is reversibly bound to $O_2$.

According to Equation 5.4, Y is the ratio of oxygen-bound Mb to total Mb molecules, and ranges from 0 to 1. The graph of Y versus $P_{O_2}$ (FIGURE 5.24) shows a curve that rises monotonically and saturates at high oxygen content. This is a typical equilibrium binding of any ligand to a receptor. Half-maximal binding occurs at about 5 mm Hg. Myoglobin stores oxygen in muscle cells and releases it when prevailing levels become depleted. This is accomplished simply through the equilibrium presented here. Myoglobin is found in humans, but it is particularly abundant in diving mammals such as whales and seals.

Hemoglobin (Hb), unlike myoglobin, has quaternary structure because it has four monomer units. Each has a bound heme group, which can bind a molecule of oxygen. In hemoglobin, the binding of oxygen to each subunit changes the subsequent binding of oxygen to another subunit. This behavior is called **cooperativity**, as opposed to independent monomer binding which is **noncooperative**. A mechanism for cooperativity is sketched in FIGURE 5.25. Here, the conformations of the low affinity monomers in hemoglobin are represented as spheres, whereas high affinity monomers are represented as squares. As oxygen binds, it alters the conformation of the monomers. More than that, neighboring subunits change their conformation even in the unbound state, as indicated, and the entire protein now binds ensuing oxygen molecules with greater affinity. Other models than the one depicted could be considered also; the important point is the increase in binding affinity that occurs with partially bound hemoglobin. This phenomenon is called **positive cooperativity**.

While it is possible to write equilibria for hemoglobin binding to multiple oxygen molecules, the overall equations require certain assumptions about *how* the cooperativity takes place. Competing models first proposed in the 1960s have still not been resolved. Experimentally, the behavior of oxygen binding to hemoglobin, as expressed as the fractional saturation (Y), is well known and is presented in FIGURE 5.26. The "S-shape" of this figure is characteristic of positive cooperativity. At low $P_{O_2}$, the saturation is very low. As $P_{O_2}$ increases, not only does Y increase, but the *rate of rise* increases as well. This is merely the graphical expression of positive cooperativity. As $P_{O_2}$ increases even further, the rate decreases as saturation is reached (as Y approaches 1.0).

$O_2$ binds Mb more avidly than Hb. This is evident from the lower value for the midpoint in the oxygen binding curve of Mb (FIGURE 5.27). The physiological role of Mb is to provide $O_2$ for intracellular use when its concentration is very low, so its tight binding to Mb is expected. Hb, the major protein of the red blood cells, is saturated with oxygen in the lung capillaries ($P_{O_2}$ of about 100 mm), and releases this oxygen in the tissue capillaries by equilibrating with the much lower $P_{O_2}$ there ($P_{O_2}$ of about 40 mm). This is the fundamental regulation of oxygen uptake and delivery, relying on the simplest form of control: the principle of equilibrium.

The binding of oxygen to Hb can be modified. For example, the **Bohr effect** is the diminished binding of oxygen to Hb in response to an increase in the concentration of protons as illustrated in FIGURE 5.28. Notice how the entire binding curve is shifted to the right under acidic conditions. For example, a greater level of anaerobic metabolism and the subsequent production of lactic acid signal the weaker binding of $O_2$ to Hb, and hence the appropriate release of oxygen. There is also an adaptation that occurs at high altitude. In this condition, the red blood cell produces the molecule 2,3-bisphosphoglycerate, which lowers the binding of $O_2$ to Hb in a manner similar to protons (Figure 5.28). This provides more oxygen to the tissues, which alleviates the problem of the lower oxygen content of the air at high altitude.

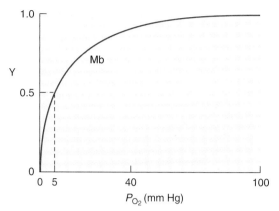

FIGURE 5.24 **The Oxygen Saturation Curve for Myoglobin.** Y represents the fraction of myoglobin bound to oxygen to the total myoglobin; $P_{O_2}$ is the partial pressure of oxygen.

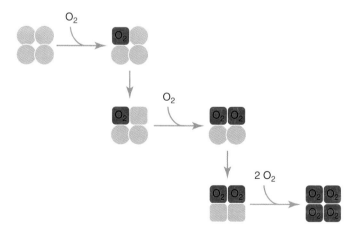

FIGURE 5.25 **Model for Cooperativity in the Binding of Oxygen to Hemoglobin.** The subunits of hemoglobin are represented as circles (low affinity) or squares (high affinity). Upon binding oxygen, the subunits switch to a new conformation that is high affinity. Subsequently, neighboring subunits are affected and their conformations switch to high-affinity sites.

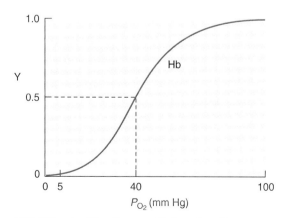

FIGURE 5.26 **The Oxygen Saturation Curve for Hemoglobin.** As a consequence of cooperativity, the binding represented as Y increases as more oxygen molecules occupy subunits of hemoglobin. Binding falls off once saturation is reached.

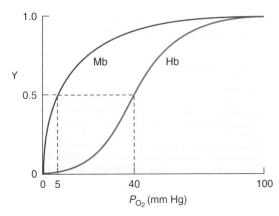

FIGURE 5.27 **Myoglobin Versus Hemoglobin.** Both saturation curves are plotted together. Myoglobin (Mb) has a much lower midpoint for binding than hemoglobin (Hb). Only Hb binding is cooperative.

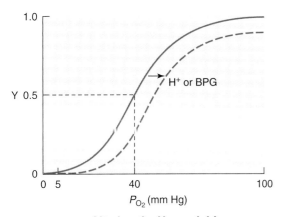

FIGURE 5.28 **Altering the Hemoglobin Saturation Curve.** Substances that decrease binding of oxygen to hemoglobin, such as protons (H+) or 2,3-bisphosphoglycerate (BPG), shift the curve to the right. The line drawn at 40 mm of Hg shows that for the same value of $P_{O_2}$, the shifted curve will be in equilibrium with a much less-saturated hemoglobin. In other words, hemoglobin will hold onto much less oxygen at the same oxygen partial pressure in the presence of the increased acidity of BPG, thereby delivering oxygen to cells.

# 5.9 Protein Purification and Analysis

The rise of biochemistry as a distinct area of chemistry coincided with advances in chemical separation and measurement methods for proteins. Accordingly, we conclude this chapter with a discussion of protein purification and analysis.

## Purification

In order to purify a protein, it must first be extracted from its natural environment with its properties intact. Subsequently, the extract is subjected to a fractionation process that divides the material into unequal portions, some of which are enriched in the protein of interest. Ultimately, this protein is obtained in a relatively pure form.

Throughout the process, a specific measure of the protein is required. For example, an enzyme is usually detected by measuring the disappearance of a substrate or the appearance of a product. All such methods—called **assays**—must also be simple to perform because they will have to be repeated hundreds of times in the course of the purification.

Several precautions are commonly taken to minimize protein denaturation during extraction and purification. These include performing experiments in a cold room; the inclusion of buffers to prevent extreme fluctuations in the pH of the protein solution; the addition of reagents to prevent oxidation of cysteine side groups; and the inclusion of protease inhibitors to prevent hydrolysis of the peptide bonds.

The extraction procedures may use physical methods, such as a blender to tear the tissue and release intracellular contents, or detergents to chemically solubilize the cell membranes. Once a solution is produced that contains the target protein, fractionation can proceed. The methods of separation of molecules are based on differences in solubility, size, and charge. As a simple example, salt precipitation is a method for the separation of proteins by differences in solubility that stems from charge differences. Different concentrations of salts (ammonium sulfate is commonly used because it preserves protein conformation) precipitate different proteins, because they vary in total charge.

The main purification is usually achieved by a form of **chromatography (BOX 5.4)**. Chromatography is a separation of components due to their different affinities towards two distinct phases: a mobile phase and a stationary phase. For example, in ion exchange column chromatography (**FIGURE 5.29**), negatively charged protein molecules added to the mobile phase are retarded by the stationary phase to different degrees and thus emerge from the column in different fractions. In size exclusion chromatography, proteins are retarded by different sized cavities in the beads making up the stationary phase. Thus, large proteins emerge from the column first. In all cases of chromatography, several modifications are possible. One is to alter the stationary phase by using very small beads, and forcing fluid flow under pressure, a method called **HPLC** (i.e., high pressure liquid chromatography or high performance liquid chromatography). The sample and carrier fluid can be converted into the gas phase in **gas liquid chromatography**. A selective binding may be achieved by attaching a molecule to the stationary phase known to strongly bind the protein; this is called **affinity chromatography**.

## Analysis

Proteins are typically measured and characterized by their interaction with light or, more generally, radiation. For example, the presence of aromatic side chains in the amino acids tyrosine and tryptophan means that proteins absorb ultraviolet radiation, which can be used to estimate the total amount of protein present. Similarly, using polarized ultraviolet light and making measurements over time provides a measure of protein conformation in solution.

The response of unpaired atomic nuclei to a magnetic field can be used in **nuclear magnetic resonance (NMR) spectroscopy**, a technique of growing importance in assessing protein structure. While NMR has long been used for small molecule identification, more sophisticated instrumentation and computer analysis has made it possible to study the structures of ever larger proteins.

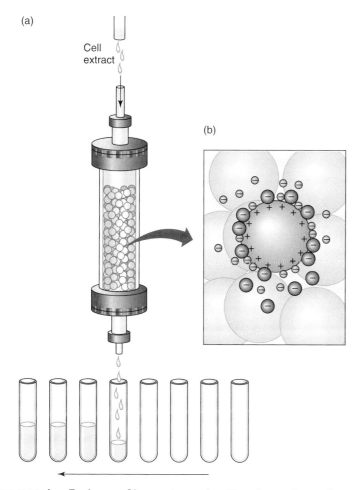

FIGURE 5.29 **Ion-Exchange Chromatography**. The chromatography column is shown as a matrix of solid material (called a gel) suspended in a buffer. (a) A cell extract representing a mixture of proteins is added to the top. Fluid containing a salt is continuously added to the column, the drain plug is removed from the bottom, and fractions are collected in tubes. Within the gel (b), differently charged proteins adhere with different strengths to the gel, altering the time it takes them to elute, thus achieving a separation. As in any chromatography, there is a stationary phase (the gel) and a mobile phase (the eluting buffer that also contains the protein mixture).

## BOX 5.4: WORD ORIGINS

### Chromatography

In 1901, the Russian botanist Mikhail Tsvet improved on prior separation methods by filling a cylinder with a support material, loading plant leaf extracts, and subsequently eluting fractions with organic solvents. Several distinct colored components emerged, leading Tsvet to name the process *chromatography*, meaning *color writing*. Modern chromatography refers to any of a wide array of methods that provide a separation between mobile and stationary phases.

While the word *chromatic* may appear to apply strictly to colors, it does have another origin. In early Greek society, it was a description of elaborate rhetoric. Also in early Greek usage, the word was applied to music, as one of three types of tetracords (the others being diatonic and enharmonic). Today, the word is retained in music, where the chromatic scale includes semitones (sharps and flats), but the diatonic scale does not. Thus, the richness of the word is fully commensurate with the richness of the chemical technique and its ongoing modifications.

Bombarding crystallized proteins with X-rays produces a pattern that, after mathematical manipulation, provides a high-resolution picture of the molecule, albeit in the solid form. **X-ray crystallography** has now been extended to increasingly large proteins and even membrane-bound proteins such as ion channels.

**Mass spectrometry** is another method borrowed from the chemical analysis of small molecules and applied to proteins. For this analysis, molecules are fragmented, vaporized into the gas phase, and the fragments are separated through a column by chromatography.

A commonly used method to determine the molecular weights and subunit compositions of proteins is **electrophoresis**. Electrophoresis is a form of chromatography in which the mobile phase is driven by an electric field imposed by electrodes. The stationary phase is a gel composed of a polymer, commonly acrylamide. Electrophoresis is usually carried out to identify various proteins present in a sample and as a test of purity. In some instances, it is used as a method of purification itself, because it performs a separation and the purified protein can be extracted from the gel.

Much of what we have been considering in this chapter about proteins in general can be applied to the discussion of enzymes. For example, the binding equilibrium behaviors of myoglobin and hemoglobin have analogies in the behavior of two classes of enzymes (the corresponding diagrams look very much the same). There is a critical distinction: binding studies reflect equilibrium behavior, but in enzyme action, the system is removed from equilibrium.

# Summary

Amino acids are the building blocks of proteins. The 20 naturally occurring amino acids found in most proteins may be divided between water soluble and hydrophobic, and can be further categorized by their acid–base properties and functional group chemistry, such as hydroxyl or thiol groups. All amino acids have multiply charged groups, with at least one amine and one carboxyl group. Thus, there are multiple charge possibilities that vary with pH. The pH value at which the net charge on an amino acid is zero called the isoelectric point. Joined together, the amine and carboxyl groups of two separate amino acids form a peptide bond. When small numbers of amino acids (e.g., <25) are linked by amide bonds (peptide bonds) the resulting molecule is called a peptide. Peptides still display "small-molecule" behavior. When larger numbers of amino acids are joined by peptide bonds, the resulting molecule assumes the behavior of a polymer and is called a protein.

The structures of proteins can be formally divided into four levels: primary (the sequence of the amino acids); secondary (a recognizable unit of structure, such as an alpha helix or beta sheet); tertiary (the three-dimensional structure of a protein strand); and quaternary (the overall structure of multi-subunit proteins). The domain is an intermediate level of structure, between secondary and tertiary, and is associated with a known biological function such as binding to a specific molecule. The overall formation of a protein into its active three-dimensional shape is called folding. While the information required to correctly fold is inherent in the primary structure, cells have proteins called chaperones that assist and accelerate the process.

An example of protein function is the binding of oxygen to myoglobin, which follows a simple equilibrium model. Hemoglobin is an analogous oxygen-binding protein, but has four separate subunits, each very similar to myoglobin. This provides more extensive control over oxygen binding, providing a broader range of oxygen release to variations in conditions at local capillary beds. Protein purification and analysis methods represented the first significant biochemical applications of the broad field of protein chemistry. The fundamental purification method is chromatography, the separation of proteins by their relative affinities towards a mobile phase and a stationary phase. Several analysis methods can be applied to proteins, including: nuclear magnetic resonance (NMR), which senses the magnetic spins of atomic nuclei; mass spectroscopy, in which proteins are fragmented into charged ions, which are then separated based on their masses; and X-ray crystallography, in which crystals of proteins are bombarded with high-energy radiation and their structures are revealed by the scattering pattern of the emerging radiation.

## Key Terms

| | | |
|---|---|---|
| α-carbon | electrophoresis | primary structure |
| affinity chromatography | fractional saturation | prosthetic group |
| α-hydrogen | gas liquid chromatography | protease |
| amide | helix breakers | proteins |
| amino acids | heme | PTB domain |
| amino end | HPLC | pyrrole |
| antiparallel | isoclectric point (pI) | quaternary structure |
| assay | loop | random coil |
| β-barrel | macromolecules | secondary structure |
| Bohr effect | mass spectrometry | SH2 domain |
| buffering regions | monomers | soft nucleophile |
| calmodulin | motif | src |
| carboxyl end | NAD binding domain | supersecondary structure |
| chaperone proteins | non-cooperative | tertiary structure |
| chromatography | nuclear magnetic resonance | through-space effect |
| cooperativity | (NMR) spectroscopy | TIM domain |
| coordination number | peptides | triosephosphate isomerase |
| domain | peptide bond | X-ray crystallography |
| EF hand | positive cooperativity | zwitterions |

## Review Questions

1. The amino acids can be organized in several different ways, and some appear in multiple categories. Suggest some different groupings, and identify those amino acids that can appear in more than one of these categories and in the groups presented in the text.

2. Draw a titration curve for glutamate, and show the major charge forms in a manner analogous to the one for lysine in the text. What is its isoelectric point?

3. A buffer is a substance that is used to ensure the pH does not change appreciably upon addition of excess acid or base. This occurs near the $pK_a$ value of a species

and is effective up to about $\pm 1$ pH unit of the p$K_a$. Use the Henderson–Hasselbalch equation to explain why this is so.

4. An unknown amino acid is found to contain a hydroxyl group and studies show it is immobile in an electric field if the pH is exactly 5.655. What is it?

5. Polyamines are multiple amine-containing molecules derived from a non-protein amino acid (ornithine) by loss of the $CO_2$ group. One of these is putrescine, which has the structure:

$$NH_2-(CH_2)_4-NH_2$$

a. Draw its titration curve, given the p$K_a$ values of 9.35 and 10.8.
b. What is its pI?
c. Polyamines have at least two known functions: binding to DNA and binding to $K^+$ channels. Based on purely electrostatic effects, explain these biological activities.

6. The molecular weight of an amino acid is roughly equal to 100. This allows a quick assessment of the number of amino acids in a protein given its molecular weight, or its molecule weight given the number of amino acids. What is the approximate number of amino acids in hemoglobin? Assuming a more accurate estimate of 115 for the molecular weight of an amino acid residue, how would this change your answer?

# References

1. Dawson, R. M. C.; Elliott, D. C.; Elliott, W. H.; Jones, K. M. *Data for Biochemical Research;* Clarendon Press: Oxford, UK, 1986.
2. Eisenberg, D. The discovery of the a-helix and b-sheet, the principal structural features of proteins. *Proc. Natl. Acad. Sci. U.S.A.* 100(20):11207–11210, 2003.
3. Hartley, H. Origin of the word "protein." *Nature* 168:244–244, 1951.
4. Irvine, R. F. Concepts of Protein Structure and Function. In *Early Adventures in Biochemistry;* Ord, M. G., Stocken, L. A., Eds.; JAI Press: Oxford, UK, 1995; p. 165.
5. Protein Data Bank, http://www.rcsb.org.
   This is a large compilation of protein structural information online.
6. Zeyar, A.; Jinyan, L. Mining Super-Secondary Structure Motifs from 3D Protein Structures: A Sequence Order Independent Approach. In *Proceedings of the 18th International Conference on Genome Informatics*, 15–26, 2007.

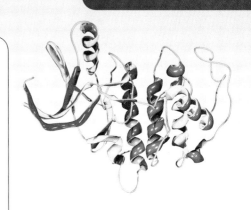

# 6

# Enzymes

Image © Dr. Mark J. Winter/Photo Researchers, Inc.

The word *enzyme* is German for *in yeast*. Near the end of the 19th century, the demonstration that a cell-free extract could cause sugar to ferment—as dramatically demonstrated by the formation of bubbles of $CO_2$—was a blow to the vitalist theory that only intact cells could carry out biological reactions. The entire extract was named *zymase*; today we use the word *enzyme* to describe a catalyst involved in a reaction.

A catalyst increases the rate of a reaction without affecting its equilibrium position. To visualize the relationship between catalysis and equilibrium, consider placing weights on a dual-pan balance (FIGURE 6.1). If weights are dropped onto both pans, they will oscillate until they settle into their final, equilibrium rest positions. If you intercede in this process just after the weights are dropped, using your hands as dampeners, the time needed to reach the final state is much shorter. Yet the heights the pans reach in their final, equilibrium (rest) state are unaffected by this intervention. The steadying hands represent the action of an enzyme, which increases the rate of approach to equilibrium (i.e., the rate of a reaction in both the forward and reverse directions), but does not affect the final state of equilibrium. An intriguing corollary is that an enzyme has no effect on a reaction at equilibrium.

Thus, the study of enzyme kinetics is that of the time course approaching but not having reached equilibrium. We first analyze reactions from energetic and chemical viewpoints and then focus on the kinetic characterization of enzyme-catalyzed reactions.

# 6.1 A Brief Look at Enzyme Energetics and Enzyme Chemistry

In any reaction, converting substrate to product requires intermediate states. Those intermediates are always less stable—and thus of higher energy—than either the substrate or the product. Theoretically, with

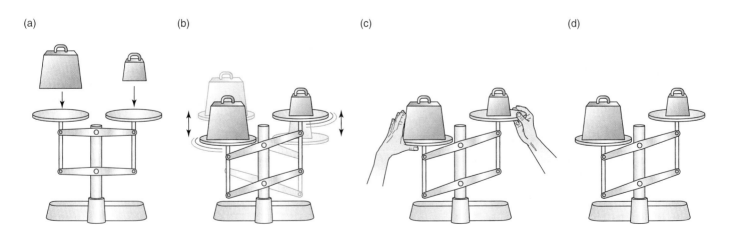

(a)　　　(b)　　　(c)　　　(d)

**FIGURE 6.1 Equilibrium Balance.** Equilibrium is modeled by a dual-pan balance. (a) Weights are dropped on the balance, and (b) it oscillates about its true balance state. (c) A pair of hands steadies the balance, and (d) it achieves an equilibrium point, which is the same point that it would reach without the hands. The hands symbolically represent enzyme action.

## BOX 6.1: Activation Energy and Murphy's Law

The concept of an activation energy can be made more intuitive by dispelling a common myth known as "Murphy's law," which states that "if anything can go wrong, it will." While mentioning Murphy's law commonly generates a cynical acquiescence, it is in fact wrong. A more accurate statement is: "most things can go wrong, but don't."

The more stable state of a building is a pile of bricks rather than the very orderly arrangement of the intact structure, yet we hardly fear an imminent collapse while sitting at our desks. This is because it would take a wrecking ball—a large investment of extra energy—to reduce the building to that more stable arrangement. Similarly, virtually all reactions that can proceed to a more stable form nonetheless require a tremendous input of energy to get them going. In the molecular world, this means a greater movement of molecules (i.e., rotations, vibrations, and translations) is needed before a reaction can occur. Thus, even though the final state may be more stable than the initial state, activation energy is always required. Murphy was apparently an incautious investigator who rushed to publication.

a single intermediate, we can speak of *the* transition state, although invariably more than one exists. The difference in energy level between the substrate and the transition state is called the energy of activation. Energy input is needed to move reactants up to the transition state's energy level (**BOX 6.1**).

Nearly all enzymes are proteins and are able to impart the necessary energy by binding interactions with the substrate. This binding lowers the activation energy, which is the essence of catalysis (**FIGURE 6.2**). We symbolize the enzyme as **E**, the substrate as **S**, and the intermediate enzyme–substrate complex as **ES**. The enzyme binds and precisely

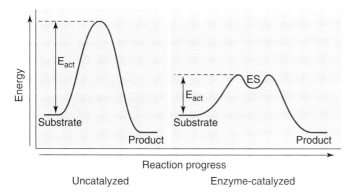

**FIGURE 6.2 Energy of Activation in Uncatalyzed and Catalyzed Reactions.** On the left is a high-energy intermediate in the uncatalyzed reaction between substrate (S) and product (P). On the right, two effects of enzyme-catalyzed reactions are evident. Most importantly, the highest energy state is much lower than the uncatalyzed reaction. In addition, there is an extra "hump" showing that the new path forged in the presence of the catalyst involves more steps.

positions the substrate. Because one or a few types of substrate bind an enzyme, only those molecules will react at an appreciable rate.

The chemical transformations of the substrate on the enzyme surface proceed as they would in nonenzymatic processes, that is, through oxidation, proton transfer, or electron rearrangement. However, enzymatic reactions are orchestrated with precision. The orientation of the substrate with ancillary groups—or even other substrates—benefits from the selective binding to the surface of the enzyme, a region called the **active site**. The intermediates symbolized here (i.e., ES) will take on chemical forms. We are strictly interested in how the reaction rate changes as the concentration of the substrate is varied.

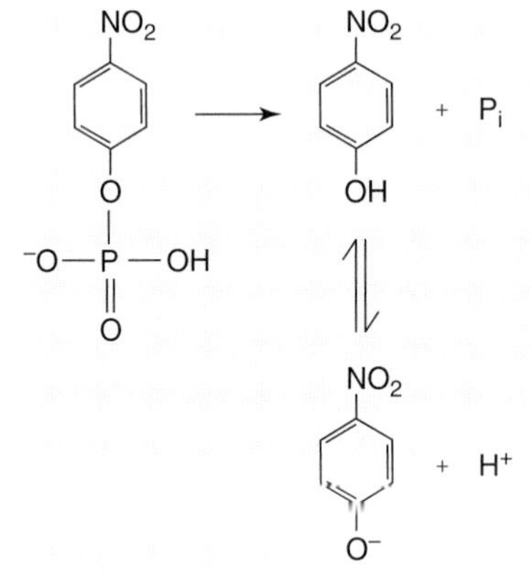

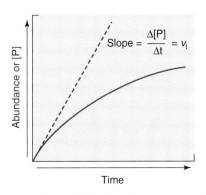

**FIGURE 6.3 Phosphatase Action on *Para*-Nitrophenolphosphate**. The phosphatase enzyme catalyzes removal of the phosphate. Subsequently, the *para*-nitrophenol equilibrates with its anion form, which has a bright yellow color that is readily detected.

## 6.2 The Enzyme Assay and Initial Velocity

**Enzyme activity** is the number of moles of substrate converted to product per unit time (e.g., moles per second). The laboratory measurement of enzyme activity is called an **assay**. As an example, consider the case of the enzyme alkaline phosphatase. This enzyme is attached to the external surface of cells and catalyzes the dephosphorylation of a variety of phosphorylated substrates. Assay of the enzyme activity uses the artificial substrate *p*-nitrophenylphosphate (FIGURE 6.3). This compound is colorless, but the dephosphorylated product, *p*-nitrophenol, is in equilibrium with its anion, which is **bright** yellow. The increase in color can be followed continuously and related to product concentration using a spectrophotometer (BOX 6.2). The resulting time course of product formation is shown in FIGURE 6.4. Note that the rate of product formation steadily decreases with time. This is the result of a decrease in substrate concentration and a corresponding increase in product concentration with time.

### BOX 6.2: The Spectrophotometer and Beer's Law

One of the simplest analytical methods is the absorption of light energy using an instrument called a spectrophotometer. Light energies from the visible and near ultraviolet ranges are isolated by a prism or a grating, passed through a sample, and then detected by a sensitive phototube. By comparing the light that passes through the sample with the same beam of light that bypasses the sample, the amount absorbed by the sample (the *absorbance*) is determined. This is a convenient measurement because it is both simple and informative. While the method is only occasionally useful for identifying functional groups in molecules (the spectrum of light at different frequencies forms broad bands), it is very useful for quantifying the amount of light absorbed because, according to Beer's law, the absorption is directly and linearly related to concentration. With a standard curve of known concentrations of a molecule, a linear plot is invariably obtained, from which it is a simple matter to determine the concentration of an unknown. For enzymatic reactions that produce (or consume) an absorbing species, the absorbance can be measured as a function of time.

**FIGURE 6.4 Initial Velocity**. The enzymatic time course shows that the rate falls off with time, as substrate is depleted and product accumulates. The most rapid rate (the one approximated by the dotted line) is the one that occurs near zero time. The slope of this line is the initial velocity ($v_i$), the parameter used in enzyme kinetics.

Reaction velocity is computed as the ratio of a change in concentration to a change in time. By convention, enzyme kinetics uses the portion of the curve indicated in Figure 6.4 with a broken straight line, which represents the **initial velocity**, abbreviated $v_i$. Using this initial portion of the curve avoids the complications of a continuous substrate concentration. While the true initial velocity strictly applies only as substrate concentration approaches zero, in practice, the first 10% of the reaction usually shows a nearly linear slope.

Using the initial velocity measurements as indicated in Figure 6.4 it is possible to characterize enzymatic reactions by varying the concentrations of substrate and of enzyme. In order to understand the results, we need a mechanism by which the enzyme functions. It happens that a very simple mechanism illustrates the key points and is broadly applicable to the majority of enzymes.

# 6.3 A Simple Kinetic Mechanism

The simplest possible reaction is S→P, where S is the substrate and P is the reaction product. For enzyme catalysis, a simple mechanism consists of three steps: (1) the enzyme binds substrate, (2) the enzyme transforms the substrate into product, and (3) the enzyme releases the product. The entire process may be written as two chemical reactions:

$$E + S \rightleftarrows ES \tag{6.1}$$

$$ES \rightarrow E + P \tag{6.2}$$

In Equation 6.1, **E** and **S** reversibly combine to form the enzyme–substrate complex, **ES**. In Equation 6.2, the complex **ES** irreversibly breaks down to form P and regenerate **E**. The net reaction—the sum of these two—is just S → P, the same as the nonenzymatic reaction. Note that in the course of the enzymatic reaction, the enzyme appears in two forms: **E** (free enzyme) and **ES** (enzyme bound to substrate or just "bound enzyme").

We can also rewrite Equations 6.1 and 6.2 in a single line and introduce rate constants for each chemical step:

$$E + S \underset{k_r}{\overset{k_f}{\rightleftarrows}} ES \overset{k_{cat}}{\rightarrow} E + P \tag{6.3}$$

The rate constants are proportionalities between substrate concentrations and reaction rate. The rate constants in Equation 6.3 are:

$k_f$ the forward rate constant for the formation of ES from E and S
$k_r$ the reverse rate constant for the formation of E and S from ES
$k_{cat}$ the rate constant for the breakdown of ES to E and P

The $k_{cat}$ rate constant, which leads to product formation, is also known as the **catalytic rate constant** indicated by the subscript *cat*.

## Assumptions

Equation 6.3 is our model for enzyme kinetics. It is possible to derive an equation from this model that relates the initial velocity $v_i$ to the substrate concentration S. To accomplish this, we make three assumptions. First, and most critically, is the **steady-state assumption**. This presumes that the concentration of the enzyme–substrate complex **ES**, in Equation 6.3 is constant with time. This condition is usually true in experimental enzyme assays and in living cells. There are specialty situations where this assumption does not hold.

The second assumption is **enzyme conservation**. That is, the total amount of enzyme, $E_{tot}$, is constant for the course of the reaction and can be expressed as the sum of all enzyme forms. In our simple mechanism there are just two forms, **E** and **ES**. Therefore,

$$[E]_{tot} = [E] + [ES] \tag{6.4}$$

The third and last assumption is that the initial velocity $(v_i)$ is expressed by an equation that leads to product formation. In our mechanism, this is the reaction determined by the catalytic rate constant $(k_{cat})$. Thus, we can express this assumption as:

$$v_i = k_{cat}[ES] \tag{6.5}$$

## The Michaelis–Menten Equation

Using the assumptions just discussed, we can derive an equation that relates velocity to substrate concentration (detailed in the Appendix). Leonor Michaelis and Maud Menten resolved the problem by postulating steady-state conditions. Steady-state conditions turn out to hold for virtually all enzymes after just a fraction of a second of reaction. They also apply broadly within living cells. In terms of our model, a steady-state condition means that the rate of formation of **ES** is equal to the loss rate of breakdown of **ES**. This reduces the mathematics required for elementary algebra, and we need only to obtain an expression for initial velocity, $v_i$, in terms of S and constants:

$$v_i = V_{max}[S]/([K_m] + [S]) \tag{6.6}$$

where

$$V_{max} = k_{cat}[E]_{tot} \tag{6.7}$$

and

$$K_m = (k_r + k_{cat})/k_f \tag{6.8}$$

Equation 6.6 is called the *Michaelis–Menten equation*. $V_{max}$ is the maximum velocity of the enzyme and, according to Equation 6.7, is proportional to the total enzyme concentration ($[E]_{tot}$). $K_m$ is the ratio of the rate constants leading to the breakdown of the ES complex

$(k_r + k_{cat})$ to the rate constant leading to the formation of the ES complex ($k_f$). We can consider $K_m$ a **steady-state constant**. As we will appreciate, $K_m$ is distinct from equilibrium constant. In the denominator of the Michaelis–Menten equation (Equation 6.6), $K_m$ is added to [S]. Clearly, $K_m$ and [S] must have the same units, indicating that $K_m$ is a concentration.

## 6.4 How the Michaelis–Menten Equation Describes Enzyme Behavior

A typical plot of initial velocity as a function of substrate concentration is shown in FIGURE 6.5. The Michaelis–Menten equation (Equation 6.6) correctly predicts the form of this equation. This is not a coincidence because a plot of the experimental finding (Figure 6.5) inspired the model and the resulting equation. Note that the curve in Figure 6.5 looks the same as the equilibrium binding curve of myoglobin. Here, however, it represents something very different: a steady state, with net flow of substrate to product.

The curve approaches two asymptotes that are indicated in Figure 6.5 as straight dashed lines. These lines approach the extremes of substrate concentration: zero and infinity. As [S] approaches infinity, the initial velocity ($v_i$) approaches the maximum velocity ($V_{max}$). This asymptote is a horizontal line, and the value of $V_{max}$ corresponds to the intersection of this line with the initial velocity axis. As [S] approaches zero, a second asymptote to the curve is illustrated in Figure 6.5. This is a straight line that intersects at the origin, and has a slope of $V_{max}/K_m$. This slope is the ratio of the two constants that appear in the Michaelis–Menten equation. Equations for both asymptotes can be derived by setting [S] to infinity or zero and solving the simplified Michaelis–Menten equation (Equation 6.6). Together, the terms $V_{max}$ and $V_{max}/K_m$ not only completely characterize the plot of Figure 6.5,

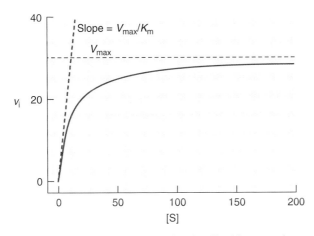

**FIGURE 6.5 Michaelis–Menten Plot.** A Michaelis–Menten plot graphs the initial velocity ($v_i$) versus substrate concentration ([S]). The curve approaches the asymptote $V_{max}$ at high [S] and the asymptote line with a slope of $V_{max}/K_m$ at low [S].

but also completely describe how the velocity changes with substrate concentration. In summary, the equations for the straight lines at the extremes of $v_i$ are as follows:

$$v_i = V_{max} \qquad\qquad [S] \to \infty \qquad (6.9)$$

$$v_i = (V_{max}/K_m)[S] \qquad [S] \to 0 \qquad (6.10)$$

# 6.5 The Meaning of $K_m$

Because $K_m$ is a substrate concentration, suppose that the prevailing substrate concentration, [S], is exactly equal to $K_m$ at some instant. If we substitute [S] for $K_m$ in the Michaelis–Menten equation, then:

$$v_i = V_{max}[S]/(K_m + [S]) = V_{max}[S]/([S]+[S]) = V_{max}[S]/2[S] \quad (6.11)$$

or, simply

$$v_i = \tfrac{1}{2}V_{max} \qquad\qquad (6.12)$$

This velocity point is plotted in FIGURE 6.6, and the projection on the substrate axis provides the value of $K_m$. From this, we see that the value of $K_m$ can also be defined as the substrate concentration that corresponds to the half-maximal velocity point.

If we consider a substrate in its cellular environment, it is commonly found that the concentration of this substrate is very close to the value of $K_m$. This is not surprising, because this is the region of concentration that provides the greatest capacity for an increase or decrease in velocity as a function of [S]. Thus, the measured $K_m$ values provide an estimate of intracellular concentrations.

There is another important use for $K_m$. If an enzyme can use different substrates—as in the case of phosphatases—then their relative $K_m$ values can be used to differentiate between them. Low values of $K_m$ indicate a tighter binding affinity to the enzyme. Caution must be exerted in this application, because $K_m$ is *not* an equilibrium constant, and *affinity* is an equilibrium concept. The actual affinity of a substrate in terms of our model is given by $K_s$:

$$K_s = k_r/k_f \qquad\qquad (6.13)$$

Recall from Equation 6.8 that the derived value of $K_m$ is $K_m = (k_r + k_{cat})/k_f$. In order to actually *equate* $K_m$ with $K_s$, we would have to set $k_{cat}$ to zero, in which case we would merely have the binding equilibrium:

$$E + S \underset{k_r}{\overset{k_f}{\rightleftarrows}} ES \qquad\qquad (6.14)$$

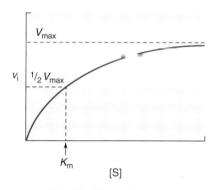

FIGURE 6.6 **The $K_m$.** The substrate concentration $K_m$ is reached by extrapolating the corresponding velocity at half of the $V_{max}$. This is the effective midpoint of the Michaelis–Menten curve.

instead of the Michaelis–Menten equation:

$$E + S \underset{k_r}{\overset{k_f}{\rightleftharpoons}} ES \overset{k_{cat}}{\rightarrow} E + P$$

This is often approximately correct because dissociation back to E and S proceeds far more rapidly than formation of E and P (that is, usually $k_{cat} \ll k_r$). Because we have replaced $K_m$ with the equilibrium constant $K_s$, the interpretation is simplified. However, there is a danger in overusing this substitution, which should become clearer in our study of reversible inhibition.

# 6.6 Reversible Inhibition

The most common means by which enzymes are regulated is through reversible inhibition. Cellular regulators, as well as most drugs are reversible inhibitors. Thus, it is important to have a clear understanding of how these agents work.

Reversible inhibitors bind to the active sites of enzymes (**BOX 6.3**), and they are distinguished by their ability to bind different forms of an enzyme. By definition, a **reversible inhibitor** is one that binds to and dissociates from an *enzyme form*, reaching equilibrium. In our simple model, there are two forms: **E** and **ES**. We can represent the interaction between an inhibitor **I** and either of these forms as the following equilibria, which are characterized by the following equilibrium constants:

$$EI \rightleftharpoons E + I \qquad K_i = [E][I]/[EI] \qquad (6.15)$$

$$ESI \rightleftharpoons E + I \qquad K_i' = [ES][I]/[ESI] \qquad (6.16)$$

Reversible inhibition can be categorized as competitive, anticompetitive, or mixed. **Competitive inhibition** is defined as the binding of the

## BOX 6.3: Inhibition and Active Sites

The active site of an enzyme is where the catalytic event occurs. The active site includes the portion that binds directly to the substrates, as well as the intermediates that form as the reaction progresses. Thus, the active site is dynamic over the course of the reaction. While inhibitors affect events at the active site, kinetics provides no information as to where an inhibitor molecule actually binds to the enzyme. So while some models suggest that inhibitors also bind at the active site, to prove this requires other, direct chemical data, such as protein modification methods. The different types of reversible inhibitors differ only in which form of the enzyme they bind, not on whether they attach to the active site.

reversible inhibitor (**I**) to the free enzyme form only (Equation 6.15). **Anticompetitive inhibition** is defined as the binding of **I** to the **ES** complex only (Equation 6.16). **Mixed inhibition** is defined as the binding of **I** to both **E** and **ES** forms (both Equations 6.15 and 6.16).

For each type of inhibitor, we will establish three points. Most critically, we will develop an intuitive picture of how the inhibitor, through binding the enzyme forms, leads to the overall inhibition of $v_i$. We will also show how the inhibitor affects the velocity–substrate curve; this reduces to effects at the extremes of very high and very low substrate concentrations. Lastly, we will show how the Michaelis–Menten equation is modified by the inhibition, and how this modification can be understood in terms of effects on two critical parameters: $V_{max}$ and $V_{max}/K_m$.

## Competitive Inhibition

Competitive inhibitors often bear a chemical structure similarity to the substrate. This stems from the fact that both the inhibitor **I** and the substrate **S** bind to the same enzyme form, that is, the free enzyme **E**. Conceptually, competitive inhibition is easy to picture because everyday examples are abundant. In ecology, organisms compete for food and territory; in academics, students compete for grades; in finance, businessmen compete for money.

Competitive inhibition can be modeled by adding the equilibrium reaction (Equation 6.15) to our standard model, producing:

$$E + S \rightleftarrows ES \rightarrow E + P$$

$$+$$

$$I$$

$$\downarrow\uparrow K_i$$

$$EI \tag{6.17}$$

An equation to describe the interaction quantitatively can be derived exactly as we have done for the Michaelis–Menten equation, with the addition of a new equilibrium, and an enzyme conservation equation that includes the dead-end form, **EI**. The resulting equation is similar to the Michaelis–Menten equation, except that the term $(1 + [I]/K_i)$ modifies the $K_m$ term:

$$v_i = \frac{V_{max}[S]}{K_m\left(1 + \dfrac{[I]}{K_i}\right) + [S]} \tag{6.18}$$

**FIGURE 6.7** shows how a competitive inhibitor affects the velocity–substrate curve. As $[S] \rightarrow \infty$, the inhibition disappears. This can also be shown with Equation 6.18, where the $[S]$ term overwhelms the modified $K_m$ term, and the equation becomes simply:

$$v_i = V_{max} \tag{6.19}$$

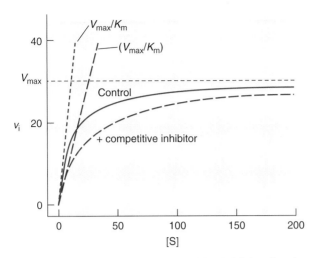

**FIGURE 6.7 Competitive Inhibition.** A competitive inhibitor has its greatest influence at low substrate concentration. Thus, a competitive inhibitor lowers $V_{max}/K_m$ (slope of the line intersecting the origin) but does not affect $V_{max}$.

We conclude that a competitive inhibitor is ineffective at saturating substrate concentrations. This behavior is also apparent from an examination of our model: as **I** and **S** compete for **E**, the concentration of **S** overwhelms **I** and wins the competition.

Competitive inhibition is most effective at very low substrate concentrations. Recall from Equation 6.10 that the uninhibited velocity at low substrate concentration is $v_i = V_{max}/K_m$ [S]. In terms of this equation, the effect of a competitive inhibitor is to lower $V_{max}/K_m$. Because $V_{max}/K_m$ is the slope of the line (Equation 6.10) that approximates the rate equation at low [S], we also know that this slope is decreased by a competitive inhibitor.

## Anticompetitive Inhibition (Uncompetitive)

Anticompetitive inhibitors do not bind the free **E** form of the enzyme and do not structurally resemble the substrate. Rather, this class of inhibitors binds to the **ES** enzyme form. To gain an intuitive understanding, imagine a bear attempting to remove honey from a hive with a narrow opening. The bear can move his empty paw in and out of the opening with no difficulty. Yet, once his paw is filled with honey inside the hive, he can't remove it. Only the paw–honey complex is blocked; if the bear had not attempted to remove the honey, he would not be "inhibited." Alternatively, consider a sting operation. A suspect is offered a lucrative deal involving contraband. When she accepts, undercover agents arrest her. Only when she associates with an illegal transaction is she at risk. Had she not become associated with it, she would not be "inhibited." In both cases, it is the bound form that is subject to inhibition.

More formally, the interaction is characterized by the equilibrium shown in Equation 6.16, where the **ESI** complex forms from the components **ES** and **I**. When this is added on to our standard mechanism (Equation 6.3), the anticompetitive model becomes:

$$E + S \rightleftarrows ES \rightarrow E + P$$

$$+$$

$$I$$

$$\downarrow\uparrow K_i'$$

$$ESI \qquad\qquad (6.20)$$

The velocity–substrate equation for this model is:

$$v_i = \frac{V_{max}[S]}{K_m + [S]\left(1 + \dfrac{[I]}{K_i'}\right)} \qquad (6.21)$$

FIGURE 6.8 illustrates the influence of an anticompetitive inhibitor on the velocity–substrate curve. In the region where $[S] \rightarrow 0$, we observe no inhibition. The simplified equation under this condition is Equation 6.10, $v_i = V_{max}/K_m\,[S]$, in which case there is no effect on the $V_{max}/K_m$ term. We interpret this to mean that unless the enzyme binds the substrate, no **ES** complex is formed. Because the inhibitor only binds **ES**, then when $[S]$ is low, so too is $[ES]$, and there is little inhibition.

Inhibition appears strongest at the opposite extreme, where $[S] \rightarrow \infty$. Under these conditions Equation 6.21 reduces to:

$$v_i = V_{max}\big/\left(1 + [I]/K_i'\right) \qquad (6.22)$$

in which case it is $V_{max}$ that is affected by the inhibitor.

Anticompetitive and competitive inhibitions are opposite extremes, which is why we use the term "anticompetitive" instead of **uncompetitive** (**BOX 6.4**). For a competitive inhibitor, the more enzyme in the **E** form, the greater the inhibition. For an anticompetitive inhibitor, the more the enzyme is drawn into the **ES** form, the greater the inhibition.

## Mixed Inhibition (Noncompetitive)

An inhibitor is said to be **mixed** if it binds both to the free enzyme form **E** and to the bound form **ES**. In essence, it is not a new form of

## BOX 6.4: WORD ORIGINS

### Noncompetitive and Uncompetitive

The established use of the words *uncompetitive* and *noncompetitive* has done much to confuse generations of students as to the true nature of reversible inhibition. For this reason, many specialists in enzyme kinetics have offered replacements. We have emphasized *anticompetitive* in place of uncompetitive because the action of such inhibitors is the mirror image of competitive. Other names have also been proposed, such as *catalytic* inhibition, because $V_{max}$ is decreased, and this contains the catalytic constant that leads to product. The substitution of *mixed* for *noncompetitive* evokes the correct picture of both a competitive (*V/K* effect) and an *anticompetitive* (*V* effect) inhibition. Others have suggested the term *pure noncompetitive* for the case in which both effects are equally inhibited, but this is merely a special case of *mixed* inhibition.

To underscore the confusion, a deaf student once pointed out that sign language does not discriminate between *un* and *non*. There are slight differences in English; typically *non* just means "not," whereas *un* suggests "not yet"; however, these distinctions are entirely without application to enzyme kinetics.

The problem that arises here is a clash between establishing the intuitive understanding of a concept and the enforcement of a uniform vocabulary. As a compromise, we use both: the less common terms to enable comprehension, and the more widely used terms that are associated with them.

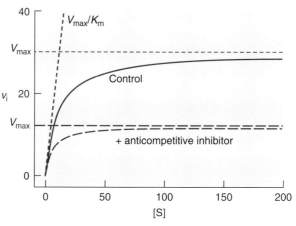

FIGURE 6.8 **Anticompetitive Inhibition.** An anticompetitive inhibitor has its greatest influence at high substrate concentration. Thus, an anticompetitive inhibitor lowers $V_{max}$ (intercept of the maximal velocity asymptote) but does not affect $V_{max}/K_m$.

inhibition at all, but merely a combination of competitive and anti-competitive inhibition. There are thus two equilibria to be added to the kinetic model:

$$E + S \rightleftarrows ES \rightarrow E + P$$

$$
\begin{array}{ccc}
+ & & + \\
I & & I \\
\downarrow\uparrow K_i & & \downarrow\uparrow K_i' \\
EI & & ESI
\end{array}
\qquad (6.23)
$$

We will consider first the situation in which the constants $K_i$ and $K_i'$ are identical. At low [S], the inhibition will be competitive, since the inhibitor binds only to the E form. This is because the concentration of ES is diminishingly small at low [S]. Thus, under this circumstance, we expect to observe a diminished $V_{max}/K_m$. At high [S], the inhibition will be anticompetitive, because the inhibitor binds only to ES form. This is because the concentration of E is diminishingly small at very high [S]. Under this circumstance, we expect to observe a diminished $V_{max}$. Indeed, at any substrate concentration, the inhibitor will reduce $v_i$, as shown by the following rate equation (see the appendix for the derivation),

$$
v_o = \frac{V_{max}[S]}{K_m\left(1 + \dfrac{[I]}{K_i}\right) + [S]\left(1 + \dfrac{[I]}{K_i}\right)}
\qquad (6.24)
$$

and by the graph FIGURE 6.9. Mixed inhibition with a single value for $K_i$ can be visualized as just removing some of the enzyme, because forms—E and ES—are equally diminished by the inhibitor, and thus so is $[E]_{tot}$.

In the more general situation, the $K_i$ and $K_i'$ values are not equal. We can make a first approximation to this inhibition behavior by realizing that the equilibrium constant representing the strongest binding will be

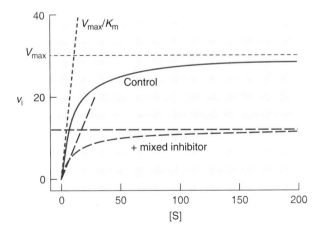

FIGURE 6.9 **Mixed Inhibition.** A mixed inhibitor that has equal inhibitor binding constants has equal effects on the two kinetic parameters, $V_{max}$ and $V_{max}/K_m$. Hence, it lowers both equally, effectively decreasing enzyme velocity at all substrate concentrations.

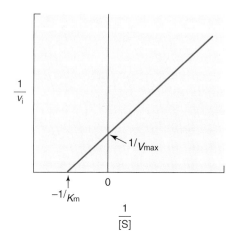

$$\frac{1}{v_i}$$

$$1/V_{max}$$

$$-1/K_m$$

$$0$$

$$\frac{1}{[S]}$$

**FIGURE 6.10 The Double-Reciprocal Plot.**
A straight line results from rearranging the Michaelis–Menten equation so that $1/v_i$ is plotted against $1/[S]$. The indicated intersections provide estimates of $V_{max}$ and $K_m$.

dominant. For example, if **I** binds much more tightly to **E** (i.e., $K_i < K_i'$), then the inhibition will resemble competitive inhibition. On the other hand, if **I** binds much more tightly to **ES** (i.e., $K_i' < K_i$), then the inhibition will resemble anticompetitive inhibition. In either case, there will be a slight inhibition for the weaker binding species, rather than the complete lack of inhibition that occurs in the competitive or anticompetitive cases. Another view of reversible inhibition is that mixed inhibition is all-encompassing, with competitive and anticompetitive inhibition as special cases in which binding to **E** or **ES**, respectively, dominates.

# 6.7 Double-Reciprocal or Lineweaver–Burk Plot

The Michaelis–Menten curve can be transformed into a straight line by rearranging Equation 6.6 and plotting $1/v_i$ versus $1/[S]$. This is known as the **double-reciprocal** or **Lineweaver–Burk plot**, and the values of $V_{max}$ and $K_m$ can be read from the intersections of its axes (**FIGURE 6.10**). Note that it is $V_{max}/K_m$ rather than $K_m$ itself that we have used to develop our understanding of enzyme inhibition. Attempting to interpret changes in $K_m$ with reversible enzyme inhibition leads to contradictions (**BOX 6.5**).

## BOX 6.5: Can $K_m$ Explain Enzyme Inhibition?

In a word, no! It is popularly thought that $K_m$ is a useful constant in enzyme inhibition because it seemingly gives the "right" answer when applied to competitive inhibition. We have seen, in some circumstances, that $K_m$ can be approximated by $K_s$, the equilibrium constant of ES dissociation. If we can continue to make this assumption for enzyme inhibition, then $K_m$ is roughly an affinity constant for substrate binding (lower values mean tighter binding). Table B6.5 shows that a competitive inhibitor increases the measured $K_m$. The temptation to conclude that this demonstrates that a competitive inhibitor decreases the affinity of the enzyme for the substrate—due to the competition—is difficult to resist.

Resist it we must, however, as is evident by the rest of the Table. For example, the anticompetitive (or uncompetitive) inhibitor causes a decrease in $K_m$. If we followed the same path, we would have to conclude that this form of inhibitor *increases* the binding of substrate to enzyme! Finally, for mixed (noncompetitive) inhibitors, when both binding constants are the same, there is no change in $K_m$. The interpretation of inhibitory behavior using affinity thus is inappropriate.

The reason for the error is that $K_m$ is not really an equilibrium constant. At times, making the approximation is reasonable, but making this assumption requires that $v_i = 0$. Inhibition analysis is intrinsically a steady-state, rather than an equilibrium process. In the Table, alterations in the true kinetic constants, $V_{max}$ and $V_{max}/K_m$, are regular and symmetric with different inhibition types.

| TABLE B6.5 Influence of Reversible Inhibitors on $V_{max}$, $V_{max}/K_m$, and $K_m$ | | | | |
|---|---|---|---|---|
| **Inhibition Type** | **Common Name** | $V_{max}$ | **Effect on $V_{max}/K_m$** | **Effect on $K_m$** |
| Competitive | Competitive | — | ↓ | ↑ |
| Anticompetitive | Uncompetitive | ↓ | — | ↓ |
| Mixed | Noncompetitive | ↓ | ↓ | — |

**CHAPTER 6** ENZYMES

In research studies, double-reciprocal plots are not used to determine kinetic constants, because the process of forming reciprocals causes the smallest values (i.e., the least experimentally certain values) to become the largest, giving the poorest data the greatest influence. However, these plots *are* widely used in inhibition studies because each type of inhibitor produces a distinct pattern. Constructing double-reciprocal plots with various amounts of inhibitor present produces lines that intersect at the $1/v_i$ axis (competitive inhibition), are parallel (anticompetitive inhibition), or intersect to the left of the $1/v_i$ axis (mixed inhibition) as shown FIGURE 6.11.

# 6.8 Allosteric Enzymes

Not all enzymes display the velocity–substrate behavior that we have examined so far. One well-known deviation is that displayed by the so-called *allosteric enzymes*. The term *allosteric* (Greek for *other place*) indicates that modifiers of the enzyme bind to a place other than the active site at which substrates are transformed to products. Their ability to influence the active site remotely, as it were, stems from conformational changes in the protein usually transmitted between separate subunits of a protein.

Allosteric enzymes are usually composed of multiple subunits which interact during the course of catalysis. The model shown in FIGURE 6.12 is one possible mechanism. Upon binding, the alteration in the subunit conformations, shown as a transformation from circles

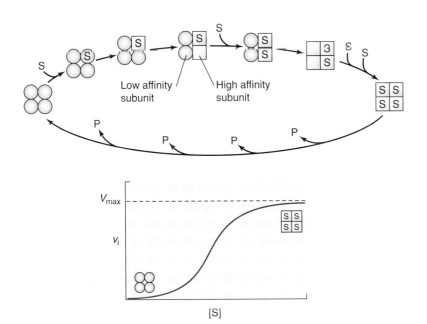

FIGURE 6.12 **Allosteric Enzymes and Cooperative Behavior.** Allosteric enzymes usually display cooperative behavior, in which binding substrate alters the ability of the enzyme to bind further substrate molecules, and is reflected in a steadily increasing velocity. Eventually the enzyme is saturated and velocity falls off to $V_{max}$, as with all enzymes.

Competitive

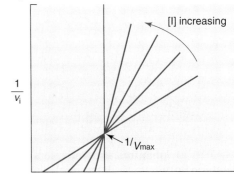

Anticompetitive (uncompetitive)

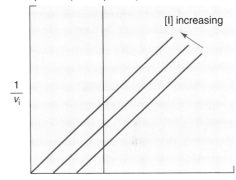

Mixed (noncompetitive)

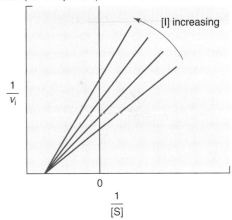

FIGURE 6.11 **Patterns of Reversible Enzyme Inhibition.** The three types of inhibition display distinct patterns when plotted in double-reciprocal form. In each case, increasing the amount of inhibitor produces new lines showing alterations in the slope (competitive), the intercept on the 1/[S] axis (anticompetitive), or both (mixed).

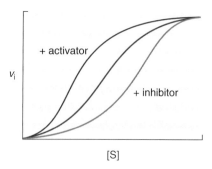

**FIGURE 6.13 Inhibitors and Activators of Allosteric Enzymes.** Modulators of allosteric enzymes cause a shift to the right (inhibitors) or left (activators), greatly extending the influence of these compounds on enzyme velocity over Michaelis–Menten type enzymes.

to squares, leads to an increased catalysis for each subunit, and thus for the enzyme as a whole. This behavior is called **cooperativity**, an idea borrowed from the equilibrium hemoglobin–oxygen binding curve. Inhibitors of cooperative enzymes shift the curve to the right because they decrease the relative effectiveness of the substrate on $v_i$, as indicated in FIGURE 6.13. Also shown in Figure 6.13 is the shift of the curve to the left, the response to an allosteric activator.

In a similar way, covalent modifications of enzymes can lead to alterations in $v_i$. A very common one is the introduction of a phosphoryl group from ATP on the side chain hydroxyl group of a serine, threonine, or tyrosine amino acid. The phosphorylated enzyme may display allosteric activation or inhibition identical to the kinetic curves shown in Figure 6.13. Returning the enzyme to the non-phosphorylated form requires a protein phosphatase, which returns the enzyme to its original activity.

# 6.9 Irreversible Inhibition

A different type of covalent modification of enzymes is irreversible inhibition. In this case, the modifying agent is covalently attached to the enzyme, and not removed, leaving the enzyme with little or no activity. As an example, the nerve gas diisopropylfluorophosphate is an irreversible inhibitor of the enzyme acetylcholine esterase, as shown in FIGURE 6.14.

Normally, acetylcholine esterase catalyzes the hydrolysis of the ester bond in acetylcholine, producing the inactive molecules acetate and choline (FIGURE 6.15). Acetylcholine is the neurotransmitter that signals muscle contraction from the connecting nerve; if it cannot be hydrolyzed, the signal remains high and muscles go into uncontrolled contraction (called contracture) and death ensues rapidly.

There are several examples of drugs that act as irreversible inhibitors, including aspirin (which inhibits prostaglandin and prostacyclin formation), penicillin (which disrupts the formation of bacterial cell walls), and prilosec (which inhibits the proton pump in the stomach, thereby reducing the amount of stomach acid).

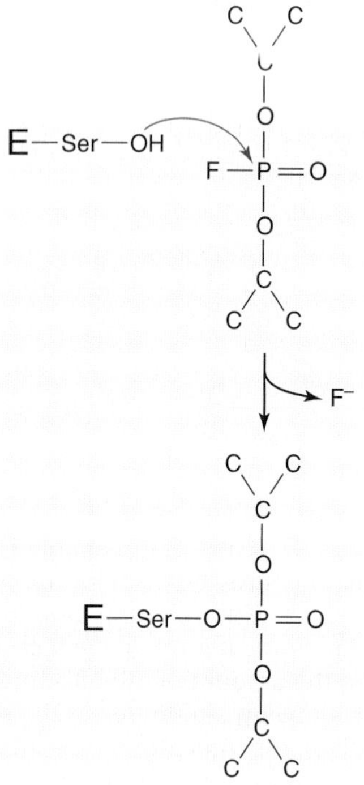

**FIGURE 6.14 Irreversible Inactivation of Acetylcholinesterase.** Acetylcholinesterase is covalently modified by the diisopropylfluoride, because the phosphoryl center provides a good site of nucleophilic attack, and the fluoride is a good leaving group. Once formed, the modified enzyme is no longer reactive toward its own substrate.

# 6.10 Enzyme Mechanisms

In our study of enzyme kinetics so far, we have suppressed the molecular nature of the reactants, using the symbols **E, S, ES,** etc. How this notation corresponds to a real mechanism is shown in Figure 6.15, using acetylcholinesterase as an example. Making the assignments in FIGURE 6.16a, the mechanism can be presented in symbolic form in Figure 6.16b.

The chemistry of enzyme mechanisms is a topic that we explore throughout this text. Here we consider two recurring aspects of the chemistry of reaction pathways: nucleophilic substitution and acid–base catalysis.

## Nucleophilic Substitution

Many reactions in biochemistry are nucleophilic substitutions. In the case of acetylcholinesterase, the first step is the attack of the hydroxyl group (which is relatively electron rich) on the carbonyl carbon (which is relatively electron poor). Thus, the electron pair migrates from the O of the serine hydroxyl to the C of the carbonyl. This reaction is shown in FIGURE 6.17. Once the electrons have been added to the C, there is now an excess of electrons, so they migrate in turn to the O of the carbonyl. Subsequently, these electrons return to the double bond of the carbonyl and the C–O bond of the substrate is ruptured.

## Acid–Base Catalysis

Often a group is made more reactive by an alteration of its acid–base status. At the simplest level, we can think of this as adding or removing a proton. In rare cases, the entire mechanism is just acid–base catalysis. It is usually a part of another mechanism, and the proton donor or acceptor is an amino acid residue contributed by the enzyme itself. In the case of acetylcholine esterase, a base from the enzyme (from a histidine residue) removes a proton from the serine hydroxyl group *during the nucleophilic attack* (**concerted**) to make the oxygen a better nucleophile. After all, if the oxygen no longer has a neighboring proton, its electron density is enhanced. However, if the proton is completely removed prior to attack, the full negative charge becomes a problem for the approach of the serine residue; the concerted mechanism (FIGURE 6.18) is a compromise.

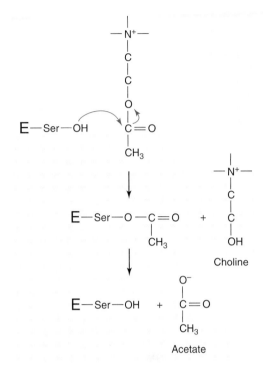

**FIGURE 6.15 Mechanism of Acetylcholinesterase.** The serine residue of acetylcholinesterase attacks the ester of the substrate acetylcholine and forms an enzyme intermediate with release of choline. Subsequently, this adduct is hydrolyzed and the free form of the enzyme is reformed and the second substrate, acetate, is released.

# 6.11 Enzyme Categories

There are a many kinds of chemical reactions in both biochemistry and organic chemistry. Just as the reactions in organic chemistry can be categorized by type, the enzymatic reactions in biochemistry can be organized into just a few general types. These divisions were originally made by the Enzyme Commission, which also extensively categorized enzymes into subtypes. We will consider here only the six major classes (Table 6.1).

(a)

E = E—Ser—OH
S = Aetylcholine
$P_1$ = Choline
$P_2$ = Acetate

(b)

$$E + S \rightleftharpoons ES \rightleftharpoons EP_1P_2 \overset{P_1}{\rightleftharpoons} EP_2 \longrightarrow E + P_2$$

**FIGURE 6.16 Matching Symbols to the Mechanism.** (a) The symbols used in enzyme kinetics are matched to the forms shown for the mechanism of acetylcholinesterase. (b) A symbolic mechanism is drawn.

**TABLE 6.1 Classification of Enzymes by the First Category of the Enzyme Commission (EC)**

| EC Type | Category | Example |
|---|---|---|
| 1 | Oxidoreductase | Lactate dehydrogenase |
| 2 | Transferase | Hexokinase |
| 3 | Hydrolase | Sucrase |
| 4 | Lyase | Aldolase |
| 5 | Isomerase | Glucose phosphate isomerase |
| 6 | Ligase | Pyruvate carboxylase |

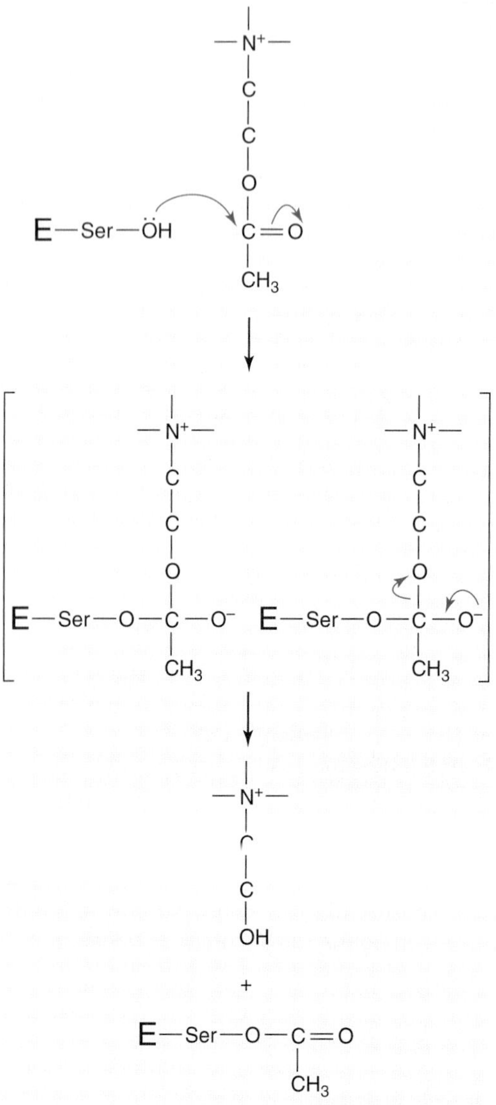

**FIGURE 6.17 Nucleophilic Substitution in Acetylcholinesterase.** Details of nucleophilic substitution are shown, a common mechanism in enzyme chemistry. The migration of electron pairs is indicated with curved arrows. An intermediate is formed first, which is released from the enzyme as electron migration leads to rupture of the C–O bond of the ester.

The first type, the **oxidoreductases**, is also known as the class of **dehydrogenases** or **reductases**. These are enzymes in which the substrate changes oxidation state in going to product. As such, there must be a mobile cofactor that removes or adds those electrons, which is commonly NAD$^+$. More enzymes are oxidoreductase than any other class.

**Transferase** enzymes catalyze reactions that move a piece of one substrate on to another. This is also a common reaction type, and includes all examples in which transfers involve the high-energy compound ATP, such as hexokinase in Table 6.1.

The **hydrolase** enzymes are formally subsets of the transferases, in which water is a substrate. Enzymes located in the exterior of cells such as lactase (Table 6.1) are often hydrolases because cofactors (other than water) are unnecessary. As a further example, all of the digestive enzymes secreted into the intestine are hydrolases.

**Lyase** enzymes catalyze the splitting of a substrate into pieces. This may seem superficially similar to hydrolases, but lyase reactions do not involve water as a substrate.

The class of **isomerase** enzymes catalyze rearrangements, such as shifting the location of a hydroxyl group. This often alters the stereochemistry of the molecule, as in the case of the enzymes **racemase** and **epimerase.**

Finally, **ligase** enzymes catalyze joining reactions that also involve a high-energy donor (usually ATP). Ligases are sometimes called **synthetases**. A similar term, **synthase**, is actually a synonym for a lyase (EC 4), with the reaction written in the opposite direction.

These six reaction types are top-level categories, and each has several subcategories. However, this overview provides us with a broad sense of the type of catalysis that exists in biology. There is one further biological event that is technically not enzyme catalysis, but has an analogous behavior: membrane transport.

# 6.12 Enzyme-Like Qualities of Membrane Transport Proteins

Many of the embedded proteins in cell membranes enable the selective exchange of small molecules. These are transporters, each of which allows a specific molecule or class of molecules to cross a membrane. For example, the glucose transporter (FIGURE 6.19) allows extracellular glucose to cross the plasma membrane of most cells and enter the cytosol.

The analogy of the transport of glucose into the cell to enzyme kinetics is strong. In fact, a plot of $v_i$ of transport against glucose concentration has exactly the same form as the Michaelis–Menten equation of Figure 6.5. Substituting extracellular glucose for **S** and intracellular glucose for **P**, we have a "reaction" in which the transporter is catalyzing a translocation, or change in location, for the glucose molecule in the same way as an enzyme catalyzes a chemical event. Even the **ES** complex in the enzyme mechanism is mirrored as a similar enzyme-transporter complex. As we learn more about enzymes and their regulation, we will find that transporters share these features. Thus, we should consider transporters as merely another form of catalysis, in which a change in space rather than a change in chemical composition occurs.

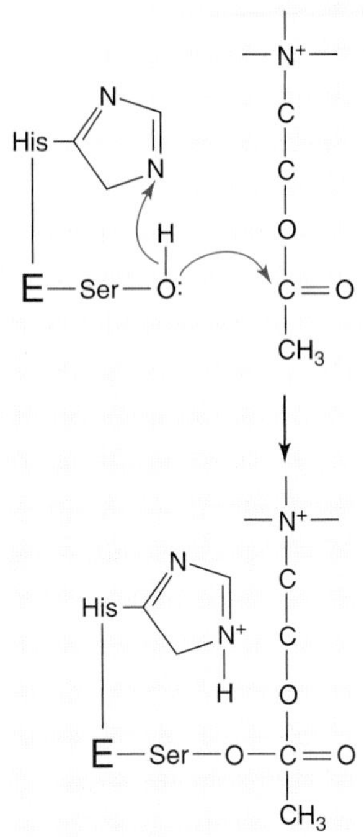

## Summary

Enzymes are measured by the reactions they catalyze. The study of the rates of conversion of substrates to products is called kinetics. The fundamental nature of catalysis is to lower the energy barrier required for reactions to proceed, called the energy of activation. Enzymes lower the energy of activation by providing binding sites for substrates and positioning groups to react with them. The rates at which enzymes catalyze reactions are measured using initial velocity ($v_i$) the beginning of the time course before appreciable substrate has been depleted or product has accumulated. In contrast to thermodynamics, kinetic studies depend upon a pathway or model for the conversion of substrates to products. The Michaelis–Menten model envisions a substrate forming an intermediate complex with the enzyme—the ES complex—and then releasing product. A key assumption in developing an equation that describes velocity as a function of substrate is that of the steady state, that is, the rate of formation of an ES intermediate is equal to the rate of its destruction. The resulting Michaelis–Menten equation has two constants, $K_m$ and $V_{max}$. $K_m$ is a substrate concentration that occurs when the initial velocity is half maximal. $V_{max}$ is an initial velocity that occurs when the enzyme is saturated with substrate. The

**FIGURE 6.18 Acid–Base Catalysis in Acetylcholinesterase.** The nucleophilic attack of acetylcholinesterase is assisted by the removal of a proton. The proton is removed by a histidine residue that is part of the enzyme.

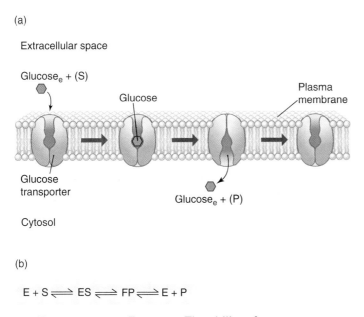

**FIGURE 6.19 Transporters as Enzymes.** The ability of transporters to "catalyze" the movement of substances from one biological space to another is directly analogous to enzyme action. The reaction is described in (a), and the symbolism analogous to other enzymes is shown in (b).

ratio, $V_{max}/K_m$, characterizes the Michaelis–Menten equation at low substrate concentration. Reversible enzyme inhibition can be readily understood by considering the extremes of substrate concentration. Thus, the parameters $V_{max}$ and $V_{max}/K_m$, respectively, are key. Inhibitors that primarily affect the enzyme at very low substrate concentration (i.e., decrease $V_{max}/K_m$) are competitive; those affecting the enzyme primarily at very high substrate concentration (i.e., decrease $V_{max}$) are anticompetitive; those that affect both $V_{max}$ and $V_{max}/K_m$ (and hence all substrate concentrations) are mixed. Some inhibitors are not reversible; they covalently bind to the enzyme and inactivate it permanently. Not all enzymes display the behavior of the Michaelis–Menten model. Another important model is that of allosteric enzymes. Allosteric enzymes have a sharper rise in their velocity–substrate curves, and modulators shift the curve to the right (inhibitors) or left (activators). This behavior arises because of subunit interactions subsequent to substrate binding. Enzymes may be categorized as oxidoreductases, transferases, hydrolases, lyases, isomerases, or ligases. Finally, membrane transport proteins are directly analogous to enzymes in that they catalyze the movement of a molecule between different biological spaces. The mathematical form of membrane transport is identical to that of enzyme kinetics.

## Key Terms

| | | |
|---|---|---|
| active site | enzyme activity | oxidoreductases |
| anticompetitive inhibition | epimerase | racemase |
| assay | hydrolases | reductases |
| catalytic rate constant | initial velocity (vi) | steady-state assumption |
| competitive inhibition | isomerases | steady-state constant ($K_m$) |
| concerted | ligase | synthase |
| cooperativity | Lineweaver–Burk plot | synthetases |
| dehydrogenases | lyase | transferases |
| double-reciprocal plot | mixed inhibition | uncompetitive inhibition |

## Review Questions

1. $K_m$ is a collection of rate constants. For our simple mechanism, how would you state in general the conditions under which $K_m$ is approximately the same as dissociation constant? What is the fundamental difference between $K_m$ and a dissociation constant (or the reciprocal, an affinity constant)?

2. The concentrations of compounds within cells are typically very close to the $K_m$ values of the enzymes that utilize these same compounds as reactants. Why do you think this is true?

3. Why isn't $V_{max}$ the same for a given enzyme under all conditions?

4. What are the units of the following constants: $k_f$, $k_{cat}$, $k_r$, $K_m$, $V_{max}$, and $V_{max}/K_m$?

5. The $E_{act}$ (energy of activation) is the highest peak in the energy diagram for an uncatalyzed reaction, but there are at least *two* peaks in the enzyme-catalyzed reaction, as shown in our model. Why must this be the case?

6. What forms of the enzyme are bound by a competitive inhibitor? A mixed inhibitor? An anticompetitive inhibitor? An irreversible inhibitor?

7. Suppose a more realistic model of enzyme kinetics was derived, with several intermediates. How would you modify your answer to Question 6?

8. If a reaction is at equilibrium, what effect would adding more enzymes have on the reaction?

9. When cells are induced to alter their state of protein synthesis, the total amount of enzyme is changed. Suppose this change leads to an increase in the amount of the enzyme by a factor of 3. How will this affect $V_{max}$? $K_m$? $V_{max}/K_m$?

10. Drugs are often enzyme inhibitors. Researchers interested in drug design find it easiest to construct a competitive inhibitor, but they are not the most desirable drugs. Why do you think that might be?

11. What is a rate constant exactly? Is there any difference between a rate constant used to express a nonenzymatic reaction and a rate constant used in enzyme kinetics?

12. In the derivation of the equations in this chapter, a steady-state assumption was used to describe the interaction of enzyme with substrate, but an equilibrium was used to describe the interaction of enzyme with inhibitors. Why couldn't a steady-state assumption also be used for inhibitors? Why couldn't an equilibrium assumption also be used for substrates?

13. Allosteric enzymes do not follow the Michaelis–Menten equation, yet the assumptions of steady state, enzyme conservation, and formation of the enzyme–substrate complex still apply. What feature makes allosteric enzymes unusual?

14. The form of the Michaelis–Menten equation and the form of the binding equilibrium of myoglobin are similar. Not only that, the form of the allosteric enzyme curve and the form of the hemoglobin curve are also similar. Discuss the reason for these similarities and how these various equations and curves are different.

# References

Cornish-Bowden, A. *Fundamentals of Enzyme Kinetics*; Portland Press: London, 1995.

Fersht, A. *Enzyme Structure and Mechanism*; W. H. Freeman and Co.: New York, 1997.

Gutfreund, H. *Kinetics for the Life Sciences*; Cambridge University Press: Cambridge, UK, 1995.

Lambert, F. L. Why don't things go wrong more often? Activation energies: Maxwell's angels, obstacles to Murphy's law. *J. Chem. Educ.* 74:947–948, 1997.

Ochs, R. S. Understanding enzyme inhibition. *J. Chem. Ed.* 77:1453–1456, 2000.

Ochs, R. S. The problem with double reciprocal plots. *Curr. Enzyme Inhib.* 6:164–169, 2010.

Ochs, R. S.; Ashby, C. R., Jr. Viewpoint: Discriminating between noncompetitive and allosteric interactions. *Synapse* 62(3):233–235, 2008.

Purich, D. L.; Allison, R. D. *Handbook of Biochemical Kinetics*; Academic Press: New York, 2000.

Segel, I. H. *Enzyme Kinetics*; John Wiley & Sons: New York, 1975.

# 7

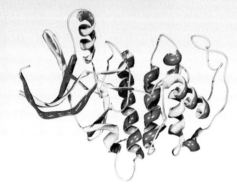

# Metabolism and Energy

**CHAPTER OUTLINE**

Image © Dr. Mark J. Winter/Photo Researchers, Inc.

etabolism is intrinsically involved with energy changes. We intuitively understand that food is chemically processed by reactions in the body to produce the energy that drives movement and maintains tissues. A deeper understanding of these processes requires a study of **thermodynamics**, the science of energy changes.

# 7.1 Origins of Thermodynamics

Thermodynamics was developed in the 19th century to answer the pressing question of the era: how to build a better steam engine. This machine was the driving force of the Industrial Revolution, which changed the world from an agrarian to a city-based economy.

Because of these historical origins, the equations of thermodynamics use terms more akin to steam engines (e.g., *heat* and *work*) than to biochemical reactions. Still, the variables in steam power—pressure, volume, and temperature—are concrete and easily grasped. These have been subsequently extended to other variables more important in chemistry, such as changes in chemical concentration (chemical potential) and changes in electrical potential (voltage).

The so-called *laws* of thermodynamics have two origins. The first is phenomenological. That is, they are laws of experience that have no proof other than plausible reasoning that has yet to be shown to have any internal contradictions. The second origin is statistical, using equations for the behavior of extremely large collections of molecules that can be averaged and show regularity. Happily, the two views arrive at identical conclusions, providing confidence in the concepts of thermodynamics.

# 7.2 First Law of Thermodynamics

### Heat and Work

The first law of thermodynamics is a conservation principle about the **internal energy** of a system. The internal energy is the energy of a **system,** as illustrated in FIGURE 7.1. A system is simply the portion of the **universe** (i.e., all that exists) that we wish to study. A boundary line divides the system from its **surroundings** (i.e., everything in the universe apart from the system under study). Both material and nonmaterial exchanges can occur across this boundary between the system and the surroundings.

The nonmaterial exchanges across the boundary of Figure 7.1 are **energy,** and experience has shown that it can be divided into two distinct forms: **heat** and **work**. These exchanges, along with a sign convention, are illustrated in FIGURE 7.2:

Heat ($q$) transferred from the surroundings to the system is positive.

Work ($w$) transferred from the surroundings to the system is positive.

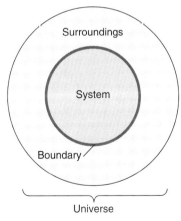

FIGURE 7.1 **Thermodynamic Definitions.** A Venn diagram of the technical meanings of some fundamental terms. The system is the object of our study. Everything that influences it is called the surroundings. The system and surroundings are separated by a boundary that defines the outer limits of the system. Practically speaking, only the immediate surroundings are significant. The universe consists of both the system and the surroundings.

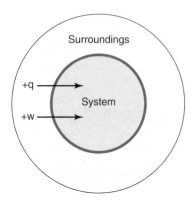

**FIGURE 7.2 Sign Conventions of Heat and Work.** Heat transferred from the surroundings to the system is positive; work transferred from the surroundings to the system is also positive. These sign conventions are different from the historical ones and have the advantage of simplifying energy exchanges of all types.

Heat and work are different forms of energy; they may be interconverted and both are measured in joules (**BOX 7.1**).

It is common to confuse *heat* with *temperature*. Temperature is a measure of how hot an object is relative to another object; it is the result of the average kinetic energy of its molecules. Heat is an energy exchange (or flow) between objects of different temperatures, always in the direction of high temperature to low temperature. The system in Figure 7.2 cannot *contain* heat; however, after the flow of heat, there will be a change in system energy, which is called the **internal energy**.

The other factor that affects the internal energy is the change in work. Work can be visualized for molecules as translational, vibrational, or rotational energy. After a change in work, there is also a change in the system's internal energy.

We are now in a position to state the first law of thermodynamics:

$$\Delta E = q + w \tag{7.1}$$

The term $\Delta E$ symbolizes the change in internal energy ($E$) as a change from one **state** to another (hence the delta symbol). This **change in state** is a rigorously defined in thermodynamics. Each thermodynamic state is characterized by **state variables**, such as temperature ($T$), pressure ($P$), volume ($V$), and number of moles ($n$). A system can be said to be in a specific state once all of those variables are defined; let us call that state-1. If we change some of these variables, we have a new state, called state-2. Moving from state-1 to state-2 might be done in different ways, but the final result is the same. Considering only the result, the change from state-1 to state-2 is **path independent**. This is the case for $\Delta E$.

Unlike $\Delta E$, both $q$ and $w$ are path dependent. The remarkable statement of the first law, however, is that the *sum* of these variables

## BOX 7.1: Joules, Calories, and Food Calories

Today energy units for reactions are expressed in units of joules, which is based on the work aspect of energy. That is, 1 joule (J) is the amount of work produced by a force of 1 newton over 1 meter (which is about the force of an apple dropped to the ground); the units are kg-m²/s². These KGS units can be easily converted to electrical ones, where 1 joule is the work of moving 1 coulomb of charge through a potential difference of 1 volt. The well-known calorie (cal) unit is based on the heat aspect of energy: it is the amount of energy needed to raise the temperature of 1 g of water 1°C (strictly speaking, the temperature is raised from 14.5°C to 15.5°C). There are approximately 4.18 joules in 1 calorie.

Food calories (Cal) are equivalent to 1000 calories or 1 kilocalorie (kcal). They are not precisely comparable, however, because nutrition databases make corrections for foodstuffs that are not entirely digestible. Apart from that, it is possible to calculate from complete combustion the energy content of foods. Because calories are so closely related to nutrition, this unit has yet to be fully displaced by the joule.

**CHAPTER 7** METABOLISM AND ENERGY

determines the internal free energy, which is itself path independent. This quality is essential for us to analyze changes in energy for reactions and pathways. A slightly different quantity than internal energy, called **enthalpy**, is used experimentally, because it is more easily measured directly in the laboratory.

## Enthalpy

Suppose that our system can only perform expansion work (called *PV* **work** where $P$ is pressure and $V$ is volume). The work involved is:

$$w = -\Delta(PV) = -P\Delta V - V\Delta P \tag{7.2}$$

where the negative sign is required by our sign convention (expansion means work is done by the system on its surroundings). If we consider only the situation where pressure is constant (i.e., $\Delta P = 0$), then

$$w = -P\Delta V \tag{7.3}$$

Under constant pressure conditions, the heat is designated $q_P$ and is path-independent. The overall equation for the first law then becomes:

$$\Delta E = q_P - P\Delta V \tag{7.4}$$

Rearranging,

$$q_P = \Delta E + P\Delta V \tag{7.5}$$

We define $q_P$ as the enthalpy, an invented word meaning "heat content." Usually, we use the term $\Delta H$ in place of $q_P$. This leads to Equation 7.6, which defines the enthalpy:

$$\Delta H = \Delta E + P\Delta V \tag{7.6}$$

It is the enthalpy change that is used to predict whether a system gives off heat (*exothermic*, $\Delta H < 0$) or takes up heat (*endothermic*, $\Delta H > 0$) from its surroundings. Actually, in biochemical systems that we are interested in, $\Delta V$ is also near zero. Thus, there is little practical difference for our purposes between internal energy and enthalpy. The data that we use are entirely derived from enthalpy values, so $\Delta H$ is retained here.

Enthalpy changes in chemical reactions involve alterations in translation, vibration, and rotation, all of which contribute to bond energy. An important property of $\Delta H$, as with all extrinsic state functions, is that they are additive. Thus, we can add up enthalpy changes for reactions, or sets of reactions, and come up with enthalpy values for reactions that cannot be determined individually.

Often, an exothermic reaction ($\Delta H < 0$) is one that is favored; that is, it reacts from left to right as the chemical equation is conventionally written. In fact, we commonly experience this correlation—most dramatically in explosions—and this produces the expectation that exothermic reactions might predict the direction for reactions. Consider three simple examples of dissolving salts in water, all of which definitely take place in the direction written, along with their associated enthalpy changes:

$$H_2SO_4 + H_2O \rightarrow \text{dissolved salts} \qquad \Delta H < 0$$

$$NaCl + H_2O \rightarrow \text{dissolved salts} \qquad \Delta H = 0$$

$$KCl + H_2O \rightarrow \text{dissolved salts} \qquad \Delta H > 0$$

All three of these events take place, and yet only the dissolution of sulfuric acid is exothermic. We must conclude that enthalpy cannot reliably predict the direction of reactions.

# 7.3 Entropy and the Second Law of Thermodynamics

Entropy ($S$) is the missing factor needed to predict the direction of chemical reactions. The second law of thermodynamics states that the entropy change in the universe for any process is greater than zero:

$$\Delta S_{univ} > 0 \tag{7.7}$$

The popular notion that entropy can be explained as "randomness" does not quite capture the meaning of the term and is also difficult to translate into a chemical driving force. In order to clarify the notion, we will explore two different views, one from **classical thermodynamics** and the other from **statistical thermodynamics**.

## Entropy as a Ratio of Heat to Temperature

The first appearance of entropy ($S$) arose from derivations used to account for the efficiencies of those same steam engines that inspired the first law. The equation that results from a model for perfect efficiency in a model engine (the Carnot engine) is:

$$\Delta S = q_{rev}/T \tag{7.8}$$

where $q_{rev}$ is the reversible heat change (the maximum that can be transferred between bodies) and $T$ is the temperature. Like $q_P$, $q_{rev}$ is a state variable. For any real processes—that is, one in which an engine actually performs work—it was found that,

$$\Delta q_{rev}/T = \Delta S_{univ} > 0$$

This is the same as Equation 7.7, the second law of thermodynamics. The fact that entropy is a ratio of heat to temperature means that, when the temperature is high, a change in heat will result in only a small increase in entropy. As the temperature decreases, a heat change will produce a greater entropy change. However, it is only by considering the statistical approach to entropy that clearer insight was achieved.

## Entropy as a Statistical Distribution of States

Ludwig Boltzmann, famous for his equation that produces a bell curve, showed that the curve for the distribution of energies of molecules had a similar shape. Using these insights, he developed the following equation for entropy:

$$S = k \cdot \ln W \tag{7.9}$$

where $S$ is the entropy, $k$ is a constant given by the molecular gas constant ($R$) divided by Avogadro's number ($N_A$), and $W$ the number of ways a system can be arranged. Boltzmann's insight was to recognize that entropy was a statistical phenomenon. An increase in the number of possible states increases the entropy. The greater *dispersion* a reaction produces in the universe, the greater the entropy. In popular belief, randomness or messiness is *equated* with entropy. More precisely, the number of arrangements that appear to be random or messy is greater than those that seem regular or neat; this is the essence of the concept of entropy.

This insight allows us to appreciate the role that entropy plays in chemical reactions. Consider the general case of a reaction that splits the molecule AB into the species A and B:

$$A - B \rightarrow A + B \tag{7.10}$$

A and B are joined together in the substrate, but separate in the products. Clearly, the products have more degrees of freedom. This is a favorable (positive) change in entropy *for the reaction*.

Unlike the entropy change of the universe, however, we cannot use the entropy of the system alone to predict the direction of a chemical reaction. It becomes possible to do so, using a combination of the entropy of the system with the enthalpy of the system.

# 7.4 Free Energy

Free energy ($G$) comes from a combination of the first and second laws of thermodynamics. The defining equation is:

$$\Delta G = \Delta H - T \Delta S \tag{7.11}$$

Unlike the second law, which applies to the universe, the free energy equation applies to the system under consideration, making it useful for reactions. A reaction is considered favorable (conventionally written in the left-to-right direction) when,

$$\Delta G < 0 \tag{7.12}$$

A reaction for which $\Delta G < 0$ is called **exergonic**. The contributions to $\Delta G$ are:

$\Delta H$: A heat term; the energy of bond formation or bond breakage

$\Delta S$: An entropy term; the probability or arrangement factor

If a reaction is exergonic and the heat term is dominant (i.e., $\Delta H$ is very large and negative), then $\Delta H$ will be the major factor in determining $\Delta G$ and we say that the reaction is **enthalpy driven**. The binding, therefore, is very favorable and it is quantitatively more important than the entropy. This is a common situation in exothermic reactions. If a reaction is exergonic and the $T\Delta S$ term is numerically dominant (i.e., $\Delta S$ is very large and positive, such as when KCl dissolves in water or a lipid bilayer forms), then the reaction is said to be **entropy driven**.

Not all reactions are exergonic. In fact, there are two other possibilities for the value of $\Delta G$. First, if the reaction is not possible under our defined conditions, then,

$$\Delta G > 0 \qquad (7.13)$$

and the reaction is **endergonic**. An endergonic reaction is not possible thermodynamically, but the reverse reaction is exergonic, because it has the same numerical value for $\Delta G$ as the forward reaction, but the opposite sign.

The other possible value for the free energy change is,

$$\Delta G = 0 \qquad (7.14)$$

In this case, neither the forward nor the reverse reaction is favored. This is another definition of equilibrium, and demonstrates that equilibrium status and free energy change are intimately related.

# 7.5 Standard Free Energy

Standard conditions are a means to compare the laboratory measurements of different observers, where temperature and pressure are set at 0°C (273 K) and 1 atm, respectively. Standard conditions applied to reactions also specify that all substrates and products have a concentration of 1 M.

Changes in enthalpy, entropy, and free energy at standard state are indicated with a superscript degree symbol (°) appended to their symbol: $\Delta H°$, $\Delta S°$, and $\Delta G°$, respectively. The standard free energy change ($\Delta G°$) is of particular interest, and can be related to the equilibrium constant ($K_{eq}$). To show this relationship, we begin with the following equation:

$$\Delta G = \Delta G° + RT \ln [C][D]/[A][B] \qquad (7.15)$$

which relates the free energy change ($\Delta G$) to the standard free energy change ($\Delta G°$) and the concentrations of the reactants ([A] and [B]) and products ([C] and [D]) for the general reaction,

$$A + B \rightleftarrows C + D \tag{7.16}$$

Recall from Section 7.4 that equilibrium exists when $\Delta G = 0$. Applying this condition to Equation 7.15 gives Equation 7.17:

$$0 = \Delta G° + RT \ln [C][D]/[A][B] \tag{7.17}$$

Rearranging slightly and adding the "eq" designation to the concentrations to emphasize that they are equilibrium concentrations,

$$\Delta G° = -RT \ln [C]_{eq}[D]_{eq}/[A]_{eq}[B]_{eq} \tag{7.18}$$

The explicit equilibrium notation in Equation 7.18 leads to an alternative definition for $\Delta G°$, because the terms inside the logarithm argument constitute the equilibrium constant. In other words, we can rewrite Equation 7.18 as:

$$\Delta G° = -RT \ln K_{eq} \tag{7.19}$$

The direct relationship between the standard free energy change and the equilibrium constant means that the substrate and product concentrations and the free energy for a reaction are directly related, too.

## 7.6 Nonstandard Free Energy Changes

Under cellular conditions, reactions are at steady state rather than at equilibrium. Nonequilibrium reactions are described by Equation 7.15:

$$\Delta G = \Delta G° + RT \ln [C][D]/[A][B]$$

If we replace $\Delta G°$ in Equation 7.15 with $-RT \ln K_{eq}$ from Equation 7.19, then we can rewrite Equation 7.15 as:

$$\Delta G = -RT \ln K_{eq} + RT \ln [C][D]/[A][B] \tag{7.20}$$

The concentration terms in this equation can be defined as,

$$Q = [C][D]/[A][B] \tag{7.21}$$

where $Q$ is the **mass action ratio**. $Q$ is the ratio of the product to the substrate for a reaction that corresponds to a specific physiological steady-state condition. Making this substitution into Equation 7.20, we arrive at,

$$\Delta G = -RT \ln K_{eq} + RT \ln Q \tag{7.22}$$

**TABLE 7.1 Reaction Free Energies in Different Situations**

| Situation | $\Delta G$ | $\Delta G°$ |
|---|---|---|
| Equilibrium | 0 | $-RT \ln K_{eq}$ |
| Standard conditions | $\Delta G°_{rxn}$ | $\Delta G°_{rxn}$ |
| Actual cellular conditions | $\Delta G_{rxn}$ | $-RT \ln K_{eq}$ |

## BOX 7.2: WORD ORIGINS

### Spontaneous

In our discussion of the direction of a reaction, some of the phrasing would have been more concise if the word *spontaneous* were used to indicate the direction that a reaction takes in nature. In fact, this word has wide popularity and is found in virtually all thermodynamic treatments from textbooks to specialty monographs. It is almost conspicuously absent in the present treatment, however, because it is a word that tends to confuse rather than clarify our understanding of thermodynamics.

The most common literature meaning for spontaneous is "of unknown origin." This is the definition behind the phrase *spontaneous generation,* a discarded idea of 19th century biology that organisms can emerge fully formed with no known precursor. Hence, if you leave fresh meat out in the open air, maggots will appear. This meaning is actually closer to "out of nowhere," another sense of the word spontaneous. There is also a strong time sense in that spontaneous events appear to happen very quickly rather than over long stretches of time. It is also used to introduce mystery, because the spontaneous is unknowable. When the word is applied to reactions, the meanings are equally rich. In one case it simply means that a reaction will occur—that is, spontaneous is synonymous with exergonic ($\Delta G < 0$). Yet another use of the word is as a synonym for *nonenzymatic,* in which reactions occur without the assistance of an enzyme are "spontaneous." It is hard to dislodge from one's mind the sense that spontaneous reactions must be faster than nonspontaneous ones. This is a confusion of kinetics (a time-based concept) with thermodynamics (an energy-based concept) in the study of reactions. The widespread use of the word *spontaneous* is one reason many students find thermodynamics difficult.

which is an expression of the nonstandard (actual) free energy change for a reaction in terms of the equilibrium constant ($K_{eq}$) and the mass action ratio ($Q$). For a reaction in cells to proceed in the forward direction, it is necessary that $\Delta G < 0$. Equation 7.22 restates this requirement as $Q < K_{eq}$. We will explore this important point further in Section 7.7.

$\Delta G > 0$ for an endergonic reaction, in which case $Q > K_{eq}$. Such a reaction will not proceed, except in the reverse direction. $\Delta G = 0$ for a reaction at equilibrium, which is equivalent to $K_{eq} = Q$.

It is worth pausing at this point to carefully consider the relationship between the equilibrium, the standard state free energy change, and the actual free energy change (Table 7.1). While $\Delta G = 0$ at equilibrium, the value of $\Delta G°$ is decidedly *not* zero, although its value can be calculated from the $K_{eq}$. Note that neither standard conditions nor equilibrium conditions provide any information about the free energy that prevails under *cellular* conditions. Notice also from Equation 7.19 that the relationship between the standard free energy change and the equilibrium constant still contains temperature ($T$) as a variable, and that this is *not* the standard temperature.

## 7.7 Near-Equilibrium and Metabolically Irreversible Reactions

Reactions that take place in cells are part of sequences called pathways. In a linear pathway, all reactions take place at the same rate and all reactions have $\Delta G < 0$ (i.e., all are exergonic).

From an energetic standpoint it is useful to divide individual metabolic reactions taking place in cells into two groups. One of these—by far the largest class—is only slightly displaced from equilibrium. We can compare $Q$ and $K_{eq}$ by using Equation 7.22. For any reaction in a pathway that is progressing in the cell, it must be the case that

$$Q < K_{eq} \tag{7.23}$$

but when $Q$ is *close* to $K_{eq}$, usually within an order of magnitude or two, we say that the reaction is **near-equilibrium**. Because most reactions fit into this category, we can consider this the typical case for cellular reactions, and so we expect to find that $Q$ approximates $K_{eq}$, the equilibrium ratio. There are two consequences of this observation. First, the reaction is sensitive to changes in substrate or product concentrations. If these are perturbed, then the same reaction can be run in reverse, because the concentrations are close to their equilibrium values. Hence, these reactions can be used for the reverse direction in other metabolic circumstances. (**BOX 7.2**) The second consequence is that these steps are rarely sites of external control, that is, of regulatory behavior. This does not mean that, *in vitro*, molecules will not be found that appear to regulate the activity of these enzymatic reactions. Instead, it means that, *in vivo*, they are not sites of pathway control. Typically, enzymes that catalyze these reactions are found in relatively large amounts, which is the means by which the cell is able to achieve near-equilibrium status.

A second case is when $Q$ is much less than $K_{eq}$. These reactions are called **metabolically irreversible**, because they are greatly displaced from their equilibrium positions. As a result, these reactions cannot run in the reverse direction under cellular conditions. Metabolically irreversible reactions are invariably the sites of metabolic control, through allosteric or covalent modification. The cellular content (i.e., protein concentration) of enzymes that catalyze metabolically irreversible reactions is relatively low. This is consistent with the role of such enzymes as bottleneck or rate-limiting points of metabolic pathways.

Determining whether a reaction is near-equilibrium or metabolically irreversible requires a comparison of $Q$ with $K_{eq}$. The best characterized pathway is glycolysis, for which complete data are available for several cell types.

# 7.8 ATP

Adenosine 5′-triphosphate (ATP) is the central energy intermediate of metabolism. In order to analyze it from a thermodynamic view, we will start by considering standard free energy values (**BOX 7.3**) for ATP hydrolysis.

**FIGURE 7.3** shows the hydrolysis reactions of ATP, and these are summarized in Table 7.2, along with $\Delta G°$ values for the reactions. It

## BOX 7.3: Standard Thermodynamic Values

The choice of conditions for a standard state reflects two distinct needs. The first is to approximate laboratory conditions. At sea level, pressure is 1 atm, and room temperature is close to 25°C, although the standard temperature is 0°C. The second need is to simplify the relationship between standard free energy and the equilibrium constant; hence, all concentrations are 1 M, so that the argument of the logarithm term becomes 1, and the logarithm itself is zero.

In what is called "biochemical standard state," however, hydrogen ion concentration is taken as $10^{-7}$ M, so that pH = 7 (which is much closer to that found in biological systems). Typically, references include a prime after the superscript zero to indicate this change. While the values used in this book also are taken from the sources that fix pH at 7, this stylistic change was not implemented because it complicates the text and focuses the student on a trivial alteration rather than the main point of standard free energy. There is another complication with making this change: it invalidates the standard free energy relationship to $K_{eq}$. It can be argued that this is just a correction, and we can take this even further and have all of the reactant concentrations set to much lower values, such as 0.1 mM, which is typically much closer to physiological values. However, this would still not produce correct numbers. Instead, it avoids the critical point that there is a distinction between standard states, equilibrium states, and actual states that must be grasped, and is far more important than having the standard states come close to actual ones.

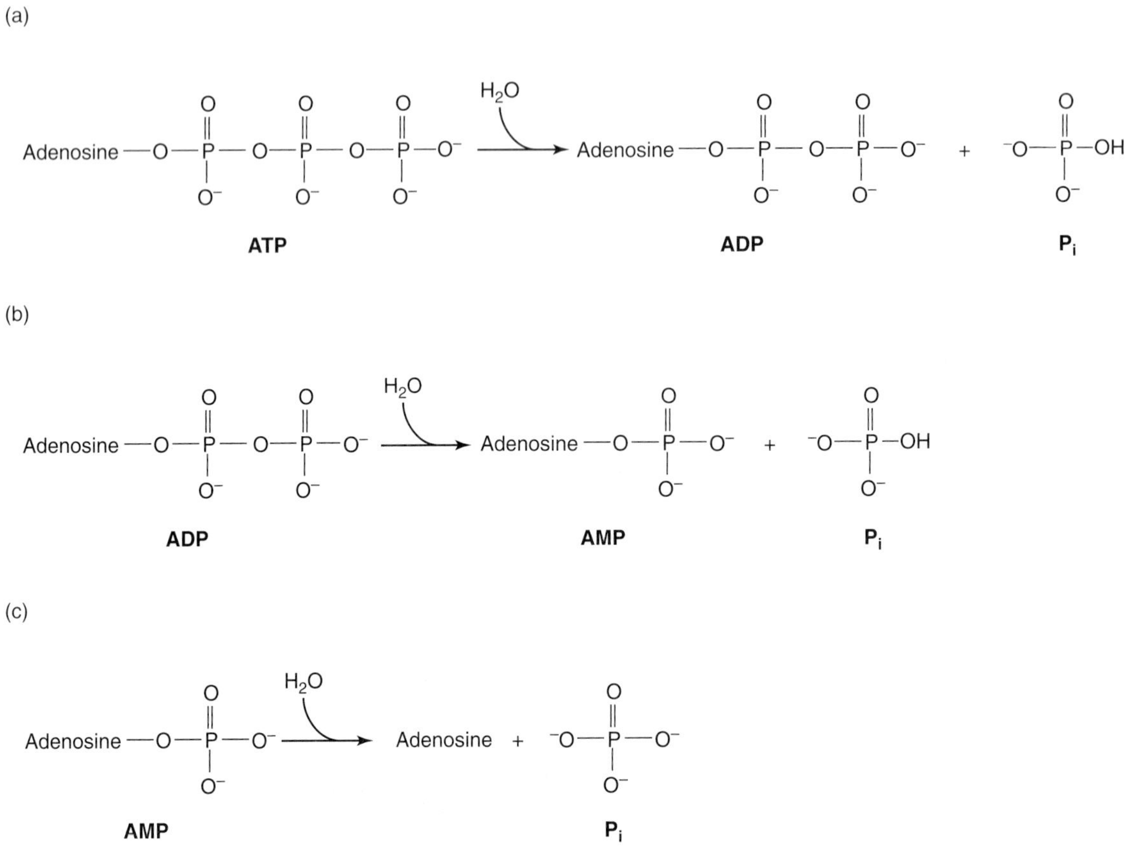

(a)

ATP → ADP + P_i

(b)

ADP → AMP + P_i

(c)

AMP → Adenosine + P_i

FIGURE 7.3 **Hydrolysis Reactions.**

### TABLE 7.2 Standard Free Energies of ATP Hydrolysis

| Reaction | $\Delta G°$, kJ/mol |
|----------|---------------------|
| ATP + $H_2O$ $\rightleftarrows$ ADP + $P_i$ | −30 |
| ADP + $H_2O$ $\rightleftarrows$ AMP + $P_i$ | −32 |
| AMP + $H_2O$ $\rightleftarrows$ Adenosine + $P_i$ | −13 |

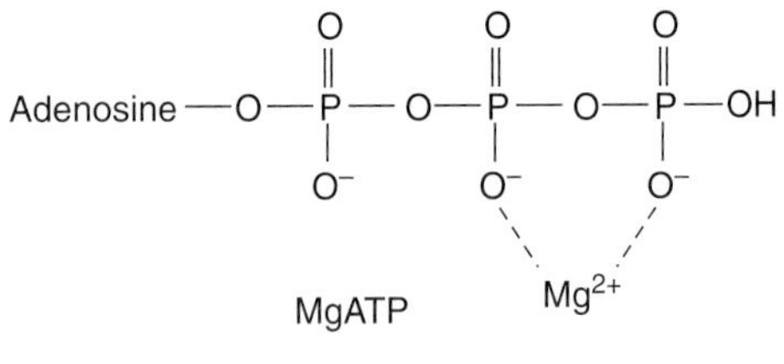

FIGURE 7.4 **MgATP.**

is clear from these values that there is a much smaller change in $\Delta G°$ for the hydrolysis of AMP to adenosine than for the hydrolysis of ATP to adenosine diphosphate (ADP) or the hydrolysis of ADP to adenosine monophosphate (AMP). In all of these reactions, there is an increase in entropy because the reactants are split into two species of products. This increased translational freedom accounts for the favorable entropy contribution. Two other effects explain the lower standard free energy change of AMP hydrolysis. First, when neighboring phosphoryl groups are separated, the repulsive interactions from the negative charges on their oxygen groups are relieved, a process that is not present in AMP hydrolysis. Second, the individual phosphate groups have more resonance possibilities in the products. This is less pronounced for AMP hydrolysis because one of the products is adenosine, which has no increased resonance in the product.

In cells, the electrical repulsion of neighboring oxygens is greatly lessened by chelation with $Mg^{2+}$ (**FIGURE 7.4**). In fact, $Mg^{2+}$ binds so strongly that all reactions involving ATP (except for the ATP translocase of mitochondria) actually use MgATP as the substrate. Calculating standard and actual free energy changes for ATP reactions is complex, in part because $Mg^{2+}$ chelation alters the values.

Two other factors must be considered before we are able to transfer these values to cellular conditions. First, measuring actual concentrations of ADP and AMP generally requires indirect analysis in living

**CHAPTER 7** METABOLISM AND ENERGY

cells because most of the ADP and AMP exist bound to proteins. It is only the free nucleotide concentration that is important thermodynamically, so measuring total values is misleading. Nonetheless, studies have been conducted to correct for these difficulties and it has been found that the cellular $\Delta G$ is about $-59$ kJ/mol in the cytosol of muscle and liver cells if the reactants are arranged to form the ratio $[ATP]/[ADP][P_i]$. The second factor to consider is that ATP hydrolysis does not actually occur physiologically. In fact, in energy transfers, it is the phosphoryl rather than a phosphate that is transferred (FIGURE 7.5). Nonetheless, it is useful to compare hydrolysis reactions and their corresponding $\Delta G°$ values, to impart ideas of inherent energy in ATP and to better understand energy coupling, our next topic.

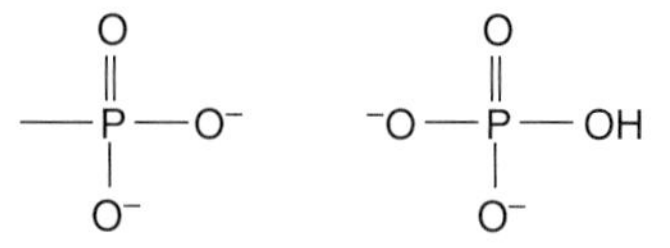

FIGURE 7.5 **Phosphoryl and Phosphate**.

# 7.9 Energy Coupling with ATP

*Energy coupling* is the joining of a process that is endergonic in isolation with one that is exergonic to produce a combined process that is exergonic overall. We will explore this concept using the asparagine synthetase reaction:

$$Aspartate + ATP + NH_3 \rightleftarrows Asparagine + ADP + P_i \qquad (7.24)$$

We will first make use of the standard free energy changes for the reactions that sum to Equation 7.24, taking advantage of the fact that state functions such as $\Delta G°$ are additive. The known value for asparagine hydrolysis is,

$$Asparagine + H_2O \rightleftarrows Aspartate + NH_3$$
$$\Delta G° = -14.2\,kJ/mol \qquad (7.25)$$

To obtain an overall $\Delta G°$ value, we write this reaction in reverse and add it to the known value for ATP hydrolysis from Table 7.3:

$$Aspartate + NH_3 \rightleftarrows Asparagine + H_2O$$
$$\Delta G° = -14.2\,kJ/mol \qquad (7.26)$$

$$ATP + H_2O \rightleftarrows ADP + P_i \qquad \Delta G° = -30\,kJ/mol \qquad (7.27)$$

| TABLE 7.3 Energy Coupling in Asparagine Synthetase | | |
|---|---|---|
| | Reaction | $\Delta G°$, kJ/mol |
| (a) | Asparagine $\rightleftarrows$ Aspartate + NH$_3$ | −14 |
| (b) | Aspartate + NH$_3$ $\rightleftarrows$ Asparagine | +14 |
| (c) | ATP + H$_2$O $\rightleftarrows$ ADP + P$_i$ | −32 |
| (b)+(c) | Aspartate + NH$_3$ + ATP $\rightleftarrows$ Asparagine + ADP + P$_i$ | −13 |

The sum of these reactions is Equation 7.24, and the sum of the $\Delta G°$ values is −15.8 kJ/mol. Thus, under *standard* conditions, the overall reaction is exergonic.

While it is tempting to conclude that ATP hydrolysis seems to drive asparagine formation, this is a mechanistic interpretation, and our calculations are purely thermodynamic. The actual mechanism of the reaction for asparagine synthetase is shown in FIGURE 7.6. Note that it involves two nucleophilic substitutions: first an acid phosphate is created on the side-chain carboxyl of aspartate; subsequently, the acid phosphate is displaced by ammonia. None of this pathway is relevant to the energy calculation, but it does provide a mechanism for how ATP bond energy is utilized to activate an otherwise stable carboxylate group to form the amide product.

The fact that we can deconstruct a reaction into one part positive $\Delta G°$ and one part negative $\Delta G°$, and that it has a negative $\Delta G°$ overall is sometimes called **energy coupling**. In ATP synthetase, ATP hydrolysis appears as part of the thermodynamic calculation, and the elements of the reaction appear in the overall reaction (i.e., ATP, ADP, and $P_i$). The energy of the phosphoryl transfer is employed to synthesize the acid phosphate intermediate as shown in the mechanism. Thus, there are two distinct meanings of energy coupling.

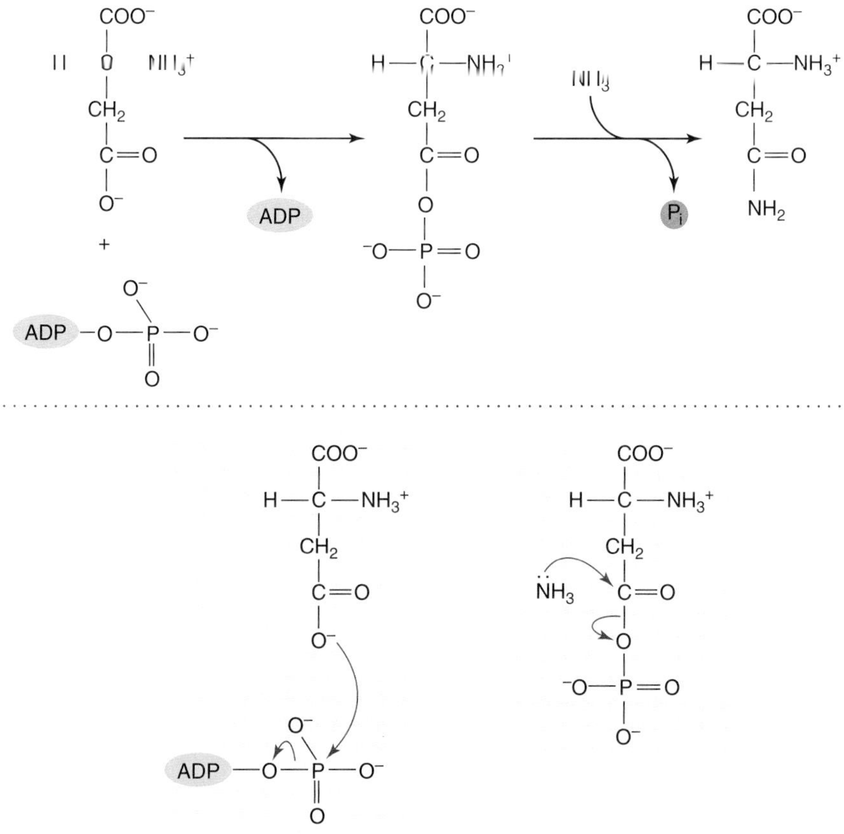

FIGURE 7.6 **Asparagine Synthetase Mechanism.** The intermediate steps above the line show the aspartyl phosphate intermediate. Below the line, intermediary steps are illustrated, along with the curved arrows indicating mechanistic electron flow.

CHAPTER 7 METABOLISM AND ENERGY

ATP can also be used in energy maintenance rather than coupling. Three examples are the creatine phosphokinase, nucleotide kinase, and adenylate kinase reactions.

## Creatine Phosphokinase

Creatine phosphokinase catalyzes the reaction of creatine with ATP to form creatine phosphate and ADP (FIGURE 7.7). The phosphoryl group of ATP is transferred to a nitrogen of creatine (metabolically derived from the amino acid arginine) to form creatine phosphate. This nitrogen is part of the resonance-stabilized guanadino group; this destabilizes the phosphoryl making creatine phosphate a very high energy compound. The creatine phosphokinase enzyme is highly expressed in muscle cells, where creatine phosphate serves as a very rapid (although limited) energy store. During periods of rest, creatine phosphate is formed from ATP. The stored concentration of creatine phosphate reaches over 40 mM, severalfold that of ATP. During contraction, a sudden increase in ATP utilization causes the reaction to run in the direction of creatine formation, replenishing ATP. In this way, the creatine phosphokinase reaction prevents a depletion of ATP in the muscle. This allows the level of ATP to remain effectively constant despite rapid changes in energy demand by the cell. Because the same reaction is used in both the forward and reverse directions under different circumstances, it is an example of a near-equilibrium reaction.

## NDP Kinase

A second reaction that involves ATP in cells is the nucleotide diphosphate (NDP) kinase, which catalyzes the reaction:

$$ATP + NDP \rightarrow ADP + NTP \qquad (7.28)$$

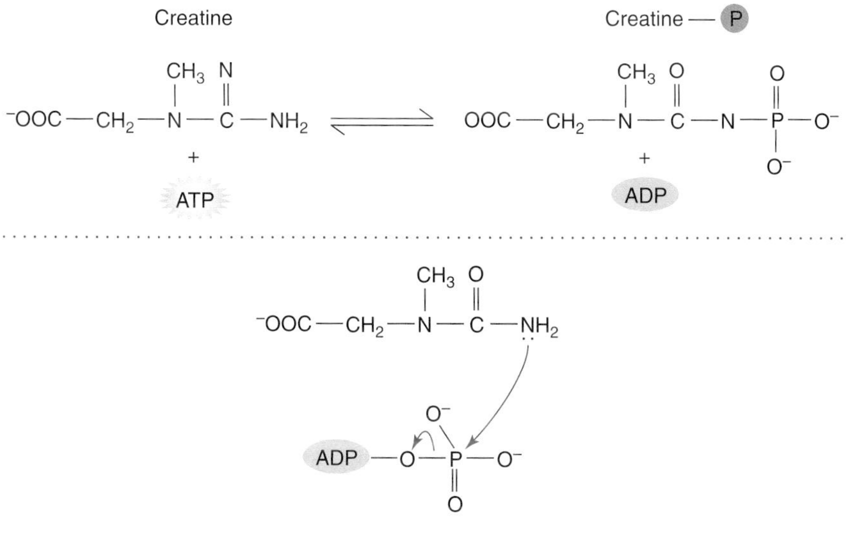

FIGURE 7.7 **Creatine Phosphokinase Mechanism.** The mechanism diagram shows the relatively simple reaction pathway that is analogous to a portion of the asparagines' synthetase reaction shown in Figure 7.6.

where the nucleotide N is a variable; it may be guanine (G), cytosine (C), etc. The reaction allows nucleotides other than ATP to serve as energy donors. The displacement from equilibrium for the NDP kinase reaction in cells remains unknown, although it is likely to be metabolically irreversible based on estimates of nucleotide concentrations. In fact, there are no known instances where this reaction is used for the purpose of synthesizing ATP.

The overall $\Delta G°$ for the reaction depicted in Equation 7.28 is approximately zero (in which case $K_{eq} \approx 1$). This stems from the fact that the reaction has essentially the same phosphate bonds in both the reactants and the products: a triphosphate (ATP/NTP) and a diphosphate (NDP/ADP). It is sometimes difficult to resist the temptation to conclude that the reaction is near-equilibrium in cells, but there is no basis for this supposition. The other intriguing aspect of NDP kinase is that it is an enzyme with somewhat relaxed specificities, accepting different nucleotides as substrate. While this broad substrate selection is true of many enzymes, in most cases their physiological substrates are limited.

## Adenylate Kinase

A third maintenance type reaction for ATP is catalyzed by adenylate kinase:

$$ATP + AMP \rightleftarrows ADP + ADP \qquad (7.29)$$

As with the NDP kinase reaction, the equilibrium constant for the adenylate kinase reaction is approximately 1, and hence its standard free energy about zero. In the adenylate kinase case, however, the reaction is near-equilibrium in cells, providing both a means of rephosphorylating AMP generated in certain activation reactions as well as producing AMP and ATP from ADP. Most ADP produced by cells is converted directly to ATP by pathways we will examine elsewhere in this text.

The mechanism of adenylate kinase involves both ATP and AMP binding simultaneously at the enzyme active site (FIGURE 7.8). Once positioned, the terminal oxygen of the AMP attacks the terminal phosphoryl of the ATP, resulting in the two ADP product molecules. The simultaneous presence of both adenine nucleotides is dramatically evident by the potent inhibitor diadenosine pentaphosphate (FIGURE 7.9), a molecule consisting of two adenine nucleotides linked with five

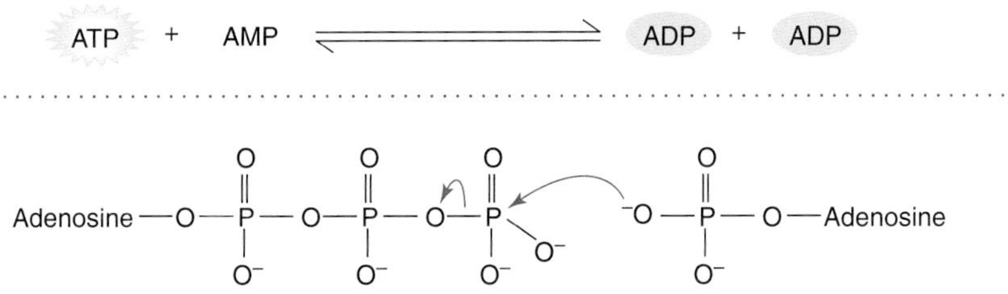

FIGURE 7.8 **Adenylate Kinase Mechanism**. In-line electron flow is made possible by simultaneous binding of ATP and AMP.

CHAPTER 7 METABOLISM AND ENERGY

phosphates. A very similar inhibitor with just one fewer phosphate, diadenosine tetraphosphate, is an order of magnitude more potent, providing strong evidence for the precise alignment of the nucleotides on the enzyme surface.

## 7.10 NADH

Electron carriers are another energy intermediate between catabolism and anabolism, and the principle electron carrier is NADH (FIGURE 7.10). In our examination of metabolic processes, and particularly in oxidative phosphorylation, we will examine the relationship between the two principal energy intermediates, ATP and NADH. In the discussion of photosynthesis elsewhere in this text, we will examine the process of trapping light energy from the sun into high-energy electrons and ATP. The present section examines NADH and the essentials of electron transfer.

Change in a redox state involves two connected processes. By definition,

> **Oxidation** is a loss of one or more electrons from a molecule or functional group.

> **Reduction** is a gain of one or more electrons from a molecule or functional group.

The two processes must occur together, because electrons lost from one entity must be gained by another. Reactions involving electron transfer are called oxidation–reduction or **redox** reactions.

While oxidation and reduction must occur together, it is possible to separate them and focus on just half of a redox reaction at a time. For example, consider the removal of two electrons from ethanol:

$$CH_3CH_2OH \rightarrow CH_3CHO + 2e^- + 2H^+ \tag{7.30}$$

In this oxidation **half-reaction**, the alcohol is oxidized to the aldehyde, and two electrons and a proton are produced. Thus, an alcohol is more reduced than an aldehyde. The other half-reaction—a reduction—consumes the electrons and a proton:

$$NAD^+ + 2e^- + H^+ \rightarrow NADH \tag{7.31}$$

The electrons are actually transferred as a **hydride ion**, H:, indicating that Equations 7.30 and 7.31 are merely *formal* statements of redox half-reactions and are not steps in a reaction mechanism. The actual reaction pathway for alcohol dehydrogenase is illustrated in FIGURE 7.11. The enzyme binds ethanol and removes a proton, attaching the oxygen of ethanol to an enzyme-bound $Zn^{2+}$ ion. Next, $NAD^+$ abstracts a hydride from the substrate. Addition of the hydride to the $NAD^+$ is facilitated by electron migration in the $NAD^+$ ring to the positively

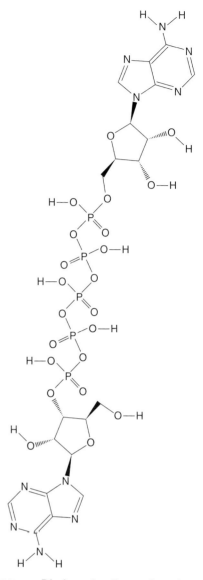

FIGURE 7.9 **Diadenosine Pentaphosphate.** The potent inhibition by diadenosine pentaphosphate provides support for the mechanism outlined in Figure 7.8.

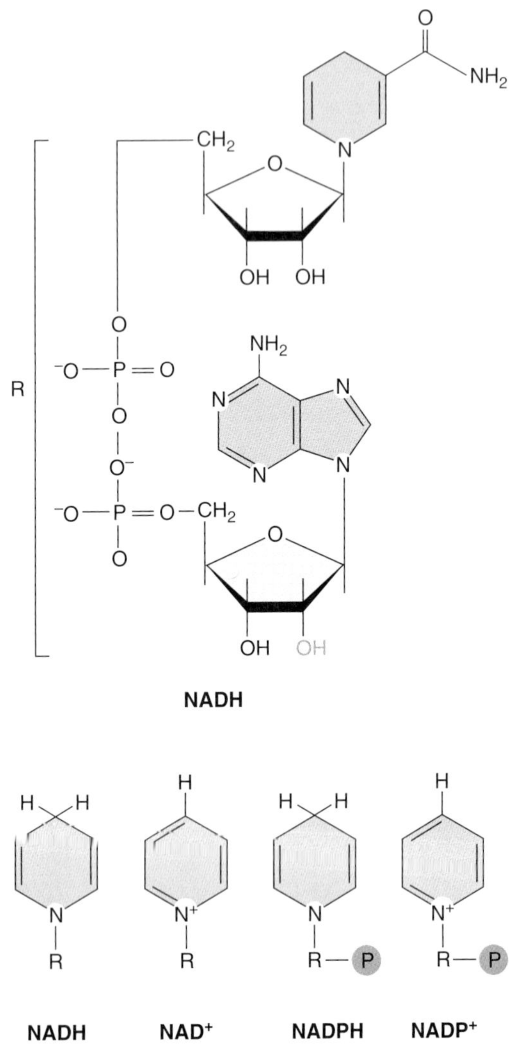

**NADH**

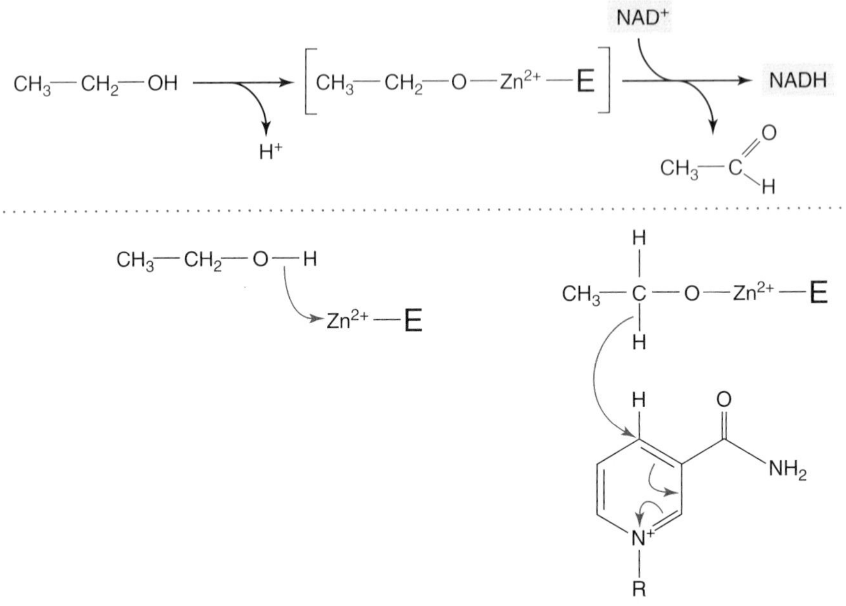

FIGURE 7.11 **Mechanism of Alcohol Dehydrogenase.** Alcohol dehydrogenase is a metalloenzyme. The tightly bound $Zn^{2+}$ helps remove the proton from ethanol and assists in the removal of the hydride to $NAD^+$.

**NADH**    **NAD⁺**    **NADPH**    **NADP⁺**

FIGURE 7.10 **NADH and Similar Nucleotides.** Nicotinamide nucleotides. $NAD^+$ and $NADP^+$ differ only in having a phosphate esterified to the 2′ hydroxyl of the adenine portion of the molecule (this position is indicated in red in the diagram).

charged nitrogen atom, an **electron sink**. Subsequently, the products (acetaldehyde and NADH) dissociate from the enzyme surface.

In an analogous way to phosphate bond energy, where ATP is not the only option, there are other electron carriers aside from NADH. For example, NADPH is also a hydride carrier, and has a molecular structure virtually identical to that of NADH. The difference is that NADPH has one an extra phosphoryl group (Figure 7.10) that confers selectivity in binding to enzymes. Most enzymes exclusively bind either NADH or NADPH. There is an interconversion that occurs in mitochondria, in a reaction catalyzed by **transhydrogenase**:

$$NADH + NADP^+ \rightarrow NAD^+ + NADPH \qquad (7.32)$$

This reaction is metabolically irreversible and, thus is a unidirectional transfer of reducing equivalents to NADPH.

# 7.11 Mobile Cofactors and the Pathway View

The energy intermediates ATP, ADP, NADH, and $NAD^+$—as well as closely related intermediates such as GTP, GDP, NADPH, and $NADP^+$—are **mobile cofactors** when they are viewed in the context of metabolic pathways. Mobile cofactors can be thought of as links between or within metabolic sequences.

Connections between reactions occur in a pathway view that do not exist when reactions are considered in isolation. The most

CHAPTER 7 METABOLISM AND ENERGY

fundamental connection—the one that defines a pathway—is that the product of one reaction is the substrate of the next. Consider the pathway shown in FIGURE 7.12. One of the interior reactions uses ATP as substrate and forms ADP as product. From the standpoint of the pathway, ATP and ADP are *not* substrates or products; they are mobile cofactors. Unlike the reactant and product of this step in the pathway, ATP and ADP are not directly connected to either the previous or the following reactions, but rather to reactions elsewhere in this pathway or in another pathway.

The other mobile cofactors, such as NADH and $NAD^+$, act in the same way; namely, they connect different pathways in *parallel*, as suggested by FIGURE 7.13. This is very different from cofactors that always stay bound to their enzymes. The chemistry of bound cofactors—often called **prosthetic groups**—may resemble mobile cofactors. However, the prosthetic groups are attached to the enzyme, and must be regenerated along with the rest of the enzyme in the course of a single catalytic cycle. Thus, they cannot connect separate pathways. If we consider the first substrate for a pathway as the **pathway substrate** and the last product as the **pathway product**, then all of the other intermediates of the pathway reach a constant concentration once the pathway has achieved a steady state. If the pathway is a linear sequence, then there is just one rate for the entire pathway, called the **pathway flux**. The steady state for the intermediates is maintained by a constant provision of pathway substrate and a constant removal of pathway product. However, the mobile cofactors are in limited supply, and so they must be regenerated by an external pathway in amounts sufficient to allow the pathway to proceed. Glycolysis exemplifies cofactor balance using both ATP and NADH.

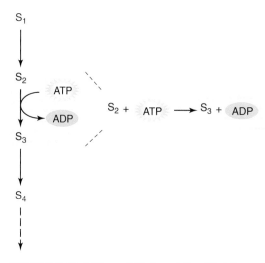

FIGURE 7.12 **Different Roles of the Mobile Cofactor ATP/ADP.** In the pathway on the left, the ATP and ADP are mobile cofactors. In the expansion on the right, they are substrate and product, respectively, of the isolated reaction.

## Summary

Energy exchanges are the central issue of metabolism. The major energy intermediate is the molecule ATP. A deeper understanding of energy requires a study of thermodynamics, which has two key laws. The first law states that the total internal energy of a system is conserved between heat ($q$) and work ($w$). Internal energy is itself a state function, defined as the sum of the heat and work ($q + w$). The second law involves entropy, which is roughly a measure of "disorder," but in actuality is a measure of the possible number of states of a system and their arrangement. According to the second law, the entropy of the universe always increases after a reaction. The free energy combines the internal energy and the entropy of the system to produce a parameter, $\Delta G$, which determines whether the reaction can take place as written. If $\Delta G$ is negative, then the reaction is exergonic, and can take place. If $\Delta G$ is positive, then the reaction is endergonic, and the reverse reaction can take place. If $\Delta G = 0$, then the reaction is at equilibrium. Under standard conditions, all reactions have substrate and product concentrations of 1 M, and can

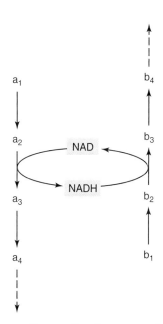

FIGURE 7.13 **Connecting Sequences in Parallel by the Pair $NAD^+$/NADH.** The sequence involving substrates $a_i$ reduces $NAD^+$ to NADH, whereas the sequence involving $b_i$ oxidizes NADH to $NAD^+$. Only with both operating is it possible to have continued flux through $a_i$.

be compared to one another *in vitro*. Under cellular conditions, the reactions are displaced from equilibrium, either slightly—in which case the reaction is near-equilibrium—or greatly, in which case the reaction is metabolically irreversible. Two energy molecules, ATP and NADH, represent mobile cofactors that connect separate pathways in metabolism. ATP is involved in several reactions that exchange high-energy phosphate: creatine phosphokinase, in which a separate store of high-energy phosphate can be produced; NDP kinase, which exchanges phosphoryl groups with nucleotides other than adenine; and adenylate kinase, which reversibly interconnects ATP, ADP, and AMP. Mobile cofactors connect reaction sequences—pathways—in parallel.

## Key Terms

| | | |
|---|---|---|
| classical thermodynamics | heat | prosthetic groups |
| electron sink | hydride ion | redox |
| endergonic | internal energy | state |
| energy | mass action ratio | state variables |
| energy coupling | metabolically irreversible | statistical thermodynamics |
| enthalpy | mobile cofactors | surroundings |
| enthalpy driven | near-equilibrium | system |
| entropy driven | path independent | thermodynamics |
| exergonic | pathway flux | transhydrogenase |
| frcc energy | pathway product | universe |
| half-reaction | pathway substrate | work |

## Review Questions

1. A reaction has an enthalpy change of 30 kJ/mol and an entropy change of 100 J/mol · K. Assuming room temperature is about 300 K, is this reaction at equilibrium? What would the free energy change be if the temperature was higher?

2. Suppose that a reaction has $\Delta G° = 0$. What conclusions can be drawn about its equilibrium constant? Can we determine if it is reversible under cellular conditions?

3. Internal energy is very similar to enthalpy. Explain why that is the case biologically, and why we still retain enthalpy in equations for free energy.

4. Redox reactions often involve the transfer of hydrogens during reactions involving organic compounds; acid–base reactions do as well. How are these processes distinct and why do both seem be involve hydrogen transfer?

5. What is the most reduced form of carbon in an organic molecule? What is the most oxidized?

6. Explain why heat is *not* a state function but heat at constant pressure is a state function.

7. Perfect interconversion between heat and work does not violate the first law of thermodynamics. Why does it violate the second law of thermodynamics?

8. One way of calculating an equilibrium constant is to use standard free energy values. Here, Equation 7.20 has been modified to use common (base 10) logs:

$$\Delta G° = -(2.3)RT \log K_{eq} \tag{7.33}$$

Use Equation 7.33 to calculate $K_{eq}$ if $\Delta G°$ is (a) 1; (b) 100; (c) 1/100.

9. Because *heat* is actually just an energy exchange due to a temperature difference, what does it mean when you open the window in winter and the cold air *comes in*? Is heat, instead, really *flowing out*?

# References

Kishore, N.; Tewari, Y. B.; Goldberg, R. N. A thermodynamic study of the hydrolysis of L-glutamine to (L-glutamate + ammonia) and of L-asparagine to (L-asparagine + ammonia). *J. Chem. Thermodynamics* 32:1077–1090, 2000.

Lawson, J. W. R.; Veech, R. L. Effects of pH and free Mg on the K of the creatine kinase reaction and other phosphate hydrolyses and phosphate transfer reactions. *J. Biol. Chem.* 254:6528–6537, 1979.

Newsholme, E. A.; Leech, A. R. *Biochemistry for the Medical Sciences;* John Wiley & Sons: New York, 1986.

Ochs, R. S. Thermodynamics and spontaneity. *J. Chem. Educ.* 73:952–954, 1996.

Veech, R. L.; Lawson, J. W. R.; Cornell, N. W.; Krebs, H. A. Cytosolic phosphorylation potential. *J. Biol. Chem.* 254:6538–6547, 1979.

# 8

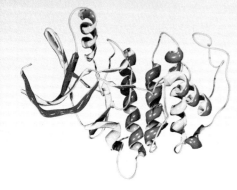

# Glycolysis

## CHAPTER OUTLINE

Image © Dr. Mark J. Winter/Photo Researchers, Inc.

**G**lycolysis is the oldest known pathway, the best studied, and the prototype for all other pathways in living cells. Most cells have the glycolytic pathway, even if they have little else. A key purpose of the pathway is the net formation of ATP. Beyond this, every intermediate in glycolysis serves as an intermediate in another pathway in most cells. Thus, glycolysis can be thought of as a metabolic crossroads (**BOX 8.1**).

In one view, the glycolysis pathway is a sequence of 10 enzymes that convert glucose to pyruvate. However, for **pathway completion**, it is necessary to balance all mobile cofactors. Both ATP and NADH are mobile cofactors formed in pyruvate production. ATP is converted to ADP by energy-utilizing reactions and NADH is reoxidized by different routes. The most common glycolytic end product is lactate.

The enzymes of glycolysis are also well-studied, and we will examine their chemical mechanisms and properties as models for those that appear later. This chapter also marks the beginning of individual steps being integrated into a pathway, and a pathway being integrated into the whole of metabolism.

## 8.1 Glucose Transport

Before glucose can be converted to pyruvate, it must first enter the cell. All of the enzymes catalyzing the steps of glycolysis exist in the cytosol. While some cells use a specialized form of glucose uptake (**BOX 8.2**), most mammalian cells use a passive glucose transport protein, abbreviated GLUT. At least 14 **isoforms**—that is, distinct proteins that have identical function—of GLUT have been discovered. These are named in chronological order of their discovery (e.g., GLUT1, GLUT2, etc.). We will examine three of them here (Table 8.1).

GLUT1, first discovered in red blood cells, is also found in many mammalian cells and is prominent in fetal tissues. A relatively low-$K_m$ transporter (about 2 mM; for comparison, blood glucose concentration is 5 mM), it provides baseline glucose uptake for most cells. The protein spans the membrane with 12 alpha helical segments and adopts a clamshell-like configuration. Glucose binding on one side induces a conformational change that flips the orientation of its binding site to the other side (FIGURE 8.1).

The rate-limiting step in transport is the return of the unoccupied receptor to the original membrane face. The basis of the continuous transfer of glucose into the cell is the relatively low concentration of intracellular glucose, which is ensured by ongoing glycolysis.

GLUT2 is found in liver, pancreas, kidney, and intestinal epithelia. The transporter operates well below saturation; its $K_m$ is above 10 mM. GLUT2 achieves near-equilibrium with extracellular glucose concentrations, so that glucose uptake responds to extracellular glucose concentrations, and the transporter becomes a route for the reverse direction in other pathways.

## BOX 8.1: WORD ORIGINS

### Zymo

*Zymo* and its variants are from Greek, meaning *leaven.* The root term refers to yeast and to ferments from yeast. When Eduard Buchner found that extracts derived from yeast degrade glucose in a cell-free fermentation, he applied the name *enzyme* to their preparation. Similarly, there are fungal diseases called *zymotic,* and the sign of an infection was named *zymosis,* which means *ferment.* The yeast extract that metabolizes glucose was later discovered to be a mixture of many components. Each of those (and thousands more) are now called an enzyme. Thus, glycolysis became the first process recognized to act independently of a cell. Embedded in the name "enzymes" are the yeasts, the organism in which the first collection of enzymes producing an observable result—namely, bubbles—was discovered.

## BOX 8.2: A Separate Class of Glucose Transporters

The glucose transporters required to supply glycolysis are not the only means by which glucose is transported into the cell. In certain epithelial cells—that is, those with distinct membranes facing different spaces—there is another type of transporter that catalyzes the uptake of extracellular glucose along with extracellular $Na^+$. These transporters are indirectly energy-linked, because there is a steep sodium gradient: $[Na^+]$ is high outside the cell and low inside. Thus, as $Na^+$ flows down its gradient, glucose enters along with it in the presence of a sodium-linked glucose transporter (SGLT). For example, the SGLT in the intestine allows the entry of glucose from the lumen into the cell. Little of the glucose is actually metabolized in the epithelial lining the intestine; instead, most is transported out of the cell into the blood to be carried to other cells by virtue of having a GLUT2 in the basolateral membrane (FIGURE B8.2). Another SGLT isoform exists in the kidneys, allowing glucose filtered into tubules to be transported into the tubule cell across an apical membrane, and transported out again using a GLUT2 in the basolateral membrane for reabsorption into the blood.

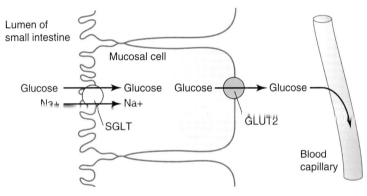

**FIGURE B8.2**

GLUT4 is found in muscle and fat cells. Like GLUT1, it is metabolically irreversible and has a relatively low Km for glucose. The unique feature of GLUT4 is insulin sensitivity, providing an increased uptake of glucose in response to an increase in blood insulin concentration. The GLUT4 protein is believed to translocate from an interior cell membrane to the plasma membrane in response to insulin.

### TABLE 8.1 Properties of Some Glucose Transporters

| Transporter | Cell Types | Near-Equilibrium | Insulin Sensitive | $K_m$ for Glucose mM |
|---|---|---|---|---|
| GLUT1 | Red cells, many others | no | no | 2 |
| GLUT2 | Liver, pancreas | yes | no | >10 |
| GLUT4 | Muscle, fat | no | yes | 1 |

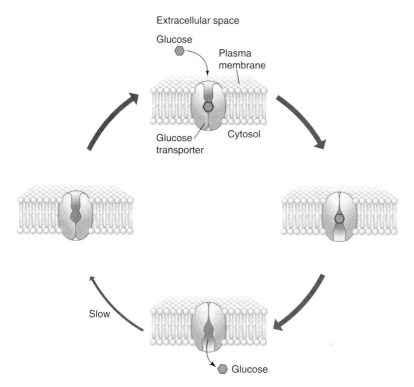

Extracellular space

Glucose

Plasma membrane

Glucose transporter

Cytosol

Glucose

Slow

**FIGURE 8.1 Glucose Transport.** The transporter undergoes a conformational change when glucose is bound. Glucose binding to the exterior causes the face of the transporter to flip to the interior, where glucose can dissociate and enter the cytosol. The slow step is the flipping of the unoccupied transporter to face the exterior once again. While the protein can flip to the exterior occupied by glucose, this would not accomplish net glucose transport.

## 8.2 From Glucose to Pyruvate

### Hexokinase

The first step of glycolysis from intracellular glucose is catalyzed by hexokinase. The reaction is a transfer of the phosphoryl group of ATP to glucose:

$$\text{Glucose} + \text{ATP} \rightarrow \text{Glucose-6-P} + \text{ADP} \tag{8.1}$$

The mechanism (FIGURE 8.2) is typical of kinase reactions, such as for adenylate kinase. As indicated in the graphic, the nucleophile—in this case, the 6-hydroxyl group of glucose—initiates a nucleophilic attack on the partially positive phosphorous of the terminal (gamma) phosphate of ATP. Electrons of the pi bond migrate to the oxygen atom, forming a pentavalent phosphate intermediate. As these electrons move back from the oxygen, they displace electrons of the P–O bond as shown, releasing ADP. In fact, any P–O bond rearrangement is possible; the others are either rapid rearrangement or a reversion to substrate. There are hundreds of kinase reactions in the cell, and they share this mechanism.

Another feature of hexokinase (also common to other kinases) is the protection of the intermediates from hydrolysis. This is accomplished by a conformational change in the enzyme upon binding substrate,

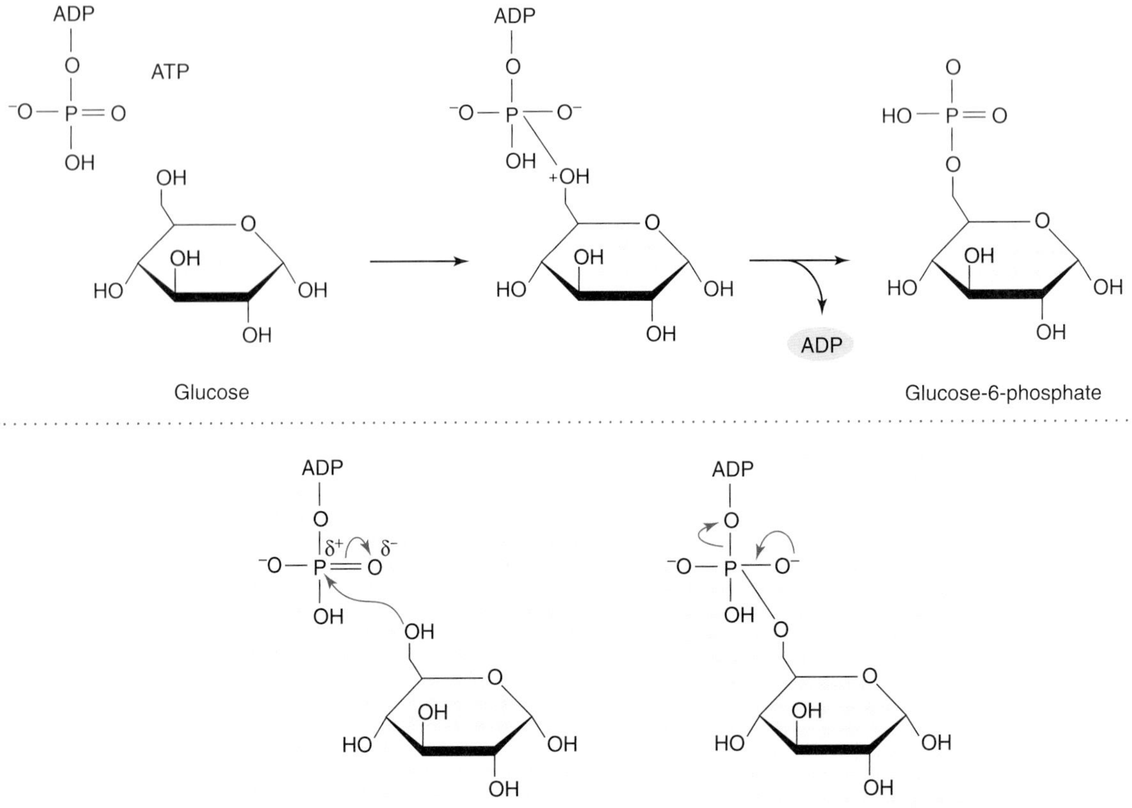

FIGURE 8.2 **Mechanism of Hexokinase**. Simultaneous binding of ATP and glucose to the enzyme provides the proximity for the nucleophilic attack of the 6-OH of glucose on the terminal phosphoryl of ATP. Electron rearrangement leads to the production of glucose-6-P and ADP.

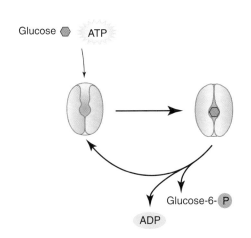

FIGURE 8.3 **Clamshell Model of Hexokinase**. The binding of glucose and ATP alters the conformation of hexokinase, excluding water from competing with the glucose hydroxyl in attacking the phosphoryl of ATP.

which isolates the active site and thus protects it from spurious water molecules (FIGURE 8.3). Note the reappearance of the "clamshell" configuration also apparent in the GLUT protein (Figure 8.1). In both cases, the two domains comprising the clamshell are connected by a hinge region. A very similar design also exists in protein kinase C.

Just as there are isoforms of the glucose transporter, there are different proteins that all catalyze the hexokinase reaction, called **isozymes**. Most hexokinase isozymes have a similar affinity for glucose, which can be estimated by their $K_m$ values, which are about 1 mM. One isozyme with a particularly high $K_m$ for glucose is called *glucokinase*. The same cells that have glucokinase also have the GLUT2 transporter. Thus, like GLUT2, glucokinase is found in liver and the beta cells of the pancreas. This pairing has a purpose: because GLUT2 allows a near-equilibrium transport of glucose across the membrane, the intracellular glucose concentration will be very high—near the blood glucose value of 5 mM. Accordingly, the $K_m$ for glucokinase is also very high, about 20 mM.

Most hexokinase isozymes are product-inhibited by glucose-6-P (i.e., the product of the hexokinase reaction). However, GLUT2 is not inhibited by physiological levels of glucose-6-P and, in accord with its near-equilibrium nature, is not a regulatory site.

Glucose-6-P is one of many sequential phosphorylated intermediates and may be considered "trapped" in the cytosol. This is not because of

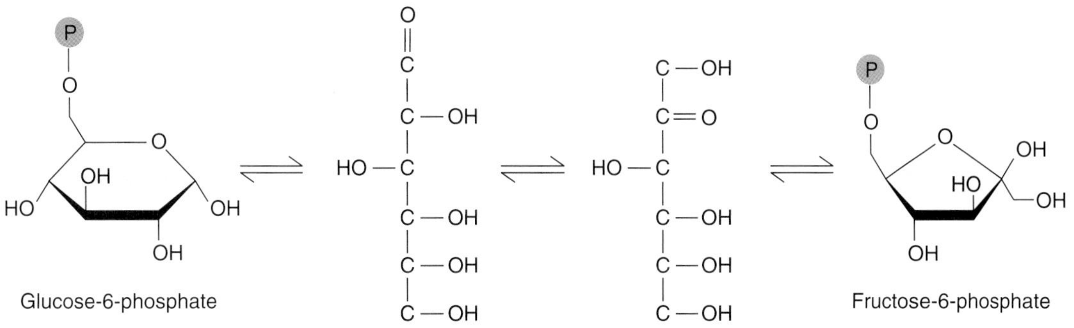

FIGURE 8.4 **Glucose Phosphate Isomerase Ring Opening.** Glucose phosphate must achieve the open-chain form before it can be isomerized to fructose phosphate.

the greater water solubility of the phosphorylated compound, but rather because the phosphate esters have no transporter that would allow them to exit the cell. The composition of every water space in the cell depends on the presence of transporters that allow them to enter and exit.

## Glucose Phosphate Isomerase

The next step in glycolysis is the isomerization of glucose-6-P to fructose-6-P. The enzyme glucose-6-P isomerase catalyzes this near-equilibrium reaction, a conversion of an aldehyde to a ketone sugar phosphate. FIGURE 8.4 shows that the reaction requires opening the ring form of glucose-6-P to reveal the aldehyde. The reaction itself is indicated in the enclosed square. The product, fructose-6-P, is in equilibrium with its ring form.

Focusing on the enclosed square in Figure 8.4, it is evident that the reaction involves "moving" a carbonyl from position one to position two. The actual mechanism is illustrated in FIGURE 8.5. Upon the

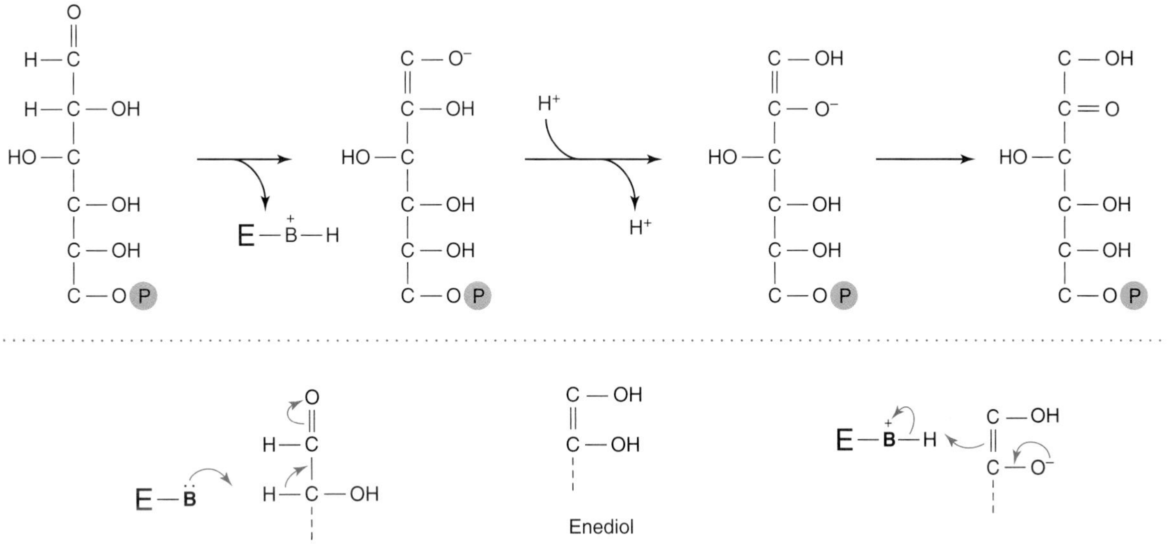

FIGURE 8.5 **Mechanism of Glucose Phosphate Isomerase.** Essentially a simple acid–base-catalyzed reaction, a proton is first abstracted from C2 of glucose-6-P by an enzyme-bound basic group, and electron rearrangement produces the symmetrical intermediate (i.e., the enediol). The proton is added back to the enediol, but at C1. Electron rearrangement then produces fructose-6-P.

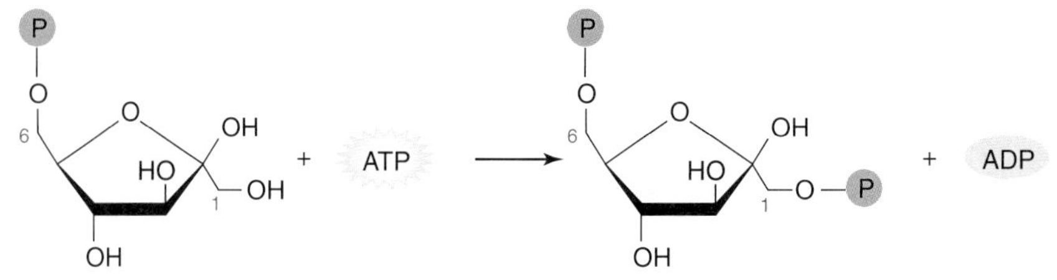

Fructose-6-phosphate  +  ATP  →  Fructose-2, 6-biphosphate  +  ADP

**FIGURE 8.6** **Phosphofructokinase Reaction.**

binding of glucose-6-P to the enzyme, an attached base (represented by **:B**) abstracts a proton from the C2 position. The ensuing electron rearrangement leads to an intermediate enediol, a **symmetrical intermediate** shown in the center of Figure 8.5. In the remainder of the mechanism, the protonated enzyme-linked base becomes positioned near C1, and donates the proton to C1 as a result of electron movement initiated from electrons on the oxygen attached to C2. Both the intermediate and the mechanism are effectively symmetrical, a strategy that we will see repeated with other isomerase-type enzymes.

## Phosphofructokinase

Phosphofructokinase (PFK) catalyzes the phosphorylation of fructose-6-P to fructose-1,6-P$_2$ (FIGURE 8.6). This phosphoryl transferase reaction is virtually identical in mechanism to the hexokinase reaction. PFK is metabolically irreversible, and critically important as the major rate-limiting step in glycolysis.

PFK is regulated allosterically by a large number of molecules *in vitro* (**BOX 8.3**). Two regulators are known to be physiologically significant: citrate (a negative modulator) and fructose-2,6-P$_2$ (a positive modulator). Both effects on the enzyme activity are indicated in FIGURE 8.7 as alterations in the S-shaped substrate–velocity curve.

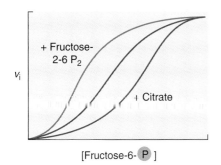

**FIGURE 8.7** **Allosteric Regulation of Phosphofructokinase.** In the presence of fructose-2,6-P$_2$, the curve shifts to the left, indicating activation. In the presence of citrate, the curve shifts to the right, indicating inhibition.

## BOX 8.3: *In Vitro* Modulators of PFK

Beyond fructose-2,6-P and citrate, a large number of other allosteric effectors of PFK are known *in vitro.* However, they are unlikely to be physiological regulators of the enzyme in the cell. It has long been known, for example, that ATP is a regulator of this kinase as well as many other reactions *in vitro.* But the fact that the concentration of ATP is essentially constant in cells means that this energy intermediate does not serve as a cellular regulator. Other candidates can be excluded for different reasons. For example, ADP activates PFK *in vitro,* and does change in concentration, but its cellular concentration is orders of magnitude lower than that needed to activate the enzyme *in vivo.* As an additional example, the concentration of fructose-1,6-P$_2$ does change in cells, is known to be a feed-forward activator of pyruvate kinase, and was once believed to also be a product activator of PFK. However, fructose-1,6-P$_2$ only mimics the true PFK activator, fructose-2,6-P, which exists at even lower concentrations.

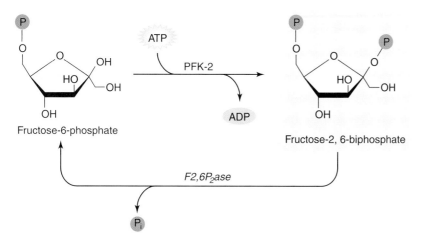

**FIGURE 8.8** **Phosphofructokinase-2 (PFK-2).** The enzyme PFK-2 is multifunctional. It catalyzes the phosphorylation of fructose-6-P to fructose-2,6-$P_2$ *and* the dephosphorylation of fructose-2,6-$P_2$ to fructose-6-P. Only one of these processes is active at any given time. PFK-2 controls the cellular level of fructose-2,6-$P_2$ and, consequently, the activity of PFK-1 (phosphofructokinase).

Citrate regulation is a key part of the **Pasteur effect,** a 19th-century observation that glucose utilization is decreased in the presence of oxygen. Citrate is an intermediate in the Krebs cycle, and an increase in its concentration inhibits phosphofructokinase and, consequently, glycolysis. This regulatory feature adjusts for the fact that oxidative metabolism is far more efficient than glycolysis in energy production.

Fructose-2,6-$P_2$ is formed from fructose-6-P via a reaction catalyzed by PFK-2 (FIGURE 8.8). This is chemically similar to the reaction catalyzed by phosphofructokinase. Also shown in this graphic is a separate enzymatic reaction that catalyzes the dephosphorylation of fructose-2,6-$P_2$, forming fructose-6-P. The two separate enzymatic reactions shown in Figure 8.8 are actually physically joined into a single protein that is a kinase *and* a phosphatase. Only one enzymatic activity prevails for a given physiological condition. For example, the presence of increased glucagon in the blood leads to the activation of the phosphatase activity in liver cells. The result is that fructose-2,6-$P_2$ is dephosphorylated, and the activity of PFK and subsequently glycolysis is decreased under these conditions. As a further example, cancer cells, which have a high rate of glycolysis, have an activated PFK-2 kinase and a high concentration of fructose-2,6-$P_2$.

### Aldolase

Aldolase catalyzes the reaction:

$$\text{Fructose-1,6-}P_2 \rightleftarrows \text{Dihydroxyacetone-P} + \text{Glyceraldehyde-P} \quad (8.2)$$

which is at near equilibrium in cells. As it converts a six-carbon to two three-carbon intermediates, this step represents the crossroads between these two stages of glycolysis.

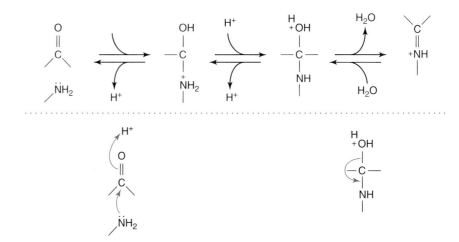

**FIGURE 8.9 Schiff Base Formation.** The reaction between an amine and a carbonyl group forms a Schiff base. The mechanism involves nucleophilic attack of the N of the amine on the C of the carbonyl, followed by protonation and electron migration to eliminate water. The mechanistic detail (below the dotted line) is shown for the forward direction, although the sequence is readily reversible, as indicated by the double arrows above the line.

The mechanism for this enzyme involves the formation of a *Schiff base*, a compound formed by the joining of a carbonyl group and an amine (FIGURE 8.9). A nucleophilic displacement leads to the N–C covalent linkage. In the case of this enzymatic mechanism, the Schiff base is formed from a lysine residue of aldolase and the carbonyl group of fructose-1,6-$P_2$.

The mechanism for aldolase is diagrammed in FIGURE 8.10. A basic residue from the enzyme abstracts a proton from the C4 hydroxyl of fructose-1,6-$P_2$. This leads to a series of electron-pair rearrangements that terminates at the positively charged nitrogen of the Schiff base, which acts as an **electron sink**. Note that one of the electron pairs which joins C3 with C4 in the substrate is cleaved during the rearrangement. This is the means by which the six-carbon sugar phosphate is split into two. The first product—glyceraldehyde-P—is released at this stage. The other electron migration starting with the Schiff base nitrogen restores the enzyme's base to its unprotonated form, with a three-carbon fragment remaining attached to the enzyme. Reversal of the Schiff base formation restores the enzyme to its original form (a requirement for every catalyst so that another round can occur), and releases the second product, dihydroxyacetone-P.

Aldolase from plants and bacteria follow a similar mechanism, but they use an enzyme-bound metal for an electron sink instead of a Schiff base.

### Triose Phosphate Isomerase

A near-equilibrium interconversion between glyceraldehyde-P and dihydroxyacetone-P, the two triose phosphates formed by the aldolase step, is catalyzed by triose phosphate isomerase. We have already

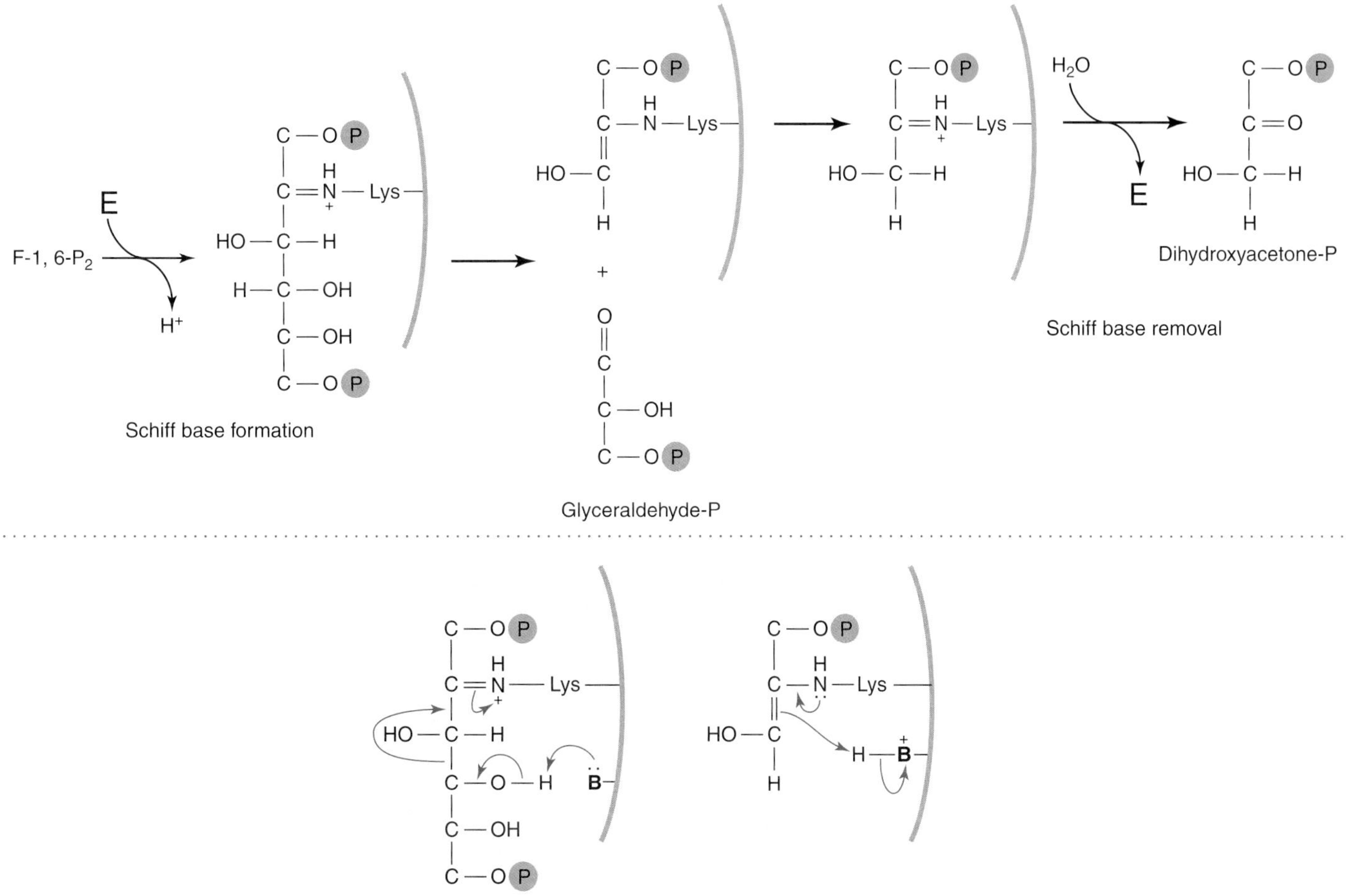

**FIGURE 8.10 Mechanism of Aldolase.** The reaction begins when a lysine group of the enzyme forms a Schiff base with the substrate. A separate basic group on the enzyme abstracts a proton from C4, leading to electron rearrangement that eliminates the C–C bond in the center of the molecule, releasing glyceraldehyde-P, the first product. Reversal of the Schiff base releases the second product, dihydroxyacetone-P.

examined the mechanism for moving a carbonyl group from one carbon to its neighbor in the glucose-6-P isomerase reaction above. The very same mechanism applies to the triose phosphate interconversion:

$$\text{Glyceraldehyde-P} \rightleftarrows \text{Dihydroxyacetone-P} \qquad (8.3)$$

The same base abstractions and enediol intermediate occurs, so no new mechanistic details need to be considered.

All subsequent reactions of glycolysis proceed from glyceraldehyde-P. This means that triose-P isomerase acts to convert all the dihydroxyacetone-P produced at the aldolase step into glyceraldehyde-P. Thus, at the conclusion of this enzymatic step, glycolysis has achieved the conversion of each glucose molecule into two molecules of glyceraldehyde-P.

## Glyceraldehyde Phosphate Dehydrogenase

The first enzyme mechanism ever solved was that of glyceraldehyde phosphate DH. This enzyme is also unique as it catalyzes the only oxidation reaction in glycolysis, and has a uniquely sensitive sulfhydryl

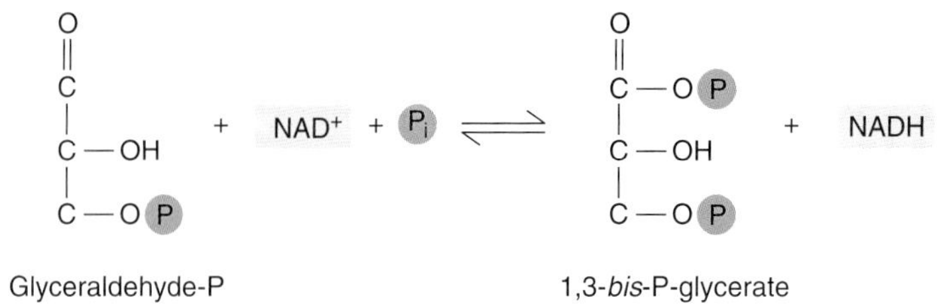

FIGURE 8.11 **Glyceraldehyde Phosphate DH Reaction**.

group (from a cysteine residue in the enzyme). The overall reaction is the conversion of glyceraldehyde-P to 1,3-*bis*-P-glycerate, with the concomitant reduction of $NAD^+$ to NADH, as shown in FIGURE 8.11. In addition to $NAD^+$ and NADH, a third mobile cofactor—inorganic phosphate ($P_i$)—is involved in the reaction. The reaction is near-equilibrium in cells, so it is not involved in cellular regulation.

The mechanism for this reaction (FIGURE 8.12) shows the role of the –SH group in the reaction. The $–S^-$ nucleophile, created by the concerted removal of hydrogen ($H^+$) by an enzyme-bound base (:B), attacks the aldehyde of the substrate. Next, the enzyme binds $NAD^+$ and an electron relay ensues between the substrate and the $NAD^+$, terminating at the positively charged nitrogen, another electron sink like the Schiff base of aldolase. Overall, this step is a **hydride** transfer (i.e.,

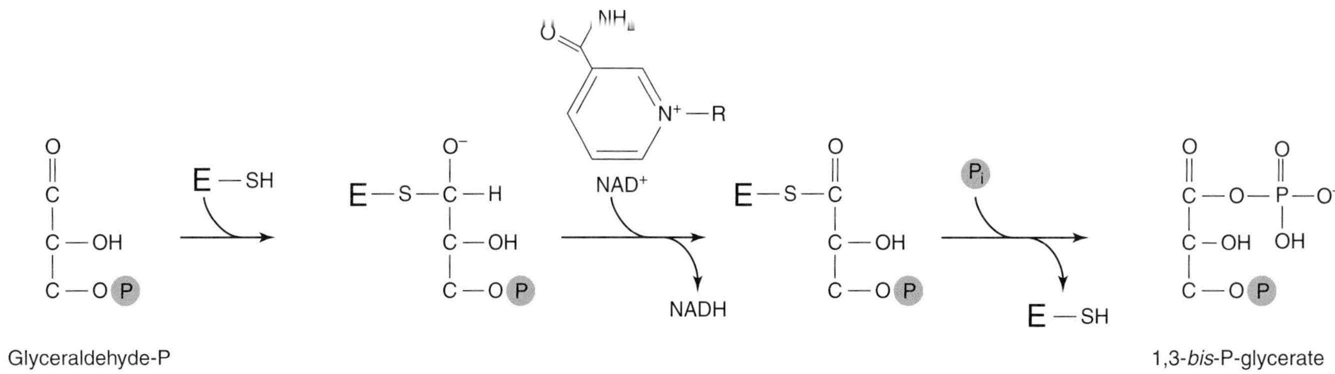

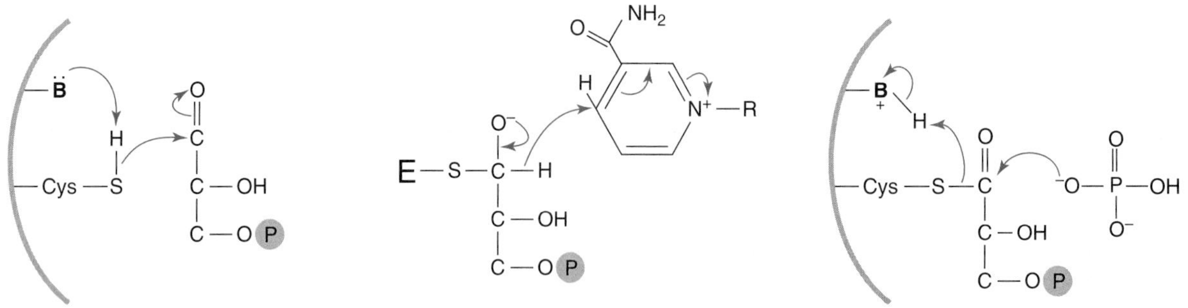

FIGURE 8.12 **Mechanism of glyceraldehyde phosphate DH**. An enzyme base (B:) removes a proton from an enzyme-linked –SH group; simultaneously, the S electrons attack the C1 of glyceraldehyde-P. This is a concerted attack. In the second step, the enzyme-bound glyceraldehyde-P reacts with $NAD^+$, because the latter is properly positioned within the enzyme to remove a hydride from the substrate and form NADH, which then dissociates. Finally, $P_i$ displaces the bound substrate from the enzyme, yielding 1,3-bis-P-glycerate.

the movement of hydrogen along with a pair of electrons) to NAD⁺. This leaves the bound triose phosphate oxidized, and the reduced cofactor NADH is released from the enzyme. Finally, an inorganic phosphate binds, displacing the thioester, creating the acid phosphate product 1,3-bis-P$_2$-glycerate, and regenerating the enzyme in its original form.

The same –SH group that participates in the glyceraldehyde-P DH catalytic mechanism can react with heavy metals and other electron deficient molecules, leading to enzyme inhibition. However, as many other –SH groups are present in most cells, it is a relatively nonselective site of inhibition and is of little analytical or clinical use in specifically blocking glycolysis.

## Phosphoglycerate Kinase

The reaction catalyzed by phosphoglycerate kinase (abbreviated P-glycerate kinase) is the first step of glycolysis in which ATP is produced (FIGURE 8.13). The reaction is near-equilibrium, and its mechanism is the same as other the kinases, such as the hexokinase we considered earlier. Kinase reactions are common in biochemistry—they number in the hundreds—and they all follow the same mechanism. While most kinase reactions are metabolically irreversible, this conclusion can only be drawn after experimental measurement, and not simply because it is energy-linked. It bears repeating that the *standard* free energy change of a reaction does not in general predict its energy status in living cells.

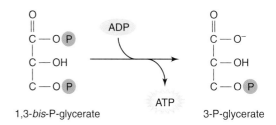

1,3-*bis*-P-glycerate       3-P-glycerate

FIGURE 8.13 **P-Glycerate Kinase Reaction**.

## Phosphoglycerate Mutase

Mutase reactions are phosphoryl transfers that cause an apparent movement of a phosphoryl group from one part of the molecule to another. These are typically near-equilibrium reactions in cells, as is the case for phosphoglycerate mutate (abbreviated P-glyceromutase; FIGURE 8.14).

The mechanism for the enzyme involves a phosphorylated histidyl residue (FIGURE 8.15). Upon binding to the substrate, the hydroxyl group in the 2-position attacks the histidyl-phosphate. This produces the intermediate, 2,3-bis-P$_2$-glycerate. Next, the newly free histidyl group of the enzyme attacks the phosphoryl in the 3-position of the intermediate compound, reforming the free histidyl-phosphate enzyme as well as the product, 2-P-glycerate. There is symmetry to this reaction, and a strong similarity to glucose phosphate isomerase, described above. In both cases, a symmetrical intermediate (compared in FIGURE 8.16) is created. Both mechanisms are also symmetrical, so the reverse reactions are mirror images of the forward reactions.

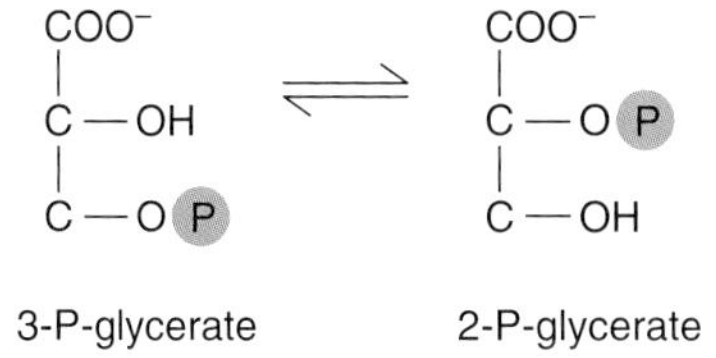

3-P-glycerate       2-P-glycerate

FIGURE 8.14 **P-Glyceromutase Reaction**.

## Enolase

Enolase catalyzes a near-equilibrium reaction that consists of a dehydration of 2-P-glycerate to P-enolpyruvate (FIGURE 8.17). The

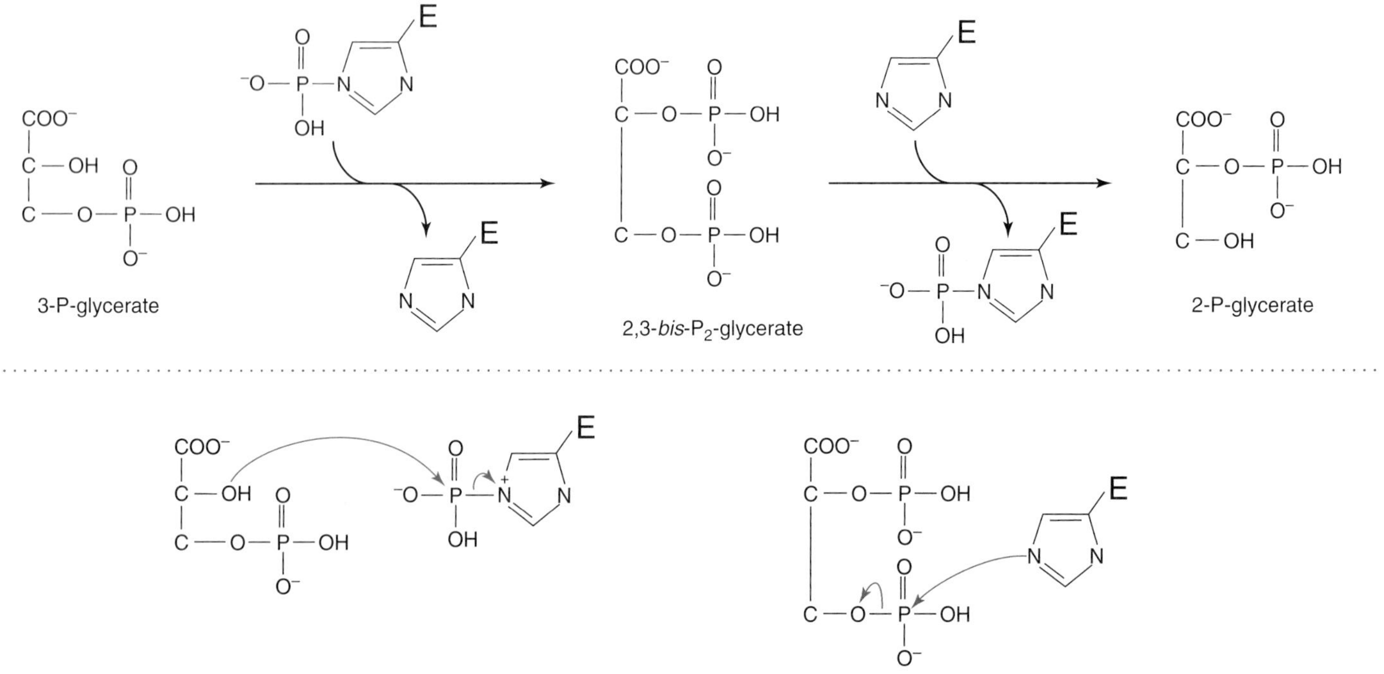

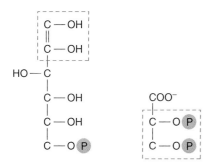

**FIGURE 8.15 Mechanism of P-Glyceromutase.** The free enzyme has a phospho-histidine residue, which is subjected to nucleophilic attack by the substrate hydroxyl group at C2. This removes the phosphoryl group from the enzyme and produces the 2,3-bis-P-glycerate intermediate. In the second phase, the histidine serves as nucleophile, attacking the phosphoryl group at C3. This restores the original enzyme form and yields the product, 2-P-glycerate.

**FIGURE 8.16 Symmetrical Intermediates: Enediol and Vicinyl Phosphates.**

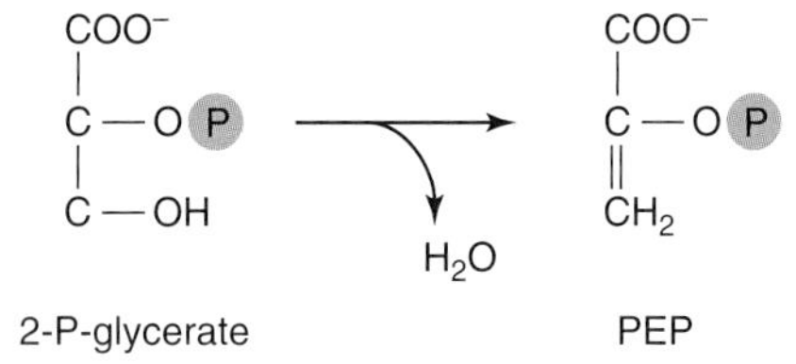

**FIGURE 8.17 Enolase Reaction.**

enzyme has a tightly attached $Mg^{2+}$ ion that binds the 3-position hydroxyl group of the substrate (FIGURE 8.18). A base from the enzyme abstracts a hydrogen from C2, creating a carbanion. The carbanion next rearranges to form the double bond and water is eliminated as shown. While enolase can be inhibited by fluoride, it is not a strong inhibitor of glycolysis, because enolase inhibition is fairly weak and nonselective.

## Pyruvate Kinase

The pyruvate kinase step is the second of the ATP-forming reactions of glycolysis (FIGURE 8.19). Its mechanism is unremarkable, because it is the same as the other kinases we have already considered. It is noteworthy as a metabolically irreversible reaction that is regulated both allosterically and by covalent modification.

The key allosteric regulator of pyruvate kinase is fructose-1,6-bis-$P_2$, the product of phosphofructokinase (PFK), the prime regulatory enzyme of glycolysis. Thus, when PFK is more active, it produces more products, which in turn activates pyruvate kinase. This is called **feed-forward activation**.

In some tissues, such as liver, pyruvate kinase is a substrate for phosphorylation. Hormonal regulation may lead to an activation of the enzyme protein kinase A, which phosphorylates and inactivates pyruvate kinase (FIGURE 8.20). A separate enzyme, a protein

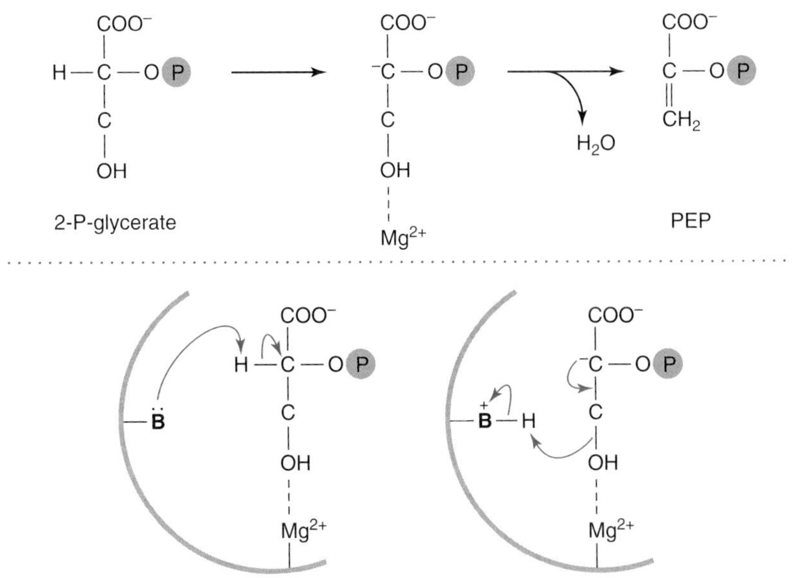

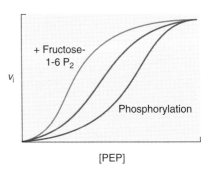

The top portion shows chemical structures of the enolase reaction (2-P-glycerate → PEP) on the left and the Pyruvate Kinase Reaction on the right.

2-P-glycerate     $Mg^{2+}$     PEP

PEP     ADP     ATP     Pyruvate

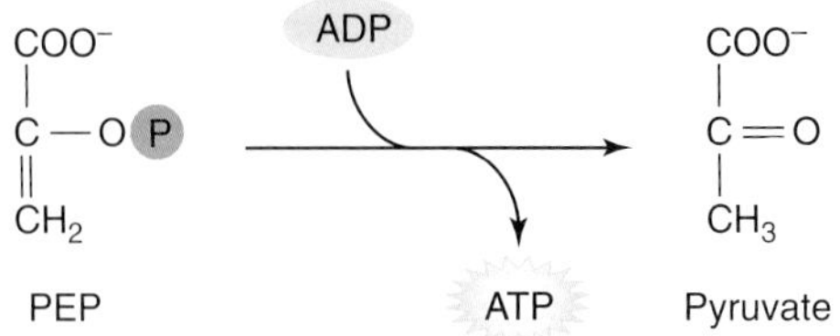

**FIGURE 8.19 Pyruvate Kinase Reaction.**

**FIGURE 8.18 Mechanism of Enolase.** An enzyme base removes a proton from C2 of the substrate, creating a carbanion intermediate. The $Mg^{2+}$ chelates the substrate and is essential for enzyme activity. In the second step, electron rearrangement from the carbanion leads to abstraction of the proton from the base, which splits the C–O bond at C3 as water is eliminated.

phosphatase, catalyzes dephosphorylation and returns pyruvate kinase to its active form.

Both the allosteric stimulation and the covalent inhibition alter the kinetic profile of pyruvate kinase (FIGURE 8.21), which resembles the strictly allosteric regulation of phosphofructokinase (Figure 8.7). Pyruvate kinase is considered a secondary regulatory site of glycolysis. As a result of feed-forward activation, it is responsive to primary control of phosphofructokinase.

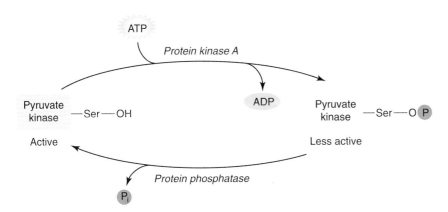

**FIGURE 8.20 Covalent Regulation of Pyruvate Kinase.** Pyruvate kinase is subjected to a phosphorylation reaction, catalyzed by protein kinase A. The phosphorylated enzyme, in turn, becomes less active. A protein phosphatase catalyzes the hydrolysis of the phosphoenzyme, with release of $P_i$.

**FIGURE 8.21 Modulation of Pyruvate Kinase Activity.** Two means of regulating the activity of pyruvate kinase exist. Covalent modification by phosphorylation causes a decrease in enzyme activity, evidenced by the shift of the curve to the right. Allosteric binding of fructose-1,6-$P_2$, the product of PFK, activates the enzyme, shifting the curve to the left. Because PFK participates in an earlier step in the pathway, its product is a feed-forward activator of pyruvate kinase.

# 8.3 Completing the Pathway

If we consider the sequence of glucose to pyruvate as the sum of its connected parts, the overall reaction is:

$$\text{Glucose} + 2\,\text{ADP} + 2\,P_i + 2\,\text{NAD}^+ \\ \rightarrow 2\,\text{pyruvate} + 2\,\text{ATP} + 2\,\text{NADH} \qquad (8.4)$$

Glucose and pyruvate represent the connected ends of the pathway, while the other molecules in this equation are mobile cofactors. The cofactors are present in limited concentrations, and must be continuously regenerated for the pathway to operate. When this is accomplished, we have a **complete pathway**; stated more succinctly, we can say that the pathway mobile cofactors are balanced.

The cofactors can be grouped into two categories:

1. Phosphorylation cofactors: ATP, ADP, and $P_i$
2. Redox cofactors: $\text{NAD}^+$ and NADH

The phosphorylation cofactors of glycolysis are regenerated by all of the reactions in the cell that utilize energy in the form of ATP breakdown. ATP is the central energy intermediate of cells. Thus, a prime function of glycolysis is to provide that energy. For example, the $\text{Na}^+/\text{K}^+$-ATPase that catalyzes the export of cellular $\text{Na}^+$ and import of $\text{K}^+$ is one reaction that turns over the phosphorylation cofactors of glycolysis. This enzyme can thus be considered to be linked in parallel to glycolysis, as are all of the energy-utilizing steps of the cell cytosol.

The redox cofactor NADH must be reoxidized to $\text{NAD}^+$ so that glycolysis can continuously operate. This can be achieved by the formation of lactate, as is discussed next. We will then consider an alternative to lactate: the conversion of pyruvate to ethanol.

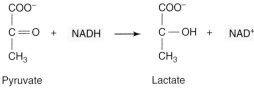

FIGURE 8.22 **Lactate DH Reaction.**

## Lactate Formation

The most common end product of glycolysis is lactate. This is catalyzed by the near-equilibrium enzyme lactate dehydrogenase (lactate DH), shown in FIGURE 8.22. The formation of lactate is a dead-end in glucose metabolism. It is not connected to any other reactions, and exits the cell for use in other cells.

There are several isozyme forms of lactate DH, so that different proteins are present, for example, in liver, muscle, and heart. This is used clinically to detect selective damage to heart cells. When these are ruptured, they release lactate DH into the bloodstream, and it can be differentiated by electrophoretic separation from other forms of this enzyme. However, the isozyme differences have no regulatory significance, because the enzyme is near-equilibrium in all of these tissues and thus has no influence on the extent to which pyruvate and lactate are interconverted.

## Ethanol Formation

Some organisms, such as yeast, can convert glucose to ethanol instead of lactate. This process is also a form of glycolysis, but it is more commonly called **fermentation.** The pathway leading from glucose to pyruvate in yeast is the same as we have documented above.

Pyruvate is converted to ethanol in two steps. First, pyruvate is decarboxylated in a reaction catalyzed by pyruvate decarboxylase:

$$\text{Pyruvate} \rightarrow \text{Acetaldehyde} + CO_2 \qquad (8.5)$$

This step is metabolically irreversible, as are most reactions that release $CO_2$. This is because $CO_2$ is a gas, and its release from the solution phase drives the reaction forward. Pyruvate decarboxylase contains the bound cofactor thiamine pyrophosphate (FIGURE 8.23a), which is derived from thiamine (**BOX 8.4**). Unlike mobile cofactors, thiamine pyrophosphate remains covalently bound to the enzyme, acting as much a part of it as an amino acid residue. The active portion of the cofactor is the proton-dissociated form, where the carbanion shown as Figure 8.23b is stabilized by resonance involving the neighboring S and N atoms.

The reaction mechanism is illustrated in FIGURE 8.24. The carbanion attacks the carbonyl of pyruvate, leading to the first intermediate. Note the presence of a positively-charged quaternary N, which serves as an electron sink in the following movement of electrons, thus leading to the loss of $CO_2$. The second intermediate is shown as two resonance forms. The charged form serves as an electron sink for the third and final electron migration that leads to release of the product, acetaldehyde.

The second step in the conversion of pyruvate to ethanol in yeast is catalyzed by alcohol DH,

$$\text{Acetaldehyde} + \text{NADH} \rightleftarrows \text{Ethanol} + \text{NAD}^+ \qquad (8.6)$$

which provides $\text{NAD}^+$ for glyceraldehyde phosphate DH and allows the pathway to continue another cycle. As a hydride transfer involving

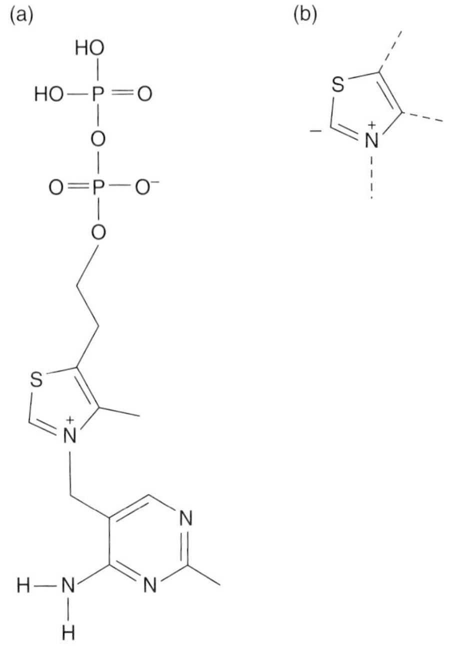

(a)  (b)

FIGURE 8.23 **Thiamine Pyrophosphate (TPP).** The structure of the entire cofactor is shown in A. The precursor is just the alcohol, that is, without the pyrophosphate attached. In B, the active thiazole portion of the cofactor is illustrated, in the dissociated carbanion form, stabilized by resonance involving the neighboring N and S atoms.

## BOX 8.4: Vitamins

Most animals have a dietary requirement for certain organic molecules that are converted *in vivo* to forms for use as enzyme cofactors. These compounds are known as **vitamins.** The name comes from the first one discovered, which was shown to be an amine that was essential for life: hence a *vital amine.* This was later renamed thiamine when it was found to contain sulfur as well. Dietary deficiency in thiamine can result from the practice of polishing rice (removing the thiamine-rich husk) in populations (particularly in Asia) where rice is a major food.

Thiamine is also called vitamin $B_1$, after similar compounds were discovered that were also essential for humans in small quantities. The fact that several such vitamins appear related (e.g., $B_1$, $B_2$, etc.) and are called "the B complex" is an accidental association only; they have only some chemical similarity, but otherwise they are unrelated.

The nicotinamide portion of $\text{NAD}^+$ is also a vitamin. Occasionally called vitamin $B_3$, it is more commonly called niacin, an invented word to avoid association with the close-sounding tobacco natural product nicotine.

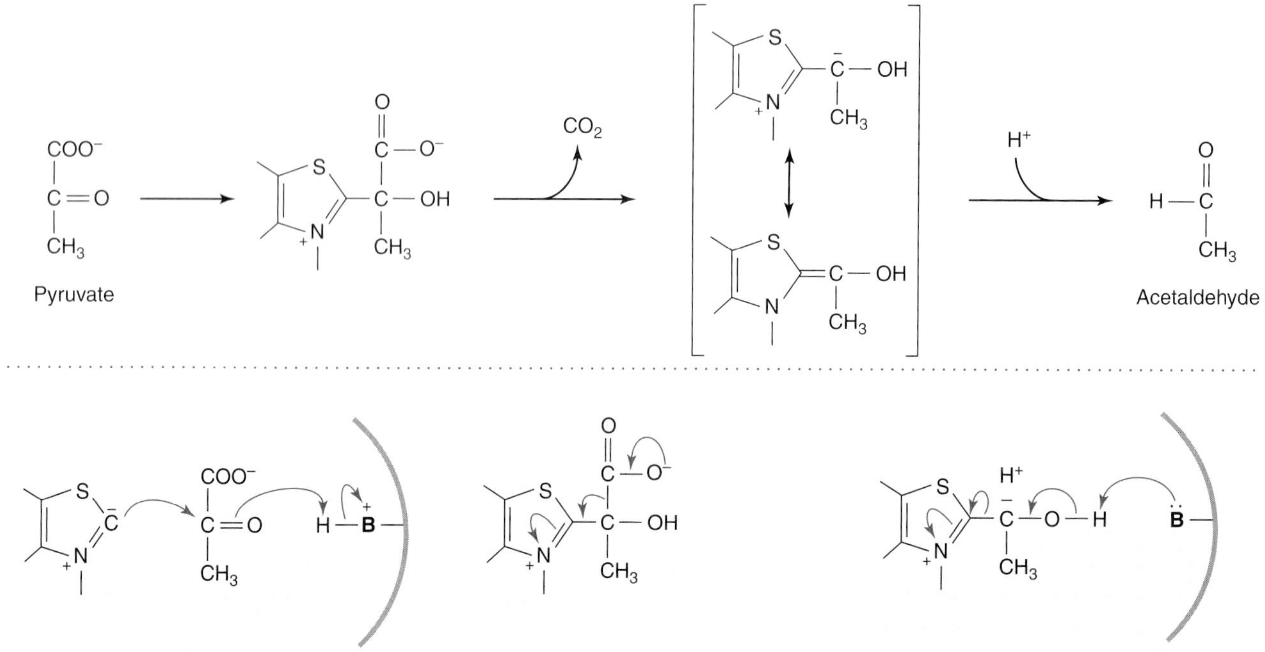

**FIGURE 8.24 Mechanism of Pyruvate Decarboxylase.** In the first step, the carbanion of the thiazole from the enzyme-bound TPP attacks the carbonyl of the substrate, leading to the thiazol-pyruvate adduct. The next step is electron rearrangement, which leads to decarboxylation and then terminates in (and is directed towards) the electron sink of the thiazole. The final step is best visualized as using the top of the two resonance forms (carbanion) of the ensuing intermediate. Electron rearrangement is initiated from a basic group on the enzyme, again utilizing the thiazole electron sink and releasing glyceraldehyde as product.

$NAD^+$, the mechanism is essentially identical to others discussed previously, such as lactate DH.

Besides yeast, alcohol DH is also present in the livers of humans (and many other animals) and is responsible for the oxidation of ethanol as well as other alcohols that are encountered there. The mammalian liver isozyme has a high $K_m$ for ethanol, and typically acts in the direction of acetaldehyde formation (i.e., in the reverse direction to yeast fermentation).

The products of yeast fermentation are thus $CO_2$ and ethanol. The first is the action of "baker's yeast," which causes bread dough to rise. The second is the basis of the centuries-old alcoholic beverage industry. In the production of beer and champagne, both fermentation products—$CO_2$ and ethanol—remain. In wine, the $CO_2$ must be drawn off during the fermentation process. Yeast is unable to survive alcohol concentrations much above 10%; hence, chemical distillation is needed for the production of high-alcohol content beverages such as whiskey and gin.

## 8.4 Energetics of Glycolysis

Balancing the phosphorylation cofactors of glycolysis requires energy-utilizing pathways, as we have observed. Thus, one function of glycolysis is to provide energy as ATP for other pathways. The net yield of energy from glycolysis is 2 ATP for conversion of 1 glucose to 2 lactate molecules.

In this section, we first examine the overall pathway of thermodynamics, extending the free energy of individual reactions to glycolysis. Next, we consider three situations in which the energetics of glycolysis is affected: a shunt pathway that exists in red blood cells, a form of arsenic poisoning, and the metabolism of fructose by the liver.

## Pathway Thermodynamics

Data for the equilibrium constants of glycolytic reactions and the concentrations of intermediates in several cell types are well known. In particular, glycolysis in red blood cells is informative because this pathway is the only means of ATP production in these cells, there are few other pathways, and there are no intracellular vesicles.

FIGURE 8.25 presents pathway data for both the standard free energy changes ($\Delta G°$) of the glycolytic reactions as well as the $\Delta G$ values measured from concentrations of compounds in red blood cells undergoing glycolysis. The diagram is a visual display of reaction energies as well as overall pathway changes. The ordinate of the plot is "G" and "G°" and the abscissa consists of the glycolytic pathway compounds in sequence.

Because only changes in thermodynamic variables are available, rather than absolute values, the points are plotted by arbitrarily setting the G° value of glucose to zero. The first G value is set lower than that one, roughly corresponding to the change in concentration from 1 M to millimolar concentrations.

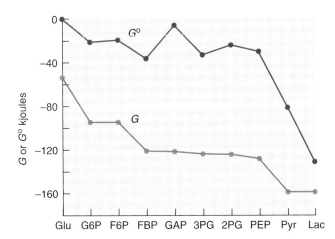

FIGURE 8.25 **Thermodynamics of Glycolysis.** The upper graph is a plot of standard free energies, and the slopes between the points representing glycolytic intermediates indicate visually the changes in free energy for each step. The $\Delta G°$ values are positive in some cases. This means that, under standard conditions, the pathway cannot go forward. The bottom graph is a plot of actual free energies. In this case, all of the slopes are negative. Three steps—hexokinase, phosphofructokinase, and pyruvate kinase—are strongly negative, corresponding to the three metabolically irreversible reactions of glycolysis. The figure emphasizes the distinction between standard and actual free energies for a pathway.

Once these arbitrary assignments are made, all of the other data plotted correspond to values derived from the known changes for each reaction. For the first $\Delta G°$, the known value was used to fix the height of $G°$ for glucose-6-P (G6P). This was then done for all of the remaining glycolytic reactions using the known $\Delta G°$ values. A similar procedure was applied to the lower line, using measured $\Delta G$ values. The line connecting any two points thus visually represents the value of the free energy change, in both direction (a positive slope shows a positive change; a negative slope shows a negative change) and quantity (a steep slope shows a large change; a shallow slope shows a small change).

The changes in $\Delta G°$ can be judged visually by the differences in slopes. Because $\Delta G°$ values can be negative or positive as we move from glucose to lactate, this pathway is not thermodynamically allowed under standard conditions. Nevertheless, the overall change in $\Delta G°$ is negative, so the overall change *is allowed thermodynamically* under standard conditions but only if we take a separate route. As indicated in the lower trace, each step has a negative $\Delta G$, which means this pathway is thermodynamically allowed. No step has a positive or a zero $\Delta G$ value. If it was positive, the reverse reaction would occur. If it was zero, the reaction would be at equilibrium and no net reaction would occur. It cannot be argued that an ensuing reaction could somehow "pull" an unfavorable one to completion. Rather, the concentrations that exist in the cell determine the status of the free energy, as we have already observed from the free energy equation.

A further important piece of information can be gleaned from Figure 8.25. Most of the slopes of the actual free energy plot are fairly shallow, visually indicating the near-equilibrium reactions. Only three are steep: hexokinase, phosphofructokinase, and pyruvate kinase. These correspond to the metabolically irreversible steps of the pathway.

## Red Blood Cell Shunt Pathway

Recall that 2,3-bis-P-glycerate decreases the affinity of oxygen for hemoglobin at high altitudes. The formation of this compound is an example of a **shunt pathway**, because it originates from one glycolysis intermediate and terminates in another.

The pathway begins with the glycolytic intermediate 1,3-bis-P-glycerate (FIGURE 8.26), the product of the glyceraldehyde-P dehydrogenase step. Red blood cells have a mutase enzyme that catalyzes the reaction:

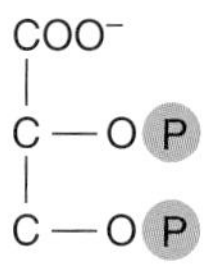

$$1,3\text{-bis-P-glycerate} \rightarrow 2,3\text{-bis-P-glycerate} \qquad (8.7)$$

The mechanism of this mutase is essentially identical to the P-glyceromutase reaction (Figure 8.15), which shuttles phosphoryl groups between the substrate and a histidyl group on the enzyme. The destruction of this regulator is catalyzed by a phosphatase reaction:

$$2,3\text{-bis-P-glycerate} \rightarrow 3\text{-P-glycerate} + P_i \qquad (8.8)$$

This phosphatase activity is accomplished by the same protein that has the mutase activity. Hence, the concentration of the regulator in the

FIGURE 8.26 **2,3-Bis-P-Glycerate**.

cell depends on a balance between two distinct activities of the same enzymatic protein, a mutase/phosphatase. This situation is similar to the protein phosphofructokinase-2/fructose-2,6-$P_2$, which catalyzes both the formation and the destruction of fructose-2,6-$P_2$.

There is an energy consequence to flow through the red blood cell shunt pathway because it bypasses the P-glycerate kinase reaction (FIGURE 8.27). Because a complete bypass of this reaction would result in zero ATP production, and as glycolysis is the only means of energy formation for red blood cells, flow through the mutase/phosphatase must be kept relatively low.

## Arsenate Poisoning

Elemental arsenic is positioned just below phosphorus in the same column (group 15) of the periodic table. This means that the outer electron configuration of As and P is the same, and we would expect the elements to have some chemical properties in common. These include the ability to form similar oxides. By analogy to $P_i$, arsenate ($AsO_4^{3-}$) can substitute in the glyceraldehyde-P DH reaction in place of phosphate, producing 1-arseno-3-P-glycerate (FIGURE 8.28). This compound, as indicated in FIGURE 8.29, is rapidly and non-enzymatically hydrolyzed to 3-P-glycerate.

The energetic consequences of arsenate on glycolysis can be appreciated by comparing Figure 8.29 with Figure 8.27, the red blood cell bypass pathway. In each case, the flow through glycolysis is uninterrupted, but ATP production is circumvented. In the case of arsenate, the extent of ATP formation by glycolysis is controlled by the amount of arsenate entering the cell. The decrease in ATP production, despite unimpeded pathway flow, is called **uncoupling,** a situation more commonly encountered in studies of mitochondrial energy production. While there are multiple explanations for arsenic toxicity, most proposed actions share the ability of arsenate to mimic phosphate in cells.

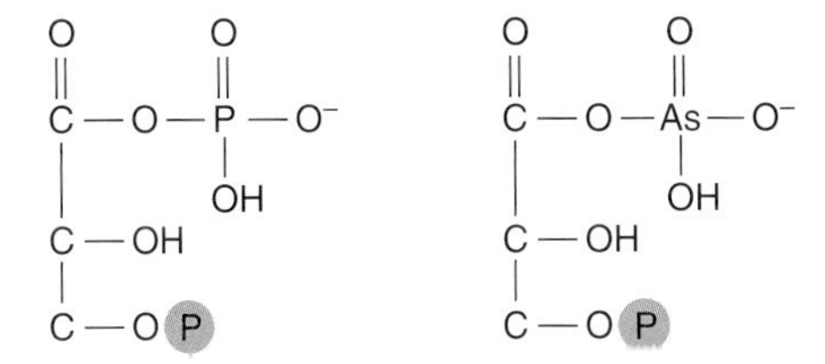

## Fructose Metabolism

Fructose is a common dietary sugar. In the past, most of the sugar in our diets came from sucrose, a disaccharide consisting of glucose and fructose. In the last few decades, most dietary sucrose has been replaced by a mixture of free glucose and fructose (**BOX 8.5**). A small amount of fructose is metabolized by many cells because it is transported by several GLUT isoforms, and because it is a weak substrate of some hexokinase isozymes. However, most of it is metabolized through a different pathway that is unique to the liver.

In liver, fructose is transported across the plasma membrane via GLUT2. Intracellular fructose is converted to fructose-1-P, catalyzed by the enzyme fructokinase (FIGURE 8.30). Next, aldolase B, a liver isozyme, catalyzes the aldol cleavage of fructose-1-P to dihydroxyacetone-P and glyceraldehyde (analogous to the aldolase reaction of glycolysis, Figure 8.10). Glyceraldehye is then a substrate for triose kinase, a third unique liver enzyme, which converts glyceraldehyde to glyceraldehyde-3-P. From this point forward, the carbon of fructose

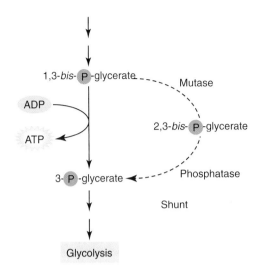

FIGURE 8.27 **The Red Blood Cell Bypass.** The multifunctional mutase/phosphatase enzyme found in red cells forms and removes 2,3-bis-P-glycerate. The latter causes a decreased affinity of hemoglobin for $O_2$, appropriate for high altitudes. Notice that flux through the shunt also bypasses the formation of ATP. To the extent that the shunt is operating, it decreases the yield of ATP from glycolysis.

1,3-*bis*-P-glycerate          1-arseno-3-P-glycerate

FIGURE 8.28 **Structures of the glycolytic intermediate 1,3-$P_2$-glycerate and the arsenate analog.**

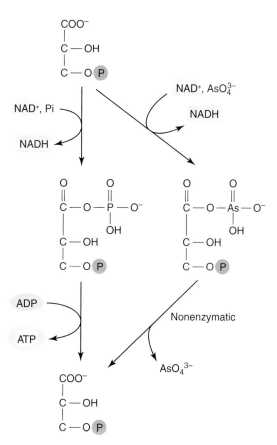

**FIGURE 8.29 Modified Glycolysis in the Presence of Arsenate.** When arsenate is present, it substitutes for $P_i$ in the glyceraldehyde-P DH reaction, producing 1-arseno-3-P-glycerate. This reverts non-enzymatically to 3-P-glycerate, bypassing P-glycerate kinase in a similar way to the red cell shunt, except that it can occur in all cells and is unregulated. The extent to which arsenate substitutes for $P_i$ depends on the concentration of arsenate.

## BOX 8.5: High-Fructose Corn Syrup

Most of our dietary sugar is derived from corn rather than the sucrose that is extracted from sugar cane and sugar beets. An industrial process enzymatically hydrolyzes cornstarch to glucose, followed by enzymatic isomerization of some of the glucose to fructose. The resulting solution, abbreviated HFCS, is cheaper to produce than sucrose due to the abundance of corn. HFCS is thus simply a mixture of glucose and fructose, and can be metabolized even more rapidly than sucrose because it obviates the need for sucrose hydrolyze to glucose + fructose. There are some indications that HFCS is nutritionally distinct from sucrose and that it may be more efficiently converted to fat. However, it is difficult to separate nutritional from sociological factors. For example, the cheaper "sugar substitute" (i.e., HFCS) is present in ever larger ("supersized") portions of beverages and fast food. It seems likely that both metabolic and situational factors contribute to obesity.

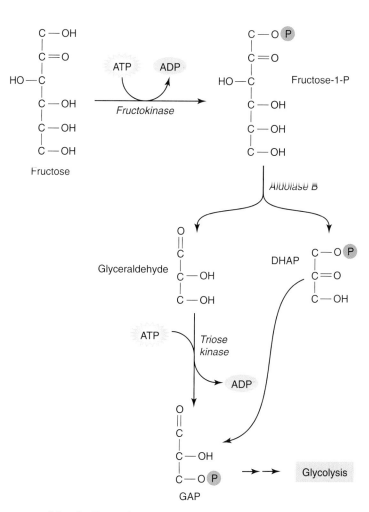

**FIGURE 8.30 Metabolism of Fructose in Liver.** Three special liver enzymes allow fructose carbon to enter glycolysis. First, fructokinase catalyzes the conversion of fructose to fructose-1-P. Second, aldolase B catalyzes the splitting of fructose-1-P into dihydroxyacetone-P and glyceraldehyde. Third, glyceraldehyde is a substrate for triose kinase, which catalyzes the formation of glyceraldehyde-P. While fructose metabolism has the same balance of ATP as glucose, the major regulatory step of glycolysis, PFK, is bypassed. This accounts for the rapid metabolism of fructose.

is metabolized using the common steps of glycolysis. Note that the ATP balance—that is, the difference between ATP input and ATP production—matches that of glucose, despite using different steps (i.e., triose kinase in place of P-fructokinase).

A major distinction between fructose metabolism and glucose metabolism is that fructose metabolism bypasses the P-fructokinase reaction, which is the rate-limiting step for glucose metabolism. This allows more rapid and relatively uncontrolled entry of fructose into the glycolytic pathway. As a further consequence, carbon from fructose can enter more rapidly into fat synthesis, contributing to obesity. A separate complication results from the ingestion of a large bolus of fructose. Phosphorylation to fructose-1-P can be so rapid that the cell's ATP supply is depleted to the point where the cell does not survive.

## 8.5 Metabolic Connections to Glycolysis

Glycolysis is a central pathway in virtually all cells, so it is extensively connected to other pathways. In our final section of glycolysis, we consider separate ways of entering glycolysis, connections to intermediates, and multiple pathway end products.

### Alternative Entry Points

We have just examined fructose metabolism from the standpoint of its effects on energetics. At the same time, we can think of fructose as a distinct entry point to glycolysis. Another common sugar and a possible starting point for glycolysis is galactose, derived from the milk sugar, lactose. In addition, glycolysis can originate from the polymer glycogen.

### Glycolytic Intermediates as Intersection Points

In keeping with its ubiquitous occurrence in cells, and its ancient origins, the intermediates of glycolysis are commonly used as intermediates of other pathways. For example, the red blood cell $2,3-P_2$-glycerate shunt described in the last section could be viewed as a pathway that begins with one intermediate of the pathway and ends with another. Another example is the pentose phosphate shunt. The formation of the regulator fructose-$2,6-P_2$ can be thought of as a pathway that begins and ends with the *same* glycolytic intermediate. The utilization of glycolytic intermediates in other pathways is more extensive than this one. In fact, *all* of the intermediates of glycolysis are used in other pathways, many of which are examined elsewhere in this text.

### Alternative Endpoints of Glycolysis

We have seen that one alternative endpoint of glycolysis is ethanol formation, but that is limited to yeast and a few other microorganisms. A significant glycolytic endpoint in mammals is the conversion of pyruvate to alanine. Virtually all cells have the pathway for the complete oxidation of pyruvate.

## Summary

Glycolysis is the best-known pathway in cells, and it serves as a model for understanding metabolic features, such as enzyme chemistry, mobile versus bound cofactors, energy links to pathways, and regulatory steps. The details of enzyme mechanisms show many of the features repeated throughout metabolism, such as nucleophilic substitution, acid–base catalysis, and electron rearrangement in isomerase reactions. The two important mobile cofactor pairs in glycolysis are ATP/ADP and $NAD^+$/NADH. The first connects glycolysis to energy production, because net ATP is produced by the pathway and is used in other cellular reactions, such as the maintenance of the plasma membrane sodium gradient. The second is the redox cofactor pair, which is balanced if glycolysis converts glucose to lactate. The most important regulatory step of glycolysis is the reaction catalyzed by phosphofructokinase, which is stimulated by fructose-2,6-$P_2$ (reflecting hormonal control), and inhibited by citrate (reflecting feedback regulation from oxidative metabolism). A secondary regulatory point, pyruvate kinase, is in part controlled by feed-forward activation by fructose-1,6-$P_2$, the product of phosphofructokinase. Because all of its intermediates intersect with other pathways, glycolysis can be thought of as the crossroads of metabolism.

## Key Terms

complete pathway
electron sink
feed-forward activation
fermentation
hydride

isoforms
isozymes
Pasteur effect
pathway completion
shunt pathway

symmetrical intermediate
uncoupling
vitamins

## Review Questions

1. Explain why fructose has the same ATP balance as glucose, even though the second phosphorylation is at the three-carbon stage in fructose metabolism, but at the six-carbon stage in glucose metabolism.

2. Pancreatic cells secrete insulin in response to glucose. Here GLUT2 plays an important role as part of the "glucose sensor." Explain.

3. Studied *in vitro*, all of the kinase reactions are inhibited, even by physiological levels of ATP. However, the concentration of ATP in cells is virtually constant. Can ATP be considered a regulator of glycolysis?

4. What is the effect of arsenate ingestion at high altitudes, where the red cell shunt mutase is more active?

5. In liver, pyruvate kinase is a substrate of protein kinase A, but so too is PFK-2. Explain how both of these will affect the rate of glycolysis.

6. Identify the reaction mechanisms that use an electron sink.

7. The P-glycerate kinase reaction is near-equilibrium, yet it leads to production of ATP, a high-energy intermediate. Explain this conundrum.

8. Both $NAD^+$, H, and TPP are enzyme cofactors; however, TPP does not rely on other reactions for its regeneration. Why is this so?

9. The product of the enolase reaction would appear from the reaction mechanism to be an enol rather than pyruvate. How can this be explained?

# References

1. DeBerardinis, R. J.; Sayed, N.; Ditsworth, D.; Thompson, C. B. Brick by brick: Metabolism and tumor cell growth. *Curr. Opin. Genet. Develop.* 18:54–61, 2008.
   A review and historical context of glycolysis in tumor cells.
2. Newsholme, E. A.; Leech, A. R. *Biochemistry for the Medical Sciences*; John Wiley & Sons: New York, 1986.
   Regulatory principles of metabolic pathways, highlighting glycolysis.
3. Taubes, G. *Good Calories, Bad Calories: Challenging the Conventional Wisdom on Diet, Weight Control, and Disease*; Knopf: New York, 2007.
   History of diets and nutritional research, including the origin of high-fructose corn syrup.
4. Walsh, C. *Enzyme Reaction Mechanisms*; W. H. Freeman: San Francisco, 1979.
   Extensive discussion of the mechanisms of major enzymes and the chemical evidence behind them.
5. Bridger, W. A.; Henderson, J. F. *Cell ATP*; John Wiley & Sons: New York, 1983.
   A monograph summarizing the findings demonstrating that ATP concentration remains constant in cells under virtually all conditions, and the role of ATP in key metabolic reactions.

# 9

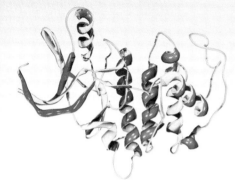

# The Krebs Cycle

## CHAPTER OUTLINE

Image © Dr. Mark J. Winter/Photo Researchers, Inc.

Thirty years after the discovery of glycolysis, investigators had identified several intermediates involved in the complete oxidation of pyruvate to $CO_2$. However, the route itself remained a mystery. Hans Krebs found that one step involved a condensation reaction involving a product of pyruvate and a four-carbon intermediate, producing citric acid, a six-carbon tricarboxylic acid. He also showed that after a series of reactions, the same four-carbon intermediate was regenerated; in other words, the pathway was a cycle. This was published in a landmark study in 1937 as the **citric acid cycle** (**BOX 9.1**). The result was not well received, however, for reasons that we explore in this chapter. In fact, many called the route the *Krebs cycle*, not as praise but to suggest that it was the personal belief of its inventor. The cyclic nature of the pathway *was* tentatively accepted, but only the part involving tricarboxylic acid intermediates other than citrate. The pathway was thus referred to as the **tricarboxylic acid cycle**. Today, all three names are used. We choose the **Krebs cycle**, as it is more pithy, vindicates the findings of a great biochemist, and conforms to the tradition of other named highlights of biochemistry, such as Michaelis–Menten kinetics, the Lineweaver–Burke plot, and the Calvin cycle.

## 9.1 A Cyclic Pathway

A series of enzymatic reactions that forms a cycle is analogous to the individual steps in an enzymatic reaction. Consider two examples of enzyme reaction mechanisms that are involved in the formation of ethanol from pyruvate by yeast, drawn in Figure 9.1 as cyclic processes. The first reaction (**FIGURE 9.1a**) shows pyruvate decarboxylase ($E_1$) and its other enzyme forms, $E_1$:pyruvate and $E_1$:acetaldehyde, as cycle intermediates. Pyruvate enters the cycle, and the products, $CO_2$ and acetaldehyde, leave. For the second reaction (Figure 9.1b), the cycle intermediates are alcohol dehydrogenase ($E_2$), and the other enzyme forms, $E_2$:acetaldehyde and $E_2$:ethanol. Acetaldehyde and $NAD^+$ enter the cycle, and NADH and ethanol leave.

Every enzymatic reaction can be written as a cycle, which means that—theoretically at least—just one molecule of the enzyme can catalyze infinitely many conversions of substrate to product. In reality, enzymes need a certain concentration of substrate for appreciable activity, and enzymes have a finite lifespan. Still, relative to the flow of substrate to product, very small amounts of enzyme are needed. The ability to write a process as a cycle is actually the essence of catalysis.

By analogy, a metabolic cycle is also catalytic, in the sense that the pathway intermediates are regenerated after each turn of the cycle. Of course, the individual enzymatic steps are themselves catalysts, and they will determine the rate of flux through the cycle. Also by analogy, the cycle has substrates and products. As shown in Figure 9.1c, acetyl-coenzyme A (acetyl-CoA) is the sole substrate of the Krebs cycle and $CO_2$ is the pathway product.

## BOX 9.1: WORD ORIGINS

### Krebs Cycle Intermediates

Many of the names for the Krebs cycle intermediates come from the organism in which the compound was first discovered. Citrate is the most transparent, derived from citrus fruits. Aconitate is a compound from aconite (monkshood), a member of the buttercup family. Succinate was first found in amber (fossilized resin); the Latin for this is *succium*. Fumarate is found in plants of the *fumaria* genus (poppy family). Malate is an acid found in abundance in the *Malus* genus (crabapple). While all of these organisms do have a Krebs cycle, the abundance of the compounds is due to specialized pathways (called **secondary metabolism**), which protect plants from predators.

The naming of other compounds is more prosaic, as they were already known to scientists at the time of their recognition as Krebs cycle intermediates, 2-ketoglutarate and oxaloacetate.

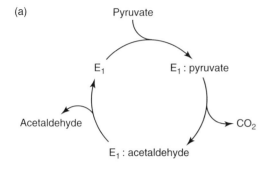

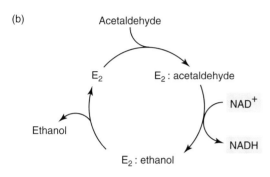

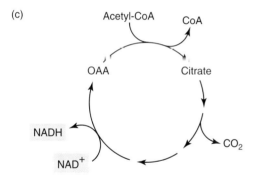

**FIGURE 9.1 Enzymatic and Metabolic Cycles.** Every catalytic event can be written as a cycle. The enzymes (a) pyruvate decarboxylase and (b) alcohol DH, the final steps of yeast fermentation, have the substrates pyruvate and acetaldehyde as inputs and include $CO_2$ and NADH as outputs. An abbreviated Krebs cycle (c) shows a similar situation, except that the intermediates of the Krebs cycle are pathway intermediates rather than enzyme-bound compounds. Like enzymatic cycles, the pathway represents a catalytic event, with a substrate as input, and $CO_2$ and reducing equivalents as output.

# 9.2 Acetyl-CoA: Substrate of the Krebs Cycle

The overall operation of the Krebs cycle oxidizes the two acetyl carbons of acetyl-CoA completely to $CO_2$, and high-energy electrons are captured in the cofactors nicotinamide adenine dinucleotide (NADH) and ubiquinone (Q). These cofactors then become substrates for the next pathway in metabolism, the electron transport chain.

The molecule acetyl-CoA is a thioester of acetate with CoA. Most significantly, it has a free -SH group at the end of the molecule, which can engage in readily reversible thioester formation. Often, the cofactor is written to emphasize the sulfur moiety, as CoA-SH.

Formation of acetyl-CoA from pyruvate requires an enzyme complex exclusively found in the mitochondria. However, pyruvate is formed through glycolytic reactions that occur in the cytosol.

Cytosolic pyruvate must cross the two membranes of the mitochondria: the **outer mitochondrial membrane** and the **inner mitochondrial membrane** (FIGURE 9.2). The outer membrane is not a barrier to molecules with a molecular weight up to about 10,000. This is achieved by the presence of the protein **porin**, which lines openings on the outer membrane. On the other hand, the inner membrane is impermeable to most molecules, so pyruvate entry requires a transport protein. This is called the **pyruvate transporter**, which co-transports pyruvate (charged −1) with a proton into the **matrix space**, the most interior portion of the mitochondria. In the matrix space, subsequent oxidation takes place.

The first step in this oxidation is the conversion of pyruvate to acetyl-CoA, catalyzed by the **pyruvate dehydrogenase complex**. An **enzyme complex** is a group of enzymes that catalyzes a metabolic sequence without releasing intermediates. The enzymes are closely associated with one another, and they act as if a single reaction was taking place. The overall reaction catalyzed by the pyruvate dehydrogenase complex is:

$$\text{Pyruvate} + \text{CoA} + \text{NAD}^+ \rightarrow \text{Acetyl-CoA} + \text{NADH} \qquad (9.1)$$

This metabolically irreversible conversion involves three separate enzymes and five cofactors. Three are mobile cofactors: CoA, $NAD^+$ and NADH, and appear in Equation 9.1. The other three are bound cofactors that stay attached to enzymes of the complex over the entire catalytic cycle: thiamine pyrophosphate, lipoic acid, and flavin adenine dinucleotide (FAD; and its reduced form, $FADH_2$; FIGURE 9.3). The bound cofactors do not appear in the overall reaction; they act as an extension of the enzyme to which they are attached. We have already encountered thiamine pyrophosphate in the discussion of pyruvate decarboxylase. It plays a nearly identical role in the pyruvate dehydrogenase complex.

The three enzymes catalyzing the conversion of pyruvate to acetyl-CoA are represented as $E_1$, $E_2$, and $E_3$ (detailed in Table 9.1). Note the presence of two other enzymes, involved in the regulation of the complex, discussed in Section 9.6.

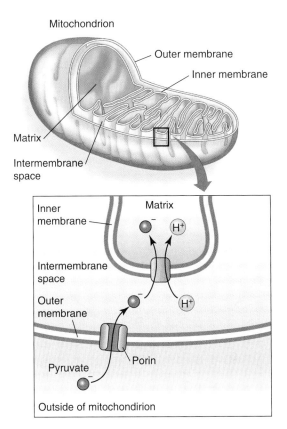

FIGURE 9.2 **Pyruvate Transport by Mitochondria.** The outer membrane of mitochondria is not a barrier for small molecules (i.e., those less than 10,000 Daltons) due to the presence of porin channels. Hence, pyruvate readily crosses into the inter-membrane space. Pyruvate is transported along with H⁺ by a protein embedded in the inner membrane. Because pyruvate has a –1 charge, no change in electrical charge of the matrix occurs as a result of this transport.

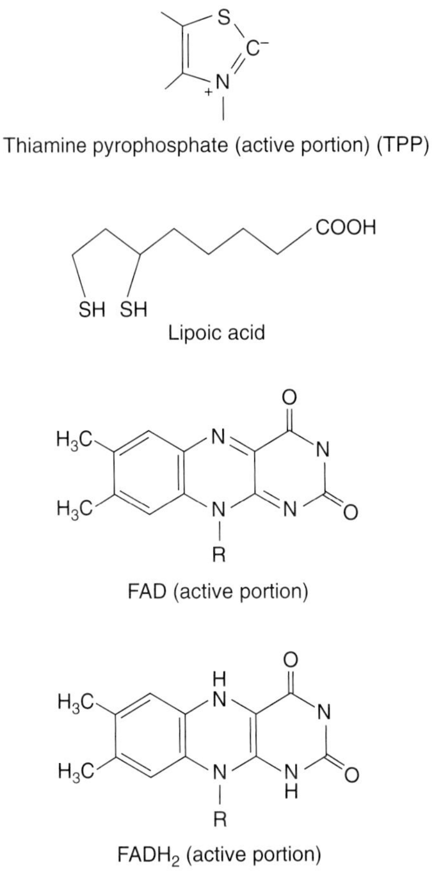

Thiamine pyrophosphate (active portion) (TPP)

Lipoic acid

FAD (active portion)

FADH₂ (active portion)

FIGURE 9.3 **Bound Cofactors of the Pyruvate Dehydrogenase Complex.** Structures of the bound cofactors of the complex show only the active portion, apart from lipoic acid.

The reaction mechanism is outlined in FIGURE 9.4. $E_1$ first reacts with pyruvate through the same mechanism as yeast pyruvate decarboxylase, except that the two-carbon fragment is not released. Rather, the central "swinging arm" of the complex—the lipoamide attached to $E_2$—forms an acetyl-thioester, regenerating the free form of $E_1$. Next, the mobile cofactor CoA-SH displaces the acetyl group from

**TABLE 9.1 Enzymes of the Pyruvate Dehydrogenase Complex**

| Name | Alternative Name or Description | Bound Cofactor | Mobile Cofactor It Reacts With |
|---|---|---|---|
| $E_1$ | Pyruvate dehydrogenase | TPP | — |
| $E_2$ | Acetyl transferase | Lipoamide | — |
| $E_3$ | Lipoamide dehydrogenase | FAD | NAD⁺ |
| PDH kinase | Kinase acting on $E_1$ | — | ATP |
| PDH phosphatase | Phosphatase acting on $E_1$ | — | — |

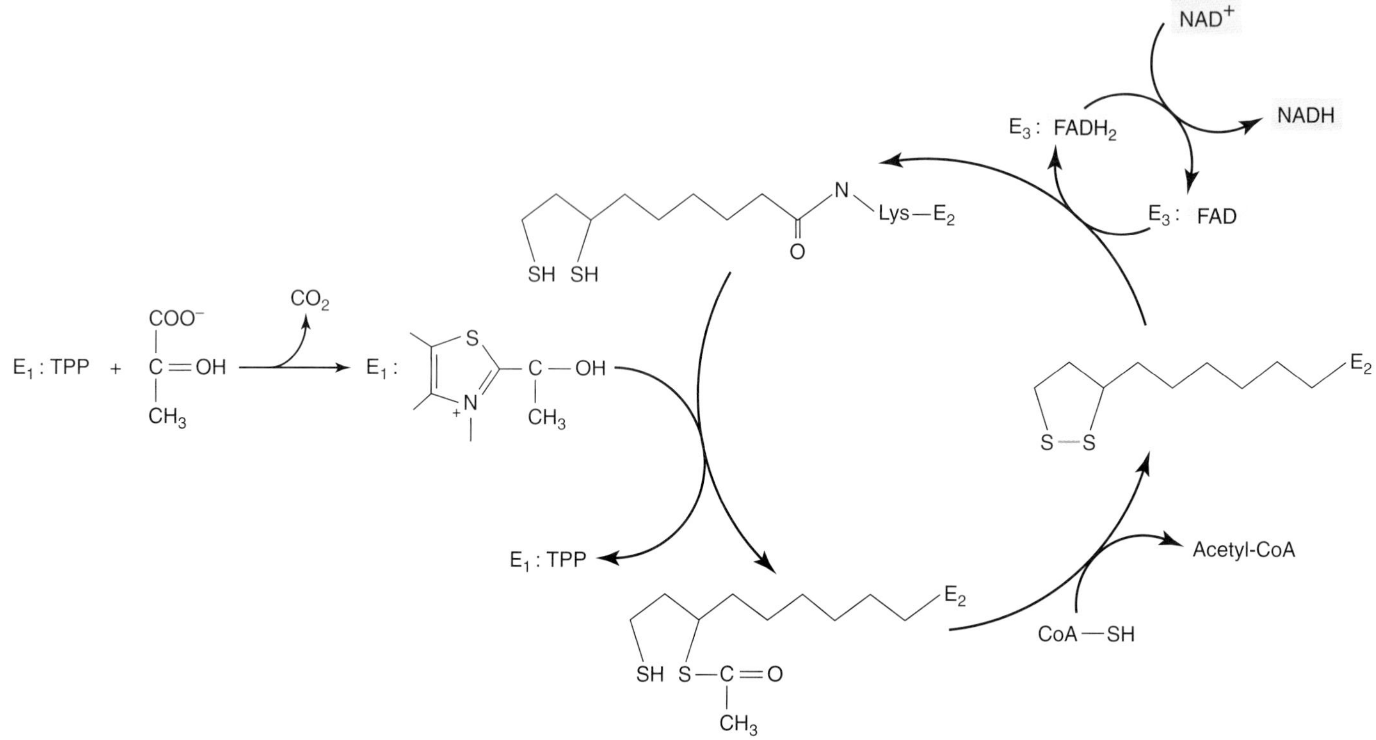

**FIGURE 9.4 "Swinging Arm" Mechanism of the Pyruvate Dehydrogenase Complex.** The first portion of the mechanism, from $E_1$ to its bound two-carbon fragment is the same as yeast pyruvate decarboxylase. However, rather than releasing this to solution, it is instead transferred to a thioester of the lipoic acid of $E_2$. Next, the "arm" of $E_2$—which is lipoic acid in an amide link to a lysine residue—reacts with CoA, releases the product acetyl-CoA, and leaves the enzyme with its lipoic acid bound in the oxidized, disulfide form. The disulfide bridge is reduced back to the sulfhydryl form through the $E_3$-bound FAD, thus producing $E_3$:$FADH_2$. $E_3$:$FADH_2$ is then reoxidized by transferring a hydride to $NAD^+$, releasing NADH, and reforming the oxidized $E_3$:FAD. Thus, every enzyme is returned to its original form so that another round of catalysis may occur.

$E_2$:lipoamide, producing the product, acetyl-CoA. This leaves the lipoamide in an oxidized (disulfide) form. The remainder of the reactions of the complex serves to regenerate the enzyme forms. The disulfide of $E_2$:lipoamide reacts with the FAD of $E_3$, producing the oxidized lipoamide (the original form of $E_2$) and reduced $E_3$, bound to $FADH_2$. Finally, the mobile cofactor $NAD^+$ reacts with $E_3$, returning it to a bound FAD form and releasing NADH.

# 9.3 Overview of Carbon Flow

FIGURE 9.5 shows the intermediates of the Krebs cycle as six-, five-, and four-carbon compounds. As input to the pathway, the two-carbon segment of acetyl-CoA (indicated as connected dark circles) condenses with a four-carbon compound to form a branched six-carbon intermediate. The release of the first carbon as $CO_2$ comes from the four-carbon (abbreviated as C4) fragment, not the incoming acetyl-CoA. Subsequently, a second $CO_2$ is released, which is also derived directly from the incoming acetyl-CoA. The resulting C4

CHAPTER 9 THE KREBS CYCLE

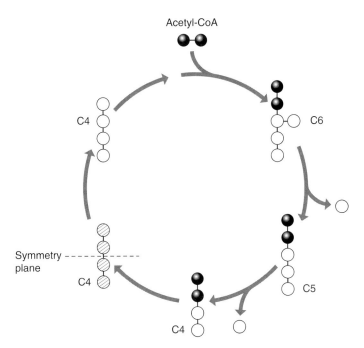

**FIGURE 9.5 Carbon Fates in the Krebs Cycle**. Acetyl-CoA is represented by the two filled circles that are the substrate of the Krebs cycle. The first reaction is with the four-carbon, open-circled substrate, to produce citrate, with the labeling pattern shown. Loss of both $CO_2$ molecules, as indicated, comes not from the input acetyl-CoA but rather from the carbon of the cycle intermediate. Conversion to a symmetrical four-carbon intermediate prior to the next round of synthesis means that half of the carbons released as $CO_2$ in the next round originated from acetyl-CoA in the previous round.

intermediate is converted to a molecule with a plane of symmetry as indicated. Thus, the top and bottom halves of the molecule are no longer distinct, indicated by the partial shading. In the second turn of the cycle, the two carbons released represent some of the carbon that entered in the *previous* turn. This pattern continues, so that in the long run, all of the incoming acetyl-CoA carbon is released as $CO_2$, but only indirectly, in subsequent turns of the cycle. This feature has confounded many radiotracer studies of the oxidation of compounds in the Krebs cycle.

Having completed an overview of carbon flow, we turn next to a detailed examination of the Krebs cycle reactions. This will allow us to understand the chemical conversions involved and the formation of mobile cofactors, which represent energy extraction.

## 9.4 Steps of the Pathway

With a cyclic pathway, it is reasonable to ask: where do we start? Because the input to the pathway is acetyl-CoA, it is customary to consider the reaction involving this metabolite as the first step.

## Citrate Synthase

The citrate synthase reaction introduces new carbon into the pathway that is ultimately released as $CO_2$:

$$\text{Oxaloacetate} + \text{Acetyl-CoA} \rightarrow \text{Citrate} + \text{CoA} \qquad (9.2)$$

While the enzyme is metabolically irreversible, no allosteric modulators are known.

In the mechanism shown in FIGURE 9.6a, extraction of a proton from the terminal methyl group of acetyl-CoA allows electrons from the methyl end of the molecule to attack the carbonyl of oxaloacetate. This nucleophilic substitution leads to the citryl-CoA intermediate, which is subsequently hydrolyzed to citrate.

Methyl hydrogens are very stable, and a proton cannot usually be removed by a basic group such as that shown in Figure 9.6a. However, the methyl carbon in acetyl-CoA is adjacent to a carbonyl. As shown in

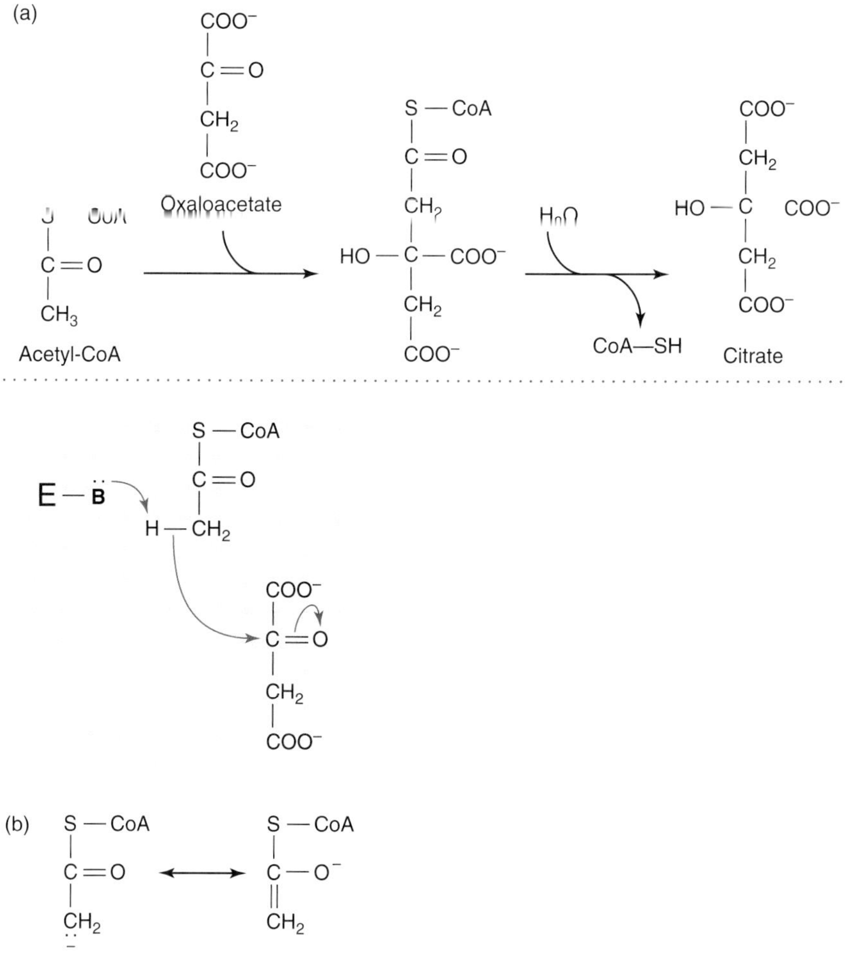

FIGURE 9.6 **Mechanism of Citrate Synthase.** (a) Condensation of acetyl-CoA with oxaloacetate is initiated by base abstraction of a methyl proton and attack on the oxaloacetate carbonyl by the resulting carbanion. The intermediate product is the CoA ester of citrate, which is hydrolyzed to the product, citrate. (b) Resonance forms of the carbanion of acetyl-CoA.

Figure 9.6b, the carbanion is resonance stabilized. Essentially, the mechanism of citrate synthase involves the development of a resonance-stabilized carbanion that undergoes an addition reaction with oxaloacetate.

## Aconitase

The aconitase reaction appears simple: citrate is dehydrated to the enzyme-bound intermediate *cis*-aconitate (hence the enzyme name) and this intermediate is rehydrated to form isocitrate. However, the enzyme mechanism is more intricate than the overall reaction would suggest. The enzyme is bound to an Fe–S cluster, shown as FIGURE 9.7. One of the $Fe^{2+}$ ions in this cluster plays a dual role in catalysis, as illustrated in FIGURE 9.8. Chelation to the carboxyl group of the substrate citrate provides an anchor, and a second chelation to the hydroxyl group provides an electron sink for the movement of electrons to form the intermediate *cis*-aconitate as shown. In the second half of the mechanism, the hydroxyl group held by the $Fe^{2+}$ attacks carbon 5, producing the product, isocitrate.

The aconitase reaction was at the center of the controversy that followed the original proposal of the Krebs cycle. Isocitrate has a chiral carbon; citrate does not. While the notion of stereochemistry itself was accepted by the 1930s, it was also common experience that reactions

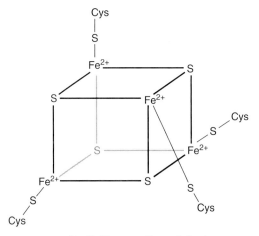

FIGURE 9.7 **Fe–S Cluster.** One of the iron–sulfur clusters is shown. The bonds between S and $Fe^{2+}$ are coordinate covalent bonds, involving the *d*-orbitals of the $Fe^{2+}$. This is the same bonding found in other chelate structures, such as heme. Only one $Fe^{2+}$ is involved in binding in the aconitase reaction. The cysteine residues shown are those of the enzyme itself. The entire cofactor is tightly bound to the protein.

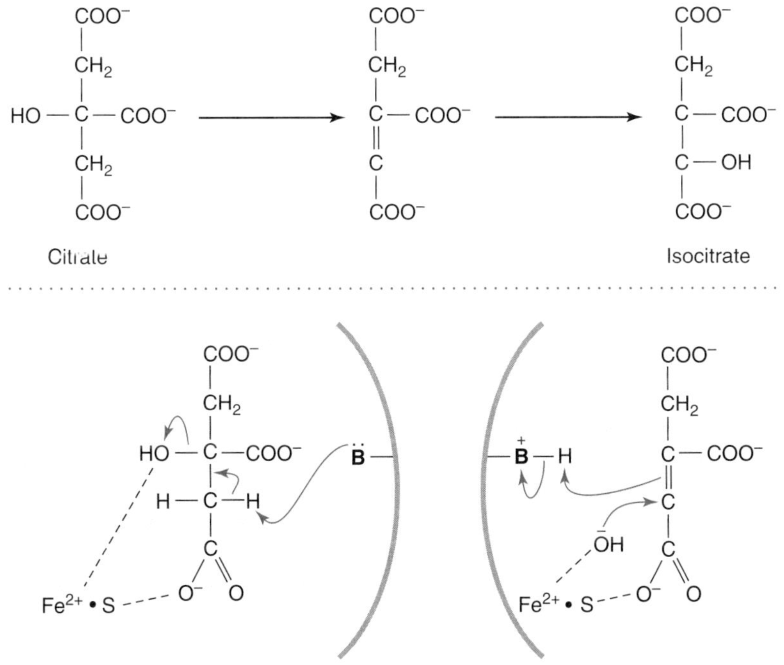

FIGURE 9.8 **Overview of the Aconitase Mechanism.** The mechanism is a dehydration/rehydration that results in the isomerization of citrate to isocitrate. First, a base abstracts a proton from citrate, and electron rearrangement leads to the double bond and removal of the OH⁻, with the assistance of the Fe–S cluster. The Fe–S cluster acts as both an anchor to the carboxylate and an electron sink. Next, the OH⁻ is returned to the bound intermediate, and the same proton that was abstracted by the enzyme is returned to form the product, isocitrate. Note that the substrate is oriented 180° to the enzyme in the second half of the mechanism relative to the first half.

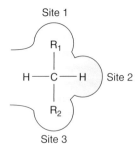

Site 1

Site 2

Site 3

FIGURE 9.9 **Three-Point Attachment for Prochiral Discrimination.** The prochiral center shown has two chemically identical substituents by definition. Yet, the tetrahedral carbon binds in three dimensions to an enzymatic site that presents one hydrogen in a chemical environment distinct from the other. Hence, stereochemical discrimination is possible.

only produce racemic mixtures from symmetrical precursors. Thus, the possibility that citrate synthase could catalyze the formation of a new chiral center in isocitrate was considered impossible.

The resolution came in the late 1940s, when Alexander G. Ogston proposed that an enzyme could provide a "three-point landing" that would distinguish two apparently identical substituents on a carbon. FIGURE 9.9 indicates how this may be achieved at an enzyme surface where two hydrogen atoms are forced into distinct chemical environments. The term **prochiral center** identifies a carbon with two identical substituents that can nonetheless be discriminated.

The stereochemical nomenclature itself had to be expanded to describe molecules like isocitrate, which cannot be easily compared to glyceraldehyde, the reference compound for the D/L system. A new system to unambiguously name chiral centers is called the *RS* system, and is described in the Appendix.

### Fluoroacetate Poisoning Involves the First Two Enzymes of the Krebs Cycle

Fluoroacetate ($CH_2FCOOH$) is the active ingredient in a poison used for many years as a rodenticide. The mechanism of its action involves not simply inhibition, but enzymatic conversion to an inhibitor. The compound first reacts to form a fluorine analog of acetyl-CoA, catalyzed by the enzyme acetate thiokinase (FIGURE 9.10). Fluoroacetyl-CoA can next react in place of acetyl-CoA in the citrate synthase reaction, in which case the product is fluorocitrate (FIGURE 9.11). Fluorocitrate is an irreversible inhibitor of aconitase. That is, aconitase catalyzes the removal of the fluoride, and a desaturated product remains enzyme-bound, thus ending all further enzyme activity.

The mechanism of aconitase with fluorocitrate as substrate has advanced the detailed understanding of the mechanism of aconitase itself. It was also part of the evidence that led scientists to accept citrate as an intermediate in the Krebs cycle.

### Isocitrate DH

In the isocitrate DH step, an oxidation is tied to the release of $CO_2$. Overall, the reaction is:

$$Isocitrate + NAD^+ \rightarrow 2\text{-}Ketoglutarate + NADH + CO_2 \qquad (9.3)$$

Like citrate synthase, this is also a metabolically irreversible reaction.

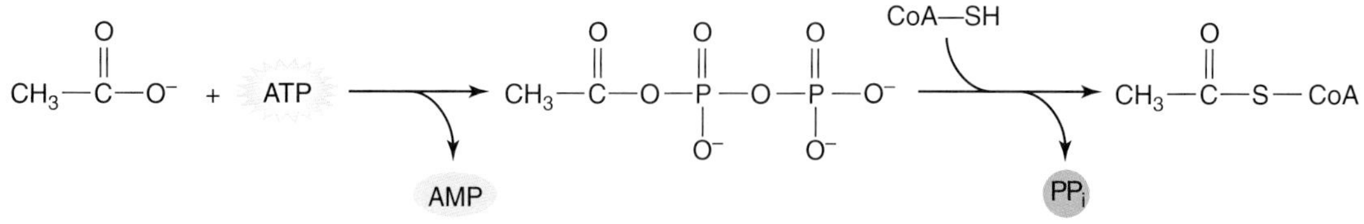

FIGURE 9.10 **Activation of Acetate: Acetate Thiokinase.** The reaction of acetate with ATP in the mitochondria forms an intermediate acid phosphate, just like the asparagine synthetase reaction. The –SH of CoA then displaces the phosphate to form the thioester, acetyl-CoA. This portion of the reaction is akin to the last portion of the glyceraldehyde phosphate DH reaction in glycolysis.

The mechanism of Equation 9.3 involves an oxidation to a chemically unstable β-keto acid intermediate followed by a decarboxylation, illustrated in FIGURE 9.12. First, $NAD^+$ accepts electrons from the hydride of the hydroxyl-containing carbon, in a manner identical to other dehydrogenases (e.g., glyceraldehyde phosphate DH), thus forming the intermediate keto acid. Electron rearrangement leads to a loss of the carboxyl group from the middle position, which forms the enol form of 2-ketoglutarate. The enol form is in equilibrium with the keto form, shown as the last step of the reaction. A similar enol–keto equilibrium occurs with pyruvate.

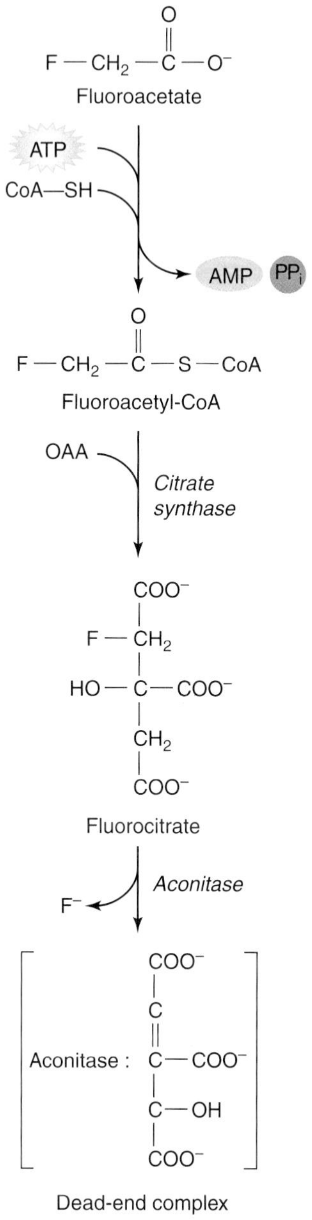

## 2-Ketoglutarate DH Complex

The other reaction that produces $CO_2$ in the Krebs cycle is the 2-ketoglutarate DH step:

$$2\text{-Ketoglutarate} + CoA + NAD^+ \rightarrow \text{succinyl-CoA} + NADH + CO_2 \quad (9.4)$$

This is actually the overall reaction of an enzyme complex that is very similar to the pyruvate dehydrogenase complex. In fact, both reactions are the same if they are written more generally as:

$$R - C(O) - COO^- + NAD^+ \rightarrow R - C(O) - S\text{-}CoA + NADH + CO_2 \quad (9.5)$$

The mechanisms for the two complexes are identical. They both use E1, E2, and E3; they both use thiamine-PP, lipoamide, and FAD in the same way; and they both employ the swinging arm mechanism. Indeed, the E2 portion of the complex is even the same enzyme in both complexes. The 2-ketoglutarate DH complex is also a metabolically irreversible step, making it an important regulatory site of the Krebs cycle, described in the following section.

## Succinyl-CoA Synthetase

The only reaction of the Krebs cycle in which a high-energy phosphate is directly formed is the succinyl-CoA synthetase step:

$$\text{Succinyl-S-CoA} + GDP + P_i \rightarrow \text{Succinate} + GTP + \text{CoA-SH} \quad (9.6)$$

This metabolically irreversible reaction is unusual because it uses guanine nucleotides rather than adenine nucleotides in many cells. Notable exceptions are mammalian brain cells and plant cells, which use ADP/ATP in place of GDP/GTP.

The mechanism of the reaction (FIGURE 9.13) resembles others we have previously considered. Like the last step of the glyceraldehyde phosphate DH mechanism (Chapter 8), the first step of this reaction is the displacement of a thioester by inorganic phosphate, forming an acid phosphate. Like the phosphoglyceromutase reaction mechanism, the high-energy phosphate is next transferred to a histidyl group of the enzyme, releasing succinate and leaving the enzyme in a

**FIGURE 9.11 Fluoroacetate Poisoning.** Fluoroacetate is a substrate for the first two mitochondrial enzymes, acetate thiokinase and citrate synthase, as shown. The resulting compound, fluorocitrate, is a suicide substrate of aconitase, thus halting the Krebs cycle.

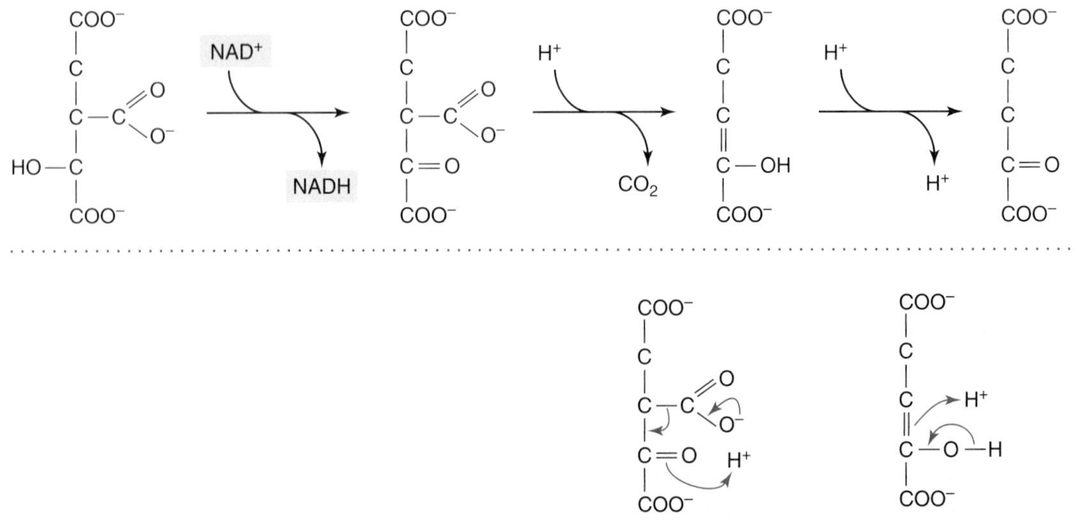

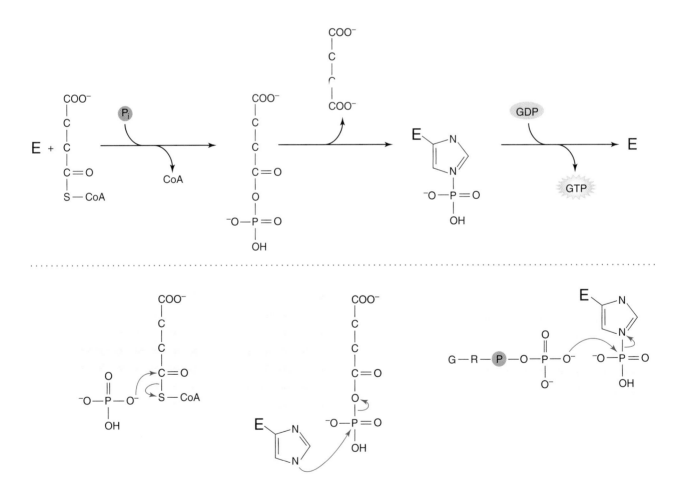

**FIGURE 9.12 Mechanism for Isocitrate Dehydrogenase.** The ketoacid intermediate is formed through the removal of a hydride from the alcohol of the substrate onto the acceptor, $NAD^+$. Next, the electron migration shown leads to loss of $CO_2$. β-Ketoacids are unstable, even in non-enzymatic reactions.

**FIGURE 9.13 Succinyl-CoA Synthetase.** This mechanism is reminiscent of two previously described ones. The first step is a phosphate displacement of a thioester, just like the terminal step of glyceraldehyde phosphate DH. The next two are phosphoryl transfers to an enzyme histidine residue, just like the phosphoglycerate mutase reaction.

phosphorylated form. Finally, the enzyme-bound phosphoryl acts as a nucleophile that attacks GDP, forming the last product, GTP, and regenerating the enzyme in its original form.

Succinyl-CoA synthetase (often called succinate thiokinase) produces a high-energy phosphate compound (GTP) that can be directly utilized as an energy intermediate. Like ATP formation in glycolysis, this process is known as **substrate-level phosphorylation.**

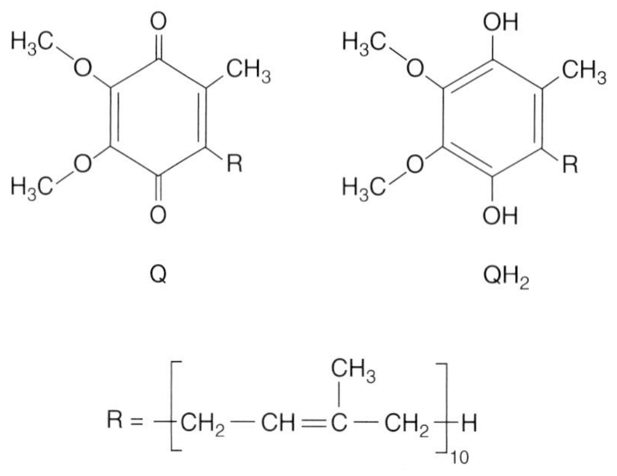

## Succinate DH

Conversion of succinate to fumarate is catalyzed by succinate DH, the only enzyme of the Krebs cycle that is membrane bound rather than soluble in the mitochondrial matrix:

$$\text{Succinate} + Q \rightarrow \text{fumarate} + QH_2 \tag{9.7}$$

Q and $QH_2$ are abbreviations for the oxidized and reduced forms, respectively, of ubiquinone (FIGURE 9.14). Succinate DH is another metabolically irreversible enzyme, making it a candidate control point, although physiological regulators are unknown. The reaction is actually another example of an enzyme complex, like the pyruvate dehydrogenase and 2-ketoglutarate complexes. In this case, however, the complex catalyzes a series of electron transfers. Electrons are removed from the interior C–C bond of succinate through intermediate, enzyme-bound electron carriers in the complex and, ultimately, to the mobile carrier Q.

Succinate DH plays another important metabolic role: It is one of five complexes of the inner mitochondrial membrane involved in oxidative phosphorylation. As such, a closer consideration of its mechanism is presented in the next chapter.

An experimentally interesting inhibitor of succinate DH is the dicarboxylate malonate (FIGURE 9.15). This inhibitor was important historically as its action elucidated the nature of the Krebs cycle itself.

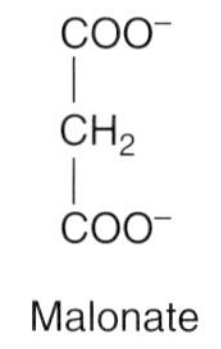

## Fumarase

The next reaction in the Krebs cycle is the hydration of fumarate to malate, catalyzed by the near-equilibrium enzyme fumarase. The reaction adds water across the carbon–carbon double bond of fumarate (FIGURE 9.16). As with the aconitase reaction (discussed above), a new stereochemical center is produced in the product that does not exist in the substrate. In the case of the fumarase reaction, the prochiral molecule (fumarate) is converted into a chiral one (L-malate) through binding to an enzyme, which is itself a chiral molecule. In another similarity to aconitase, fumarase also has an Fe–S complex bound to the enzyme for substrate positioning and hydroxylation.

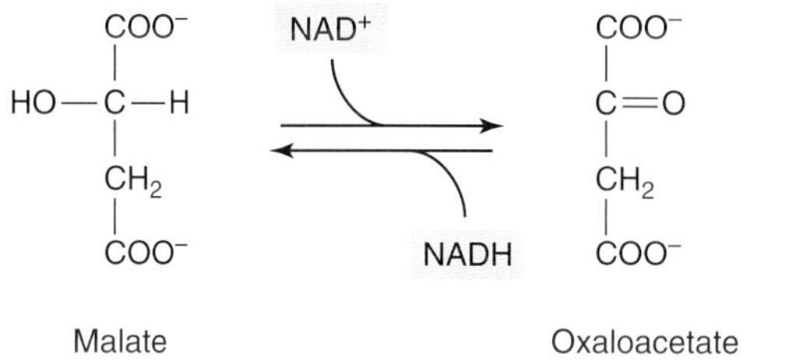

## Malate DH

The final step in the Krebs cycle is catalyzed by malate DH, another near-equilibrium reaction (FIGURE 9.17). The mechanism is the transfer of a hydride from the C–H bond of the substrate (L-malate) to the

**FIGURE 9.14 Ubiquinone.** Oxidized (Q) and reduced ($QH_2$) forms of ubiquinone are shown. The long chain substituted R group is 10 subunits of the branched five-carbon lipid. This cofactor is often called $Q_{10}$.

**FIGURE 9.15 Malonate.**

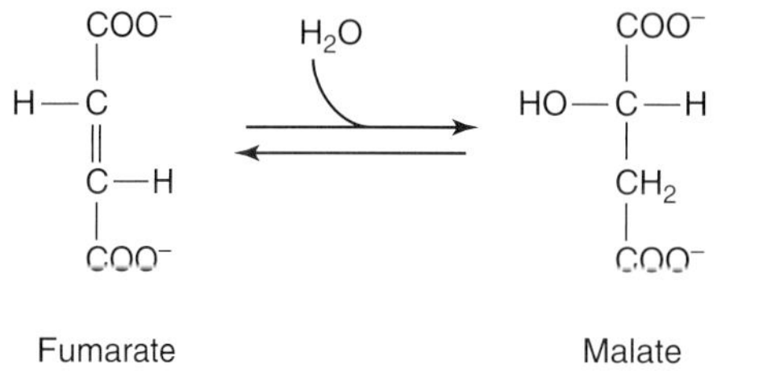

**FIGURE 9.16 Fumarase Reaction.** This hydration of fumarate to malate is similar to the aconitase reaction (Figure 9.8) in that fumarase also has an Fe–S cluster and produces a new chiral center in malate (the L-isomer).

**FIGURE 9.17 Malate DH Reaction.** The malate DH reaction is yet another redox reaction involving hydride transfer. As it is near-equilibrium, the reverse reaction also produces a new chiral center, so that enzymatic orientation—the three-point binding site to the enzyme—is involved.

NAD$^+$ ring. The reaction product oxaloacetate is achiral, but when the reaction is run in the reverse direction, which happens in pathways other than the Krebs cycle, the product is always the L-isomer of malate. Thus, we have yet another case of a reaction that produces a chiral molecule from a prochiral one.

# 9.5 Energy Balance

In order to estimate the energy available from the oxidation of acetyl-CoA by the Krebs cycle, it is necessary to make some assumptions about how much energy can be extracted by the energy cofactors (**BOX 9.2**). The approximate equivalents for calculations are:

> NADH: 3 ATP
>
> QH$_2$: 2 ATP
>
> GTP: 1 ATP

With these assumptions, the three NAD$^+$-linked dehydrogenase reactions produce 9 ATP; the single QH$_2$-generating step catalyzed by succinate DH produces 2 ATP; and the succinyl-CoA synthetase reaction produces 1 GTP, which we will consider to be roughly equivalent to 1 ATP. Indeed, in many cells, the enzyme uses ADP as substrate, thus producing ATP directly. The total is 12 ATP per acetyl-CoA oxidized. How NADH and QH$_2$ oxidations can produce ATP requires a study of electron transport chain of the mitochondria. However, it is already clear that the Krebs cycle produces prodigious amounts of ATP when glucose is fully oxidized to CO$_2$. In fact, a single glucose molecule

## BOX 9.2: Estimating ATP Energy Equivalents

The values stated in this text are the "classical" ones that have been used for many years: 3 ATPs for NADH oxidation and 2 ATPs for succinate oxidation. For metabolic analysis in nutrition and medicine and for comparing pathways, these conversion factors are universally accepted. However, there are reasons to consider more precise values that have emerged in the last two decades.

The use of exact integer values given above is a throwback to an era when it was believed that ATP was synthesized locally at three separate sites in mitochondria. When succinate was oxidized, it was believed that it skipped "site 1" and thus only produced 2 ATPs. Once it became clear that a proton gradient was the energy intermediate, fractional values—which had been observed in many labs—became a possibility. Many studies have proposed values of 2.5 and 1.5, respectively, although controversy remains concerning the precise number. In addition, there is uncertainty in assigning the value of 1 ATP for 1 GTP. The reaction must produce somewhat less ATP, because the efficiency cannot be 100%. These considerations show that the estimates of ATP equivalents, while useful, are inexact.

## TABLE 9.2 Comparison of Energy Production

| Portion of Pathway | Energy Intermediate | Equivalents of ATP |
|---|---|---|
| Glucose → Pyruvate | 2 ATP<br>2 NADH | **2**<br>**6** |
| Pyruvate → Acetyl-CoA | 1 NADH | 3 |
| 2 Pyruvate → 2 Acetyl-CoA | | **6** |
| Acetyl-CoA → 2 $CO_2$ | 3 NADH<br>1 $QH_2$<br>1 GTP | 9<br>2<br>1 |
| 2 Acetyl-CoA → 4 $CO_2$ | | **24** |
| Glucose → 6 $CO_2$ | | 38 |

produces 38 ATP via complete oxidation, compared to 2 ATP using glycolysis alone (Table 9.2).

# 9.6 Regulation

There are two levels of regulation of the Krebs cycle: by its supply of substrate, acetyl-CoA, and by its intrinsic activity. We have emphasized the production of acetyl-CoA from pyruvate, which in turn arises from glycolysis. However, as discussed elsewhere in this text, all major nutrients—fats, carbohydrates, and proteins—are converted into acetyl-CoA for oxidation by the Krebs cycle. Thus, the Krebs cycle is the final common route for the complete oxidation of foods into $CO_2$.

Regulation of pyruvate to acetyl-CoA in mammals is largely due to phosphorylation and dephosphorylation of the $E_1$ component, catalyzed by the protein kinase and phosphatase listed in Table 9.1. Control of those enzymes is illustrated in FIGURE 9.18. The pyruvate

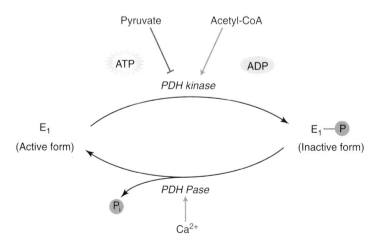

FIGURE 9.18 **Regulation of the PDH Complex.** $E_1$ (the PDH component) is phosphorylated and inactivated in a reaction catalyzed by PDH kinase. PDH kinase is inhibited by pyruvate and stimulated by acetyl-CoA. The protein phosphatase that catalyzes dephosphorylation of $E_1$–P is stimulated by $Ca^{2+}$.

dehydrogenase kinase (PDH kinase) that catalyzes phosphorylation of $E_1$ is allosterically inhibited by pyruvate and allosterically stimulated by acetyl-CoA. Because conversion to the phosphorylated form of $E_1$ causes its inactivation, pyruvate stimulates overall pyruvate dehydrogenase complex activity and acetyl-CoA inhibits it. Thus, the substrate and product of the complex itself also serve as allosteric regulators of the kinase. The protein phosphatase activity of $E_1$ (PDH phosphatase) is stimulated by $Ca^{2+}$ ions, leading to dephosphorylation of $E_1$ and thus stimulation of the overall complex.

Besides controlling the supply of acetyl-CoA to the Krebs cycle, several steps in the pathway are metabolically irreversible, and thus potentially regulatory. Citrate synthase, isocitrate DH, 2-ketoglutarate DH, succinyl CoA synthetase, and succinate DH all fit this classification. Of these, isocitrate DH and 2-ketoglutarate DH can be stimulated by an increase in the concentration of $Ca^{2+}$ ions. Control of the other enzymes is usually related to control of flow through other reactions that connect to the Krebs cycle.

There is another regulatory circuit that operates between the Krebs cycle and glycolysis. As the rate of the Krebs cycle increases, so do the concentrations of its intermediates. This leads to an increased concentration of citrate in mitochondria and the cytosol, because there is a transporter for citrate in the mitochondrial inner membrane.

An increased citrate concentration in the cytosol decreases phosphofructokinase activity and hence glycolysis. The underlying observation that glycolysis is diminished in aerobic metabolism was made in the 19th century and is known as the Pasteur effect.

## 9.7 Krebs Cycle as a Second Crossroad of Metabolic Pathways

We have just considered the reactions of the Krebs cycle as a complete pathway for the oxidation of acetyl-CoA to $CO_2$. The Krebs cycle also can be viewed as a *metabolic hub* in which one or more of its enzymes play a part in other metabolic pathways.

We will consider some of these reactions in this text because they are major routes in metabolism. For example, fatty acid synthesis uses citrate from the Krebs cycle, although a separate reaction must supply carbon at the same rate to sustain this pathway. Additionally, some amino acids can be converted to Krebs cycle intermediates by the process of transamination. Comparing the structure of the amino acid glutamate with 2-ketoglutarate (FIGURE 9.19a), you may notice that the two are similar except for the amine in one case and the keto group in the other. Similarly, aspartate can be viewed as an amine analog of the Krebs cycle intermediate oxaloacetate (Figure 9.19b).

The Krebs cycle also connects to pathways that are more elaborate and not considered in detail in this text. For example, heme biosynthesis is a pathway stemming from succinyl-CoA. As a further example,

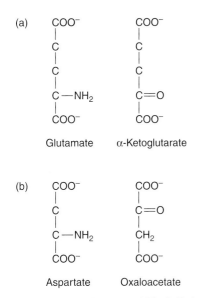

FIGURE 9.19 **Amino Acids and Their Related Keto Acids.**

the glyoxylate cycle is a pathway found in bacteria and plants that is related to the Krebs cycle. This specialty route is discussed in the Appendix.

Just like an airline hub that connects numerous stations that themselves have modest networks, the Krebs cycle intermediates are each connected to separate pathways in the cell. Thus, at any given moment, the pathway can be viewed as a separate entity or as a portion of many others. It is always important to remember that, as a distinct pathway, the Krebs cycle has just one substrate, acetyl-CoA, and one product, $CO_2$.

## Summary

All major nutrients—carbohydrates, lipids, and proteins—are oxidized first to acetyl-CoA and then to $CO_2$ through the Krebs cycle. When the precursor is carbohydrate, pyruvate is the source of acetyl-CoA, and the conversion is catalyzed by the pyruvate dehydrogenase complex. The pyruvate DH complex has three separate enzymes and five cofactors: thiamine pyrophosphate, lipoic acid, CoA, FAD, and $NAD^+$; the latter two in their reduced forms are $FADH_2$ and NADH. In the first step of the Krebs cycle, acetyl-CoA condenses with oxaloacetate to form citrate, a six-carbon intermediate. Next, aconitase catalyzes the conversion of citrate to isocitrate. Historically, this was the first example of prochirality in enzymes, in which identical substituents on the same carbon could be distinguished by binding to an enzyme surface. Isocitrate loses $CO_2$ in the next step, a reaction catalyzed by isocitrate dehydrogenase, thus forming the five-carbon molecule 2-ketoglutarate. This is followed by the other decarboxylation reaction of the Krebs cycle, catalyzed by the 2-ketoglutarate DH complex. This complex is closely analogous to the pyruvate DH complex, because it has the same three types of enzymes, the same five cofactors, and the same mechanism. Next, succinyl-CoA and GDP are converted to succinate and GTP, catalyzed by succinyl CoA synthetase. Succinate is converted to fumarate via succinate DH, the only membrane-bound enzyme of the cycle. Next, the reaction also captures reducing equivalents in the lipid-soluble mobile cofactor ubiquinone. The fumarase-catalyzed hydration of fumarate produces another new chiral center in L-malate, another example of prochiral discrimination by enzymes. The final step is the oxidation of malate to oxaloacetate, with the formation of NADH.

Overall, the Krebs cycle produces 12 ATP molecules for each input acetyl-CoA. The Krebs cycle is regulated in part by the concentration of acetyl-CoA, the pathway substrate. The pyruvate dehydrogenase complex is regulated by phosphorylation (inhibition) and by mitochondrial $Ca^{2+}$ concentration (stimulation). $Ca^{2+}$ ions also directly stimulate the 2-ketoglutarate DH complex. Aside from its role as the final oxidation pathway in cells, the Krebs cycle serves as a metabolic hub; each enzymatic step is utilized as part of other pathways in the cell.

## Key Terms

citric acid cycle
enzyme complex
inner mitochondrial
  membrane
Krebs cycle
matrix space
outer mitochondrial
  membrane

Pasteur effect
porins
prochiral center
pyruvate dehydrogenase
  complex
pyruvate transporter
secondary metabolism

substrate-level
  phosphorylation
tricarboxylic acid cycle

## Review Questions

1. One consequence of a cyclic pathway is that its intermediates are in very low, catalytic amounts. However, in the linear pathway of glycolysis, the intermediates are also in very low concentrations. How is a cyclic pathway fundamentally different?
2. The Krebs cycle has two basic strategies for oxidizing molecules and extracting electrons as mobile carriers. What are they?
3. How does fluoroacetate act as a poison?
4. Explain the notion of prochirality and where it is significant in the Krebs cycle.
5. Which carbon of acetyl-CoA attaches to oxaloacetate in the citrate synthase reaction?
6. Which step of the Krebs cycle is called *substrate-level phosphorylation*? What step does this correlate to in glycolysis?
7. Which steps of the Krebs cycle produce $CO_2$? Which produce NADH?
8. Which reaction of the Krebs cycle has the same mechanism as the pyruvate dehydrogenase complex?
9. Malonate is chemically similar to succinate, as both are dicarboxylic acids with intervening methyl subunits. Why can't malonate also be a substrate of succinate DH? What type of inhibition would you predict for malonate?

## References

1. Newsholme, E. A.; Leech, A. R. *Biochemistry for the Medical Sciences;* John Wiley & Sons: New York, 1986.
2. Owen, O. E.; Kalhan, S. C.; Hanson, R. W. The key role of anaplerosis and cataplerosis for Krebs cycle function. *J. Biol. Chem.* 277:30409–30412, 2002.
3. Tong, W. H.; Rouault, T. Metabolic regulation of citrate and iron by aconitases: Role of iron-sulfur cluster biogenesis. *BioMetals* 20:549–564, 2007.

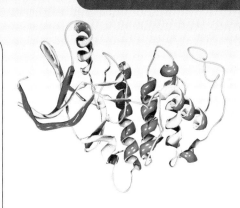

# 10

# Oxidative Phosphorylation

## CHAPTER OUTLINE

Image © Dr. Mark J. Winter/Photo Researchers, Inc.

**E**lectrons from the Krebs cycle are captured as two mobile carriers: NADH and $QH_2$. The pathway that oxidizes these compounds is called *oxidative phosphorylation* and it is responsible for most of the ATP produced in cells. A unique feature of oxidative phosphorylation is a proton gradient that acts as the energy intermediate. The mechanism is known as the **chemiosmotic hypothesis,** and it was originally proposed by Peter Mitchell in the early 1960s. It was such a break from the traditional notion of chemical energy transfer through bond exchange that it wasn't fully accepted until extensive evidence had been accumulated. We will consider some of this evidence after discussing the pathway of oxidative phosphorylation.

## 10.1 The Phenomenon

The term *oxidative phosphorylation* refers to two processes: oxidation, involving electron flow, and phosphorylation, the synthesis of ATP from ADP and $P_i$. In isolated mitochondria, as in intact cells, these processes are **coupled.** A simple experiment with mitochondria shows this coupling directly: adding ADP to the preparation causes the oxidation of substrates as well as the NADH and $QH_2$ they produce, and increases the formation of ATP. $QH_2$ actually enters as an intermediate in the process. Because the first electron donor is NADH and the final electron acceptor is oxygen, the overall equation for electron flow is:

$$NADH \rightarrow Q \rightarrow O_2 \qquad\qquad (10.1)$$

NADH is oxidized to $NAD^+$, consistent with its role as a mobile cofactor, and $O_2$ is reduced to water. Thus, there is a continuous consumption of $O_2$ and production of $H_2O$.

Another kind of coupling occurs in this process, closely related to the first: a coupling of electron and proton movement. The movement of electrons from NADH to $O_2$ takes place exclusively within the inner membrane of the mitochondria. Electrons are passed from one electron carrier to another in a series of steps collectively called the **electron transport chain.** As electrons move through the membrane, protons move *across* the membrane, from the interior of the mitochondrial matrix to the outside of this inner membrane (**FIGURE 10.1**). This movement of protons is **vectorial,** because it has both a magnitude (the number of protons transported) and a direction (from the matrix to the cytosol). We will get a sense of how this coupling works by considering some details of the molecules involved in supporting this process.

The movement of protons across the inner membrane leaves the inner membrane space negatively charged relative to the outside of the matrix. There is also a concentration gradient of protons, with more outside than inside the matrix. Hence, there is both an electrical gradient (due to charge separation) and a chemical gradient (due to differences in concentration) across the membrane. This is the energy intermediate of the entire process. The re-entry of protons occurs through a special channel linked to ATP synthesis.

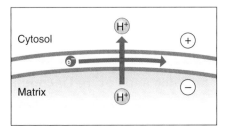

**FIGURE 10.1 Vectorial Transport.** As electrons move through the membrane, protons move across it.

We can summarize this overview by saying that *oxidation* is the stepwise removal of electrons from substrates that ultimately reduce oxygen to water, and that it creates a proton gradient as it does so. *Phosphorylation* is the process of importing protons back into the mitochondrial matrix and forming ATP from its precursors, ADP and $P_i$.

# 10.2 Mitochondrial Inner Membrane

An electron micrograph of a typical mitochondrion is shown in FIGURE 10.2. The outer membrane is seen as the dark regular outline. The inner membrane shows characteristic infoldings, which provide a much greater membrane surface area within the same volume.

A sketch of the essential topology of mitochondria is presented as FIGURE 10.3a. As with the entry of pyruvate into the mitochondria, large pores exist in the outer membrane, rendering it permeable to molecules up to 10,000 daltons. In fact, it is experimentally possible to strip off the outer membrane of isolated mitochondria and reproduce all of the events described in this chapter. Hence, only the inner membrane is critical for oxidative phosphorylation. It also follows that the space outside the inner membrane, while technically part of the inner membrane space, is equivalent to the cytosol for small molecules under 10,000 daltons. As a result, our view of mitochondrial topology can be simplified to the functional picture presented as Figure 10.3b, in which the inner membrane—usually referred to as just *the membrane*—divides the matrix from the cytosol.

The inner membrane contains a relatively high concentration of proteins, which include exchangers like the pyruvate transport, but also several protein complexes embedded in the membrane that conduct the flow of protons and electrons. We can classify carriers for the electron transport chain into three groups: molecules that carry electrons, protons, or both.

# 10.3 Carriers of Electrons, Protons, or Both

## Carriers of Electrons

Metals are pure electron carriers; the most common of these is iron, with copper a close second. The metals in the electron transport chain exist in a chelate form, either as a heme or as a complex with sulfur, such as the Fe–S complex. Both heme iron (in hemoglobin) and an Fe–S cluster (in aconitase) are not redox-active. Instead, the iron is exclusively in the $Fe^{2+}$ state. However, as electron carriers in oxidative phosphorylation, metals rapidly undergo changes in redox state, such as $Fe^{2+}$ alternating with its oxidized form, $Fe^{3+}$. Those compounds that are pure electron carriers always mediate single-electron transfers.

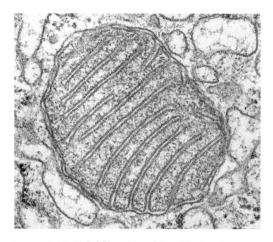

FIGURE 10.2 **A Mitochondrion Under the Electron Microscope.** (© Don W. Fawcett/ Photo Researchers, Inc.)

(a)

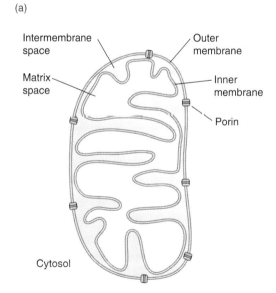

(b)

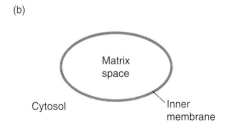

FIGURE 10.3 **Functional Spaces in Mitochondria.**

## BOX 10.1: WORD ORIGINS

### Flavins

Flavius was a family from Ancient Rome, where it meant *blonde*. The house of Flavius included emperors and writers. In the 19th century, a yellow dye derived from the bark of the North American oak tree (*Quercus velutina*) contained an active principle, originally called quiercitron, but now called quercitin, which belongs to a family of related compounds called the **flavonoids**. By the 1930s, the flavins (a subset of the flavonoids) were established as structural derivatives of isoalloxazine (the tricyclic ring portion of riboflavin). When purified, flavin-containing enzymes all have a striking golden-blond color, testimony to the tight binding of these cofactors to their apoproteins.

### Carriers of Protons

We have previously encountered pure carriers of protons in another guise: as acids or bases. Thus, aspartyl residues of proteins are proton carriers, as are bases such as histidyl residues of proteins. These side groups play an important role in establishing the proton gradient.

### Carriers of Both Electrons and Protons

The major carriers of both electrons and protons are NADH, $QH_2$, and the flavin nucleotides $FADH_2$ and $FMNH_2$ (**BOX 10.1**). NADH transfers a pair of electrons along with a proton (i.e., a hydride, H:$^-$). By contrast, $QH_2$ transfers electrons in two distinct steps. The process, illustrated in FIGURE 10.4 for the reduction of ubiquinone (Q), is fully reversible. Q accepts one electron to form the radical ion intermediate, $Q^-$, which is stabilized by resonance. A second electron is then incorporated along with two protons to form the fully reduced species ubiquinol ($QH_2$).

The flavin cofactors are unique in that they can transfer electrons *either* one at a time *or* two at a time. In a sense, they mimic the other two transfers we have just considered. In the single-electron transfer mode, there is a radical intermediate, stabilized by the aromaticity of the ring. Alternatively, flavins can undergo a two-electron, hydride ion transfer. It is this ability that makes the flavin cofactors (FAD and FMN, which have identical redox properties) *transducers* between the obligate two-electron exchange with the cofactor NADH and all of the other cofactors that transfer one electron at a time. The structure of the oxidized, semiquinone, and reduced flavins are shown in FIGURE 10.5.

## 10.4 Membrane-Bound Complexes

Five membrane-bound protein complexes participate in oxidative phosphorylation. These can be viewed as islands within the membrane sea that only communicate with each other indirectly. For example, the mobile cofactor NADH can donate electrons and protons to complex I (FIGURE 10.6). This complex releases protons to the cytosol and transfers its electrons to Q, resulting in $QH_2$ formation. Q and $QH_2$ are mobile cofactors confined to the hydrophobic space of the membrane. $QH_2$ donates its electrons to complex III, and releases protons to the cytosol. The electrons of complex III reduce cytochrome *c*, a mobile cofactor that slides along the cytosolic surface of the membrane and donates its electrons to complex IV. Finally, electrons from complex IV react with $O_2$ to form $H_2O$.

The pathway just described traces the route electrons follow from NADH to $O_2$, and it involves complexes I, III, and IV, and the intermediary mobile cofactors $QH_2$ and cytochrome *c*. Complex II also is known as succinate dehydrogenase, an enzyme that catalyzes a step in the Krebs cycle. Electrons from succinate enter this complex and they leave as $QH_2$, so that Q can be thought of as a collector of electrons from complexes I and II. As we will appreciate later, other membrane complexes also form $QH_2$, so this cofactor is a point of common intersection. Together, the first four complexes account for the movement of electrons in oxidative phosphorylation as well as the generation of the proton gradient.

Complex V uses the proton gradient to form ATP (Figure 10.6). As protons flow down the gradient, complex V synthesizes ATP. We next consider the processes of the electron pathway and proton circuits individually before we directly examine the operation of the complexes.

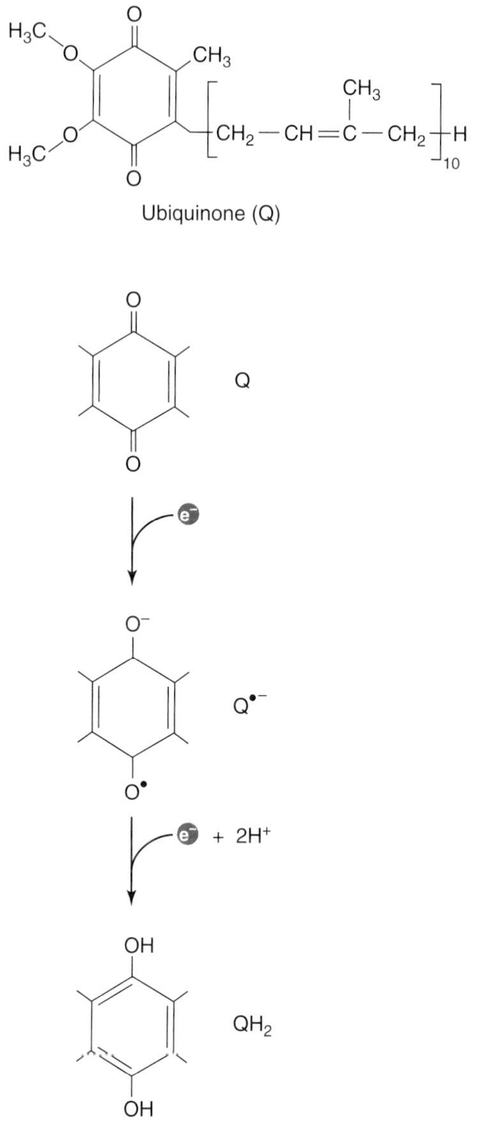

# 10.5 Electron Pathways

## Electrochemical Cells

There is a natural ordering to redox reactions that can be appreciated by studying electrochemical cells. In Chapter 7, we noted that a redox reaction can be formally split into an oxidation half-reaction and a reduction half-reaction. This can also be done experimentally. Consider the case of electron transfer between the metals Zn and Cu. Suppose a solid bar of each metal is dipped in a separate **half-cell**, each in contact with a salt solution of its metal. In FIGURE 10.7, the Zn is in equilibrium with $ZnSO_4$ and the Cu is in equilibrium with $CuSO_4$. There are two connections between the cells. First, a wire connects the two bars (with an in-line meter to measure current flow), allowing electrons to pass between them. Second, a salt bridge connects the two solutions, allowing ion flow to complete the circuit with minimal mixing between the solutions. These two connections represent the ways in which current can travel: as a flow of electrons (as in the wire) or as a flow of ions (as in the solutions and salt bridge).

The instant the connections are complete, Zn is converted to $Zn^{2+}$ (that is, the bar partially dissolves) and electrons leave the bar to the wire, where they flow to the copper half-cell. At the Cu half-cell, $Cu^{2+}$ ions react with the electrons entering the bar to form Cu metal (that is, the bar is expanded, or *plated*). The flow of electricity is completed as an equivalent number of $SO_4^{2-}$ ions move through the salt bridge from the Cu to the Zn half-cell. The meter records a potential of 1.1 V overall, which is the $\Delta\epsilon°$ for the overall oxidation of Zn coupled to the reduction of Cu:

$$Zn + Cu^{2+} \rightarrow Zn^{2+} + Cu \qquad (10.2)$$

Note that there is directionality in the Zn/Cu electrochemical cell. Zn is oxidized and $Cu^{2+}$ is reduced. Each of these events can be written separately as half-reactions corresponding to just one cell. Although half-reactions can be written as either oxidations or reductions, by general agreement scientists write half-reactions as reductions. In order to compile a list of these **reduction potentials**, we need a reference. The reference standard is the hydrogen reduction potential, which is defined as zero:

$$2H^+ + 2e^- \rightarrow H_2 \rightleftarrows \Delta\epsilon° = 0V \qquad (10.3)$$

FIGURE 10.4 **Ubiquinone.** The full structure of the oxidized form is shown above, followed by stepwise reduction. The intermediate is both negatively charged (an anion) and a free radical.

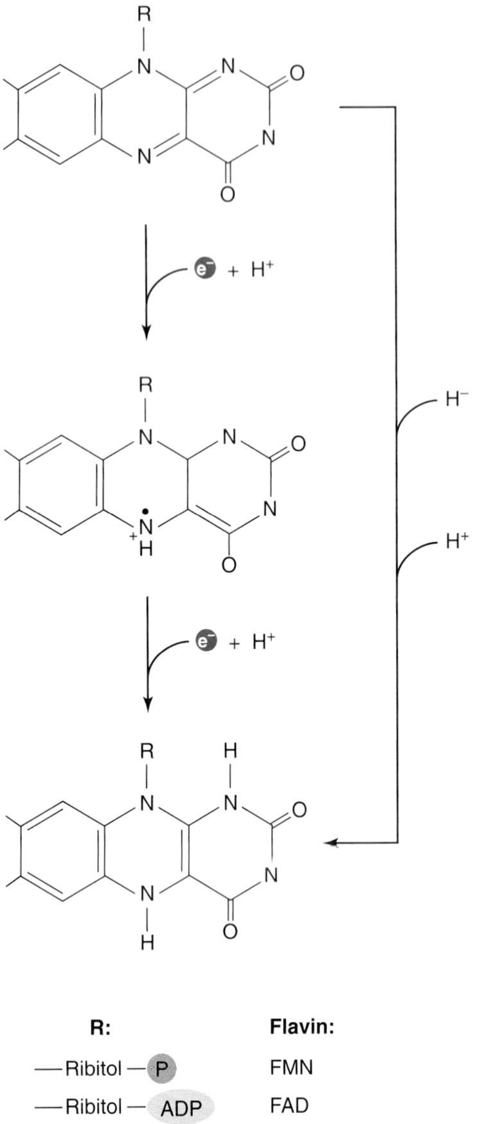

**FIGURE 10.5 Redox states of the flavin nucleotides.** The oxidized flavin is shown moving stepwise through single electron steps, first to the radical intermediate and then to the fully reduced form, which accepts protons. Alternatively, the molecule can accept a hydride ion and a proton and directly convert to the fully reduced form. As both transforms are reversible, the flavin can accept an electron pair and donate single electrons, a critical function in the electron transport chain.

Every half-reaction can be compared with this reference standard, and a table of reduction potentials can be constructed, such as the one in Table 10.1. Note that the values, as is customary in biochemistry, are corrected for a slightly different standard, where pH = 7.

## Sequence of Electron Flow

The pathway of connected redox reactions that comprises the respiratory chain (FIGURE 10.8) follows the order of increasing reduction potential (Table 10.1). Both NADH and succinate feed electrons into Q, through either complex I or complex II, respectively. Electrons from Q pass through complex III and are removed by cytochrome *c*. Finally, cytochrome *c* donates electrons through complex IV to form $H_2O$. This electron path can be viewed as a steadily oxidizing flow. NADH can be considered to be the pathway substrate because most of the electrons arise from NADH. $H_2O$ is the common termination point, and is thus the pathway product.

It may seem surprising that these redox potentials—which are *standard* state values—are consistent with the sequence of mitochondrial redox carriers. For most metabolic pathways, standard thermodynamic values bear no relation to the actual ones (e.g., glycolysis). However, it *is* possible to use standard values for the respiratory chain. This is not because the concentrations are all near 1 M (standard conditions), but because they are in nearly 1:1 ratios. This is due to the fact that the components are mostly in complexes. Thus, they act *as if* the concentrations were near standard conditions. Because the pH is taken to be 7.0 rather than the usual standard of 0, this single outlier has been removed. Finally, the ratio of $NAD^+$ to NADH happens to have a value of about 1 in the mitochondrial matrix. This is not the case in the cytosol, where the $NAD^+$ to NADH ratio is two orders of magnitude greater. For these reasons, the standard values serve as reasonable approximations to the actual ones.

Experimental support for this sequence comes from the use of selective inhibitors of the respiratory complexes. As $O_2$ is the final acceptor, this end of the pathway can be called the *oxidized side*, or just the oxygen side. Inhibitors are known for each of the complexes of respiration. These are also indicated in Figure 10.8. Rotenone, used commercially as a rat poison, specifically inhibits complex I. Malonate inhibits complex II (succinate dehydrogenase), albeit weakly (**BOX 10.2**). Antimycin is a classical inhibitor of complex III and cyanide ($CN^-$) inhibits complex IV.

The logic of inhibitor actions on the pathway is straightforward: wherever an inhibitor blocks the respiratory chain, electron carriers closer to the oxygen side will be more oxidized, as electrons are removed ultimately to oxygen, and those carriers on the NADH side will be more reduced. The concentrations of the redox carriers can be measured because they absorb light differently in their reduced and oxidized forms. In fact, the original names of the major cytochromes were based on the appearance of their peaks in a spectrum, where they were labeled a, b, and c. These names have been modified to

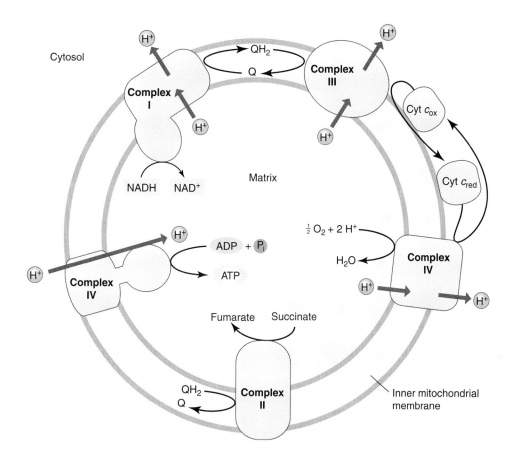

**FIGURE 10.6 Schematic of Oxidative Phosphorylation.** Oxidative phosphorylation is illustrated as a function of five complexes and the communicating cofactors. Protons are the intermediate between redox extraction (complexes I–IV) and ATP synthesis (complex V). The cofactors are shown to be of three types. NADH is water soluble, Q is soluble in the hydrophobic interior of the membrane, and cytochrome *c* moves along the outer leaflet of the membrane.

incorporate cytochromes discovered later, such as cytochrome $c_1$, cytochrome $b_{560}$, and cytochrome $b_{566}$.

## Energetics of Electron Flow

Analogous to the diagram of glycolysis in Chapter 8, we can plot the sequence of mobile electron carriers against their increasing redox potentials (FIGURE 10.9). The figure shows that the change in redox potential is substantial for complexes I, III, and IV (quantified as vertical bars in the figure), and nearly zero for complex II.

The bars in Figure 10.9 correspond to values of $\Delta\epsilon°$, which have a direct relationship to the standard free energy change through the relation given in Equation 10.4:

$$\Delta G° = -nF\ \Delta\epsilon°$$

$$(10.4)$$

where *n* is the number of electrons and *F* is the faraday constant (96,500 joules/coulomb). $\Delta G°$ and $\Delta\epsilon°$ are closely related because free energy is a chemical potential and the redox change is an electrical

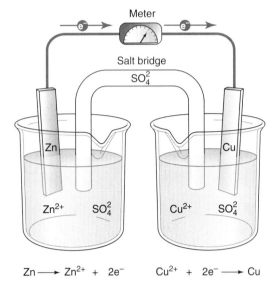

$$Zn \longrightarrow Zn^{2+} + 2e^- \qquad Cu^{2+} + 2e^- \longrightarrow Cu$$

**FIGURE 10.7 The Electrochemical Cell.** Measuring redox changes is possible by physically isolating Zn oxidation from $Cu^{2+}$ reduction as shown. The cells are joined electrically through metal rods and a wire, and in solutions by a salt bridge, represented by a tube. The complete circuit functions by electrons carrying the current through the wire and rods, and by ions carrying the current through the solutions and salt bridge.

## TABLE 10.1 Reduction Potentials for Mitochondrial Electron Transport Chain Mobile Cofactors

| Half-Reaction | $\epsilon°$ (mV) |
|---|---|
| NADH → NAD$^+$ | −0.32 |
| Succinate → Fumarate | +0.03 |
| QH$_2$ → Q | +0.04 |
| Cytochrome $c$ (red.) → Cytochrome $c$ (ox.) | +0.23 |
| O$_2$ → ½ H$_2$O | +0.82 |

potential. The faraday constant is a conversion factor between chemical and electrical units. A positive reduction potential corresponds to a negative free energy change. The $\Delta G°$ and $\Delta \epsilon°$ values for complexes I–IV are calculated assuming a two-electron transfer ($n = 2$) and listed in Table 10.2.

Experimentally, the free energy change ($\Delta G$) of ATP synthesis has a value of approximately 44 kJ/mol. If we compare this with the $\Delta G°$ values in Table 10.2, we would conclude that complexes I and IV have a free energy value that is sufficient to drive ATP synthesis, but complex III is not. This is a classical interpretation, but an incorrect one. The three complexes are not physical sites of ATP formation but merely separate contributors to the proton gradient. The proper analysis is to sum all three free energy changes. This would suggest that the electron transport chain produces more than enough energy for the synthesis of 3 ATP molecules, or 2 if we start from the level of Q (from complex II). The exact numbers are still in dispute, because it depends on the ratios of electrons flowing per protons produced. Most investigators have calculated that fewer than three ATP molecules are actually synthesized from NADH. For our treatment, we will simply assume that three ATPs are made for each NADH, and two are made for each QH$_2$ oxidation, with the understanding that these are estimates.

The key issue is not the precision in the numbers of ATP molecules synthesized. Rather, it is how electron movement through the respiratory chain creates a proton gradient. We will thus examine features of each respiratory complex, and subsequently take up the issue of how the protons move back into the mitochondrial matrix to drive the formation of ATP.

## 10.6 Mechanisms for Electron and Proton Flows Through the Mitochondrial Membrane

We began our study of mitochondrial energy production with the overall result: a proton gradient is produced across the inner mitochondrial membrane and re-entry of protons drives the synthesis of ATP. The first part of this pathway—the production of the proton gradient—involves four of the five complexes of this membrane, along with the mobile cofactors that connect them. In this section, we will see how the flow of both electrons and protons together leads to the creation of a proton gradient. We already have considered a general picture of these flows (Figure 10.6). We now examine these events in some detail.

### Complex I: Proton Pump

The first complex is also known as NADH-Q reductase. It accepts electrons and protons in the form of a hydride ion from NADH, donates electrons to the mobile carrier Q, and pumps protons across the membrane in the process. Mammalian NADH-Q reductase is composed of

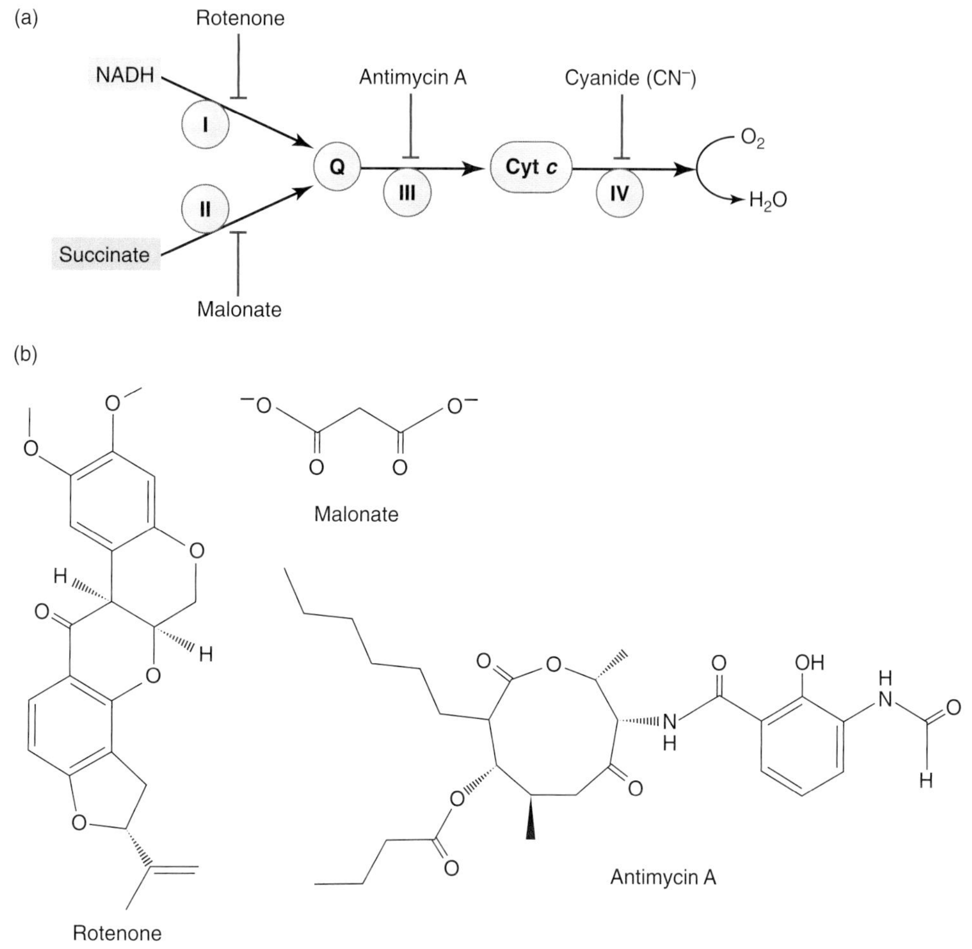

FIGURE 10.8 **Sequence of Electron Transport.** (a) The order of electron movement was experimentally determined using inhibitors of each complex, which are illustrated. (b) Structures of rotenone, malonate, and antimycin A.

## BOX 10.2: A New Class of Complex II Inhibitors

Complex II is distinct from the other three mitochondrial complexes because it does *not* contribute to the proton gradient, and because it does *not* have selective inhibitors that block it. Malonate is a classic inhibitor that was used to help elucidate the Krebs cycle. However, it is less potent and selective than the inhibitors of the other complexes. Recently, a series of compounds was reported that are structural analogs of ubiquinone, called Atpenins (e.g., atpenin A5, FIGURE B10.2). Atpenins, originally isolated from a fungus (genus *penicillium*), ironically have antifungal activity. Compared to malonate, atpenins are over 1000-fold more potent inhibitors of complex II.

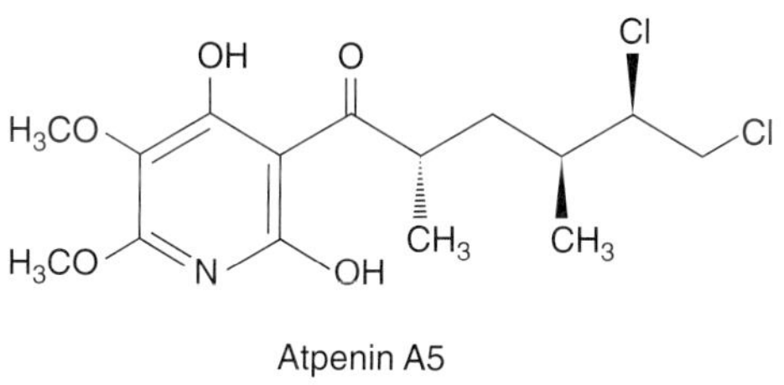

Atpenin A5

FIGURE B10.2 **Structure of Atpenin A5.**

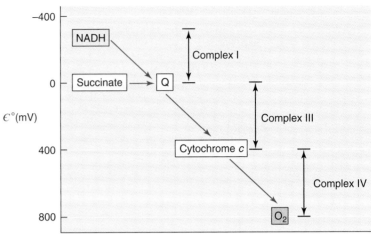

**FIGURE 10.9 Redox Sequence of Electron Flow.** The progress of electron transport is plotted against the standard reduction potential. Three changes in reduction potential, for complexes I, III, and IV, are large and contribute to energy capture. Free energy is directly proportional to changes in redox potential.

**TABLE 10.2 Reduction Potentials and Free Energies of Complexes I to IV**

| Complex | ΔG° (kJ/mol) | Δε° (mV) |
|---|---|---|
| I | −70 | 0.36 |
| II | ~ 0 | 0.01 |
| III | −37 | 0.19 |
| IV | −116 | 0.60 |

a large number of subunits—about 45 protein chains—and contains the bound flavin FMN, in addition to several Fe–S proteins. Structural studies so far have provided only an outline of how complex I functions, but its operation is well understood.

Complex I can be divided into three general portions (**FIGURE 10.10**): the N-module, the input domain that directly reacts with NADH; the P-module, through which protons are transported; and the Q-module,

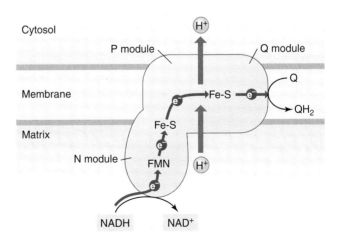

**FIGURE 10.10 Complex I: NADH-Q Reductase.** There are three modules to this complex, called N, P, and Q. The N-module is the site of electron entry, where NADH reduces the flavin FMN. The P-module is where proton pumping occurs. The Q module is where electrons, along with protons from the matrix, attach to Q. Most of the electron transport is sequentially passed to several Fe–S proteins, some of which are indicated.

which reacts with Q. The N-module contains the bound FMN along with several Fe–S proteins, so that electron flow is from NADH to FMN to Fe–S. Here, the transducer role of the flavin is apparent, as FMN accepts a pair of electrons in the form of a hydride, but donates them one at a time to an Fe–S cluster. Protons take a separate route, passed to side chains of the protein subunits until they are finally released through the P-module.

The nature of "proton pumping" can be visualized as a conformational change that situates a proton on the matrix face first and then the cytosolic side. FIGURE 10.11 shows that the process is analogous to the action of the glucose transporter. Prior to passage of high-energy electrons through the complexes, a proton binding site faces the matrix. After the conformational change, the site faces the cytosol. Reaction at the Q-module takes electrons from the Fe–S clusters and protons from the matrix space to produce $QH_2$, the reduced form. Note that both Q and $QH_2$ are mobile entirely within the hydrophobic portion of the membrane.

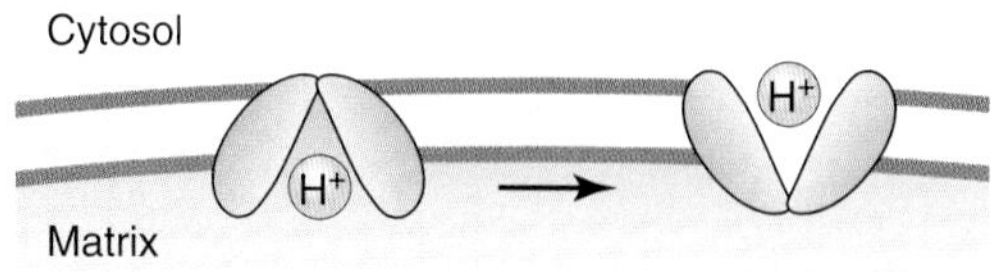

FIGURE 10.11 **Proton Pumping Analogous to Glucose Transport.** Like the glucose transporters, proton pumping can be viewed as binding to a protein that faces one water space or the other in a membrane. In electron transport, the energy driving the conformational change is from electron transport through the protein.

## Complex II: Succinate Dehydrogenase

Complex II is synonymous with succinate dehydrogenase, also a step in the Krebs cycle. Thus, this reaction sequence plays a dual role in central metabolic pathways. The sequence of reactions is analogous to complex I: a hydride is transferred to a flavin cofactor, and the electrons are then passed to a series of Fe–S complexes and finally to Q. In the case of succinate dehydrogenase, the hydride comes from succinate, the flavin is FAD, and no protons are translocated in the overall process.

Complexes I and II can be viewed as converging on $QH_2$ formation, as do other complexes within the mitochondrial membrane. These include glycerol phosphate oxidase, which we will examine at the end of this chapter, and electron transferring flavoprotein, an enzyme of fatty acid oxidation. All of these reaction sequences produce $QH_2$, which is oxidized exclusively by complex III.

## Complex III: Loop Mechanism

Complex III is also known as Q-cytochrome $c$ reductase, because it takes electrons from $QH_2$ and reduces cytochrome $c$. Unlike the other complexes of the respiratory chain, electron flow is partially cyclic in complex III. Understanding this mechanism requires that we first examine the reduction states and the movement of Q within the membrane.

Q undergoes exclusively single-electron transfers, so the reduction of Q to the fully reduced $QH_2$ takes place in two steps, as shown in FIGURE 10.12. The intermediate is similar to the flavin single-electron intermediate in that it is resonance stabilized. However, it is distinct in that it contains both a free radical and a negatively charged ion, called a **radical anion**. The molecule is mostly hydrophobic, but the negative charge is fairly localized; thus, $Q^{\bullet-}$ is an amphiphile. The negative charge locks $Q^{\bullet-}$ at either the matrix membrane face or the

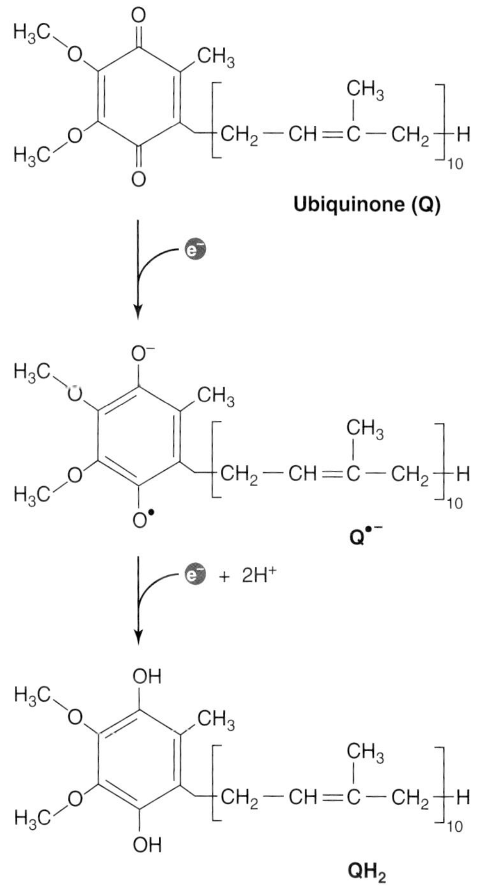

FIGURE 10.12 **Forms of Ubiquinone.** The three redox states of ubiquinone are shown. The cofactor accepts or donates electrons strictly one at a time, but the intermediate state has both a negative charge and a free radical (i.e., a radical ion, $Q^{\bullet-}$). Both protons add to $Q^{\bullet-}$ to produce $QH_2$.

cytosolic membrane face. In both situations, the molecule is anchored to the membrane's water-phase edge. The other forms of the molecule, Q and $QH_2$, are entirely hydrophobic and mobile within the plane of the membrane. Hence, the interconversion of forms turns an anchored molecule into a mobile one over the course of the reaction.

Electron flow diverges in complex III. The electrons that arrive at the Fe–S cluster after step (B) in FIGURE 10.13 are donated to cytochrome $c_1$ and they leave the complex when they are subsequently transferred to cytochrome $c$, the mobile cofactor that connects complex III with complex IV. Other electrons move through a cyclic path, from the cytosolic face at step (B) through the membrane via cytochromes $b_{566}$ and $b_{562}$.

Proton flow is carried by the mobile cofactor $QH_2$, reacting at the matrix face in (F) of Figure 10.13 and released at the cytosolic face in (B).

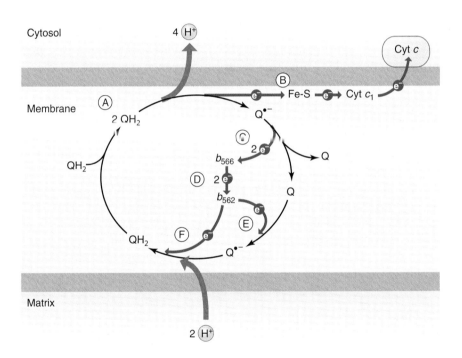

FIGURE 10.13 **The Q cycle.** This is actually a bi-cycle, because forms of Q and electrons cycle through the membrane. First, $QH_2$ (A) dissociates, releasing protons to the cytosol and electrons to an Fe–S protein (B) (and subsequently cytochromes $c_1$ and $c$), leaving $Q^{\cdot-}$ at the cytosolic membrane leaflet. Next (C), electrons from the radical ion are passed to cytochrome $b_{566}$, producing Q. The electrons are then passed to cytochrome $b_{562}$ (D). In one step, Q can accept electrons from cytochrome $b_{562}$ to form $Q^{\cdot-}$ at the matrix membrane leaflet (E). Further reduction requires electrons from another round of $QH_2$ oxidation, providing another electron from cytochrome $b_{562}$, which reduces the matrix-bound $Q^{\cdot-}$ to $QH_2$ (F). The Q from this second step contributes to the free Q pool shown to the far right of the figure. Overall, half of the electrons cycle through this complex, and the other half move through it from $QH_2$ to cytochrome $c$.

CHAPTER 10 OXIDATIVE PHOSPHORYLATION

These movements are keys to the overall mechanism of electron and proton movement through complex III, known as the **Q cycle**. The Q cycle has a certain analogy to the Krebs cycle in that more than one turn of the cycle is necessary to completely balance the flows. The overall reaction for protons only is:

$$QH_2 + 2H^+_{matrix} \rightarrow Q + 4H^+_{cyto} \tag{10.5}$$

The overall reaction for electrons only is:

$$QH_2 + 2 \cdot \text{cytochrome } c_{ox} \rightarrow Q + 2 \cdot \text{cytochrome } c_{red} \tag{10.6}$$

The scheme, illustrated in Figure 10.13, is known as a **loop mechanism**. If we start with $QH_2$ (A), note that a dissociation of protons occurs at the cytosolic face, with separate routes for the products (B). While the protons enter the cytosol, electrons are donated to an Fe–S cluster, leaving the semiquinone $Q^{\cdot-}$ with its hydrophilic charged portion locking it onto the cytosolic face. Loss of the second electron (C) from this pool of $Q^{\cdot-}$ produces fully oxidized Q, which is now mobile in the membrane interior, and electrons, which are accepted by cytochrome $b_{566}$. Passage of electrons through the membrane is achieved by transferring them to cytochrome $b_{562}$ located within the membrane (D). Q can accept an electron from cytochrome $b_{562}$ only at the matrix face (E), producing the second pool of $Q^{\cdot-}$ that has its charged portion locked onto the matrix face. Another electron from a second cytochrome $b_{562}$ (F), along with protons from the matrix, produce $QH_2$ once again. $QH_2$ is mobile in the hydrophobic membrane phase, and completes the cycle by diffusing back to the cytosolic face.

When it was first proposed, the Q cycle was highly controversial, but it is now the clearest example of how protons can move across the membrane and electrons can move through it. One convincing piece of evidence in favor of this cycle was the observation that two pools of $Q^{\cdot-}$ are present in the mitochondria. These data were gathered using **electron paramagnetic resonance**, a technique that identifies and measures free radicals. Two different signals are observed for the same chemical structure due to the very different electrical environments: one positively charged (the cytosolic side) and one negatively charged (the matrix side).

## Complex IV: Pump and Annihilation

The final complex of electron transport is also known as cytochrome $c$ oxidase. Like complex I, the pumping of protons is tied to conformational changes in protein subunits as high-energy electrons are passed through the complex. FIGURE 10.14 shows the path of electron flow from cytochrome $c$. Electrons are donated first to $Cu_A$, a **dinuclear copper center**, which contains two copper ions joined by sulfur atoms (FIGURE 10.15a), a structure analogous to the iron–sulfur complexes. The path of electron flow continues through cytochrome $a$, then cytochrome $a_3$, which is associated with another copper ion, $Cu_B$, in which the ion is chelated to three histidine groups (Figure 10.15b). Overall, four electrons flow through the complex and two water molecules are

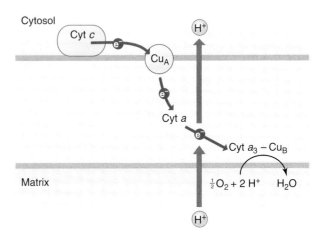

**FIGURE 10.14 Complex IV: Cytochrome *c* Oxidase.** Electrons from cytochrome *c* are donated to $Cu_A$, a cofactor pictured in Figure 10.15. Subsequently, electrons are passed to cytochrome *a*. As electrons move next to cytochrome $a_3$, a conformational change drives proton pumping, in a way similar to complex I. Electrons are passed to the cofactor $Cu_B$, a structure also shown in Figure 10.15, and lastly to water. Protons consumed in water formation also contribute to the proton gradient, a mechanism called annihilation.

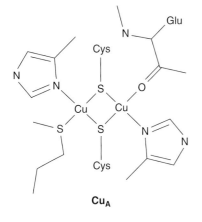

**FIGURE 10.15 Copper Centers.** Structures of copper centers participating in electron transfer in complex IV.

formed from $O_2$. The water synthesis step—the terminal stage of electron transport in the chain—involves oxygen bound alternatively between $Cu_B$ and cytochrome $a_3$ in an interior portion of the protein complex. Note that the $Cu_B$ ions as well as cytochrome $a_3$ have dual roles: binding oxygen (like the non-redox-active hemoglobin) and undergoing redox state transitions, as electrons are moved through complex IV.

Complex IV has two mechanisms for creating the proton gradient. One of these is a pump mechanism, similar to that involved in complex I. As electrons move down their energy gradient between cytochromes *a* and $a_3$, they induce a conformational change in proteins, pumping protons as in Figure 10.11. The other mechanism is an **annihilation** of protons specifically from the matrix side. These protons go into the synthesis of water. Because a loss of protons from the matrix is equivalent to a gain of protons to the cytosol, this further contributes to the proton gradient.

## Complex V: ATP Synthesis

The movement of protons in the direction of cytosol to matrix is catalyzed by complex V, which includes a passageway for protons and an active site for the synthesis of ATP from ADP and $P_i$. The complex is also known as the $F_1F_o$ ATP synthase. This stems from the fact that one portion of the complex—called $F_1$ (factor one)—binds adenine nucleotides and is responsible for ATP synthesis. The other—called $F_o$ (oligomycin sensitive factor)—conducts protons through the membrane.

Early micrographs of the complex showed that $F_o$ is anchored to the membrane, and $F_1$ seemed to stick out into the matrix like a lollypop.

The determination of the mechanism for complex V was, in large part, the work of Paul Boyer. A series of enigmatic findings had to be resolved. First, the binding of ATP to the complex was found to display **negative cooperativity**. Thus, for the three ATP binding sites, the first has a dissociation constant of less than 1 nM; the second, 1 µM; and the third, 30 µM. In contrast to binding, enzymatic activity exhibited just the *opposite* behavior—namely, as more ATP molecules were bound, the activity increased. This situation could be explained if the limiting step in the overall reaction was the *release* of the nucleotide. Somehow, the enzyme complex would have to undergo a conformational change that would release ATP, and this had to be related to the entry of protons.

Boyer proposed that a protein segment of the complex might move from one of the three ATP binding sites to the other, in concert with the proton entry in order to effect the conformational change that elicits ATP release. This mechanism was called **binding change**. Further development of this mechanism used an analogy from the operation of a motor (FIGURE 10.16). The two parts are the moving **rotor** and the stationary **stator**. In this depiction, the movement is effected entirely by magnetism; repulsion of like magnetic poles causes the rotor to spin with respect to the stator. Real motors use electromagnetic induction, but the principle is the same.

The detailed subunit structure of complex V is shown in FIGURE 10.17a. Structures that correspond to the stator are shaded in red, while those corresponding to the rotor are shaded in green.

### The stator

Three αβ dimers are arranged as a cylinder, attached to a stalk made of *b* subunits. This is attached, in turn, to a membrane-embedded *a* subunit. The cylinder of αβ subunits comprises most of the knob structure of the previously mentioned $F_1$ that appears in electron micrographs. Adenine nucleotides bind to the β subunit of the αβ dimers.

### The rotor

A ring of membrane-bound γ subunits is attached to a stalk containing ε and *c* subunits. The γ subunit can contact αβ dimers sequentially, which changes the conformation of the β subunits. This conformational change is key to the reaction as indicated in Figure 10.17b. Each of the three stages of ATP synthesis is represented by one of the three states of the β subunit: the binding of ADP and Pi to the subunit; the formation of the bond to synthesize ATP, and the release of ATP. The moving rotor supplies the conformational change to drive each of these states in turn.

The entry of protons through the membrane occurs at the interface between the stator and rotor, specifically at the junction between the α and *c* subunits, as shown in Figure 10.17. This is the cause of rotation of the γ-subunit that moves among the three αβ dimers. In this way, the entry of protons is indirectly coupled to ATP synthesis.

Complex V is often studied in the reverse direction. In this case, any added ATP is split into ADP and $P_i$, and drives a proton gradient in reverse. It also causes a spinning of the rotor but only in the reverse direction.

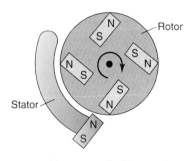

FIGURE 10.16 **Essence of a Motor.** Two essential components of a motor are the stator and the rotor. In this simplified diagram, repulsion of fixed magnets is the driving force for rotation. In a real motor, electromagnetic induction is used to drive the rotor.

(a)

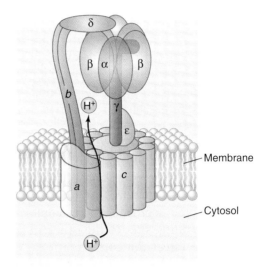

(b) Top view    (c) Side view

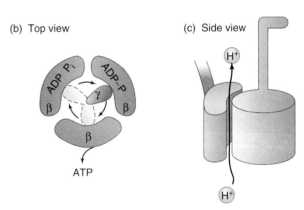

**FIGURE 10.17 Complex V: ATP Synthase as a Molecular Motor.** The component structure of the ATP synthase (a) is colored to correspond to the motor of Figure 10.16, where the stator is in red and the rotor in green. Various protein subunits are labeled with both Greek (exterior portions) and Roman letters (largely membrane-embedded portions) by convention. The rotor tip is hidden by the three pairs of $\alpha\beta$ proteins of the stator. The diagram is opaque so that it is partly visible. A top view of the edge of the rotor ($\gamma$) rotating and interacting intermittently with the ATP-binding portion of the stator ($\beta$) shows the action of the enzyme; as each interacts, ATP dissociates, which is the rate-limiting step of the overall reaction. In the side view (c), it is apparent that stator and rotor closely interact in the membrane segment, and as the protons pass between these, it drives the rotor.

# 10.7 Mitochondrial Membrane Transport

Several other inner membrane proteins are essential for the operation of the mitochondria. Many are transporters that provide the molecules necessary for communication between the cytosol and mitochondria. We begin with one that is needed to bring ATP synthesized in the matrix out to the cytosol.

## Adenine Nucleotide Translocase

As most of the ATP synthesized by mitochondria is needed in the cytosol, it must be exported. The adenine nucleotide translocase takes fully charged $ATP^{4-}$ from the matrix side and $ADP^{3-}$ from the cytosolic side (FIGURE 10.18a). The result is a net movement of negative charge out of the mitochondria, meaning that a portion of the membrane potential is used to drive the transport in one direction, making the translocase metabolically irreversible. This is the only metabolic step involving ATP that does *not* involve a chelate with $Mg^{2+}$. Because the ATP/ADP exchange causes an alteration in charge, it is an example of an **electrogenic** transporter.

## Phosphate Exchange

Inorganic phosphate can use different exchange proteins in the inner mitochondrial membrane. Two are shown in Figure 10.18b and c. The $P_i^-/H^+$ cotransporter directly draws on the pH portion of the

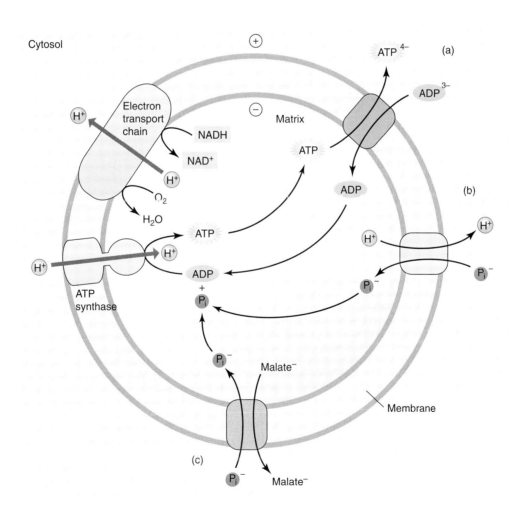

FIGURE 10.18 **Mitochondrial Transporters.** Three transport systems of the mitochondria are illustrated: (a) ATP/ADP exchange is electrogenic, driven by the membrane potential; (b) $P_i/H^+$ exchange is driven by the concentration differences in $H^+$ ion (pH); and (c) malate/$P_i$ exchange is driven exclusively by changes in concentrations of the components.

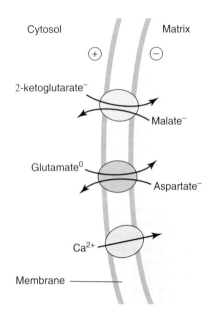

Cytosol      Matrix

2-ketoglutarate⁻

Malate⁻

Glutamate⁰

Aspartate⁻

Ca²⁺

Membrane

**FIGURE 10.19 Further Mitochondrial Transporters.** Three more transporters within the mitochondrial inner membrane are illustrated. 2-Ketoglutarate/malate exchange responds only to concentration changes. Glutamate/aspartate exchange, on the other hand, is electrogenic, driven by membrane potential. Ca²⁺ uptake is a uniport, which is also electrogenic and driven by the membrane potential.

mitochondrial proton motive force when $P_i$ is taken up by mitochondria. This exchanger, however, is near-equilibrium and can operate in the opposite direction. Similarly, there is a near-equilibrium exchange of $P_i$ for malate. The compounds are both exchanged in the $-1$ form. For both $P_i$ exchangers, there is no change in membrane potential (i.e., they are electroneutral, not electrogenic). This is the most common type of exchange across the inner membrane.

### Other Transport Proteins

Other mitochondrial membrane transporters can also be classified as either electroneutral or electrogenic. For example, the malate/$P_i$ exchanger (Figure 10.18c) and the 2-ketoglutarate/malate exchanger (FIGURE 10.19) are electroneutral, near-equilibrium transport reactions. The glutamate/aspartate exchanger and the Ca²⁺ uptake uniport (Figure 10.19) are both electrogenic, metabolically irreversible steps. Like the adenine nucleotide transporters, these draw upon the membrane potential to drive glutamate into the matrix and aspartate out to the cytosol, making the matrix more positive. Similarly, Ca²⁺ uptake into the matrix is electrogenic, bringing with it two uncompensated charges. While these steps cannot be reversed, they are not likely to be regulatory steps either, because they respond only to concentration changes.

## 10.8 Coupling of Oxidation and Phosphorylation

The idea of coupling oxidation with phosphorylation predates the chemiosmotic hypothesis that explains how mitochondria make ATP. The early experiments were direct and compelling: when mitochondrion was mixed with substrates that could supply NADH and $P_i$, the addition of ADP caused a large increase in oxygen consumption from the media. This can be explained on the basis of the proton gradient as the energy intermediate, as shown in Figure 10.18.

ADP enters via the adenine nucleotide exchanger, and $P_i$ enters through its transporter, so substrates are available for the $F_1F_o$ATP synthase. With adenine nucleotides bound, H⁺ enters and drives the rotor to release the synthesized ATP. The entry of protons draws down on the proton gradient, which causes more protons to be exported and electrons to flow through the respiratory chain. This process is called **respiration**. As we have already seen, the movements of electrons and protons through the respiratory chain are entirely interlinked. The coupling of oxidation with phosphorylation, on the other hand, is indirect, through the proton gradient. This is dramatically illustrated by what happens when this coupling link is broken—a situation called **uncoupling**.

# 10.9 Uncoupling

The link between oxidation and phosphorylation can be broken in a way that allows respiration to continue without the synthesis of ATP. Historically, compounds that caused this phenomenon of uncoupling were difficult to categorize, because there were many of them and they varied in structure. Because earlier hypotheses about mitochondrial energy generation centered on a specific chemical intermediate, it was hard to see how a large array of different compounds could selectively interfere with the hypothetical intermediate.

The mystery was resolved by the chemiosmotic hypothesis. If the intermediate is a proton gradient, and protons normally enter mitochondria via the $F_1F_o$ ATPase, then a proton **ionophore** would provide an alternative and easier path for protons. An ionophore is a molecule that allows a charged species to cross a membrane. A mechanism for the uncoupling of a weak acid ionophore is diagrammed in FIGURE 10.20. In its protonated form, HA crosses the membrane and dissociates in the matrix into $H^+$ and $A^-$. The $A^-$ returns to the exterior, where it can combine with a proton that has been driven out of the mitochondria by the electron transport chain, form HA, and complete the cycle by entering the matrix again. The ability of the $A^-$ ion to cross the membrane is facilitated by two features: hydrophobicity and resonance delocalization of the negative charge. The fact that the charge is negative will allow it to be driven across the membrane by the established membrane potential. One common molecule that satisfies these criteria is dinitrophenol (Figure 10.20b). It is also possible to have uncouplers that are weak bases (conjugate acids), so that they enter the matrix in their positive form when protonated and leave uncharged. In fact, currently popular uncouplers used experimentally are of exactly this type, such as the compound abbreviated CCCP (Figure 10.20c).

A large number of drugs are hydrophobic, because this helps them enter cells for therapeutic purposes. Many of these are also weak acids so they may act as uncouplers. It is standard practice to test potential drugs for their potency as uncouplers as part of a toxicology screen. The classical uncoupler dinitrophenol was marketed in the 1930s as a diet drug before its full toxic qualities were realized (**BOX 10.3**).

# 10.10 Superoxide Formation by Mitochondria

Oxygen can abstract electrons from intermediate sites in the electron transport chain to produce superoxide ($O_2{}^{\cdot-}$), an oxygen molecule with an extra electron. Superoxide is extremely reactive, and can destroy a number of biological molecules.

For many years, it was believed that oxidative damage was a critical cellular hazard, and that it was associated with premature aging. However, extensive studies performed in 2009 found no correlation between oxidative stress and longevity across a number of species. This is likely due, in large part, to cellular systems that inactivate superoxide

(a)

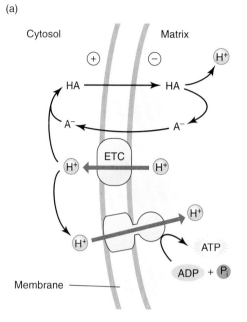

(b)

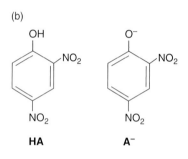

(c)

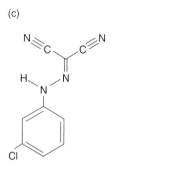

FIGURE 10.20 **Uncoupling.** The general process is illustrated in (A). The proton produced by the respiratory chain reacts with an anion symbolized by $A^-$ to form the weak acid, HA. HA crosses the membrane without a carrier, delivering the proton back to the matrix after dissociation. The anion, $A^-$, also can cross the membrane, driven by the electrical gradient (i.e., the membrane potential). The structures of the classical inhibitor DNP (dinitrophenol) in HA and $A^-$ forms are shown in (B). CCCP (carbonyl cyanide *m*-chlorophenyl hydrazone), a more modern inhibitor that is commonly used experimentally, is shown in (C).

## BOX 10.3: The Once and Future Uncoupling Diet

During World War I, dinitrophenol (DNP) was used as an explosive (FIGURE B10.3). This action, like the structurally similar trinitrotoluene (TNT), is the result of a rapid release of its components into gases; in particular the nitro groups are particularly unstable. A group of French munitions workers exposed to DNP were found to experience weight loss. A pharmacologist studied low doses of DNP as a diet drug in the 1930s, and then marketed it as "Formula 281" (as in "too weighty one"). The product was extremely popular. However, even slight excesses of DNP led to cataracts, blindness, and death. There can be little doubt about its effectiveness: by allowing respiration and attenuating ATP formation, uncouplers such as DNP permit overeating without weight gain. Despite the clear toxicology of uncouplers, DNP has emerged as a diet pill once again thanks to the Internet and counterculture viewpoints.

There is, in fact, a possible future for the uncoupler diet. Brown adipose tissue (or just "brown fat") has long been known to exist in animals that hibernate, and in small amounts in newborn humans. This fat is brown due to the abundance of mitochondria. These mitochondria contain a protein called UCP-1, or uncoupling protein-1. Under the appropriate stimulus (involving epinephrine induction of UCP-1 protein), a controlled increase in uncoupling occurs. This generates heat, which can arouse animals from hibernation. Its function in newborn humans is unknown. However, in 2009 it was discovered that adult humans also have brown fat. Thus, there is the possibility that a genetic manipulation of brown fat may become a willpower-free diet. Until then, we are left with the only diet that actually works: eat less; exercise more.

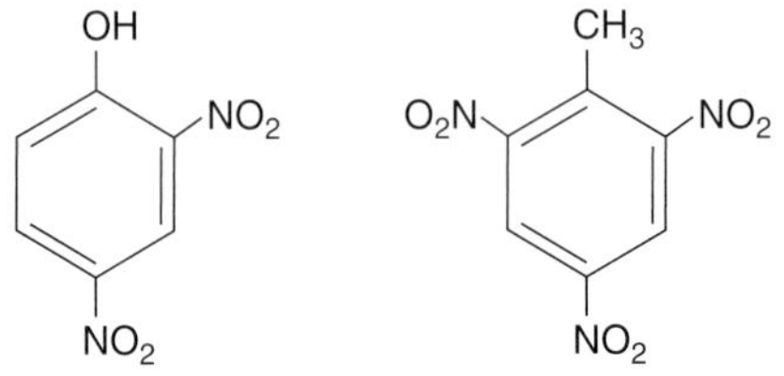

FIGURE B10.3 **Structures of DNP (Dinitrophenol) and TNT (Trinitrotoluene).**

and related species. First, superoxide dismutase catalyzes the formation of the more stable hydrogen peroxide ($H_2O_2$). Subsequently, peroxidase catalyzes the conversion of hydrogen peroxide to water. Similar inactivation enzymes also exist within the mitochondrial matrix space.

There are two sites in the respiratory chain where superoxide can be formed, both of which are inferred from the presence of cytosolic and mitochondrial detoxification systems. Matrix superoxide can be formed at complex I; however, like the mechanism of complex I itself, details of this process remain unclear. Cytosolic superoxide is formed from the reaction of the cytosolic-facing radical ion of Q via the reaction:

$$Q^{\bullet -} + O_2 \rightarrow Q + O_2^{\bullet -} \tag{10.7}$$

CHAPTER 10 OXIDATIVE PHOSPHORYLATION

In both cases, electrons drawn into superoxide formation result in fewer electrons flowing through the electron transport chain. It turns out, however, that only a small percentage of electrons end up in superoxide.

# 10.11 Control of Mitochondria

The primary function of mitochondria is to produce ATP for use by cytosolic reactions. Most of that ATP is converted to ADP. It is fitting, then, that the concentration of ADP is the principal means of controlling further ATP formation by mitochondria. ADP limits the production of ATP, which in turn controls the number of protons that enter the matrix. As a consequence, this also determines the number of protons that exit the matrix and, consequently the rate of electron flow. This level of control is a fundamental one, establishing a connection between energy demand, reflected in ADP concentration, and mitochondrial energy supply, reflected in ATP formation.

It should be noted that the *total* ADP content of cells greatly exceeds the *free* solution concentration that regulates mitochondrial energy production. This is because most ADP exists bound to cytosolic proteins. A number of techniques have estimated free ADP concentration to be about 25 µM, whereas the total cytosolic concentration is an order of magnitude greater.

In one sense, the adenine nucleotide translocase, which exchanges cytosolic ADP for mitochondrial matrix ADP, also can limit oxidative phosphorylation, because this reaction is metabolically irreversible. However, the exchanger activity is linked to membrane potential, so it is not related to cellular energy utilization. Rather, it is attenuated by other changes that depress membrane potential.

Another transporter that utilizes membrane potential is the mitochondrial $Ca^{2+}$ uniporter (introduced in Section 10.7, Other Transport Proteins). Its activity largely depends upon changes in cytosolic $Ca^{2+}$ released by the endoplasmic reticulum in response to external cellular regulation, such as electrical depolarization or hormonal activation. In turn, increased mitochondrial $Ca^{2+}$ activates the Krebs cycle and pyruvate oxidation. While very high concentrations of $Ca^{2+}$ could decrease the membrane potential enough to limit ADP uptake, this is a pathological rather than a physiological condition. This is due to cytosolic ADP concentration being more than an order of magnitude greater than cytosolic $Ca^{2+}$ concentration.

# 10.12 How Mitochondria Can Utilize Cytosolic NADH

There is no mitochondrial transporter that directly ports NADH (or any nicotinamide nucleotide) directly into the matrix. Yet, the conversion of glucose to pyruvate generates cytosolic NADH. We now examine two pathways (commonly called **shuttles**) that *indirectly* oxidize NADH by transferring just its electrons, or **reducing equivalents**.

## Glycerol Phosphate Shuttle

The simpler of the shuttles illustrates the general pattern: two redox reactions utilize the same pathway intermediates in two metabolic spaces, except in opposite directions. For the **glycerol phosphate shuttle**, the first of these is the cytosolic glycerol phosphate dehydrogenase. This catalyzes the near-equilibrium reaction:

$$\text{Dihydroxyacetone-P} + \text{NADH} \rightleftarrows \text{Glycerol-P} + \text{NAD}^+ \quad (10.8)$$

The second reaction, which is metabolically irreversible, is catalyzed by glycerol phosphate oxidase, an enzyme complex located in the inner membrane:

$$\text{Glycerol-P} + Q \rightarrow \text{Dihydroxyacetone-P} + QH_2 \quad (10.9)$$

The reactive site for glycerol-P is on the cytosolic face of the membrane.

Because both of the reactions operate at the same time, it becomes a short cyclic pathway, which serves to move reducing equivalents from cytosolic NADH to the membrane resident mobile carrier, $QH_2$. The sequence is illustrated in FIGURE 10.21, which also shows the key components of the glycerol phosphate oxidase complex. The latter is strikingly similar to succinate dehydrogenase. Like complex II, glycerol phosphate oxidase does not contribute to the proton gradient.

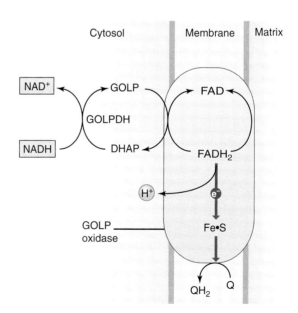

**FIGURE 10.21 The Glycerol Phosphate Shuttle.** Reducing equivalents from NADH are brought into the mitochondria by the operation of two enzymes. The cytosolic GOLPDH (glycerol phosphate dehydrogenase) reduces DHAP (dihydroxyacetone phosphate), forming GOLP. Subsequently, the GOLP reacts with the GOLP oxidase complex in the mitochondrial membrane at the cytosolic face, and electron transport ensues until electrons are ultimately captured in $QH_2$.

**CHAPTER 10** OXIDATIVE PHOSPHORYLATION

In essence, the oxidation of cytosolic NADH through the glycerol phosphate shuttle bypasses complex I and the protons it pumps. Only a portion (about ⅔) of the energy of NADH can be extracted as a result. A second shuttle, called the malate/aspartate shuttle, while more elaborate, provides a greater energy yield.

## Malate/Aspartate Shuttle

The **malate/aspartate shuttle** shares the same features we have just noted in the glycerol phosphate shuttle for importing reducing equivalents: an overall cyclic pathway with at least one metabolically irreversible step to drive the flow in just one direction, and duplicate reactions in separate metabolic spaces, running in opposite directions.

The malate/aspartate shuttle uses two sets of enzymes rather than one, because the intermediate oxaloacetate, like NADH, has no transporter in the mitochondrial inner membrane. The pathway is illustrated in FIGURE 10.22. Note that it involves two isozymes of malate DH: the cytosolic enzyme, which oxidizes NADH, and the mitochondrial enzyme, which reduces NAD+. This brings NADH into the mitochondria indirectly; thus, the malate is imported in exchange with 2-ketoglutarate (as in Section 10.7, Other Transport Proteins). Because oxaloacetate itself cannot be transported out of the mitochondria, it is converted to aspartate through the aspartate aminotransferase reaction:

$$\text{Oxaloacetate} + \text{Glutamate} \rightleftarrows \text{Aspartate} + \text{2-Ketoglutarate} \quad (10.10)$$

Aspartate can exit the mitochondria in exchange for glutamate. As discussed in Section 10.7, Other Transport Proteins, this step is electrogenic and hence metabolically irreversible. In fact, it is the only step of the entire shuttle that *is* metabolically irreversible, and it

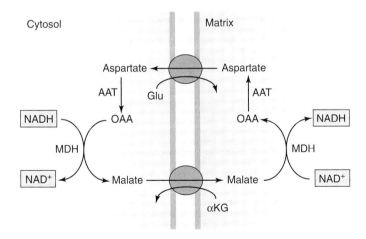

FIGURE 10.22 **The Malate–Aspartate Shuttle.** Reducing equivalents from NADH are stored as malate through the malate dehydrogenase (MDH) reaction. Once inside the mitochondria, malate reacts with a separate malate dehydrogenase, producing mitochondrial NADH. Two aspartate aminotransferase enzymes (AAT) are needed to recycle the oxaloacetate because oxaloacetate has no mitochondrial transporter.

accounts for the unidirectionality of the shuttle. A cytosolic isozyme of aspartate aminotransferase catalyzes the reconversion of aspartate to oxaloacetate.

The malate/aspartate shuttle is more elaborate than the glycerol phosphate shuttle. For example, rather than just two reactions, the malate/aspartate shuttle requires six. The cycle is neatly balanced by the fact that glutamate and the 2-ketoglutarate needed at the transamination steps are appropriately transferred in the correct directions in the operation of the pathway. Despite the apparent complexity, this is by far the more common route for the delivery of reducing equivalents into the mitochondria.

We will see some of the same reactions of the malate/aspartate shuttle utilized again, as well as the concept of shuttles applied more generally, in gluconeogenic pathway and in lipid and nitrogen metabolism. However, before we study additional pathway connections, we will first discuss photosynthesis, a pathway that, like oxidative phosphorylation, extracts energy and can be understood on the basis of the same chemiosmotic hypothesis.

## Summary

Oxidative phosphorylation is the major pathway for the extraction of energy from food. There are two linked, vectorial flows: electrons from NADH to $O_2$, and protons from the mitochondrial matrix to the cytosol. These flows establish a proton gradient that leaves the mitochondrial matrix with a lower concentration of protons and a negative electrical charge. The return of protons to the matrix is coupled, in turn, to the synthesis of ATP, the major energy currency for the cell.

Five protein complexes are responsible for these activities, and they are known as complexes I through V (as well as having other more descriptive names). The first four are electron carriers and the fifth catalyzes ATP synthesis. Each of the complexes is embedded in the mitochondrial inner membrane and is physically isolated from the other four. Communication occurs exclusively through mobile cofactors. NADH, Q, and cytochrome $c$ connect complexes I through IV. This flow can be depicted as follows:

$$NADH \rightarrow [Complex\ I] \rightarrow Q \rightarrow [Complex\ III]$$
$$\rightarrow cyt.\ c \rightarrow [Complex\ IV] \rightarrow O_2$$

Additionally, electrons from succinate enter the pathway via:

$$succinate \rightarrow [Complex\ II] \rightarrow Q$$

and the flow continues as in the first sequence, bypassing complex I.

Protons are produced by a variety of mechanisms: a pump in complexes I and IV; a loop in complex III, and by annihilation (i.e., the removal of matrix protons by forming water) at complex IV. The electrical and chemical gradient of these protons is known as the proton motive force, which is the intermediate for energy generation. Complex V, also known as ATP synthase, conducts protons back into the mitochondrial matrix space and drives the synthesis of ATP from ADP and $P_i$ through an unusual mechanism of rotating a portion of the complex, one of the few examples of molecular motors.

The linking of substrate oxidation, ultimately to $H_2O$, with ATP formation is called coupling, and it is expressed in the very name *oxidative phosphorylation*. Some substances are uncouplers, that is, they allow oxidation to proceed in the absence of ATP formation. Uncouplers are hydrophobic, weak acids that provide an alternative to proton re-entry to the mitochondrial matrix. Beyond the protein complexes, there are membrane proteins of the inner mitochondria that allow selective exchange of a number of other substances, such as ADP, ATP, $P_i$, malate, aspartate, and glutamate. Molecules that are not themselves transported, such as NADH, can use indirect means—in some cases making use of these transporters—to allow the entry of the electrons from NADH (reducing equivalents) for use in mitochondrial oxidation.

The involvement of protons as the key energy intermediate is known as the chemiosmotic hypothesis. It accounts not only for mito-chondrial energy production but also for the capture of light energy in photosynthesis.

## Key Terms

annihilation
binding change
chemiosmotic hypothesis
electrogenic
electron paramagnetic
  resonance
electron transport chain

flavonoids
glycerol phosphate shuttle
ionophore
loop mechanism
malate/aspartate shuttle
negative cooperativity
Q cycle

radical anion
reducing equivalents
reduction potentials
respiration
shuttles
uncoupling
vectorial

## Review Questions

1. What is a proton motive force?
2. What are the complexes of the inner mitochondrial membrane?
3. Which step of the Krebs cycle is also part of the electron transport chain?
4. How are the respiratory complexes related to the respiratory chain?
5. Suppose you were presented with the following sequence of electron flow:

   NADH → Fe–S → FMN → b → Q

   Is this sequence possible? Explain.
6. What are the three ways that a proton gradient can be formed?
7. How does the loop mechanism work?
8. What is the role of flavin nucleotides in electron transport?
9. How is ATP synthesized?
10. How do uncouplers work?
12. Brown fat, rich in mitochondria, has recently been discovered in adult humans. This tissue expresses a protein that causes uncoupling. If the tissue can be increased, it would cause weight reduction. Why?
13. What is the mechanism for the regulation of ATP production by mitochondria?
14. Both the dissociated uncoupler ($A^-$) and the ubiquinone radical ion $Q^{\cdot-}$ are delo-calized molecules. Yet, $A^-$ crosses the membrane and $Q^{\cdot-}$ does not. Explain.

# References

1. Lanza, I. R.; Nair, K. S. Functional assessment of isolated mitochondria in vitro. *Methods Enzymol.* 457:349–372, 2009.

2. Nicholls, D. G.; Ferguson, S. J. *Bioenergetics;* Academic Press: New York, 2002.

3. Schafer, G.; Penefsky, H. S. *Bioenergetics. Energy Conservation and Conversion;* Springer-Verlag: Berlin, 2008.

4. von Ballmoos, C.; Wiedenmann, A.; Dimroth, P. Essentials for ATP synthesis by F1F0 ATP synthases. *Annu. Rev. Biochem.* 78:649–672, 2009.

5. Miyadera, H.; Shiomi, K.; Ui, H.; Yamaguchi, Y.; Masuma, R.; Tomoda, H.; Miyoshi, H.; Osanai, A.; Kita, K.; Omura, S. Atpenins, potent and specific inhibitors of mitochondrial complex II (succinate-ubiquinone oxidoreductase). *Proc. Natl. Acad. Sci. U.S.A.* 100:473–477, 2003.

6. Pérez, V. I.; Van Remmen, H.; Bokov, A.; Epstein, C. J.; Vijg, J.; Richardson, A. The overexpression of major antioxidant enzymes does not extend the lifespan of mice. *Aging Cell* 8:73–75, 2009.

7. Salmon, A. B.; Richardson, A.; Pérez, V. I. Update on the oxidative stress theory of aging: Does oxidative stress play a role in aging or healthy aging? *Free Radical Biol. Med.* 48:642–655, 2010.

8. Oxford English Dictionary (online edition) for definitions of *flavin* and *riboflavin.*

9. Allchin, D. To err and win a Nobel Prize: Paul Boyer, ATP synthase and the emergence of bioenergetics. *J. History Biology* 35:149–172, 1999.

   A nonspecialist view of oxidative phosphorylation, showing both the history and how errors made by Boyer led to his Nobel prize winning discovery of the mechanism of ATP synthase.

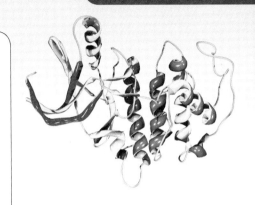

# 11

# Photosynthesis

## CHAPTER OUTLINE

Image © Dr. Mark J. Winter/Photo Researchers, Inc.

The **carbon cycle** is the global interchange of $CO_2$ between animal production and plant utilization. It is a topical issue: currently much more $CO_2$ is being produced than is being utilized, so this cycle is out of balance. This likely began with the increased burning of fossil fuels at the onset of the industrial revolution in the 19th century. At that time, it was postulated that an increase in atmospheric $CO_2$ might cause an increased global temperature. A steadily increasing global $CO_2$ concentration was experimentally confirmed in the middle of the 20th century.

Broadly speaking, organisms that consume organic compounds and produce $CO_2$ are **auxotropic**. $CO_2$ is converted back to organic compounds by **photosynthetic** organisms. Our focus will be on plants. A second cycle involves shuttling oxygen between its production by plants and its consumption by animals.

Photosynthesis has two natural divisions: the **light reactions**, which include oxygen production and photochemical events, and the **dark reactions** through which $CO_2$ is converted to organic compounds, such as sugars. While we discuss them separately, both can proceed simultaneously in plants.

# 11.1 Light and Dark Reactions

The overall reaction of photosynthesis is:

$$CO_2 + H_2O + \text{light} \rightarrow CH_2O + O_2 \tag{11.1}$$

The unique feature of Equation 11.1 is the incorporation of light energy to drive carbohydrate production (i.e., $CH_2O$) from $CO_2$. A slightly more detailed view of this process exposes the roles of the mobile cofactors ADP, $P_i$, ATP, $NADP^+$, and NADPH:

$$ADP + P_i + NADP^+ + H_2O \rightarrow O_2 + ATP + NADPH \tag{11.2}$$

$$CO_2 + ATP + NADPH \rightarrow CH_2O + ADP + P_i + NADP^+ \tag{11.3}$$

Equation 11.2 summarizes the overall **light reactions**, whereas Equation 11.3 summarizes the overall **dark reactions**. We have already encountered ATP, ADP, and $P_i$, mobile cofactors that represent the transfer of phosphate bond energy. The pair of mobile cofactors, $NADP^+$/NADPH, was introduced in Chapter 7 as virtually identical in structure and function to $NAD^+$/NADH. These pairs have the same standard redox potential, but very different cellular redox potentials. In photosynthesis, NADPH represents electron-rich energy in a manner similar to that of NADH in mitochondria.

Many of the individual light reactions are closely analogous to those of the mitochondria. In fact, the same chemiosmotic hypothesis (i.e., production of a $H^+$ gradient across a membrane) explains the generation of energy in both cases. There are other remarkable analogies between these processes, with the principle distinction being the input of light energy for photosynthesis.

Dark reactions can and do proceed in the light; they are distinct in that do not require light. These pathways accomplish **carbon fixation,** by which plants (and other photosynthetic organisms) incorporate $CO_2$ into carbohydrates. Because both light and dark reactions take place within the **chloroplast,** the organelle unique to plant cells, we begin by examining this structure.

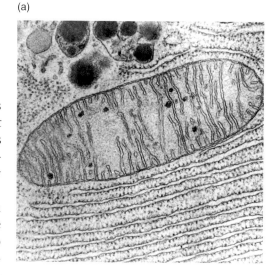
(a)

# 11.2 Chloroplasts

While mitochondria are found in both plants and animals, chloroplasts occur only in photosynthetic organisms. We have noted previously that mitochondria are evolutionary descendents of bacteria. Chloroplasts are descendents of cyanobacteria. Chloroplasts retain their own (limited) genetic machinery and can carry out photosynthesis successfully in isolation.

FIGURE 11.1 presents three views of a chloroplast. The micrograph (Figure 11.1a) reveals extensive inner membranes, some of which are in dense stacks, called **grana.** The simplified diagram of Figure 11.1b shows the arrangement of the membranes and spaces of the chloroplast. Two external membranes separate the cell cytosol from the **stromal** space. The innermost membrane is called the **thylakoid membrane.** The space within the thylakoid membrane is called the **lumen.** The thylakoid membrane is analogous to the mitochondrial inner membrane because, in photosynthesis, all of the membrane-bound proteins involved in light absorption and electron transport are embedded in this membrane. Figure 11.1c shows the functional essentials for photosynthesis. The thylakoid membrane contains embedded proteins that allow electron flow and produce a proton gradient between the lumen and the stroma. In addition, the light-absorbing molecules that drive photosynthesis are also embedded within the thylakoid membrane.

(b)

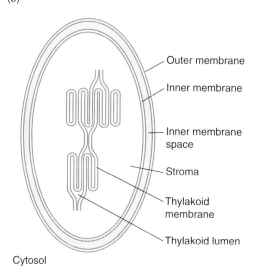

## Orientations: N and P Sides of the Membrane

The chemiosmotic hypothesis applies broadly to mitochondria, chloroplasts, and bacteria. The biological distinctions between these organelles can obscure their mechanistic unity; that is, a proton gradient is formed across a membrane, leaving one side more deficient in hydrogen ions. Thus, the mitochondrial inner membrane separates the matrix from the cytosol, the thylakoid membrane separates the

(c)

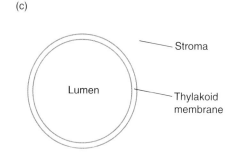

FIGURE 11.1 **Three Views of the Chloroplast.** (a) The electron micrograph shows extensive membrane folding in some regions (grana) interior to the organelle. (© Keith R. Porter/Photo Researchers, Inc.) (b) Spaces and membranes of the chloroplast are indicated diagrammatically. The thylakoid membrane is in-folded. (c) Functional diagram of the chloroplast for the light reactions. Only the thylakoid membrane and its inner (lumen) and outer (stromal) spaces are required.

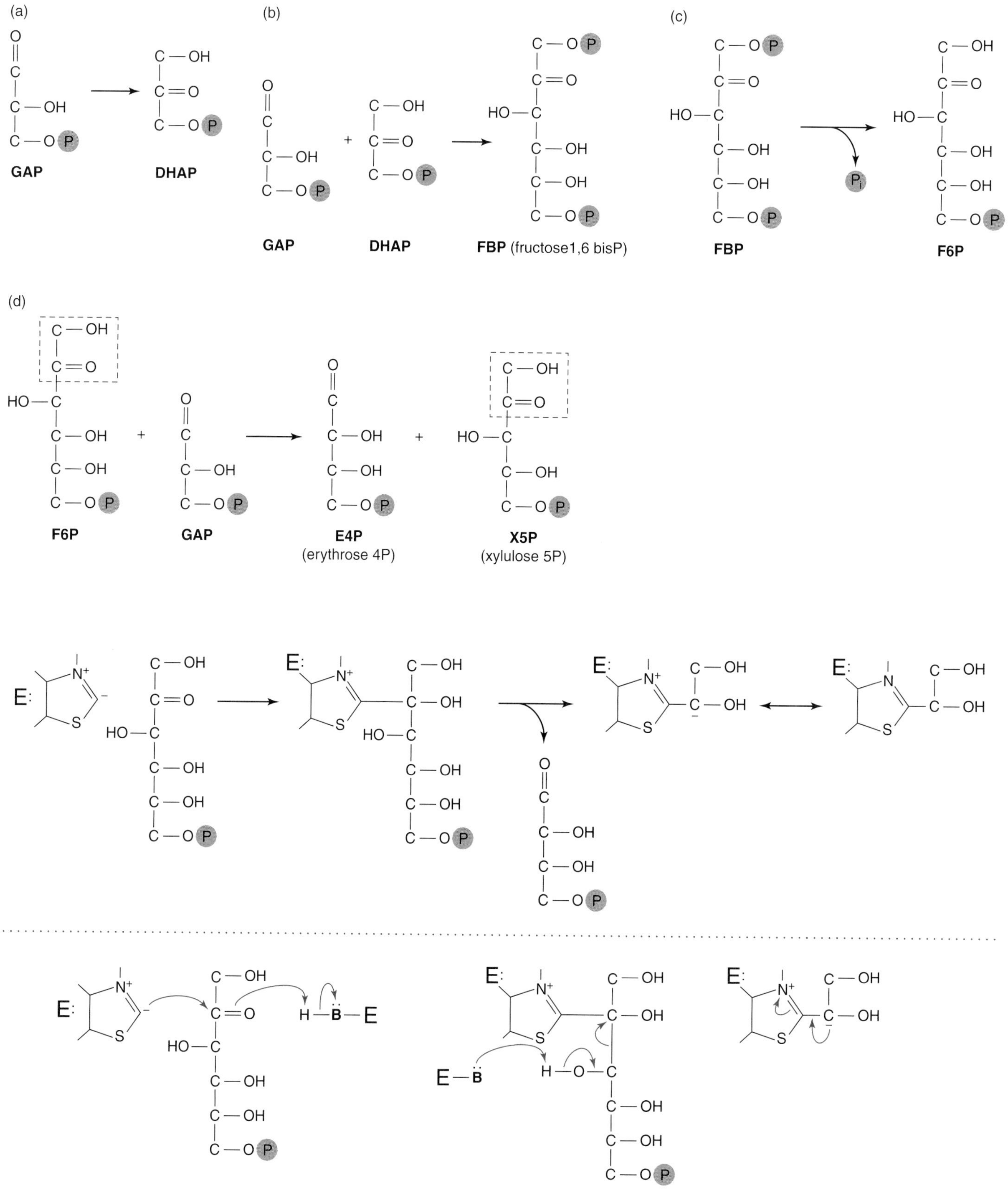

FIGURE 11.16 **The Calvin cycle reactions from GAP to RuBP.** (a) Triose isomerase. (b) Aldolase. (c) Fructose bisphosphatase. (d) Transketolase and mechanism. The enzyme-bound thiamine facilitates removal of a two-carbon fragment of the substrate. In the second half (not pictured), the fragment is donated to the carbonyl of the acceptor molecule (here, GAP). *(Continues)*

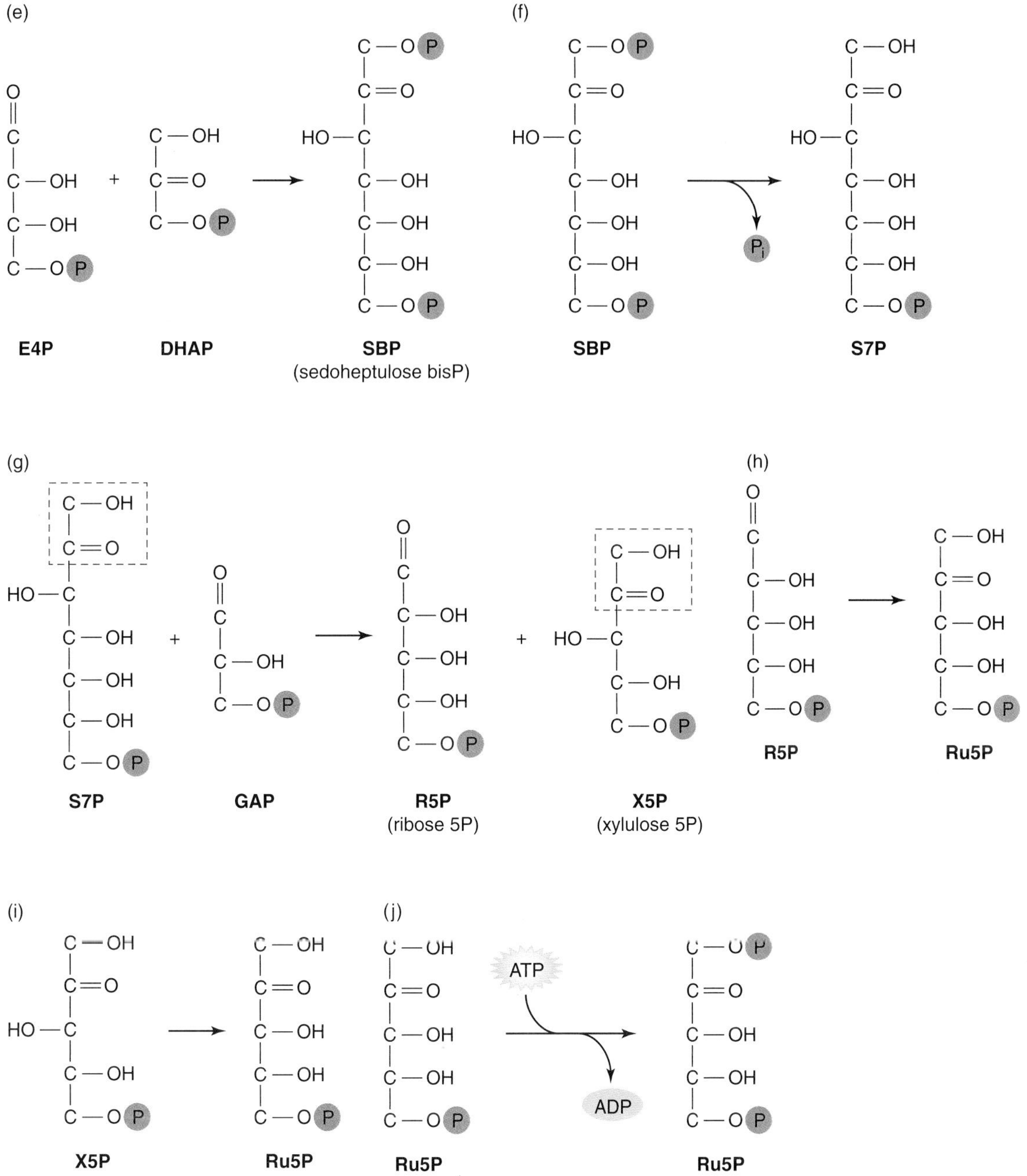

**FIGURE 11.16** *(Continued)* **The Calvin cycle reactions from GAP to RuBP**. (e) Aldolase. (f) Sedoheptulose bisphosphatase. (g) Transketolase; sedulose-7-P and GAP as substrates. (h) Ribose-P isomerase. (i) Xylulose-P epimerase. (j) Ribulose-P kinase.

electron rearrangement that splits the C–C bond between C2 and C3. This causes the release of E4P and leaves the two-carbon fragment attached to the thiamine cofactor of the enzyme. As indicated, this has a resonance form, which contributes to its stability. The remainder of the mechanism involves binding to the second substrate, GAP, attacking the carbonyl with the anion form, and ultimately releasing the free enzyme and X5P, the joined five-carbon product. With this reaction, we can appreciate the versatility of the thiamine cofactor, which participates in a decarboxylation (pyruvate decarboxylase), the transfer to a separate nucleophile (pyruvate dehydrogenase complex E1), and the carbon shuffling reaction shown here.

(e) The joining of erythrose-4-P with DHAP is also catalyzed by an aldolase enzyme; the product is the seven-carbon sugar sedoheptulose-1,7-bisP (SBP).

(f) The next reaction, the hydrolytic removal of one of the phosphoryl groups of SBP to form S7P, is catalyzed by a phosphatase. This is analogous to the phosphatase reaction of step 3.

(g) A second transketolase reaction, using S7P as a reactant with GAP, produces R5P and X5P.

(h) An epimerase, an enzyme encountered in lactose metabolism, catalyzes the conversion of Xu5P to the ketone Ru5P.

(i) An isomerase, with the same mechanism as the glycolytic isomerase enzymes, catalyzes the conversion of the R5P to the ketone Ru5P.

(j) A kinase catalyzes the conversion of Ru5P to the RuBisCo substrate, RuBP.

Having considered the individual reactions, we can now step back and see that the Calvin cycle incorporates $CO_2$ into PGA at the RuBisCo reaction. Ru5P, the substrate for the reaction, is regenerated as all but one of the PGA molecules produced, which undergo a complex dance of transketolase, aldolase, and rearrangement steps. The remaining PGA molecule is the pathway product of the Calvin cycle, derived from three input $CO_2$ molecules. PGA, which represents "fixed carbon" from atmospheric $CO_2$, is itself converted in plants largely to sucrose and starch.

# 11.7 Variations in $CO_2$ Handling: C3, C4, and CAM Plants

Most plants fix $CO_2$ by the process described in Section 11.6. Because the product of the fixation reaction has three carbons (PGA), these plants are also known as **C3 plants**. However, there is a built-in inefficiency, because RuBisCo also catalyzes its reaction with $O_2$ in place of $CO_2$ (FIGURE 11.17). $O_2$ can add to the enediol intermediate of RuBP, producing PGA and P-glycolate. While the $K_m$ for $O_2$ is much greater than the $K_m$ for $CO_2$, the far greater concentration of atmospheric $O_2$ (21% in air vs. 0.04% for $CO_2$) means that a significant amount of $O_2$ can substitute for $CO_2$. In a few cases in which the climate magnifies the loss of carbon flow through the Calvin cycle, there are adaptations that minimize this loss by the presence of additional reactions.

For example, in plants originally found in tropical climates, such as sugar cane and corn, the first intermediate incorporating $CO_2$ is a four-carbon intermediate rather than a three-carbon intermediate; these are called **C4 plants**. An example of this pathway is illustrated in FIGURE 11.18. Photosynthesis is divided between two cell types in these plants; the bundle sheath cells and the mesophyll cells. The **bundle**

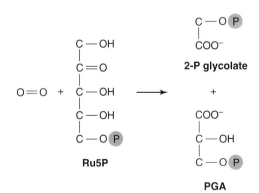

**FIGURE 11.17 RuBisCo Reaction with $O_2$ in Place of $CO_2$.**

**CHAPTER 11** PHOTOSYNTHESIS

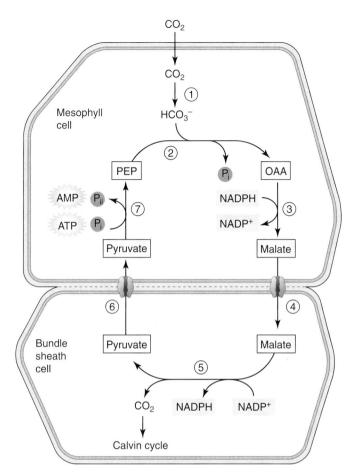

**FIGURE 11.18 The C4 Pathway in Plants.** The pathway involves two separate cells—the mesophyll and bundle sheath—connected by transporters that span both cell membranes. Upon entry to the mesophyll cell, $CO_2$ is hydrated to $HCO_3^-$ (1) in a carbonic anhydrase–catalyzed step, and then incorporated into oxaloacetate (OAA), catalyzed by PEP carboxykinase (2). Following reduction of OAA, catalyzed by an NADPH-linked malate DH (3), malate is transported (4) into the bundle-sheath cell. Malate is converted to pyruvate and $CO_2$ through a reaction catalyzed by malic enzyme (5). The $CO_2$ is the Calvin cycle substrate, and the pyruvate is transported (6) back to the mesophyll cell and converted to PEP by pyruvate dikinase (7).

**sheath** cells are arranged in an inner ring around the vascular system of the plants. An outer ring contains mesophyll cells. The Calvin cycle enzymes are located exclusively in the bundle sheath cells. After $CO_2$ enters the mesophyll cell from the exterior, it is converted to $HCO_3^-$, a step catalyzed by carbonic anhydrase. Next, PEP carboxykinase catalyzes the reaction of $HCO_3^-$ with PEP to form oxaloacetate (OAA). As we will appreciate when we complete this pathway, this carboxylation reaction is fundamentally different from RuBisCo in that $CO_2$ is ultimately released in the pathway. OAA is converted next to malate in a reaction catalyzed by an $NADP^+$-linked malate dehydrogenase. Malate is transported to the bundle sheath cell by a transport protein that spans both cell membranes (a **plasmodesmata**). Once in the bundle sheath cell, malate undergoes decarboxylation to pyruvate and $CO_2$, catalyzed by malic enzyme. This reaction shares a mechanism with iso-citrate dehydrogenase of the Krebs cycle. The pyruvate is transported back to the mesophyll cell (through a plasmodesmata), and converted to PEP in a reaction catalyzed by a dikinase. The regeneration of PEP completes the cycle, which functions in producing augmented levels of $CO_2$ for RuBisCo so that the reaction with $O_2$ is minimized. Reactions similar to this shuttle are in the pathways of gluconeogenesis and fatty acid synthesis.

A further variation exists in certain plants that thrive in extremely hot, dry climates. As these plants are succulents of the Crassulaceae family, they are called **CAM plants**, for Crassulacean *a*cid *m*etabolism. Similar to C4 plants, an intermediate carboxylation occurs, using the same enzymes, PEP carboxylase and malate dehydrogenase, to convert PEP to oxaloacetate. In CAM plants, however, the separation of the parts of the cycle is accomplished in time rather than in space. At night, the cell admits air containing $CO_2$ and stores malate, allow-ing entry through openings of pores to the cell called stomata. In the daytime, the stomata close, and malate is decarboxylated to generate $CO_2$ for the Calvin cycle.

Beyond these specializations, there are numerous minor pathways in specific plant species as well as in microorganisms, such as bacte-ria and yeast. The product of these pathways, known as **secondary metabolites**, have no obvious functional except perhaps for their pro-tection against consumption by predators. There is medical interest in these pathways, because their products often selectively kill pests with little adverse effect on humans. As a result, many drugs have the products of secondary metabolic pathways as their natural origins.

# 11.8 Pathway Endpoints: Sucrose and Starch

GAP that arises from $CO_2$ fixation can be converted to either sucrose or starch. Sucrose is formed by transporting GAP out of the chloroplast and into the cytosol, followed by a series of enzymatic steps leading to a condensation of an activated glucose moiety and fructose-6-P. Starch is formed by the sequential condensation of a distinct activated glucose moiety with pre-formed glucose polymers. These "activated glucose" moieties are nucleotide sugars.

Both sucrose and starch can be used as energy precursors in plants for the generation of ATP by pathways discussed elsewhere in this text. The two have roles analogous to glucose and glycogen in animals, in the sense that sucrose is a mobile energy sugar that can be transported between cells through the plant vascular system, while starch is a stor-age form of energy that is local to cells.

In the ensuing chapters of this text, we consider the remaining criti-cal pathways of mammals: other carbohydrate routes, fat metabolism, and nitrogen metabolism. Nonetheless, the pathways taken by plants can be considered as a superset of those for mammals. The closest photochemical reaction in humans is in the visual cycle, where the

## BOX 11.2: Plants and the Visual Cycle

The first part of the visual cycle in humans and other animals is similar to photosynthesis. For both processes, an embedded pigment bound to a membrane protein absorbs light, and the energy of that absorbed light drives a reaction. In the visual cycle, the reaction is the isomerization of a *cis* double bond in retinal to the *trans* configuration. In turn, this changes the conformation of the bound protein opsin, which transmits the light through a series of steps involving membrane polarization and neurotransmitter release from the retina to the optic nerve.

There are other similarities. The origin of human retinal is dietary carotene (see the structures in FIGURE B11.2a), an accessory plant photosynthetic pigment. Retinal is bound to carotene through its aldehyde bond, forming a Schiff base to a lysine residue of the protein opsin; this adduct is known as **rhodopsin**. Not only does the retinal portion alter the conformation of opsin after its conversion to the trans form, but the opsin influences the light absorption properties of the bound retinal. In the cone cells of the retina, three different opsin proteins exist, each of which senses specific bands of the light spectrum, accounting for sensitivity to blue, green, and red light.

Finally, the membrane arrangement of the light-absorbing pigment in the eye resembles the grana stacks. Rhodopsin exists in dense stacks in internal membranes of the outer segment of the rod cells (FIGURE B11.2b).

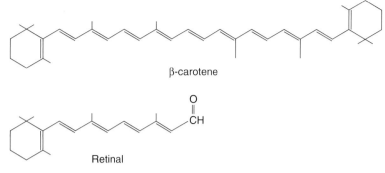

β-carotene

Retinal

FIGURE B11.2a **β-Carotene and Retinal**.

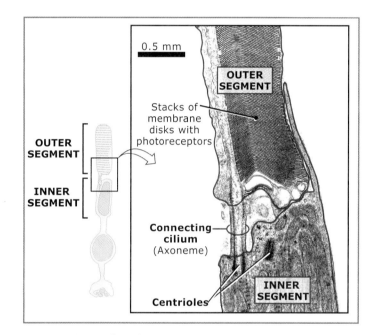

FIGURE B11.2b **A Rod Cell**. On the left, a drawing of a rod cell showing the inner and outer segments. On the right is an electron micrograph showing a detail of the rhodopsin stack in the outer segment. (Reprinted from *Histology of the Human Eyes*, M. J. Hogan, J. A. Alvarado, and J. E. Weddell, p. 425. Copyright 1971, with permission from Elsevier.)

key molecule employed is carotene, one of the accessory pigments of photosynthetic organisms (**BOX 11.2**).

## Summary

Photosynthesis can be divided into two separate processes: the light reactions, in which energy is formed, and the dark reactions, in which energy is used to convert $CO_2$ to sugars. The intermediate energy forms are ATP and NADPH and they are employed in the first portion of the dark reactions. Photosynthesis takes place in the plant chloroplast, a specialized organelle that is not found in animal cells. Like the mitochondria, the chloroplast arose evolutionarily from bacteria. The pathway for energy formation involves light absorption by chlorophyll molecules that is used to excite electrons and move them through a redox chain in close analogy to mitochondria (run in reverse), from $H_2O$ to NADPH. The generation of a proton gradient is consistent with the chemiosmotic hypothesis, which also accounts for mitochondrial and bacterial energy formation. The use of the energy molecules ATP and NADPH in the Calvin cycle drives the conversion of $CO_2$ into sugar ($C_6H_{12}O_6$); the Calvin cycle converts three $CO_2$ molecules into one of PGA (P-glycerate). In the carbon-fixing reaction, ribulose-bis-P-carboxylase (RuBisCo) catalyzes the incorporation of $CO_2$ into ribulose-5-P, producing two PGA molecules. PGA is recycled back to ribulose-5-P using several enzymatic reactions, most of which are very similar to ones encountered previously. An exception is transketolase, which catalyzes two separate steps of the pathway, in which carbon fragments are reshuffled to create all five-carbon molecules to complete the cycle. The PGA produced by the Calvin cycle is converted to sucrose for transfer to other parts of the plant, or to starch for energy storage. Variations exist in some plants to concentrate the $CO_2$, as in the C4 plants, or to prevent water loss, as in the CAM plants. Numerous other variations also exist within distinct plant species to make a wide array of distinctive molecules, some of which have significant medical uses.

## Key Terms

| | | |
|---|---|---|
| antennae | exciton transfer | photosystem I (PS I) |
| auxotropic | dark reactions | photosystem I (PS II) |
| bundle sheath | fructose bisphosphatase | plasmodesmata |
| C3 plants | grana | reaction center |
| C4 plants | light reactions | rhodopsin |
| Calvin cycle | pentose phosphate shunt | secondary metabolites |
| CAM plants | pheophytins | stroma |
| carbon cycle | photosynthetic | thylakoid membrane |
| carbon fixation | photon | transketolase |
| chloroplast | photosystem (PS) | |

## Review Questions

1. The two absorption maxima of light were given as 400 and 700 nm. What is the maximum amount of energy that can be incorporated to drive electron flow?
2. While all plants would benefit from the use of the C4 route, most do not use it. What steps of the C4 pathway require extra energy formation by the plant? What is one possible effect of global warming?
3. Other accessory pigments in plants account for the autumnal colors, mainly orange and reds. One of these was identified in this chapter as carotene. What are some others? Why should those molecules persist in leaves while chlorophyll, which is typically far more abundant, does not?
4. What is the principal advantage of energy transfer by exciton transfer in the antenna complex versus direct absorption at the reaction center?
5. In terms of overall pathway consequence, distinguish $CO_2$ incorporation by RuBisCo from $CO_2$ incorporation by PEP carboxylase in C4 plants.

# References

1. Hall, D. O.; Rao, K. K. *Photosynthesis,* 6th ed.; Cambridge University Press: Cambridge, UK, 1999.
2. Morton, O. *Eating the Sun. How Plants Power the Planet*; Harper Collins: New York, 2008.
3. Fromme, P.; Jordan, P.; Krauss, N. Structure of photosystem I. *Biochim. Biophys. Acta* 507:5–31, 2001.
4. Mullineaux, C. W. Function and evolution of grana. *Trends Plant Sci.* 10:521–525, 2005.
5. Walker, D. A. The Z-scheme—down hill all the way. *Trends Plant Sci.* 7:183–185, 2002.

# 12

## Carbohydrate Pathways Related to Glycolysis

**CHAPTER OUTLINE**

Image © Dr. Mark J. Winter/Photo Researchers, Inc.

The pathways discussed in the present chapter will reveal the richness of carbohydrate metabolism, showing new connections to the glycolytic pathway. Together with the principal pathways of lipid and nitrogen metabolism, we have a set of reactions collectively called **intermediary metabolism**.

A short bypass of glycolysis forms 2,3-bisphosphoglycerate. Fructose, after three metabolic steps, enters intermediate steps of glycolysis. The complete oxidation of glucose can be considered a continuation of glycolysis as pyruvate progresses through the pyruvate dehydrogenase complex, the Krebs cycle, and oxidative phosphorylation.

Here we consider several pathways that are connected to glycolysis. FIGURE 12.1 shows several glycolytic intermediates as points of intersection for five routes of carbohydrate metabolism. The first two, **glycogen synthesis** and **glycogenolysis**, will serve as a model for understanding regulation in metabolism. **Galactose catabolism** intersects glycogen metabolism in its use of uridine nucleotides. **Gluconeogenesis**, essentially the reverse of glycolysis, converts molecules such as lactate and amino acids to glucose. Finally, the **pentose phosphate shunt** serves the dual purpose of providing five-carbon sugars and generating NADPH; it bears a strong similarity to the Calvin cycle of photosynthesis.

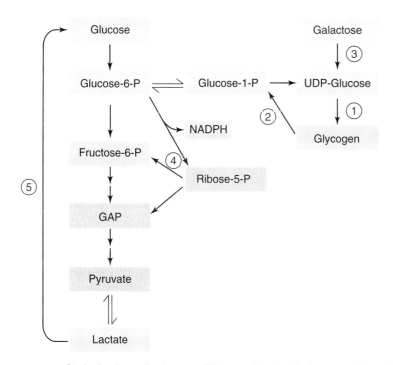

FIGURE 12.1 **Carbohydrate Pathways.** The synthesis of glycogen (1) and its breakdown (2) are connected to glycolysis through glucose-6-P, as is the catabolism of galactose (3). Also stemming from glucose-6-P is the pentose phosphate shunt (4), which provides two molecules for the cell: the pentose phosphate ribose-5-P and NADPH. Gluconeogenesis (5) is the synthesis of glucose from certain precursors, such as the lactate illustrated here.

## 12.1 Glycogen Metabolism

Glycogen represents a cell's readily available energy store. These glucose polymers can have molecular weights up to 100 million. Most of the glycogen macromolecule has "straight chain" links ($\alpha 1 \rightarrow 4$ linked glucosyl units), with about tenfold fewer "branch points" ($\alpha 1 \rightarrow 6$ linkages). Thus, most glycogen metabolism involves adding or removing the $\alpha 1 \rightarrow 4$ linked glucosyl units from the nonreducing end. The other features in its metabolism involve the branch points. At the core of the glycogen particle is a protein molecule called **glycogenin.** We begin by describing the formation and breakdown of these immense molecules, and then examine the regulation of glycogen metabolism as a model for controlling processes in cells.

Glycogen breakdown begins with the splitting of $\alpha 1 \rightarrow 4$ links and, subsequently the $\alpha 1 \rightarrow 6$ branch points. Glycogen is much more extensively branched than starch, the plant storage sugar, so more $\alpha 1 \rightarrow 4$ linked glucosyl units are available for breakdown in glycogen. Glycogen breakdown thus provides rapid formation of glucose units for energy.

Glycogen is a glucose *buffer*. Its function in liver is unique: it can provide the rest of the body with glucose in times of need, such as in fasting when no glucose is provided by the diet, and in times of stress, such as escaping from a predator. The glycogen broken down in muscle can only be used for energy in the muscle cell itself, because free glucose is not released. Thus, rapid muscle contraction stimulates the immediate breakdown and utilization of glucosyl units from glycogen. Other tissues have glycogen in far smaller amounts, and their function is similar to that of muscle: glucose is not released, but glucosyl units can be provided to glucose-6-P and subsequent pathways.

With this broad background established, we turn to an examination of the separate processes of glycogen synthesis and glycogen breakdown.

### Glycogen Synthesis

The incorporation of glucose-6-P into glycogen proceeds by the pathway outlined in FIGURE 12.2. The first step is the formation of glucose-1-phosphate (glucose-1-P), catalyzed by phosphoglucomutase. This near-equilibrium enzyme has the same mechanism as the phosphoglyceromutase of glycolysis; that is, the phosphate groups are exchanged through the intermediate formation of histidyl-phosphoryl groups on the enzyme.

The next step is the formation of uridine diphosphate (UDP)-glucose, or *activated glucose*, because the nucleotide portion is a highly favorable leaving group. This is similar to the *activated acetate* represented in the Krebs cycle substrate acetyl-coenzyme A (CoA). For the present activation step, the molecule uridine triphosphate (UTP) reacts with glucose-1-P, catalyzed by a transferase enzyme (FIGURE 12.3). The mechanism is a direct nucleophilic substitution: one of the oxygen anions of the phosphate of glucose-1-P attacks the

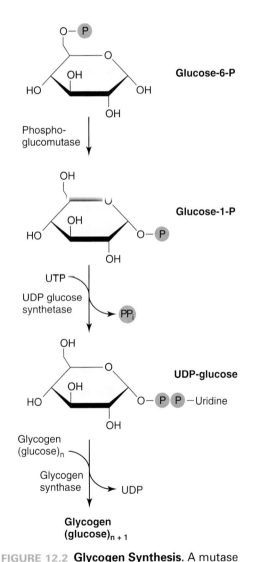

**FIGURE 12.2 Glycogen Synthesis.** A mutase catalyzes the formation of glucose-1-P, followed by the activation step, the formation of UDP-glucose. The key step in chain formation is catalyzed by glycogen synthase, which adds the glucose portion of UDP-glucose to an existing chain of glucose residues.

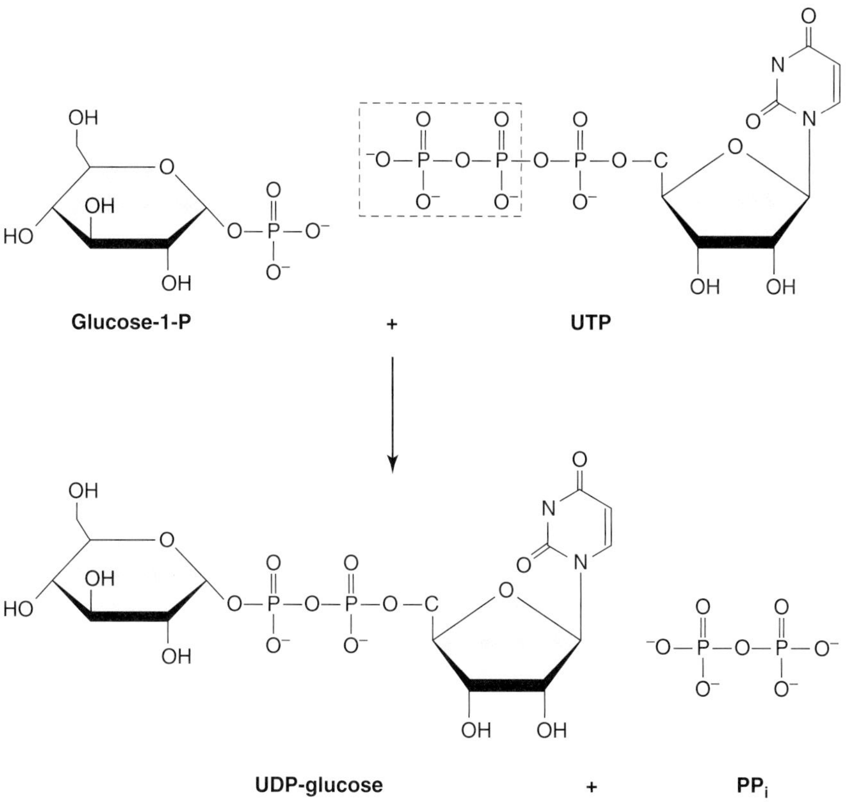

FIGURE 12.3 **Formation of UDP-Glucose.**

α-phosphoryl of UTP. The other reaction product, inorganic pyro-phosphate (PP$_i$) is rapidly hydrolyzed due to the presence of a very active pyrophosphatase in cells. The formation of *activated glucose* is thus driven by the splitting of two high-energy phosphate groups. The pyrophosphatase maintains the overall formation of UDP-glucose far displaced from equilibrium, and leaves cellular levels of PP$_i$ at very low concentrations.

The rate-limiting step in glycogen synthesis is catalyzed by **gly-cogen synthase**. This reaction is drawn in different levels of detail in FIGURE 12.4. The overall reaction is shown in Figure 12.4a, in which all hydroxyl groups are indicated, as well as the new incoming sugar ring (in red) and the hydroxyl group at the nonreducing end of the initial glycogen polymer. An abbreviated representation of the sugar ring is presented in Figure 12.4b, similar to the sugar polymer. Intermediate steps of the mechanism are drawn in Figure 12.4c. Displacement of the UDP portion of UDP-glucose produces a glucose carbocation that is subsequently attacked by the 4′-OH group of the nonreducing end of glycogen. A key piece of evidence supporting this mechanism is inhibition of the enzyme by gluconolactone; its structure, alongside the glucose carbocation, is shown in Figure 12.4d. The right-hand portions of both molecules are flat, whereas the left-hand portions are identical, which explains why gluconolactone binds to the active site of glycogen synthase. Both assume a six-membered ring structure in the half-chair configuration.

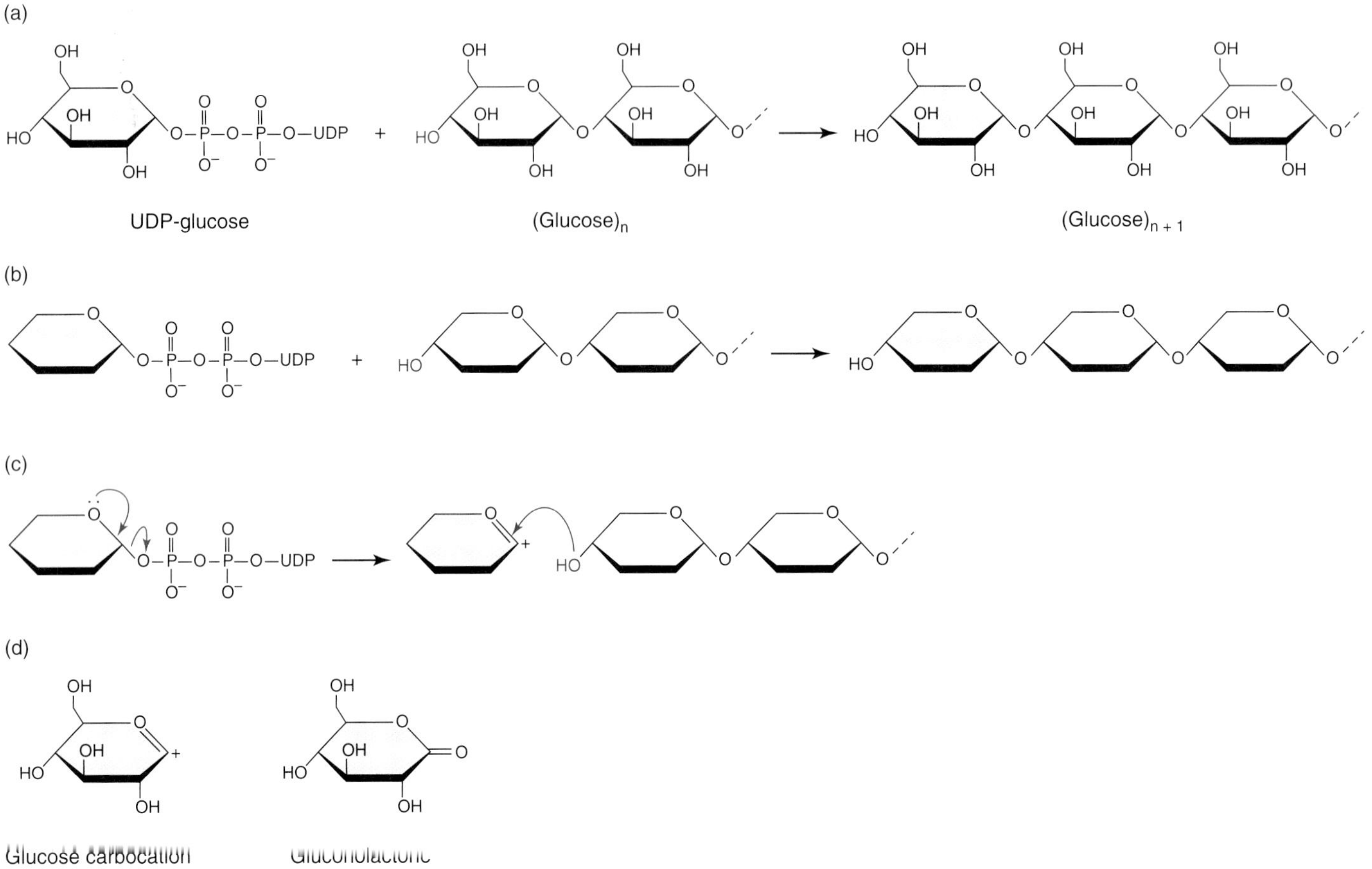

**(a)**

UDP-glucose    +    (Glucose)$_n$    ⟶    (Glucose)$_{n+1}$

**(b)**

**(c)**

**(d)**

Glucose carbocation          Gluconolactone

**FIGURE 12.4 Glycogen Synthase.** (a) Reaction structures. (b) Abbreviated reaction structures. (c) The mechanism involves electron rearrangement from a lone pair of electrons on the ring oxygen and release of UDP to form the carbocation intermediate, which is subsequently attacked by the 4 hydroxyl of the nonreducing end of glycogen. (d) The intermediate glucose carbocation and the enzyme inhibitor gluconolactone are both forced into a half-chair configuration.

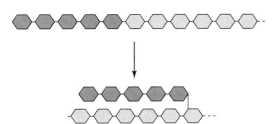

**FIGURE 12.5 Branching Enzyme.**

After about 10 glucosyl units have been added to the nonreducing end, a reaction catalyzed by **glycogen branching enzyme** (FIGURE 12.5) removes a string of at least six glucosyl units from the reducing end to an interior 6-hydroxy position. This creates a new interior α1→6 bond, known as a *branch point*.

Complete *de novo* synthesis of the glycogen molecule requires a separate protein called **glycogenin**, which serves as both an enzyme and a scaffold (FIGURE 12.6). Glycogenin is a dimer in which two active sites with their acceptor tyrosine residues are arranged head to toe. UDP-glucose donates glucose residues first to the tyrosine hydroxyls, and then a small chain of glucosyl residues are incorporated. Subsequent buildup of the polymer involves glycogen synthase and the branching enzyme, as above. A complete glycogen molecule is depicted in FIGURE 12.7. Most glycogen synthesis, however, involves the incorporation of glucosyl units into an already substantial glycogen molecule. This is because the glycogen molecule rarely breaks down completely. We next consider the major reactions involved in that process.

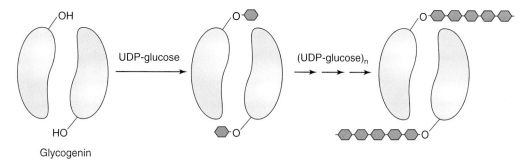

FIGURE 12.6 **Actions of Glycogenin**. Glycogenin dimers are the scaffold for initiating the glycogen chain as well as the catalysts for the incorporation of the first glucosyl units.

## Glycogenolysis

Virtually all glycogen breakdown is accomplished by a single reaction: the removal of glucose residues from the nonreducing end. This reaction is catalyzed by **glycogen phosphorylase:**

$$(\text{glucose})_n + P_i \rightarrow (\text{glucose})_{n-1} + \text{glucose-1-P} \qquad (12.1)$$

The enzyme has a bound cofactor, **pyridoxal phosphate,** derived from dietary pyridoxine (vitamin $B_6$) in humans (FIGURE 12.8). Because of the vast amount of muscle tissue and correspondingly large quantity

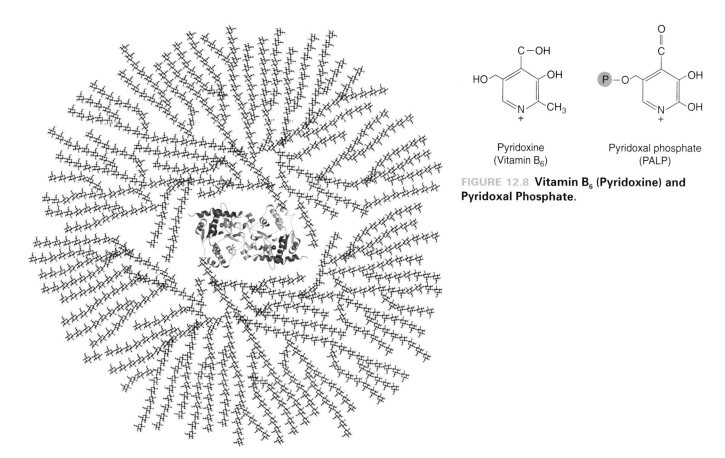

FIGURE 12.8 **Vitamin $B_6$ (Pyridoxine) and Pyridoxal Phosphate.**

Pyridoxine
(Vitamin $B_6$)

Pyridoxal phosphate
(PALP)

FIGURE 12.7 **The Glycogen Particle.** (Courtesy of Mikael Häggström.)

of glycogen phosphorylase in muscle, most of the body's vitamin $B_6$ is incorporated into this enzyme. The cofactor forms a covalent bond between the aldehyde of the ring and a lysine residue of the enzyme (a Schiff's base). The phosphate group acts as an acid–base catalyst.

The enzyme mechanism (**FIGURE 12.9**) involves the same glucose carbocation intermediate as glycogen synthetase. Accordingly, gluconolactone (Figure 12.4d) also inhibits glycogen phosphorylase activity. The carbocation is formed after protonation and electron rearrangement, leading to cleavage of the 1→4 bond. Inorganic phosphate then attaches itself (as a nucleophile) to the carbocation to form the product, glucose-1-P. The enzyme-bound pyridoxal phosphate forms an ionic bond to the inorganic phosphate substrate, alternately deprotonating and protonating it.

Glycogen phosphorylase undergoes several catalytic cycles until it stalls at about four residues from an α1→6 bond. At that point,

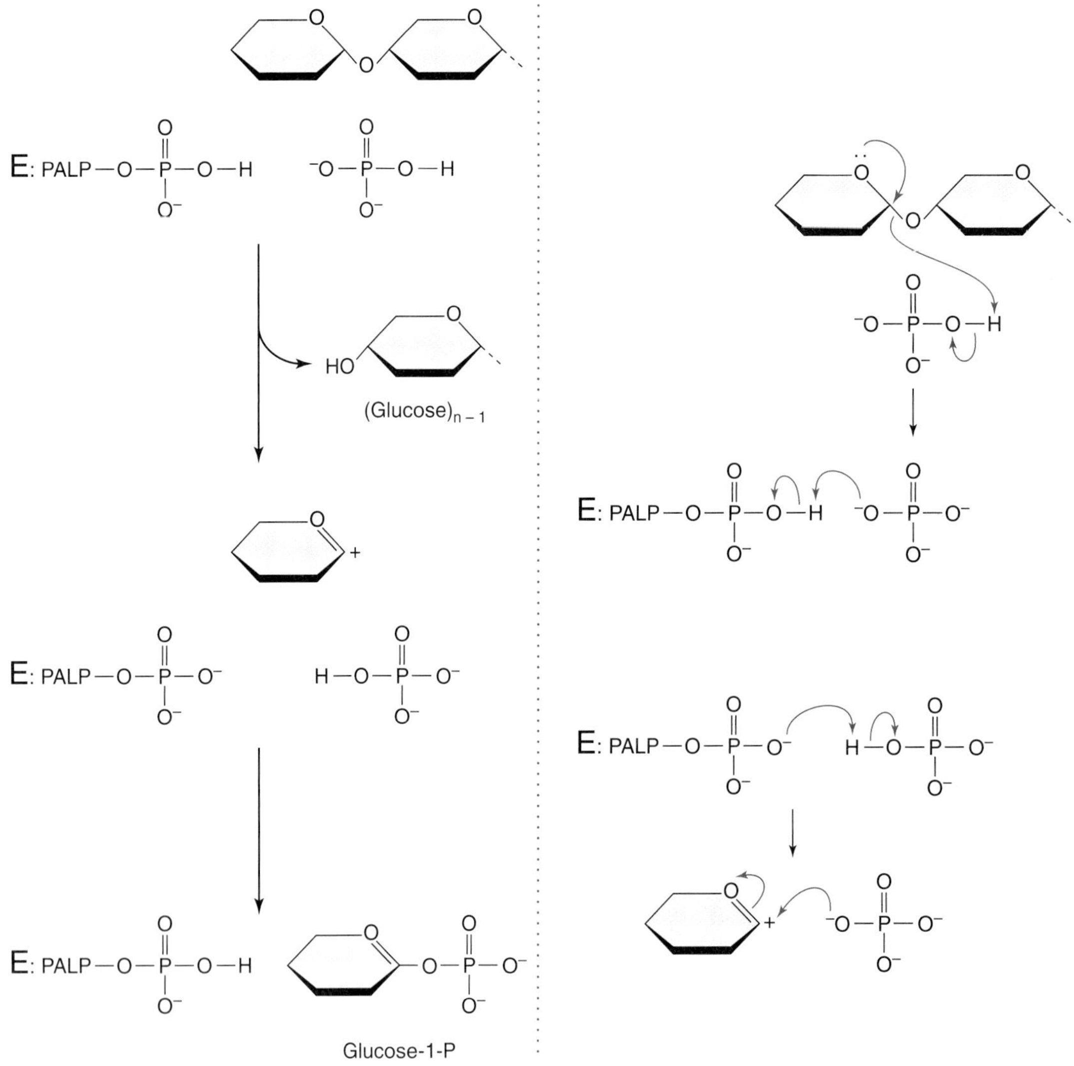

FIGURE 12.9 **Mechanism of Glycogen Phosphorylase**. The pyridoxal phosphate serves to position the $P_i$ substrate and participate in acid–base catalysis. Note the carbocation intermediate is attacked by the $P_i$.

debranching enzyme catalyzes the removal of a polymer containing all but one glucosyl subunit in the $\alpha 1 \rightarrow 6$ bond. The last glucosyl unit of the branch point is removed by a reaction catalyzed by glucosidase (FIGURE 12.10), and a free glucose molecule is the product. Thus, the overall reaction of glycogen breakdown produces mostly glucose-1-P, with a small amount of free glucose reflecting the number of branch points.

Most of the well-established genetic defects due to deranged glycogen metabolism involve enzymes of glycogen synthesis or degradation, with few exceptions (BOX 12.1). Normal control of glycogen metabolism ultimately comes down to the regulation of glycogen synthase and glycogen phosphorylase. We next consider this issue in physiological situations.

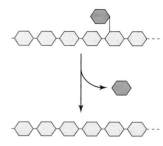

FIGURE 12.10 **Glucosidase.** This hydrolysis reaction removes the branch point, producing a glucose molecule.

## BOX 12.1: Glycogen Storage Diseases

The **glycogen storage diseases** are a classic set of inborn errors of metabolism—genetic defects that have a metabolic consequence deleterious to glycogen metabolism. Because these are well-established disease states, the missing gene is known in each case, and usually its consequences understood. Some of the defects are not direct deficiencies of glycogen enzymes themselves, but rather in enzymes of connected pathways, such as glycolysis and gluconeogenesis, that cause abnormalities of glycogen metabolism.

The original classification used Roman numerals, which were assigned in order of their discovery. Major storage diseases are listed in Table B12.1.

Consider type IV and type V in detail. Type IV, the loss of branching enzyme, leaves the glycogen polymer unbranched and much less soluble. Because branching enzyme is more abundant in liver than in muscle, its deficiency is more obvious in the liver, which becomes oversized (*hepatomegaly*) and scarred (*cirrhotic*). In type V, the deficiency in muscle phosphorylase makes many types of exercise difficult because the glycogen cannot be readily broken down as an energy source. This is particularly important for intense *anaerobic* exercise (using fast muscle). The more sustained type of aerobic exercise is supported by lipid metabolism.

### TABLE B12.1 Types of Storage Diseases

| Type | Deficiency | Pathway Lesion |
|---|---|---|
| I | Glucose-6-Pase (Ia); glucose-6-P transporter (Ib) | Gluconeogenesis, glycogenolysis (liver) |
| II | Lysosomal glucosidase | Lysosomal catabolism |
| III | Glycogen debranching enzyme | Glycogenolysis |
| IV | Glycogen branching enzyme | Glycogen synthesis |
| V | Phosphorylase | Glycogenolysis (muscle) |
| VI | Glycogen phosphorylase | Glycogenolysis (liver) |
| VII | Phosphofructokinase | Glycolysis (muscle) |

## Physiological Context of Glycogen Metabolism

Glycogen represents readily available energy, providing cells with glucose on a minute-to-minute basis. The control of glycogen pathways has been well studied and is a model for the regulation of other pathways. While at least trace amounts of glycogen are found in many animal cells, muscle and liver account for virtually all of the body's store of this polymer. Hence, we will consider the role of glycogen in just these two tissues.

Muscle (in particular, *fast* or *glycolytic* muscle) uses glycogen as a fuel to support its contraction. Both muscle contraction and glycogenolysis are triggered by a rise in cytosolic $Ca^{2+}$. $Ca^{2+}$ stored in the muscle endoplasmic reticulum is released to the cytosol in response to nerve-directed depolarization of the muscle's plasma membrane. Contraction of the muscle involves the movement of two protein complexes relative to each other: actin and myosin. The myosin converts ATP to ADP and $P_i$ to drive this process. This accounts for most of the energy utilization in contracting muscle. It is thus appropriate that the rise in $Ca^{2+}$ concentration also serves to activate glycogen breakdown to metabolically support ATP formation.

Synthesis of glycogen is under the control of **insulin**, the hormone that is secreted into the bloodstream during feeding conditions. Due to the large amount of skeletal muscle in the body, skeletal muscle uptake of glucose and its incorporation into glycogen represents the major depository of glucose after feeding.

Liver serves as a *glucose buffer* for the entire body, releasing or taking up glucose from the blood to maintain a relatively constant level. The breakdown of liver glycogen provides free glucose to the blood for use by all of the cells in the body. A key trigger for glycogen breakdown is **glucagon**, the hormone that rises during fasting conditions. Liver also can release glucose from glycogen during conditions that elevate cytosolic $Ca^{2+}$, such as adrenal glands' release of **epinephrine** during a stress response. As in muscle, insulin stimulates the synthesis of liver glycogen in the fed state.

In the following sections, we will examine hormonal regulatory systems of glycogen metabolism imposed by insulin in the muscle, and by glucagon and epinephrine in the liver. Next, we will examine *gluconeogenesis*, the synthesis of glucose from non-carbohydrate precursors such as lactate. Finally, we will construct a broader picture of carbohydrate metabolism in the physiological contexts of the feeding–fasting and resting–exercise transitions.

## Regulation of Glycogen Metabolism by Glucagon

When glucose concentrations drop, which happens routinely between meals, the α-cells of the pancreas release glucagon to the blood. The target of this 29-amino acid peptide is the glucagon receptor of the liver. Glucagon does not enter the cell in order to exert its effects; rather, it binds to the exterior face of the receptor, which then undergoes a conformational change. The glucagon receptor is a protein that spans the membrane. In its glucagon-bound state (symbolized as a square in FIGURE 12.11), the cytosolic portion of the receptor can bind a **G-protein** and catalyze an exchange of GDP for GTP on the G-protein surface. The name G-protein indicates that a guanine nucleotide is always

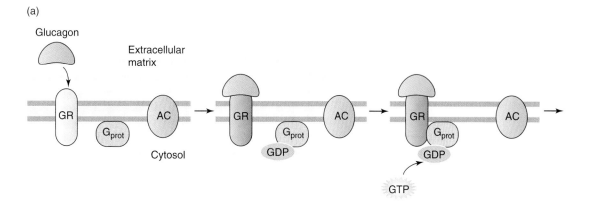

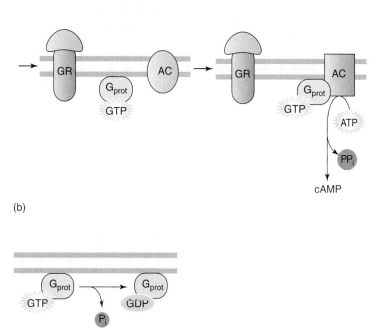

**FIGURE 12.11** **Glucagon and the G-Protein.** (a) G-protein ($G_{prot}$) shuttles between the glucagon receptor (GR) and the effector, adenylate cyclase (AC). $G_{prot}$ becomes activated when GR is occupied by glucagon, enabling $G_{prot}$ to exchange its bound GDP for GTP. The $G_{prot}$:GTP species is the active form, stimulating AC, which catalyzes cAMP formation. (b) $G_{prot}$ has an endogenous enzymatic activity that hydrolyzes bound GTP constantly, releasing $P_i$ and leaving bound GDP (inactive form). The $G_{prot}$:GTP form can be regenerated under the conditions of (a) above.

bound to it: either GDP or GTP. GTP can only displace GDP on the G protein when the hormone receptor is bound to glucagon. Once the G-protein is in its GTP-bound state, it dissociates from the glucagon receptor, slides along the surface of the membrane, and binds a separate membrane-embedded protein, adenylate cyclase. Only the GTP-bound form can bind and activate this enzyme, which catalyzes the reaction:

$$ATP \rightarrow cAMP + PP_i \tag{12.2}$$

$PP_i$ is rapidly hydrolyzed (as in UDP-glucose formation in Section 12.1, Glycogen Synthesis), ensuring metabolic irreversibility. As long as the G-protein has GTP bound, it can continue to activate adenylate cyclase and drive the production of cAMP. However, G-protein has an enzymatic activity of its own: a GTPase, which catalyzes:

$$GTP : G\text{-protein} \rightarrow GDP : G\text{-protein} + P_i \qquad (12.3)$$

The GDP-bound form of the G-protein can no longer activate adenylate cyclase, so this represents signal termination. However, as long as glucagon is bound to its receptor, the GDP:G-protein can slide along the membrane and be converted again to the active, GTP-bound form. This system ensures that the effector—adenylate cyclase—is activated only when the hormone is bound to the receptor.

There are over 20 distinct G-proteins known that transduce signals in a similar way in cells. At least one type of cancer is linked to a defect in the GTPase activity of a G-protein. The *ras* oncogene results from an inactive GTPase, which leads to a state of constant activation and hyperactive cell replication.

Overall, the result of glucagon action is an increase in cAMP concentration in the cytosol. cAMP is a signal molecule that acts by binding to protein kinase A (PKA), an enzyme composed of regulatory and catalytic subunits (FIGURE 12.12). In the resting (inactive) state, PKA is a tetramer. When the concentration of cAMP is elevated, it binds the regulatory subunits, releasing active catalytic monomers and dimeric regulatory subunits.

The reaction pictured in Figure 12.12 is reversible. Thus, PKA can revert to an inactive form if cAMP concentration is lowered and dissociates from the regulatory subunits. Subsequently, regulatory and catalytic subunits combine, thereby inactivating PKA. This inactivation will occur as glucagon levels drop and the G-protein reverts to its inactive form. In addition, the constant presence of the enzyme **phosphodiesterase**, removes cAMP by catalyzing its hydrolysis:

$$cAMP + H_2O \rightarrow AMP \qquad (12.4)$$

This provides signal termination, like the GTPase of the G-proteins.

As depicted in FIGURE 12.13, active PKA catalyzes the phosphorylation of two proteins in glycogen metabolism. First, glycogen synthase (GS) is converted to an inactive form, reducing the rate of glycogen formation. Second, the enzyme **glycogen phosphorylase kinase** is converted to its active form. Glycogen phosphorylase kinase acts on glycogen phosphorylase itself, catalyzing its phosphorylation and

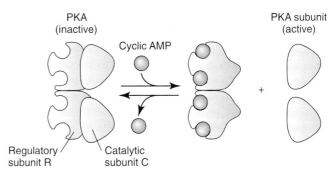

FIGURE 12.12 **Activation of Protein Kinase A**. Elevated concentrations of cAMP bind the regulatory subunit (R) of protein kinase A (PKA) and release it from the catalytic subunit (C), which is the active form of PKA.

CHAPTER 12 CARBOHYDRATE PATHWAYS RELATED TO GLYCOLYSIS

activation, and leading to increased glycogen breakdown. Thus, the effect on glycogen breakdown is indirect, involving an intermediate enzyme phosphorylation. As we will appreciate in the next section, this provides a separate opportunity for regulatory control.

## Regulation of Glycogen Metabolism by Epinephrine

Whereas glucagon always leads to an increased concentration of cAMP, epinephrine produces different responses in different target tissues. This is the result of distinct plasma membrane-embedded receptor proteins. In muscle, epinephrine binds to β-adrenergic receptors coupled to similar G-proteins of the type we have just considered and leads to an increased cAMP and PKA activation.

However, in the liver of some species—including humans—epinephrine binds to the α-adrenergic receptor, which binds a separate G-protein. This G-protein activates a distinct effector system: the enzyme phospholipase C. Phospholipase C catalyzes the hydrolysis of a phosphatidylinositol-4,5-$P_2$ in the inner leaflet of the plasma membrane via the reaction (FIGURE 12.14):

$$PIP_2 \rightarrow IP_3 + DAG \qquad (12.5)$$

Both products of this reaction are regulatory molecules. Inositol trisphosphate ($IP_3$) is a water-soluble molecule that binds a protein on the endoplasmic reticulum membrane surface. Diacylglycerol (DAG) remains in the lipid phase of the membrane bilayer, activating a protein at the cytosolic surface.

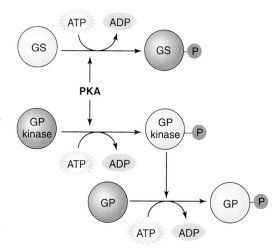

FIGURE 12.13 **Protein Kinase A Regulation of Glycogen Metabolism.** Protein kinase A (PKA) leads to the inactivation of glycogen synthase (GS) and the activation of glycogen phosphorylase (GP), with both enzymes in their phosphorylated state. PKA directly catalyzes GS phosphorylation; however, GP is not a substrate for PKA. Rather, an intermediate kinase, GP kinase, is a substrate for PKA. Once GP kinase is phosphorylated, this enzyme catalyzes the phosphorylation of GP.

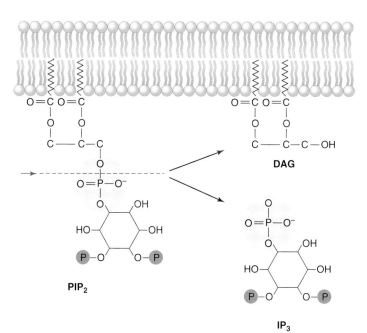

FIGURE 12.14 **Phospholipase C and the Formation of $IP_3$ and DAG.** The phospholipid $PIP_2$, phosphatidylinositol 4,5-bisphosphate, is a substrate for phospholipase C, the reaction shown here. One product, diacylglycerol (DAG) remains in the membrane; the other, inositol triphosphate ($IP_3$), is released to the cytosol.

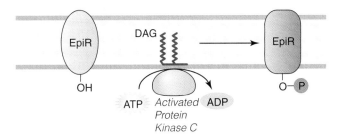

**FIGURE 12.15** **Action of Protein Kinase C**. Protein kinase C (PKC) is activated by the lipid-bound modulator diacylglycerol (DAG). As shown in the top part, PKC associates with the inner plasma membrane. Once active, as shown in the bottom part, PKC catalyzes the phosphorylation of its targets, such as the membrane-bound epinephrine receptor (EpiR). The phosphorylated EpiR is inactive.

$IP_3$ binds to and activates a calcium channel in the endoplasmic reticulum, allowing $Ca^{2+}$ to move from the lumen of the endoplasmic reticulum to the cytoplasm. One consequence of a rise in cytosolic $Ca^{2+}$ concentration is the activation of glycogen phosphorylase kinase. This enzyme is activated by $Ca^{2+}$ regardless of its phosphorylation status. Phosphorylase kinase in turn activates glycogen phosphorylase, so the latter enzyme is indirectly stimulated by the rise in $Ca^{2+}$. $Ca^{2+}$ has no effect on glycogen synthase, so it selectively activates glycogen breakdown in response to a rise in epinephrine.

Epinephrine also can lead to a rise in DAG through the activation of phospholipase C, although most DAG is derived from other phospholipid substrates, such as phosphatidylcholine. Moreover, the rise in DAG is slower than the rise in $Ca^{2+}$ concentration. In liver, for example, an increase in cytosolic $Ca^{2+}$ concentration occurs in less than one minute, but it can take at least 10 times as long for an appreciable increase in DAG concentration. DAG is an activator of **protein kinase C (PKC)**, which has multiple targets in the cell. One of these is the epinephrine receptor itself (**FIGURE 12.15**), where it serves to dampen its response to epinephrine binding. In this way, epinephrine binding to its receptor eventually leads to inactivation of that receptor, a phenomenon known as **down regulation**.

## Regulation of Glycogen Metabolism by Insulin

A rise in blood glucose level as a result of a meal triggers the release of *insulin* from the pancreatic β-cells. This 50-amino acid peptide hormone is an anabolic signal. In both muscle and liver, insulin leads to increased glycogen synthesis. The most prominent and well-studied effect is on muscle tissue.

Insulin stimulation of glycogen stores ultimately leads to the activation of glycogen synthase. A key proximal regulator of glycogen synthase is the enzyme **glycogen synthase kinase 3 (GSK3)**, so-named for a region of phosphorylation sites on glycogen synthase. The overall scheme is shown in **FIGURE 12.16**. Note that insulin binds its receptor,

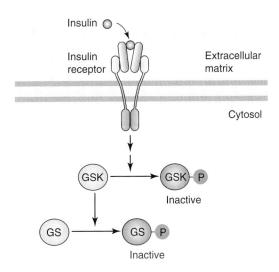

**FIGURE 12.16** **Summary of Insulin Stimulation of Glycogen Synthase**. Insulin binds the insulin receptor (IR), followed by a series of intermediate kinases and binding proteins, ultimately leading to the phosphorylation and inactivation of glycogen synthase kinase 3 (GSK), and thereby lifting the inhibition of glycogen synthase (GS).

ultimately leading to a phosphorylation and inactivation of GSK3. This lifts the inhibition that GSK3 tonically exerts on glycogen synthase, and increases the activity of glycogen synthesis.

A more detailed view of insulin action is presented in FIGURE 12.17. Step (1) shows insulin binding the exterior portion of the insulin receptor. This triggers a conformational change in the intracellular portion of the insulin receptor and activates the intrinsic kinase activity of the insulin receptor [step (2)]. Reminiscent of glycogenin discussed above, the insulin receptor is both enzyme and substrate. The activated insulin receptor catalyzes phosphorylation of **insulin receptor substrate 1 (IRS-1)** specifically at tyrosine residues; this is step (3).

The phosphorylated tyrosine residues of IRS-1 can bind to proteins that contain **Src homology-2 (SH2) binding domains** (FIGURE 12.18). This occurs in step (4) of Figure 12.17. **Phosphatidylinositol phosphate 3-kinase (PI3K)** has a regulatory subunit that contains SH2 binding domains, through which it binds the phosphorylated IRS-1. With its regulatory

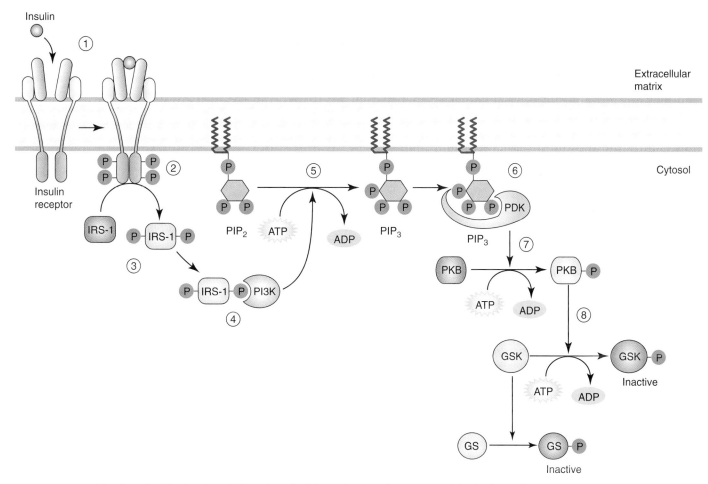

FIGURE 12.17 **The Insulin Mechanism.** When insulin binds the insulin receptor (1), the latter becomes a tyrosine kinase (2), phosphorylating itself as well as the insulin receptor substrate (IRS-1) (3). Phosphorylated IRS-1 binds targets such as the phosphatidylinositol 3-kinase (PI3K) (4), which catalyzes the formation of phosphatidyl 3,4,5-trisphosphate ($PIP_3$) (5). $PIP_3$ serves as a binding site for phospholipid dependent kinase (PDK) (6), which catalyzes the phosphorylation and activation of protein kinase B (PKB) (7). PKB catalyzes phosphorylation and inactivation of glycogen synthase kinase 3 (GSK) (8). This lifts GSK inhibition of glycogen synthase, leaving it less phosphorylated, and activated.

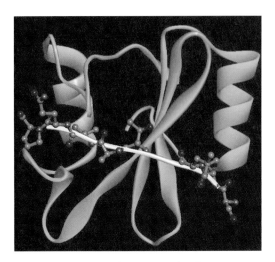

**FIGURE 12.18 The SH2 Domain.** The ribbon diagram shows the coil–sheet–coil pocket structure of this domain, with the phosphotyrosine residue shown in relation to it. (Lee, C.H., Kominos, D., Jacques, S., Margolis, B., Schlessinger, J., Shoelson, S.E., Kuriyan, J. 1994. Crystal structures of peptide complexes of the amino-terminal SH2 domain of the Syp tyrosine phosphatase. *Structure* 2:423–438. Protein Data Base #1AYA; prepared by Molly Steinbach.)

subunit bound, the catalytic subunit of PI3K becomes activated. The reaction it catalyzes is the phosphorylation of a membrane lipid [step (5)]:

$$\text{Phosphatidylinositol-P}_2\left(\text{PIP}_2\right) + \text{ATP}$$
$$\rightarrow \text{Phosphatidylinositol-P}_3\left(\text{PIP}_3\right) + \text{ADP} \quad (12.6)$$

As shown in Figure 12.17, the inositol head group of the lipid product has three phosphates that serve to anchor the next protein, **phospholipid-dependent kinase (PDK)**, which is step (6). In step (7), PDK catalyzes the phosphorylation and activation of protein kinase B (PKB). PKB, in turn, catalyzes the phosphorylation of GSK3 [step (8)]. This phosphorylation inactivates GSK3, leaving glycogen synthase less phosphorylated, and hence with greater activity. The result is an increase in glycogen synthesis.

The insulin receptor is one member of a class of **receptor tyrosine kinases**, which also includes the insulin-like growth factor 1 (the Growth Hormone mediator) and Epidermal Growth Factor. All have the same pattern. First, a hormone binds the receptor at the cell exterior, leading to a conformation change in the receptor protein that activates its intrinsic kinase activity. This leads to the phosphorylation of one or more intracellular proteins, and typically membrane modifications. These latter events produce a binding site for other proteins, which then become activated. The first of these events permits a *transduction* of a hormone such as insulin, which acts at the cell exterior, yet exerts alterations in the cell interior. The intracellular steps transmit the signal to specific targets, dictated by the extracellular hormonal milieu. Other proteins with SH2 domains besides PI3K can attenuate insulin action. This is not the only means of down-regulating insulin. Protein kinase C (which mediates adrenergic receptor inactivation as described above) also phosphorylates and inactivates the insulin receptor.

## Regulation of Glycogen Metabolism by AMP Kinase

The AMP-dependent protein kinase (AMPK) is an enzyme found in virtually all cells. Because the activator AMP is elevated under periods of energy deprivation, this type of control is distinct from hormonal regulation, which involves cell-to-cell signaling. The generation of AMP requires the breakdown of cellular ATP to ADP, followed by the action of adenylate kinase:

$$\text{ADP} + \text{ADP} \rightleftarrows \text{AMP} + \text{ATP} \quad (12.7)$$

The reaction in Equation 12.7 is responsible for increasing AMP concentration when extensive amounts of ADP are produced from ATP. This occurs when ATP utilization outstrips ADP rephosphorylation, for example during intense muscle contraction, or when energy supplying pathways such as mitochondria are inadequate as in anoxia. AMPK itself is the substrate of a protein kinase; in fact, phosphorylation of AMPK is essential for its activation. Binding AMP to AMPK makes it a better substrate for this upstream kinase. Once activated, AMPK phosphorylates and activates glycogen synthase kinase 3 (GSK3), the same enzyme that is inhibited by insulin. In this way, AMPK inhibits the synthesis of glycogen. Thus, in situations where energy generation is depressed, the accumulation of ADP and then AMP will activate AMPK and decrease the synthesis of glycogen in both liver and muscle.

# 12.2 Gluconeogenesis

The pathway for forming glucose from non-carbohydrate precursors occurs almost exclusively in the liver. Kidney has the complete pathway, but contributes little to total body blood glucose. Gluconeogenesis is also found in non-mammalian species, such as yeast and some plants (**BOX 12.2**).

In mammalian liver, lactate, glycerol, and many amino acids (such as alanine) are converted to glucose. These precursors take separate pathways initially, but they ultimately converge. We will consider the steps from lactate to glucose in detail here. The prime importance of this pathway is to maintain blood glucose as glycogen (the first line of supply) is depleted.

Most of the gluconeogenic reactions are reversals of the near-equilibrium steps of glycolysis. Only the metabolically irreversible steps—namely, pyruvate to phosphoenolpyruvate (PEP), fructose-1,6-$P_2$ to fructose-6-P, and glucose-6-P to glucose—involve specific gluconeogenic enzymes. We start with the first reaction of the pathway, catalyzed by lactate dehydrogenase.

## Lactate Dehydrogenase as a Gluconeogenic Enzyme

Lactate dehydrogenase catalyzes the last step of glycolysis and the first step in gluconeogenesis:

$$\text{Lactate} + \text{NAD}^+ \rightleftarrows \text{Pyruvate} + \text{NADH} \qquad (12.8)$$

In the gluconeogenic direction, the *standard free energy change* is extremely unfavorable, with $\Delta G° = +46$ kJ/mol. However, this has no metabolic significance. Under cellular conditions, the reaction is near-equilibrium, like most gluconeogenic steps. Hence, there is no regulation of this reaction apart from the substrate and product concentrations, all of which are close to their equilibrium values.

Like lactate DH, the other near-equilibrium steps in gluconeogenesis use the same enzymes as in glycolysis. Their direction is reversed in all cases due to the changes in the ratio of substrates to products, which allows the enzymes to be used for flow in either direction as metabolic circumstances dictate. Our focus in gluconeogenesis will be on the reactions required to enable flow in the reverse direction to the metabolically irreversible steps of glycolysis. This first is the conversion of pyruvate to PEP.

## Pyruvate to PEP

The conversion of pyruvate to PEP is the rate-limiting step in gluconeogenesis. The reverse process in glycolysis requires just one enzyme

## BOX 12.2: WORD ORIGINS

### What's Really New About Gluconeogenesis?

Two parts of the term *gluconeogenesis* indicate that it is new: both the direct Latin *neo*, and the concept of beginnings in *genesis*. Genesis also could be viewed as *generated*, but linked with *neo*, it strongly conveys the meaning of origins. But is it really new?

The **Cori cycle** is a combination of two pathways that span different tissues in the body. Glucose, once ingested, is stored largely in liver and muscle. Not only that, its metabolism by these tissues is quantitatively of paramount significance. Muscle takes up most of the glucose in the body by virtue of its mass, and uses it to support contraction. Some of the carbon is completely oxidized, and some is released from the muscle to the blood as lactate and carried to the liver. In the Cori cycle, carbon is shuttled between these two organs, as suggested in FIGURE B12.2.

Thus, gluconeogenesis can be viewed from this perspective as the regeneration of glucose carbon that was converted to lactate by the muscle in a continuing cycle. This led biochemist Mitchell Halperin to suggest the pathway be renamed *glucopaleogenesis*.

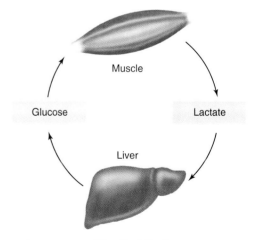

FIGURE B12.2 **The Cori Cycle.**

(pyruvate kinase). However, gluconeogenesis requires two enzymes, two high-energy phosphates, and the shuttling of substrates between the cytosol and mitochondria.

The first metabolically irreversible enzyme of gluconeogenesis is **pyruvate carboxylase**, which exists exclusively in the mitochondrial matrix. Cytosolic pyruvate is transported into the mitochondria, using the pyruvate transport system. The overall reaction catalyzed by pyruvate carboxylase is:

$$\text{Pyruvate} + CO_2 + ATP \rightarrow \text{Oxaloacetate} + ADP + P_i \qquad (12.9)$$

The incorporation of $CO_2$ requires the bound cofactor **biotin** (FIGURE 12.19), a vitamin that is taken into mammalian cells and covalently attached to the enzyme through an amide linkage to a lysine residue. Also indicated in Figure 12.19 is the position on biotin where $CO_2$ becomes transiently attached.

The mechanism is analogous to that of the pyruvate dehydrogenase complex, in that a "swinging arm" moves a molecular fragment between active sites. In the case of pyruvate carboxylase, there are just two active sites, residing in separate portions of the same enzyme. A biotin-lysine arm becomes carboxylated at one site, then moves to another, where the carboxyl group is added to pyruvate (FIGURE 12.20a). In the first site $CO_2$ is attached to biotin in a reaction assisted by ATP:

$$CO_2 + ATP + \text{Enz-biotin} \rightarrow \text{Enz-biotin-}CO_2 + ADP + P_i \qquad (12.10)$$

The mechanism is shown in Figure 12.20b. $CO_2$ (as its hydrate, bicarbonate) reacts with ATP to form a carboxyphosphate intermediate. One of the nitrogens of biotin acts as a nucleophile, attacking the carboxyphosphate and forming a carboxy-biotinylated enzyme.

Once the swinging arm moves the carboxyl-biotin portion to the second active site, the sequence shown in Figure 12.20c takes place.

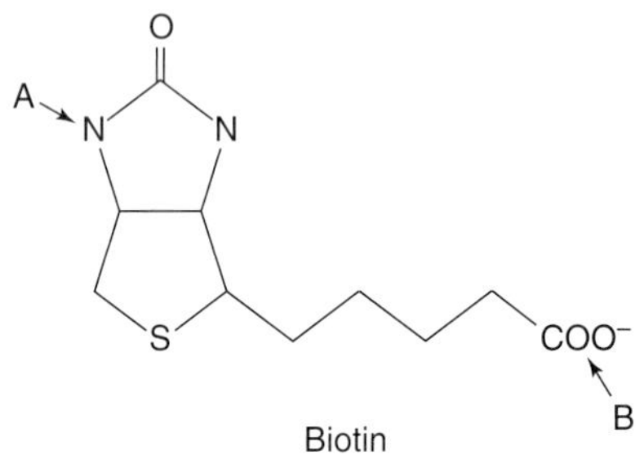

FIGURE 12.19 **Biotin.** The N atom at (A) is the point of attachment of bicarbonate. The carboxyl group at (B) forms an amide bond to a lysine residue, anchoring the cofactor to pyruvate carboxylase.

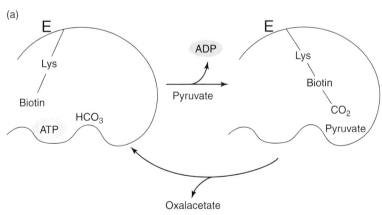

FIGURE 12.20 **Mechanism of Pyruvate Carboxylase.** (a) Overall mechanism: The biotin arm of the enzyme reacts with ATP and $HCO_3^-$ at one site, attaching $CO_2$ to the biotin, then moves the other, where carboxylation of pyruvate occurs. (b) Carboxylation of biotin: ATP and $HCO_3^-$ react to form carboxyphosphate, which is attacked by a nitrogen of the biotin to produce carboxybiotin. (c) Pyruvate carboxylation: A concerted electron rearrangement between pyruvate and carboxybiotin produces oxaloacetate and the enol form of biotin. The enol form is in equilibrium with the keto form of biotin. *(Continues)*

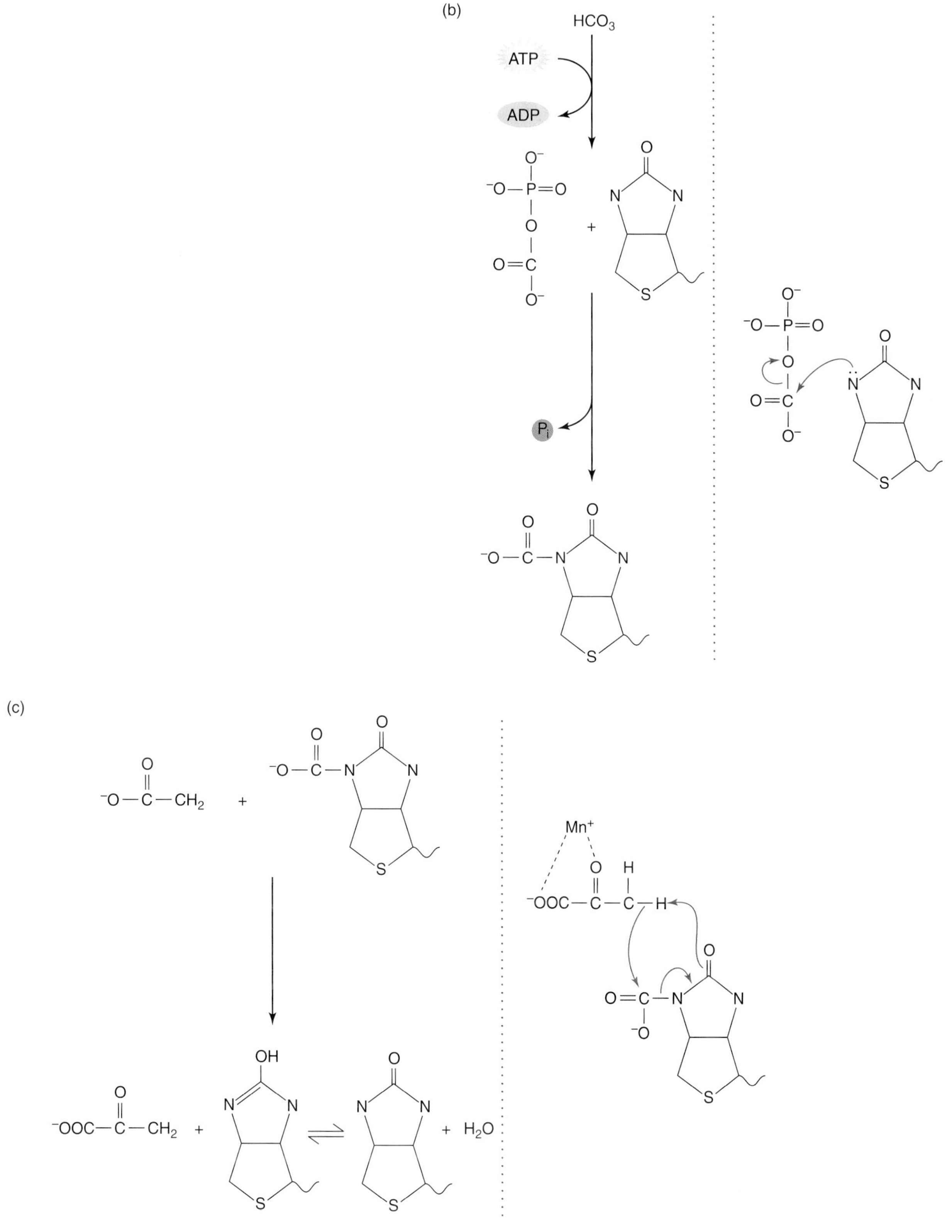

(b)

(c)

FIGURE 12.20 (Continued)

The series of concerted electron rearrangements shown attach $CO_2$ to pyruvate, producing oxaloacetate and the *enol* form of the attached biotin. This enol–biotin tautomer is in equilibrium with the original cofactor form as illustrated in the figure. Once biotin is restored to its original form, the enzyme can engage in another cycle of catalysis.

**Phosphoenolpyruvate carboxykinase** (usually abbreviated **PEPCK**) is the next step, catalyzing the reaction:

$$Oxaloacetate + GTP \rightarrow PEP + GDP + CO_2 \qquad (12.11)$$

The electron rearrangements that decarboxylate and phosphorylate the oxaloacetate are illustrated in FIGURE 12.21. The two substrates are brought in close proximity on the enzyme with the assistance of a $Mn^{2+}$ ion that attracts the negative charges of the carboxylate of oxaloacetate and the phosphate of GTP.

While there are no known allosteric modulators of PEPCK, this enzyme is widely considered the rate-limiting step of gluconeogenesis under a variety of physiological conditions. Its activity is regulated by genetic control of PEPCK protein production rather than the minute-to-minute modulation of enzyme activity, such as those described for glycogen metabolism. In brief, glucagon leads to increased PEPCK protein content, whereas insulin leads to diminished PEPCK protein content, thereby setting the pace of gluconeogenesis.

Note that $CO_2$ is incorporated into pyruvate carboxylase, only to be released by PEPCK. Hence, there is no *net* incorporation of $CO_2$ in gluconeogenesis, in contrast to the Calvin cycle in plants. When pyruvate carboxylase was first discovered, this distinction between *net* incorporation (plants) and using $CO_2$ as a temporary inclusion (animals) was unknown, which made the discovery controversial (**BOX 12.3**).

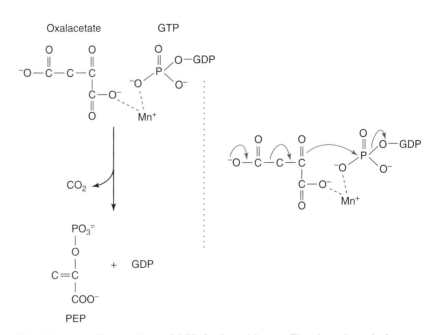

FIGURE 12.21 **Mechanism of PEP Carboxykinase.** The decarboxylation of oxaloacetate is linked to phosphorylation. Oxaloacetate and GTP are positioned appropriately on the enzyme, with the assistance of the metal $Mn^{2+}$, enabling electron rearrangement to form the products.

## BOX 12.3: $CO_2$ Fixation in Animals?

The incorporation of $CO_2$ in the pyruvate carboxylase reaction was first discovered in the early 1930s when isotopically labeled carbon became available to researchers. Incorporation of $CO_2$ into stable products in animal tissue was met with some surprise by the scientific community. Even then it was well understood that a major difference between plants and animals was whether it was possible to stably incorporate $CO_2$ (plants) or not (animals). Plants and some bacteria can synthesize all of their carbon from $CO_2$ by photosynthesis in specialized organelles (chloroplasts in plants), but these do not exist in animals. Skepticism over the pyruvate carboxylation findings was expressed by the reaction to the publication, by Harland Wood and Chester Werkman, as the "wouldn't work" reaction.

Only later did the pattern emerge in animal metabolism: whenever $CO_2$ is incorporated in a reaction such as pyruvate carboxylase, it is followed very shortly by a corresponding decarboxylation, such as PEPCK. The purpose is to energetically drive the latter reaction, because escape of $CO_2$ to the gas phase contributes to making reactions endergonic.

### Indirect Transport of Oxaloacetate from Mitochondria to Cytosol

In the sequence of pyruvate conversion to PEP, the production of oxaloacetate is catalyzed in the mitochondria by pyruvate carboxylase. However, a substantial fraction of PEPCK, which catalyzes the subsequent conversion of oxaloacetate to PEP, occurs in the cytosol. No transporter exists to permit oxaloacetate to cross the inner mitochondrial membrane. Instead, oxaloacetate is indirectly transported by the conversion to other molecules that do have transporters in the inner mitochondrial membrane. The two separate conversions that are possible are:

1. Reduction to malate
2. Transamination to aspartate

Both malate and aspartate have transport proteins in the inner mitochondrial membrane. Thus, they act as surrogates of oxaloacetate. After malate or aspartate cross the inner mitochondrial membrane and enter the cytosol, they are converted back to oxaloacetate. This is achieved by separate mitochondrial and cytosolic isozymes of the relevant enzymes.

Malate DH is present in mitochondria (also serving as part of the Krebs cycle):

$$Oxaloacetate + NADH \rightarrow Malate + NAD^+ \qquad (12.12)$$

The malate is then transported to the cytosol, where a separate malate DH catalyzes the reverse reaction:

$$Malate + NAD^+ \rightarrow Oxaloacetate + NADH \qquad (12.13)$$

Both malate DH enzymes catalyze near-equilibrium reactions. The direction of the reaction is thus dictated by the relative concentrations of substrates and products established by the production of oxaloacetate (by pyruvate carboxylase in the mitochondria) and by its removal (by PEPCK in the cytosol).

Alternatively, oxaloacetate can be converted to aspartate in a mitochondrial reaction catalyzed by aspartate aminotransferase:

$$\text{Oxaloacetate} + \text{Glutamate} \rightarrow \text{Aspartate} + \text{2-Ketoglutarate} \quad (12.14)$$

Aspartate crosses the inner mitochondrial membrane on its own transporter and is the substrate of the cytosolic enzyme, aspartate aminotransferase:

$$\text{Aspartate} + \text{2-Ketoglutarate} \rightarrow \text{Oxaloacetate} + \text{Glutamate} \quad (12.15)$$

Like the malate DH isozymes, the aspartate transaminases are also near-equilibrium.

The pathways for oxaloacetate transfer are illustrated in FIGURE 12.22. Note that, with malate as the intermediate, NADH is effectively transported from mitochondria to cytosol. However, lactate DH also generates NADH, and this is reoxidized later at glyceraldehyde-P DH: the reverse of the glycolytic sequence. Hence, gluconeogenesis from lactate uses aspartate as the intermediate, because the use of malate would cause an overproduction of NADH. More oxidized precursors for gluconeogenesis, such as pyruvate itself, use malate as an intermediate. The availability of two separate transport systems thus provides metabolic flexibility to meet the redox state needs for gluconeogenesis from different substrates.

### Fructose-1,6-P$_2$ to Fructose-6-P

The bypass of the phosphofructokinase step is catalyzed by fructose-1,6-P$_2$ phosphatase. The reaction is:

$$\text{Fructose-1,6-P}_2 \rightarrow \text{Fructose-6-P} + \text{P}_i \quad (12.16)$$

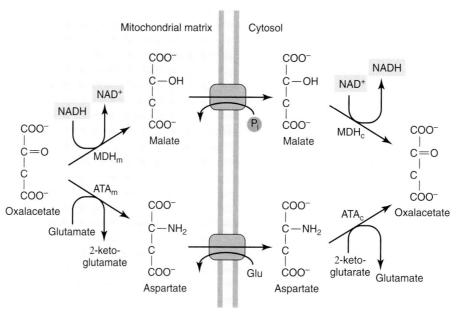

FIGURE 12.22 **Indirect Transport of Oxaloacetate Between Mitochondria and the Cytosol**. In the mitochondrial matrix (mito), oxaloacetate forms either malate (malate DH) or aspartate (aspartate aminotransferase). Both malate and aspartate, unlike oxaloacetate, have exchangers in the mitochondrial inner membrane, allowing them to appear in the cytosol. Isoforms of the same enzymes re-form oxaloacetate in the cytosol.

and it is metabolically irreversible. The mechanism is a typical hydrolysis, in which a nucleophilic attack by water on the 1-phosphate bond of the substrate releases inorganic phosphate ($P_i$).

The enzyme is allosterically activated by citrate and is inhibited by fructose-2,6-$P_2$. Both of these modulators have opposing effects on phosphofructokinase, for which citrate is an inhibitor and fructose-2,6-$P_2$ is an activator. Together these two enzymes constitute a **substrate cycle**, previously described as a *futile cycle* because their net result is simply to split ATP:

$$PFK: \text{ATP} + \text{Fructose-6-P} \rightarrow \text{ADP} + \text{Fructose-1,6-P}_2 \qquad (12.17)$$

$$FBPase: \text{Fructose-1,6-P}_2 \rightarrow \text{Fructose-6-P} + P_i \qquad (12.18)$$

*Net reaction:* $\text{ATP} \rightarrow \text{ADP} + P_i$

While it is true that the simultaneous operation of such opposing reactions does lead to a net loss of ATP energy, the advantage is that control of either gluconeogenesis or glycolysis becomes more sensitive when both directions are subject to regulatory modulation.

## Glucose-6-P to Glucose

The conversion of glucose-6-P to glucose is catalyzed by the enzyme **glucose-6-phosphatase**, which is actually a combination of a transport system and a catalytic phosphatase, involving the endoplasmic reticulum (FIGURE 12.23). Glucose-6-P is first transported into the lumen of the endoplasmic reticulum. Subsequently, a relatively nonspecific phosphatase (Pase in Figure 12.23) catalyzes the hydrolysis to glucose. The other product is $P_i$, which enters the cytosol using an exchange transport protein. As illustrated in the figure, this transporter carries glucose-6-P into the lumen of the ER as it carries $P_i$ into the cytosol. Finally, GLUT7 is a specific transporter for the entry of glucose into

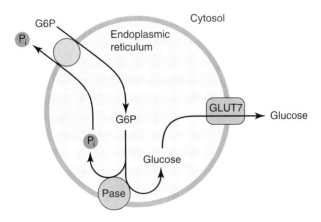

**FIGURE 12.23 Glucose-6-Pase.** This enzyme system has a transporter to bring G6P into the endoplasmic reticulum (ER), where a membrane-bound phosphatase (Pase) catalyzes glucose formation, and a transporter, GLUT7, moves glucose back into the cytosol.

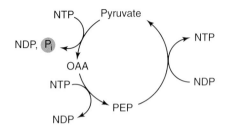

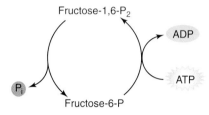

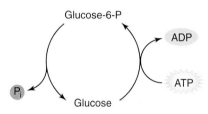

**FIGURE 12.24 Substrate Cycles.** Three substrate cycles catalyze a net hydrolysis of ATP (or NTP) with the simultaneous activity of gluconeogenesis and glycolysis. This provides fine control of both pathways.

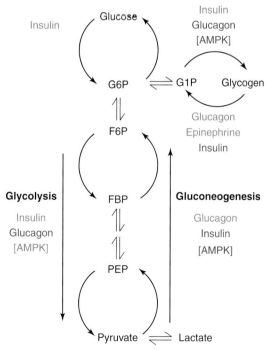

**FIGURE 12.25 Regulation of Carbohydrate Pathways.**

the cytosol. Glucose can exit the cell using the same GLUT2 plasma membrane transporter that is used for hepatic glucose uptake.

Regulation of the glucose-6-Pase system is exerted through increased protein formation (i.e., by control of genetic expression) of the catalytic subunit during starvation, directed by glucagon, and through its suppression in the fed state, directed by insulin. Together with glucokinase, glucose-6-Pase forms the following substrate cycle:

$$GK: \text{Glucose} + \text{ATP} \rightarrow \text{Glucose-6-P} + \text{ADP} \quad (12.19)$$

$$Glu\cos e\text{-}6\text{-}Pase: \text{Glucose-6-P} \rightarrow \text{Glucose} + \text{P}_i \quad (12.20)$$

*Net reaction:* $\text{ATP} \rightarrow \text{ADP} + \text{P}_i$

In addition to the two substrate cycles already described, the reactions between PEP and pyruvate constitutes a third substrate cycle, if we ignore the nucleotide specificities involved (i.e., we substitute NTP for either ATP or GTP):

$$PK: \text{PEP} + \text{NDP} \rightarrow \text{Pyruvate} + \text{NTP} \quad (12.21)$$

$$PC: \text{Pyruvate} + \text{NTP} \rightarrow \text{Oxaloacetate} + \text{NDP} + \text{P}_i \quad (12.22)$$

$$PEPCK: \text{Oxaloacetate} + \text{NTP} \rightarrow \text{PEP} + \text{NDP} \quad (12.23)$$

*Net reaction:* $\text{NTP} \rightarrow \text{NDP} + \text{P}_i$

Regulation of any step of this cycle can regulate gluconeogenesis. Thus, short-term control (activation) of gluconeogenesis can be exerted through PKA-dependent phosphorylation and the inactivation of pyruvate kinase. All three substrate cycles are shown in **FIGURE 12.24**.

## Pathway Integration: Glycolysis, Glycogen Metabolism, and Gluconeogenesis

We can gain biochemical insight into physiological process by examining how pathways are integrated. Here we examine two common events: the feeding–fasting transition and the resting–exercise transition. We will confine our discussion to the principal routes of carbohydrate metabolism: glycolysis, gluconeogenesis, and glycogen metabolism.

### The Feeding–Fasting Transition

During a meal, glucose moves from the digestive tract to the bloodstream, rising there to a peak of about 12.5 mM. The elevated blood glucose concentration triggers insulin release from the pancreas, which in turn activates glycogen synthesis in the liver and muscle. In addition, there is active glycolysis in liver and adipose tissue, and suppression of liver gluconeogenesis. Glycogenolysis is depressed by insulin in both the liver and muscle. In muscle, insulin stimulation of glucose transport is critical for the incorporation of the sugar into glycogen. These actions, summarized in **FIGURE 12.25**, counteract the rise in blood glucose. The carbohydrate pathways prominent in the fed state are glycogen synthesis and glycolysis.

| TABLE 12.1 Insulin, Glucagon, and Epinephrine: Origins and Targets | | |
|---|---|---|
| **Hormone** | **Origin** | **Target** |
| Insulin | Pancreatic β-cell | Liver, muscle, others |
| Glucagon | Pancreatic α-cell | Liver |
| Epinephrine | Adrenal gland | Liver, muscle, others |

After a meal (the **postprandial state**), glucose concentration drops, decreasing insulin release from the pancreatic β-cells, and stimulating glucagon release from the pancreatic α-cells. This results in the activation of glycogen breakdown and gluconeogenesis in the liver, and the inhibition of glycolysis, releasing glucose to the blood. The decrease in insulin concentration reduces glucose uptake by muscle. Epinephrine is released from the adrenal glands in response to a low blood glucose concentration. This signaling originates in the central nervous system, carried to the periphery via nerve fibers that synapse on the adrenal glands. Epinephrine stimulates glycogen breakdown in the liver. The glucose phosphate formed in the cell cannot enter glycolysis because this pathway is suppressed under fasting conditions; glucose is released to the blood. The three hormones discussed are summarized in Table 12.1.

### The Resting–Exercise Transition

The skeletal muscle, by virtue of its mass, is a major consumer of glucose even in the resting state. During exercise, skeletal muscle can increase its energy utilization by 100-fold. Accordingly, dramatic events in metabolic pathways are needed to support this increased energy utilization.

The regulation of muscle metabolism is largely intrinsic; the increased AMP that results from the extensive adenine nucleotide turnover activates AMPK and drives glucose uptake, inhibits glycogen formation, and stimulates glycolysis. During exercise, blood flow is shunted away from the liver and toward muscle, diminishing the metabolic role of liver in this circumstance. In the post-exercise state, blood flow returns to the liver, and excess lactate accumulated by muscle glycolysis is converted to glucose by liver gluconeogenesis. An extreme example is found in the alligator (**BOX 12.4**).

## BOX 12.4: Alligators and the Resting–Exercise Transition

The rapid and powerful muscle contraction of the alligator can be sustained for only several minutes. It is an extreme example of glycolysis supplied by glycogen breakdown in white muscle (fast or glycolytic muscle). This anaerobic metabolic response characterizes the similar brief, yet powerful muscle movements of the weight lifter.

In the alligator, lactate produced by muscle glycolysis can be converted to glucose by the liver (see the Cori cycle, Figure B12.2). However, the process of gluconeogenesis is relatively slow, and the long resting period of the alligator reflects this restoration process. A similar pattern applies to the more ancient but closely related species, the crocodile. (For a humorous poem on the habitat differences of the two, read the poem *The Purist* by Ogden Nash.)

A further effect of exercise is both intrinsic and hormonal: a rise in the concentration of intracellular $Ca^{2+}$. Epinephrine increases cytosolic $Ca^{2+}$ concentration, which stimulates breakdown of liver glycogen. In muscle, $Ca^{2+}$ concentration is elevated during the contraction cycle itself, which also serves to activate glycogen breakdown.

Two other pathways connected to glycolysis proceed at more modest rates: the pentose phosphate shunt and the more specialized route of galactose metabolism.

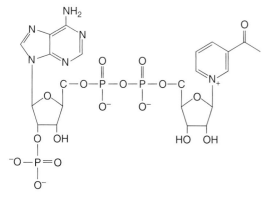

FIGURE 12.26 **NADP⁺**

## 12.3 The Pentose Phosphate Shunt

Like glycolysis, the pentose phosphate shunt occurs in virtually all cells in the body. This pathway has two distinct contributions to the cell. First, it partially oxidizes glucose-6-P in a reaction that also reduces NADP⁺ (FIGURE 12.26) to NADPH. These pyrimidines nucleotides are virtually identical to NAD⁺ and NADH, apart from an extra phosphate group that is located far from the nicotinamide ring. *In vitro* the redox states of the NAD⁺/NADH couple are identical to the NADP⁺/NADPH couple, so they have indistinguishable $\Delta E°$ values. In the cell, however, the NADP⁺/NADPH couple is far more reduced—by a factor of up to five orders of magnitude—than the NAD⁺/NADH couple. The distinction of the extra phosphate group between these cofactor pairs enables binding discrimination between different dehydrogenase enzymes. The NADPH functions as an electron donor for biosynthetic reactions.

The other function of the pentose phosphate shunt is the production of distinct sugar phosphates, most notably *ribose phosphates* used in the synthesis of nucleotides. The products of the pentose phosphate shunt are intermediates in glycolysis. Cells can readily control the relative amount of pentose or NADPH by means of a simple equilibrium, which we will reconsider once we have examined the overall pathway.

### Oxidative Stage

All of the reactions of the oxidative stage of the pentose phosphate shunt are metabolically irreversible. The overall process is:

Glucose-6-P + 2 NADP⁺
$$\rightarrow \text{Ribulose-5-P} + CO_2 + 2\ \text{NADPH} \quad (12.24)$$

The first step (FIGURE 12.27) is catalyzed by glucose-6-P DH, which is allosterically inhibited by NADPH. The mechanism (hydride transfer) is similar to other dehydrogenases (e.g., lactate DH). The product is the cyclic ester 6-phosphogluconolactone, which is hydrolyzed to the open chain 6-P-gluconate in a lactonase-catalyzed reaction. A second NADPH is produced in the 6-P-gluconate DH reaction. The mechanism is another example of oxidative decarboxylation, just like the mechanisms of the isocitrate DH reaction in the Krebs cycle and the malic enzyme in lipid metabolism.

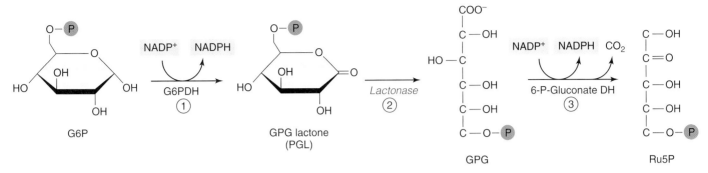

**FIGURE 12.27 Pentose Phosphate Shunt: Oxidative Reactions**. Two dehydrogenases, in reactions (1) and (3), produce NADPH. Hydrolysis of the lactone product of reaction (2) is catalyzed by a lactonase.

## Nonoxidative Stage

All of the enzymes in the nonoxidative stage of the pentose phosphate shunt catalyze near-equilibrium reactions. Starting from ribulose-5-P, there are two reactions that we have already encountered in the previous chapter, one catalyzed by an epimerase and one by an isomerase (FIGURE 12.28). Both are important in converting these molecules back into glycolytic intermediates, with the added assistance of enzymes **transketolase** and **transaldolase**. Some of the isomerase product, ribose-5-P, can be incorporated into ribonucleotides, as indicated in the Figure 12.28.

Transketolase and transaldolase catalyze reactions that remove a portion of one substrate and add it to another. Both overall reactions are illustrated in FIGURE 12.29. In the transketolase reaction, cleavage occurs between the carbonyl and the α-carbon; in the transaldolase reaction, cleavage occurs between the carbonyl and the β-carbon. In each case, the fragment is added to the aldehyde group of an acceptor. We have already come across the transketolase mechanism in our study of the Calvin cycle. In the pentose cycle, two transketolase reactions make up the final two reactions of the nonoxidative stage (FIGURE 12.30a and c). The transaldolase reaction has a mechanism is similar to aldolase, an enzyme of glycolysis (Chapter 8). The first part of both the aldolase and the transaldolase mechanisms is the same up to the release of the aldehyde. In aldolase, the enzyme-bound fragment is released by hydrolysis, whereas in transaldolase, the enzyme-bound fragment condenses with a new aldehyde (the second substrate), forming a new aldol product. There is just one transaldolase in the pentose phosphate shunt. The first three carbons of one substrate, sedoheptulose-7P, add to glyceraldehyde-3-P to form the aldol fructose-6-P, a glycolytic intermediate (Figure 12.30b).

Overall, the nonoxidative process converts the ribulose-5-P formed in the prior steps to the two intermediates of glycolysis, fructose-6-P and glyceraldehyde-P. To help visualize the nonoxidative stage of the pentose phosphate shunt, consider the pseudo–three-dimensional view of FIGURE 12.31a. Shown next to this are similar reactions of the Calvin cycle (Figure 12.31b). For both pathways, all of the intermediates at the vertices of the cube are identical.

For the pentose phosphate shunt, Ru5P carbon enters at X5P (epimerase) and at R5P (isomerase). These two undergo a transketolase

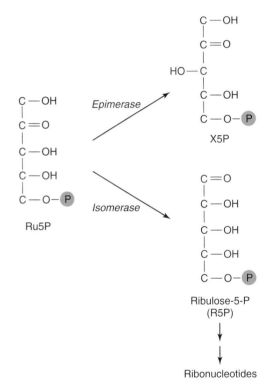

**FIGURE 12.28 C5 Interconversions in the Pentose Phosphate Shunt.** These reactions are the same as those of the Calvin cycle. The ketone xylulose P (X5P) and the aldose ribose 5P (R5P) are both intermediates of the nonoxidative portion of the pathway. R5P can additionally be utilized to form the pentose part of nucleotides.

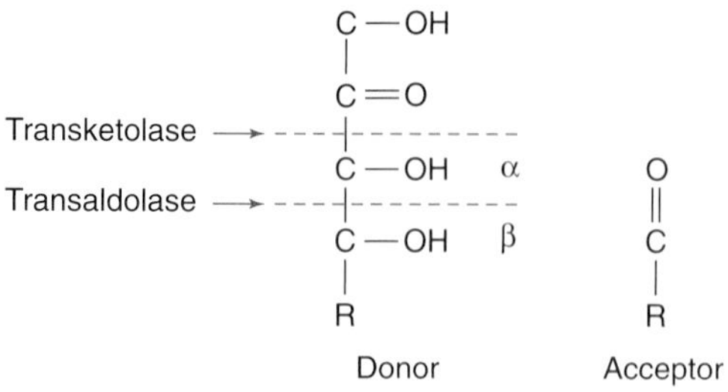

FIGURE 12.29 **Schematic Representation of Transketolase and Transaldolase Reactions.** The points of cleavage of each of these reactions are indicated. One is α to the carbonyl, the other is β to the carbonyl. In each case, the acceptor molecule is an aldehyde.

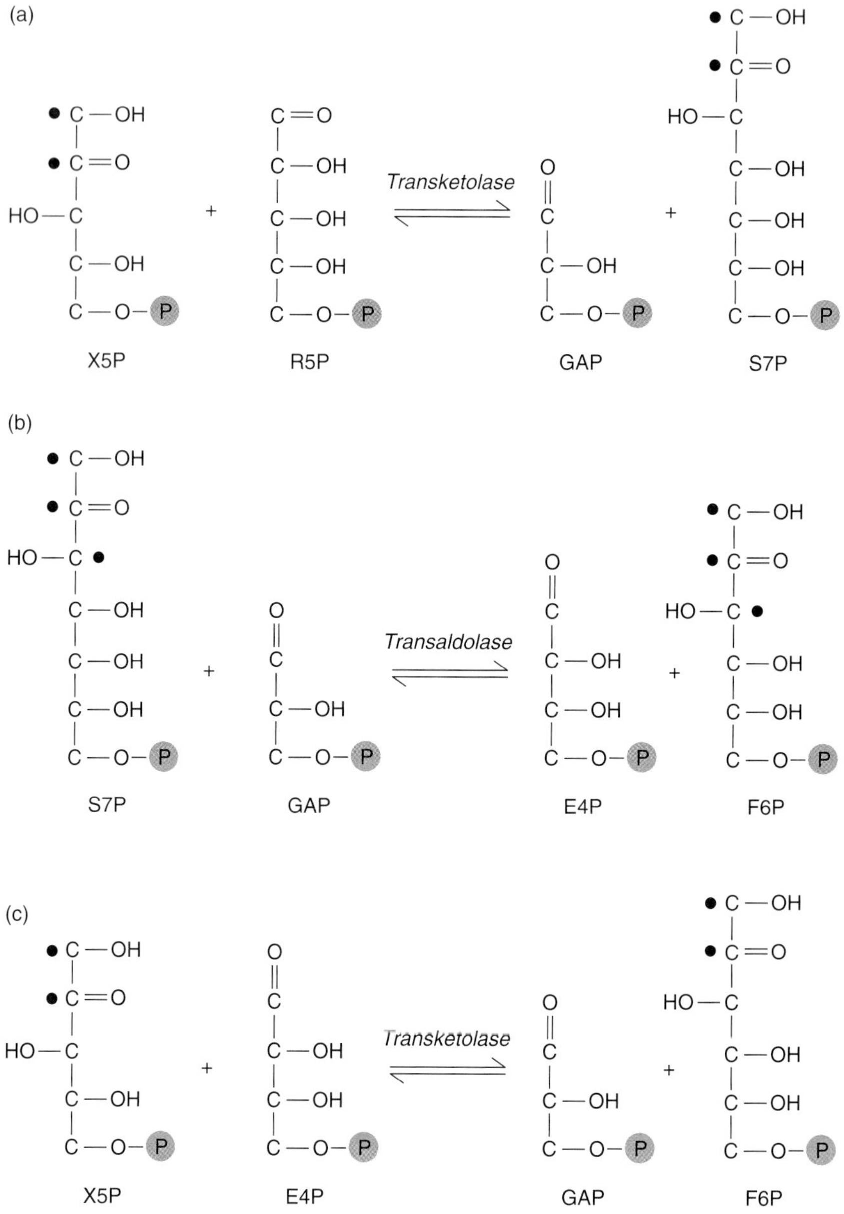

FIGURE 12.30 **Transaldolase and Transketolase Reactions of the Pentose Phosphate Shunt.** Dots indicate carbons transferred to the aldehyde acceptor in each reaction.

reaction, and their products undergo a transaldolase reaction (the bottom faces of the cube). Finally, the top face shows E4P (produced in the transaldolase reaction highlighted in red) reacting with X5P in a second transketolase reaction, producing the products GAP and F6P that enter glycolysis.

For the Calvin cycle segment, GAP is the sole input and X5P and R5P are the outputs. As indicated in Figure 12.31b, reactions involving DHAP (isomerase and aldolases) are required to generate three other intermediates that participate in transketolase reactions. The appearance of DHAP in the center of the bottom right square makes

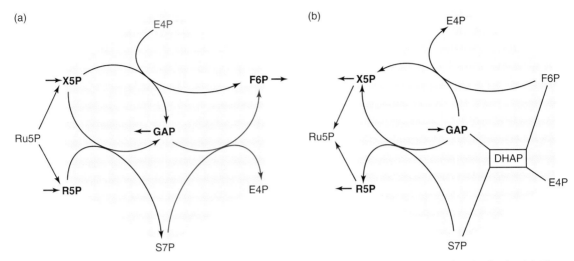

**FIGURE 12.31 Three-Dimensional Pathway Views of the Pentose Phosphate Shunt and the Calvin Cycle.** (a) The nonoxidative reactions of the pentose phosphate shunt. The transaldolase reaction is indicated with red arrows. (b) The Calvin cycle reactions are shown to be closely analogous, except that no transaldolase reactions occur. Aldolase conversions involving the intermediate dihydroxyacetone P (DHAP) occur instead. The reaction directions are the reverse of the pentose phosphate shunt.

this pathway distinct. Note that the Calvin cycle does not have a transaldolase reaction. Otherwise, the reactions directly correspond to the pentose phosphate shunt, but in the reverse direction.

Of the two pathways, the pentose phosphate shunt is more broadly distributed biologically, occurring in both plants and animals. It also displays metabolic flexibility: the oxidative and nonoxidative stages proceed at different rates depending upon the metabolic needs for NADPH or ribose. We next consider how this is accomplished.

### Distribution Between NADPH Production and Ribose-5-P

Cellular demand fluctuates between requirements for NADPH (for several biosynthetic pathways such as lipid synthesis) and ribose-5-P (for nucleotide synthesis). The relative needs can be accommodated due to the near-equilibrium nature of the nonoxidative segment of the pentose phosphate pathway. If demand for ribose carbon increases, more will be drawn from the pathway, in which case less will be returned to glycolytic intermediates. If relatively more NADPH is needed, then more carbon can be returned to glycolytic intermediates and less ribose-5-P is removed from this segment of the pathway. This simple means of endogenous control underscores the self-regulating nature of near-equilibrium reactions.

## 12.4 Galactose Utilization

Galactose is derived from lactose, the glucose-galactose disaccharide found in milk. After hydrolysis of lactose to glucose and galactose by the intestinal enzyme lactase, the galactose is uniquely metabolized by

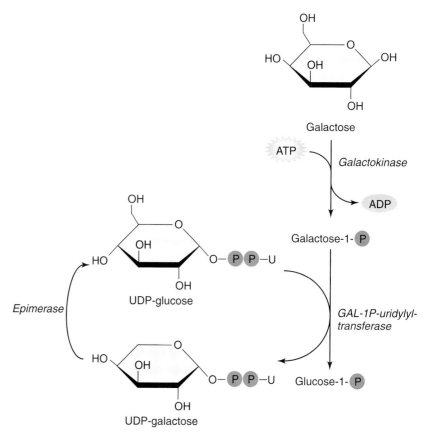

FIGURE 12.32 **Galactose Catabolism**. The phosphorylation is catalyzed by liver galactokinase, after which a uridylyltransferase catalyzes the exchange of sugars on the UDP carrier. Interconversion of UDP-galactose with UDP-glucose is catalyzed by an epimerase.

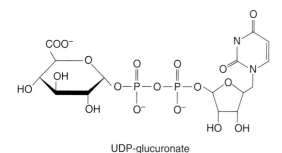

FIGURE 12.33 **UDP-Glucuronate**.

the liver through a pathway that intersects with glycogen metabolism via a UDP-sugar intermediate.

Galactose catabolism is outlined in FIGURE 12.32. First, galactose is converted to galactose-1-P, catalyzed by the liver-specific galactokinase (analogous to fructose metabolism in the liver). In the next step, the galactose moiety exchanges with the glucose moiety of UDP-glucose, catalyzed by uridylyltransferase. The products of this reaction are glucose-1-P and UDP-galactose. Finally, an epimerase converts UDP-galactose to UDP-glucose, recycling the substrate for the uridylyltransferase. As a result, the pathway produces net glucose-1-P and *catalytic* amounts of UDP sugars.

While galactose metabolism is a specialized route, UDP-glucose can be otherwise modified, such as through oxidation of its C6 to the acid, forming UDP-glucuronide (FIGURE 12.33). This is an important *activated glucuronide* used in many drug conjugation reactions, by which foreign substances in the body (**xenobiotics**) are converted in the liver to sugar conjugates that are thereby more water soluble and can be eliminated by the kidney.

# Summary

Pathways connected to glycolysis include glycogen synthesis, glycogen breakdown, gluconeogenesis, the pentose phosphate shunt, and galactose utilization. Most of the glycogen in mammals (and, consequently, most of its metabolism) occurs in liver and muscle. The synthesis of glycogen involves two key enzymatic steps: conversion of UDP-glucose to glycogen, catalyzed by glycogen synthase, and the formation of branches, catalyzed by branching enzyme. Complete synthesis of the glycogen molecule, a rare event, involves the small protein glycogenin, both as scaffold and enzyme for the first few glucosyl units. The key enzyme of glycogenolysis is glycogen phosphorylase, which catalyzes the removal of glucosyl units from the nonreducing end of the polymer to form glucose-1-P. Debranching enzyme catalyzes the relocation of chains from the $\alpha 1 \rightarrow 6$ branch points to the nonreducing ends of glycogen, leaving a single glucosyl unit in $\alpha 1 \rightarrow 6$ linkage. A glucosidase activity then removes this unit to form free glucose. Several control systems target two critical enzymes of glycogen metabolism: glycogen synthase and glycogen phosphorylase. In liver, glucagon leads to protein kinase A activation and phosphorylation (and inactivation) of glycogen synthase, and phosphorylation (and activation) of glycogen phosphorylase kinase, which in turn phosphorylates (and activates) glycogen phosphorylase. Epinephrine can increase intracellular $Ca^{2+}$, activating both glycogen phosphorylase kinase and glycogen phosphorylase activity. Insulin, through a receptor tyrosine kinase, leads to a series of binding and phosphorylation steps that inactivate glycogen synthase kinase, which lifts a tonic inhibition of glycogen synthase. A rise in AMP, triggered by increased ATP utilization, activates the AMP kinase, which phosphorylates and inactivates glycogen synthase. These regulatory systems appear widely in metabolic pathways, including gluconeogenesis. That pathway is essentially glycolysis in reverse, with three unique segments that are metabolically irreversible. First, pyruvate is converted to PEP using transport steps that link mitochondrial pyruvate carboxylase (pyruvate $\rightarrow$ oxaloacetate) to cytosolic PEP carboxykinase (oxaloacetate $\rightarrow$ PEP). Second, fructose-1,6-$P_2$ is converted to fructose-6-P, catalyzed by fructose bisphosphatase. Finally, the glucose-6-Pase system, which involves transport between the cytosol and the endoplasmic reticulum, converts glucose-6-P to glucose. All three of these segments constitute metabolic cycles, in which both glycolytic and gluconeogenic enzymes are simultaneously active. This incurs a net loss of a high-energy phosphate, but enables more exquisite metabolic control, as every enzyme of the cycle can serve as a control point. The pentose phosphate shunt is a side branch of glycolysis from glucose-6-P. One portion of this pathway—the oxidative arm—produces NADPH and pentose phosphates. The NADPH is used in reductive synthesis reactions. The pentose phosphates, primarily ribose-5-P, can be used for nucleotide synthesis or can be converted through reactions similar to those of the Calvin cycle to glycolytic intermediates. The assimilation of galactose, derived from the milk sugar lactose, is a liver pathway that converts galactose to UDP-galactose and then to UDP-glucose for entry into glycolysis.

## Key Terms

biotin
Cori cycle
down regulation
epinephrine
galactose catabolism
glucagon
gluconeogenesis
glucose-6-phosphatase
glycogen branching
  enzyme
glycogen storage diseases
glycogenin
glycogenolysis
glycogen phosphorylase
glycogen phosphorylase
  kinase

glycogen synthase
glycogen synthase kinase
  3 (GSK3)
glycogen synthesis
G protein
insulin
insulin receptor substrate
  1 (IRS-1)
intermediary metabolism
pentose phosphate shunt
phosphatidylinositol
  phosphate 3-kinase
  (PI3K)
phosphodiesterase
phosphoenolpyruvate
  carboxykinase (PEPCK)

phospholipid-dependent
  kinase (PDK)
postprandial state
protein kinase C (PKC)
pyridoxal phosphate
pyruvate carboxylase
ras oncogene
receptor tyrosine kinases
Src-homology 2 (SH2)
  binding domains
substrate cycle
transaldolase
transketolase
xenobiotic

## Review Questions

1. Prior to certain physical exercise, such as sprinting, a nutritionist recommends a diet heavy in carbohydrates. What is the rationale?
2. Explain why a transaldolase enzyme is necessary in the pentose phosphate shunt, but not in the Calvin cycle.
3. Which control systems are active during muscle contraction? During starvation?
4. How does insulin lead to the activation of glycogen synthase? Is there an insulin effect on glycogen phosphorylase?
5. Early studies with isotopic labeling showed that muscle could release glucose, which raised the possibility that muscle undergoes gluconeogenesis. What was the reason behind this conclusion?
6. How is the relative formation of pentose phosphate shunt products NADPH and ribose-P regulated?
7. What steps of glycogen metabolism are inhibited by gluconolactone?

## References

1. Chen, S. Y.; Pan, C. J.; Nandigama, K.; Mansfield, B. C.; Ambudkar, S. V.; Chou, J. Y. The glucose-6-phosphate transporter is a phosphate-linked antiporter deficient in glycogen storage disease type Ib and Ic. *FASEB J.* 22:2206–2213, 2008.
2. Hutton, J. C.; O'Brien, R. M. Glucose-6-phosphatase catalytic subunit gene family. *J. Biol. Chem.* 284:29241–29245, 2009.
3. Van Schaftingen, E.; Gerin, I. The glucose-6-phosphatase system. *Biochem. J.* 362:513–532, 2002.
4. Sillero, A.; Selivanov, V. A.; Cascante, M. Pentose phosphate and Calvin cycles. Similarities and three-dimensional views. *Biochem. Mol. Bio. Ed.* 34:275–277, 2006.
5. Taylor, S. S.; Buechler, J. A.; Yonemoto, W. cAMP-dependent protein kinase: Framework for a diverse family of regulatory enzymes. *Ann. Rev. Biochem.* 59:971–1005, 1990.

# 13

# Lipid Metabolism

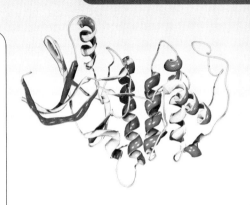

Image © Dr. Mark J. Winter/Photo Researchers, Inc.

**M**ost of the energy for sustaining animal activity is derived from lipids. As an extreme example, migratory birds can travel a thousand miles on their lipid stores, enough for more than a week of nonstop flight. Apart from energy, all cells require membranes, so they need to synthesize phospholipids. In the present chapter, we examine the pathways for utilizing lipid for energy, and the synthesis of the major lipid molecules: fatty acids, triacylglycerols, phospholipids, and cholesterol. In addition to their energy storage and barrier roles, lipids also serve as important signal molecules, such as inositol lipids. Some further examples are explored as well as an extension of the interaction between lipid and carbohydrates as energy sources. We begin our study of the major lipid pathways by examining the fate of dietary lipids.

## 13.1 Absorption of Dietary Lipids

There are two routes of lipid assimilation in mammals: endogenous and exogenous. The endogenous pathway is the biosynthesis of fatty acids from excess dietary carbohydrates and protein. The exogenous pathway is dietary lipid.

Most dietary lipids are triacylglycerols; the remainder is mostly phospholipid (representing membrane material of ingested foodstuffs), cholesterol, and its derivatives. After food is conducted into the stomach, the churning action of the organ's muscles produce a liquefied suspension called **chyme** (FIGURE 13.1). Once chyme enters

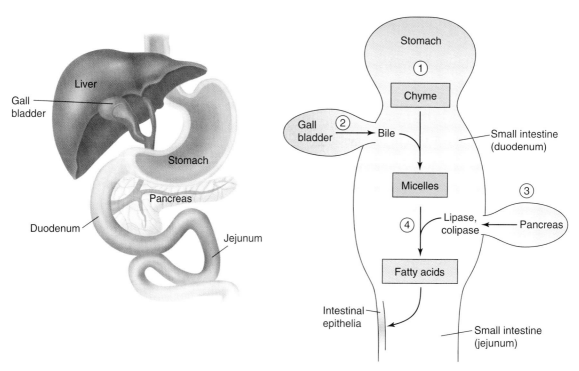

**FIGURE 13.1 Lipid Absorption: View From the Intestine.** (1) Stomach reduces food to a liquid suspension called chyme. (2) Bile stored in the gall bladder enters the intestine, emulsifying the fat. (3) Pancreas secretes lipase and colipase, which lead to (4) hydrolysis of triglycerides and uptake of fatty acids by the intestinal epithelia.

**CHAPTER 13** LIPID METABOLISM

the duodenum (the first portion of the small intestine), the lipid portion mixes with **bile,** which is a mixture of bile salts, phospholipids, and cholesterol. Bile is produced by the liver and is conducted to the small intestine through the bile duct. Along the way, bile also flows into a blind pouch, the gall bladder. This can be rapidly expelled for delivery of a large amount of bile during digestion of a particularly fatty meal. The mixture of bile with lipid forms smaller lipid droplets as the bile salts coat the outside of the particles, a process called **emulsification**.

The pancreas secretes several enzymes and proteins into the duodenum, including pancreatic lipase and co-lipase. Co-lipase assists the binding of the lipase to the suspensions of bile-salt-coated lipid particles. Lipase catalyzes the hydrolysis of the ester bonds in triacylglycerols, releasing fatty acids. Fatty acids are passively transported (**BOX 13.1**) across the plasma membrane of the intestinal epithelia. Once inside these cells, triacylglycerols (and cholesterol esters) are resynthesized (FIGURE 13.2). The cells also synthesize a protein called an **apolipoprotein**, which along with phospholipid forms the exterior of a spherical particle called a **chylomicron**. Chylomicrons leaving the intestinal epithelia enter the lymphatic system. Lymphatic vessels terminate at the vena cava, at which point chylomicrons enter the bloodstream. Once in the blood, the triacylglycerols in the chylomicrons are hydrolyzed back to fatty acids and taken into adipocytes. The hydrolysis reaction is catalyzed by esterases covalently attached to the walls of blood vessels near adipocytes. Fatty acids entering into the adipocytes reform triacylglycerols. Adipocytes are thus the storage site for fatty acids, in the form of triacylglycerols.

## BOX 13.1: Fatty Acid Binding Proteins

While fatty acid uptake into cells is described here as passive, there is extensive evidence for a large number of different proteins that may transport fatty acids across membranes, and may serve as intracellular binding proteins. Evidence in favor of this hypothesis comes largely from genetic knockout experiments or gene insertions and the subsequent observations of the effects on fatty acid metabolism. While these proteins may have some influence on transport or storage, controversy remains as to whether this is their true function. For one thing, the ability to bind a fatty acid is not a selective property of a protein. In fact, virtually all globular proteins bind fatty acids. Even proteins with fatty acid binding as their specific role can be replaced, as in the case of a group of people found to have no serum albumin and yet have no obvious clinical problems. In that case, a compensating increase in the blood γ-globulin protein served to bind blood fatty acids. In many cases, fatty acid transport across the plasma membrane is not saturable and, even when it is, this analysis is complicated by the binding of fatty acids to proteins in the cell exterior and the limited further metabolism of the fatty acid in the cell. Some fatty acid binding proteins have been found to have other functions. Whether this is ultimately found to be true of all of them remains an open question.

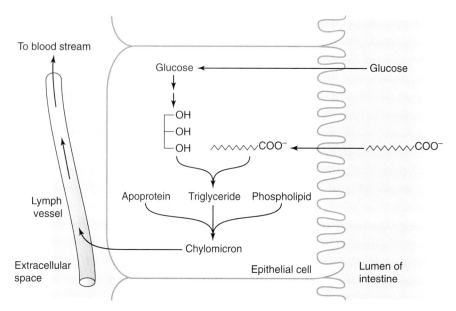

FIGURE 13.2 **Chylomicron Synthesis in Intestinal Epithelia.** Fatty acids are esterified in the epithelia and form chylomicron particles after being coated with apoprotein and phospholipids. The chylomicron exits the endothelial cell into the extracellular space and is transported through the lymphatic circulation until it enters the bloodstream.

# 13.2 Fatty Acid Oxidation

In the fed state, the blood concentration of free fatty acids is low (0.01 mM), but in the fasted state, this value can rise 100-fold. As blood glucose falls, the brain triggers adrenal glands to release epinephrine. Epinephrine binds to receptors in the adipocytes, which leads to the hydrolysis of stored triacylglycerols (and cholesterol esters), thus forming free fatty acids through a stimulation of hormone-sensitive lipase (FIGURE 13.3). The fatty acids, bound to the protein albumin in the blood, reach tissues for oxidation, such as muscle, heart, and liver.

Friedrich Wohler, famous for producing urea in the laboratory (1828) and dispelling organic mysticism, had observed even earlier (1824) that fatty acids could be completely oxidized to carbon dioxide and water in animals. Hence, fatty acid oxidation is actually the earliest pathway to be established. In an early form of metabolic labeling, Fritz Knoop in 1904 fed dogs fatty acids of various chain lengths having a phenyl group attached at the methyl end. From the urine products, a pattern was clearly established: carbon fragments are released in pairs from the carboxyl end of the chains.

The pathway for fatty acid oxidation can be conceptually divided into three steps (FIGURE 13.4): *activation,* the conversion of a fatty acid to an acyl-coenzyme A (acyl-CoA); *transport,* importing acyl-CoA into the mitochondria; and *β-oxidation,* the conversion of fatty acyl-CoA to acetyl-CoA. The activation step occurs in the cytosol, while β-oxidation occurs within the mitochondrial matrix. The fatty acids are mixtures of saturated and unsaturated molecules, and

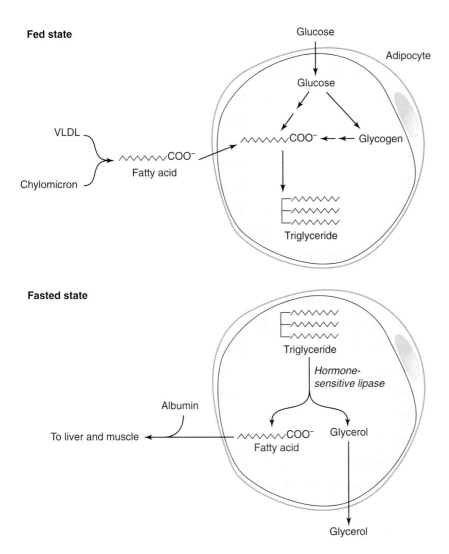

**Fed state**

Glucose

Adipocyte

Glucose

VLDL

Glycogen

∿∿∿∿COO⁻

Chylomicron

Fatty acid

Triglyceride

**Fasted state**

Triglyceride

*Hormone-
sensitive lipase*

Albumin

To liver and muscle

∿∿∿∿COO⁻

Glycerol

Fatty acid

Glycerol

**FIGURE 13.3** **The Adipocyte in Fed and Fasted States.** In the fed state (top), the adipocyte synthesizes fatty acids from glucose and receives them from chylomicrons (or VLDL, representing liver synthesized fat). The fatty acid is stored as triacylglycerol. In the fasted state (bottom), triacylglycerol is hydrolyzed in the adipocytes. The released fatty acid travels in the blood bound to albumin, and is delivered to muscle and liver for oxidation.

usually have 14, 16, or 18 carbons. We will present the pathway for the oxidation of palmitate, and briefly consider modifications needed to oxidize unsaturated fatty acids and those with different chain lengths.

## Activation

The activation step converts the fatty acid to the CoA ester:

$$R\text{-}COO^- + CoA\text{-}SH + ATP \rightarrow R\text{-}C(=O)SCoA + AMP + PP_i$$

$$(13.1)$$

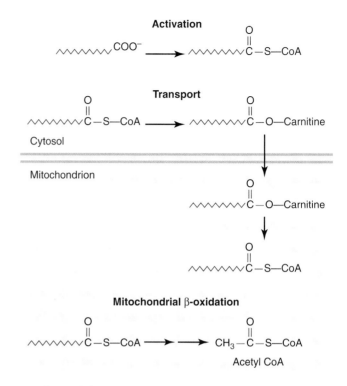

**FIGURE 13.4 General Steps in Fatty Acid Oxidation.**

This reaction is catalyzed by fatty acyl-CoA synthase, and is metabolically irreversible, assisted by rapid pyrophosphate hydrolysis, catalyzed by pyrophosphatase:

$$PP_i \rightarrow 2P_i \tag{13.2}$$

Pyrophosphatase is also employed in uridine diphosphate (UDP)–glucose formation for glycogen synthesis. The mechanism of fatty acyl-CoA synthase is analogous to asparagine synthetase. Here, the thiol of the CoA cofactor (explicit when written as CoA-SH) serves as the nucleophile that displaces the acid phosphate intermediate.

### Transport

Movement of fatty acyl-CoA into the mitochondria is indirect because no transport proteins exist for CoA or its esters in the mitochondrial inner membrane. A shuttle similar to that for NADH is needed. Fatty acid oxidation uses the **carnitine shuttle** for this purpose. First, fatty acyl-CoA is converted to its carnitine ester in a reaction catalyzed by carnitine acyl transferase I (CAT I; FIGURE 13.5). The hydroxyl group of carnitine initiates an attack on the carbonyl of the acyl-CoA, releasing CoA-SH and producing fatty acyl carnitine. Carnitine, named for its discovery in meat (muscle tissue), is present in high concentration in muscle and liver. Animals both synthesize carnitine and obtain it from the diet, so it usually not limiting for fatty acid oxidation.

In the next step of the carnitine shuttle, a mitochondrial transport protein catalyzes the exchange of cytosolic acyl-carnitine for

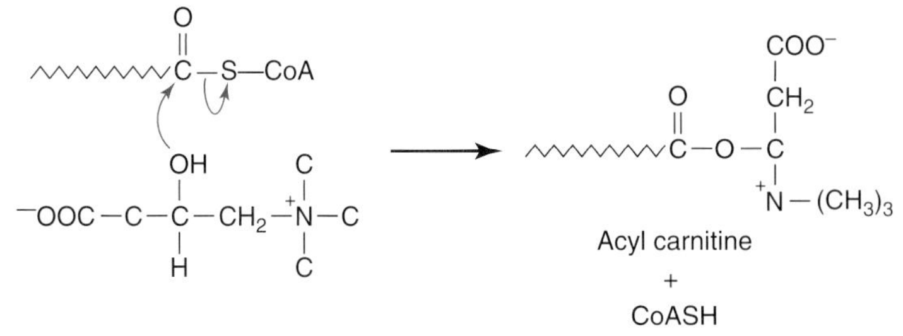

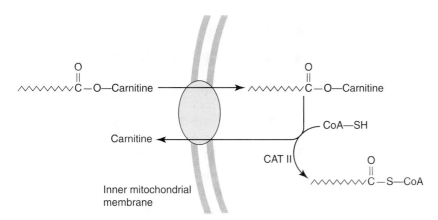

FIGURE 13.5 **Carnitine Acyl Transferase I (CAT I).**

mitochondrial carnitine. Once acyl-carnitine enters the mitochondrial matrix, an isozyme of carnitine acyl transferase, CAT II, catalyzes the conversion of fatty acyl-carnitine to fatty acyl-CoA. Thus, the CAT II reaction runs in the reverse direction to CAT I (FIGURE 13.6). The carnitine released by CAT II is recycled by exchange transport back to the cytosol, where it can participate in the CAT I reaction.

CAT I is the key regulatory step in the cellular process of fatty acid oxidation. The regulator is malonyl-CoA, which allosterically inhibits CAT I. As we will see below, malonyl-CoA is formed in the process of fatty acid synthesis in the liver, so there is a coordination of synthesis and oxidation of fatty acids.

## β-Oxidation

The mitochondrial fatty acid oxidation steps remove two carbons at a time from the chain of the fatty acid, at a position β to the carbonyl. As a result, this sequence, which is step 3 of the overall process, is known as β-oxidation, having the overall reaction

$$(\text{Fatty acyl})_n -\text{CoA} \rightarrow (\text{Fatty acyl})_{n-2} -\text{CoA} + \text{Acetyl-CoA} \quad (13.3)$$

The pathway is illustrated in FIGURE 13.7, along with the steps of the Krebs cycle that are analogous to the first three reactions of β-oxidation. The Krebs cycle sequence from succinate to oxaloacetate

FIGURE 13.6 **Import of Fatty Acids into Mitochondria.** An exchange protein catalyzes the import of acyl-carnitine to the mitochondria and the export of carnitine to the cytosol. The carnitine is the product of CAT II.

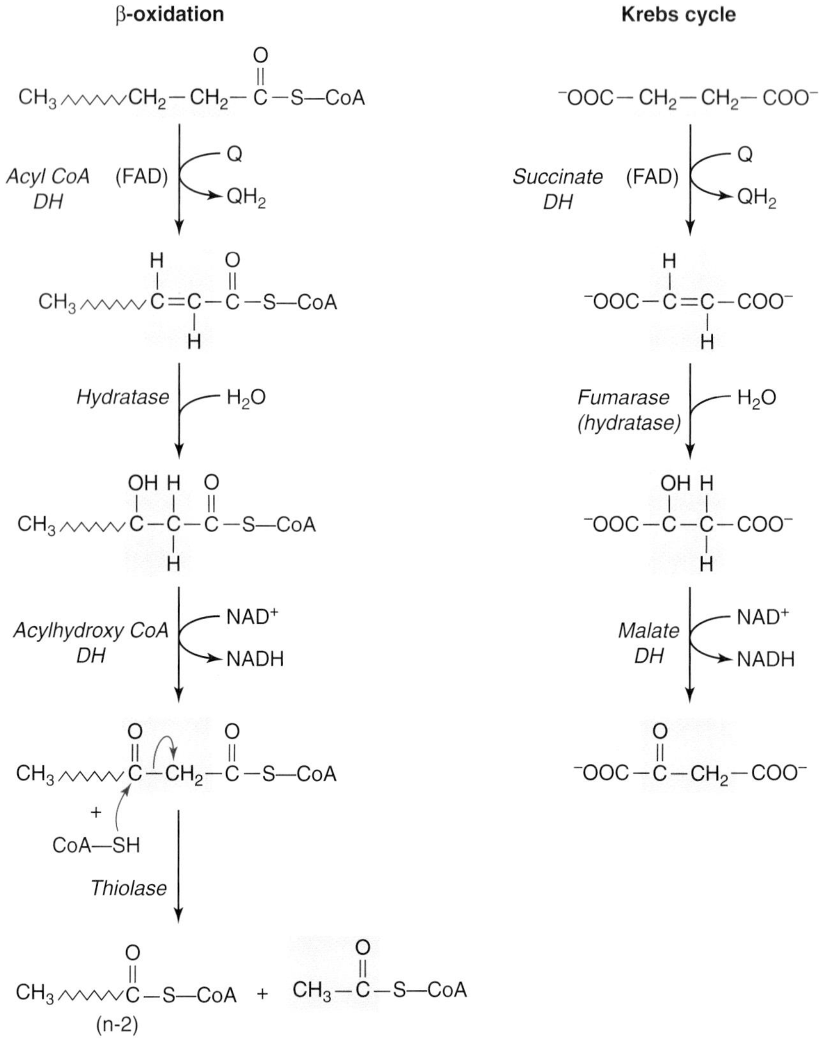

**β-oxidation**

**Krebs cycle**

$$CH_3 \text{—}\!\!\!\!\sim\!\!\!\!\text{—} CH_2 \text{—} CH_2 \text{—} \overset{\overset{\displaystyle O}{\|}}{C} \text{—S—CoA}$$

Acyl CoA DH  (FAD)  Q → QH_2

$$^-OOC\text{—}CH_2\text{—}CH_2\text{—}COO^-$$

Succinate DH  (FAD)  Q → QH_2

$$CH_3\text{—}\!\!\!\!\sim\!\!\!\!\text{—}\overset{\overset{\displaystyle H}{|}}{C}=\underset{\underset{\displaystyle H}{|}}{C}\text{—}\overset{\overset{\displaystyle O}{\|}}{C}\text{—S—CoA}$$

$$^-OOC\text{—}\overset{\overset{\displaystyle H}{|}}{C}=\underset{\underset{\displaystyle H}{|}}{C}\text{—}COO^-$$

Hydratase — H_2O

Fumarase (hydratase) — H_2O

$$CH_3\text{—}\!\!\!\!\sim\!\!\!\!\text{—}\overset{\overset{\displaystyle OH}{|}}{C}\text{—}\underset{\underset{\displaystyle H}{|}}{\overset{\overset{\displaystyle H}{|}}{C}}\text{—}\overset{\overset{\displaystyle O}{\|}}{C}\text{—S—CoA}$$

$$^-OOC\text{—}\overset{\overset{\displaystyle OH}{|}}{C}\text{—}\underset{\underset{\displaystyle H}{|}}{\overset{\overset{\displaystyle H}{|}}{C}}\text{—}COO^-$$

Acylhydroxy CoA DH  NAD^+ → NADH

Malate DH  NAD^+ → NADH

$$CH_3\text{—}\!\!\!\!\sim\!\!\!\!\text{—}\overset{\overset{\displaystyle O}{\|}}{C}\text{—}CH_2\text{—}\overset{\overset{\displaystyle O}{\|}}{C}\text{—S—CoA}$$

$$^-OOC\text{—}\overset{\overset{\displaystyle O}{\|}}{C}\text{—}CH_2\text{—}COO^-$$

\+

CoA—SH

Thiolase

$$CH_3\text{—}\!\!\!\!\sim\!\!\!\!\text{—}\overset{\overset{\displaystyle O}{\|}}{C}\text{—S—CoA} \quad + \quad CH_3\text{—}\overset{\overset{\displaystyle O}{\|}}{C}\text{—S—CoA}$$
(n-2)

**FIGURE 13.7 Mitochondrial Steps in Fatty Acid Oxidation.** The mitochondrial steps, known as β-oxidation, are shown on the left side. Along the right side are similar reactions that occur in the Krebs cycle. Only the last reaction, catalyzed by thiolase, is distinctive.

involves reduced ubiquinone ($QH_2$) production, a hydration, and NADH production. In the first step of β-oxidation, catalyzed by fatty acyl-CoA dehydrogenase, electrons are removed from the two methylene carbons adjacent to the carbonyl and transferred to the mobile cofactor $QH_2$. The complex catalyzing this transfer contains the fixed electron carriers $FAD^+$ and Fe–S protein. This process is analogous to the succinate dehydrogenase complex, which receives electrons from two adjacent methylene carbons that are transferred first to FAD, then to an Fe–S protein, and finally to $QH_2$. Both enzyme complexes consist of integral membrane proteins and interact with the mobile cofactor Q.

A hydratase catalyzes the second reaction, which is the addition of water across a double bond, with the same mechanism as fumarase. Both reactions are stereospecific, producing exclusively the L-isomer. The hydroxyl group is subsequently oxidized to produce a ketone in a reaction catalyzed by acyl-hydroxy dehydrogenase. The

**CHAPTER 13** LIPID METABOLISM

mechanism—removal of a hydride ion to form NADH—is identical to the malate dehydrogenase reaction in the Krebs cycle.

The product of the preceding reactions, a ketoacyl-CoA, is attacked by the sulfur of a CoA-SH molecule in a thiolase-catalyzed reaction. This yields acetyl-CoA, and a shortened ($n − 2$) acyl-CoA. From this point, the newly shortened acyl-CoA repeats the four steps just described. While the steps are repeated, different isozymes exist for the first enzyme (acyl-CoA DH) with distinct specificities for long-, medium-, and short-chain acyl-CoA molecules.

## Ancillary Enzymes

Ancillary enzymes are needed to oxidize unsaturated, very long chain, or other fatty acids. Unsaturated fatty acids such as oleate ($18\Delta^9$) are abundant in mammalian blood, and can be oxidized by a slight modification of the pathway described (**BOX 13.2**). After three rounds of β-oxidation (FIGURE 13.8), an intermediate is formed that has a *cis* double bond *beta* to the carbonyl. This species is a substrate for an isomerase that catalyzes conversion to a *trans*-double bond *alpha* to the carbonyl. The remaining oxidation steps proceed as before.

The rare odd-numbered fatty acid chains require ancillary reactions for their complete oxidation (see the Appendix).

Finally, chains that are relatively short or very long have a different catabolic fate. Fatty acids having eight or fewer carbons (*medium chain* fatty acids) bypass the cytosolic activation step because they are not substrates for the activation step in the cytosol, and are

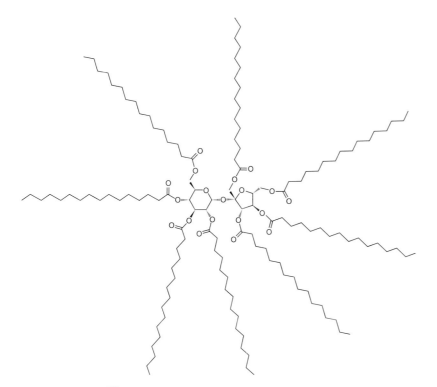

FIGURE B13.2 **Olestra.**

## BOX 13.2: WORD ORIGINS

### Oleate

The stem of *oleate* is from the Greek *olea,* or olive, a source of this fatty acid. Triacylglycerols of the olive flesh consist mostly of oleic acid, an 18:1 fatty acid. The related term *oleaginous* is used as a rough synonym for anything oily. *Oleacious* is more strictly defined as being a member of the family Oleacea (of which *Olea* is a subcategory, a genus); these include lilac and tropical trees. Surprisingly, the *oleo* part of *oleomargarine* has an unusual origin: It was the name of the first margarine, produced in the 1860s from a beef fat extract subjected to a churning process. Since the 1950s, margarine has been produced by the partial chemical saturation of plant oils, including oleic acid, but also several others, such as peanut, soybean, and cottonseed oils. The high heat required for the process causes rearrangement of some *cis* double bonds to the *trans* configuration. Currently there is a health concern over consumption of margarine because eating *trans* fats is correlated with cardiovascular disease.

More recently, Proctor and Gamble produced the eponymous *olestra* (FIGURE B13.2) as a fat substitute. Chemically, olestra resembles triacylglycerols because it is a multiple fatty acid ester, but the backbone is sucrose rather than glycerol. Note that not all of the hydroxyl groups of sucrose are substituted in commercial preparations of olestra. Olestra is not broken down in the intestine by esterases, and it is not incorporated as a dietary lipid, so it is *calorie free*. However, use of olestra is controversial, because many consumers suffered intestinal disorders arising from undigested lipids passing through their intestinal tract. An additional problem is that olestra decreases the absorption of fat-soluble vitamins. Nonetheless, the enormous increase in the incidence of obesity and consequent diabetes has spurred the development of this and undoubtedly many future new fat substitutes for consumers.

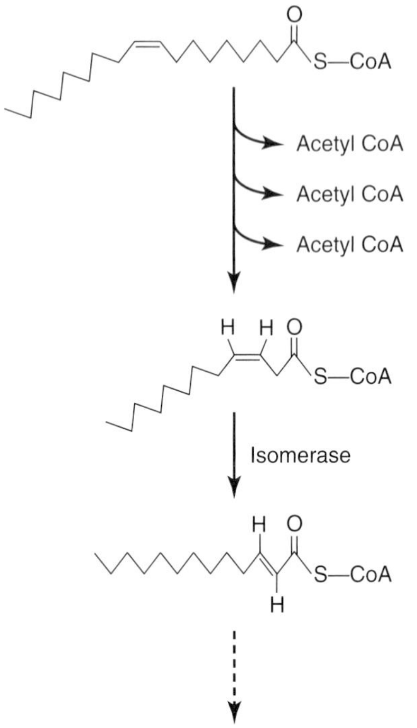

**FIGURE 13.8 Oleate Oxidation.** After three rounds of β-oxidation, oleate becomes a 3-*cis* fatty acyl-CoA. This is a substrate for an isomerase that leads to the 2-*trans* fatty acid that can proceed through the remainder of the usual steps for fatty acid oxidation.

activated directly in the mitochondrial matrix. As they do not require a transport step to enter mitochondria, these short-chain fatty acids are carnitine independent and escape malonyl-CoA control of their oxidation. Very long chain fatty acids, such as erucic acid ($22\Delta^{13}$), derived from canola oil, are partially oxidized (a few rounds of β-oxidation) in the peroxisomes, a separate organelle from the mitochondria. The oxidation reactions are not energy linked, but rather reduce $O_2$ to $H_2O_2$, which is then converted to $H_2O$ through the action of peroxidases in the peroxisome. The shorter fatty acids produced in this way are medium chain fatty acids, activated within the mitochondria, are not substrates of CAT I, and thus are not regulated by malonyl-CoA.

# 13.3 Ketone Body Metabolism

Ketone bodies are produced exclusively in the liver as a *spillover pathway* during excess production of acetyl-CoA, usually arising from fatty acid oxidation. These molecules can also be derived from the oxidation of certain amino acids, which are thus known as **ketogenic** amino acids. Typically, a rise in ketone bodies in the blood parallels an increase in blood fatty acid concentration.

The first stage in ketone body synthesis is the formation of the acetyl-CoA carbanion (**FIGURE 13.9a**). In the next step, the acetyl-CoA anion displaces CoA from a separate molecule of acetyl-CoA to form acetoacetyl-CoA (Figure 13.9b). In the subsequent 3-hydroxy-3-methylglutaryl (HMG)-CoA synthase step, another acetyl-CoA carbanion attacks the carbonyl of acetoacetyl-CoA, resulting in the tertiary alcohol, HMG-CoA. The HMG-CoA degrades to acetyl-CoA and acetoacetate through a lyase-catalyzed reaction (Figure 13.9c). Acetoacetate, the first of the ketone bodies, reacts with NADH in a reaction catalyzed by β-hydroxybutyrate dehydrogenase (a near-equilibrium enzyme), to form the product β-hydroxybutyrate, the second ketone body (**FIGURE 13.10**). The liver has no reactions that further metabolize either acetoacetate or β-hydroxybutyrate, and these are both transported out of the cell.

Once in the bloodstream, acetone—a third ketone body—arises from nonenzymatic decarboxylation of acetoacetate (Figure 13.10). This decarboxylation reveals the inherent instability of β-ketoacids. When ketone bodies are elevated, it is possible to detect *acetone breath* due to the solubility of acetone in the gas phase (**BOX 13.3**).

Oxidation of ketone bodies (**FIGURE 13.11**) occurs in the mitochondria. First, β-hydroxybutyrate is oxidized to acetoacetate, a reaction catalyzed by β-hydroxybutyrate dehydrogenase, an isozyme of the liver form. The next step is absent from liver, which explains why liver cells cannot oxidize ketone bodies. This is the transferase-catalyzed reaction in which succinyl-CoA reacts with acetoacetate, forming succinate and acetoacetyl-CoA. This can be viewed as bypassing the succinyl-CoA synthetase step of the Krebs cycle, analogous to bypassing P-glycerate kinase of glycolysis in red blood cells to

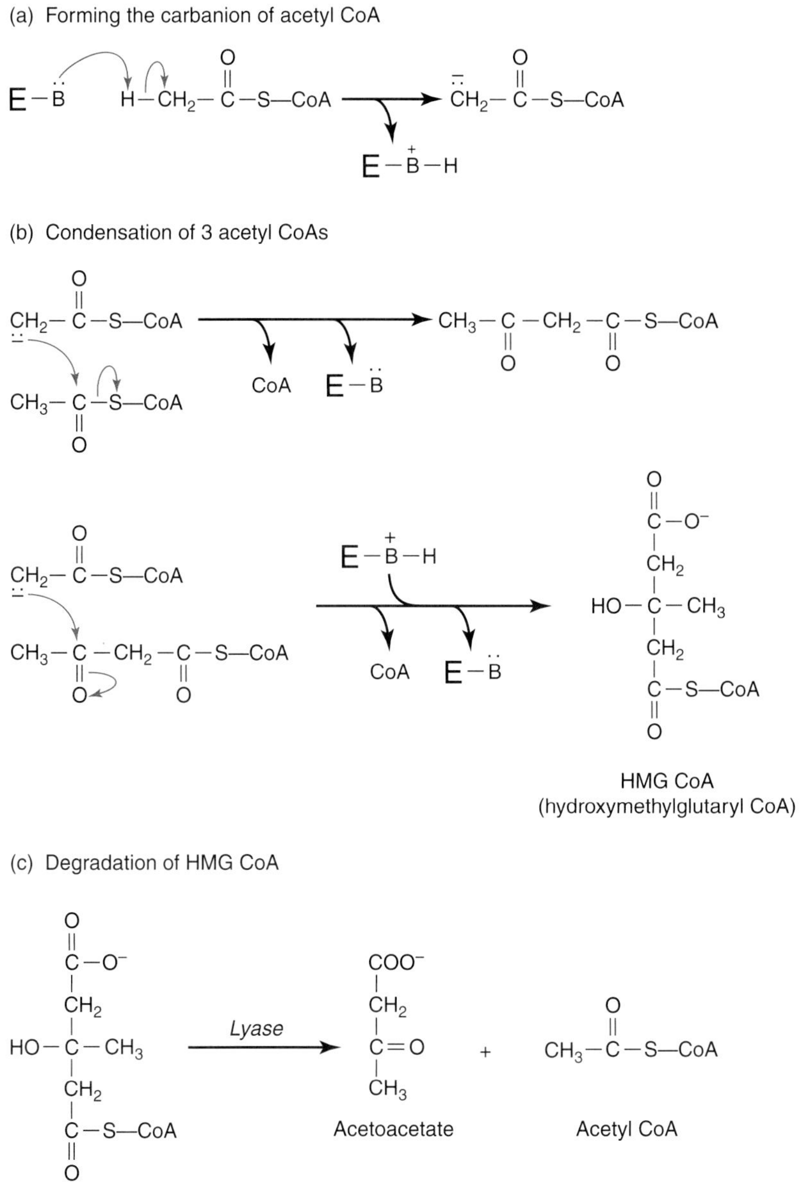

## (a) Forming the carbanion of acetyl CoA

## (b) Condensation of 3 acetyl CoAs

HMG CoA
(hydroxymethylglutaryl CoA)

## (c) Degradation of HMG CoA

FIGURE 13.9 **Ketogenesis I: Formation of Acetoacetate.**

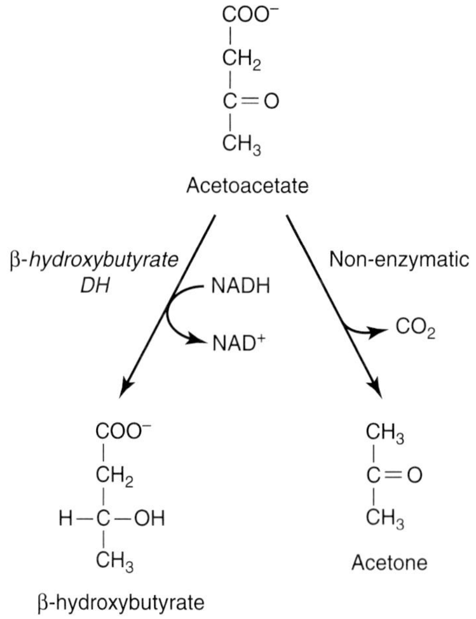

FIGURE 13.10 **Ketogenesis II: Formation of β-Hydroxybutyrate Acetone from Acetoacetate.**

produce 2,3-bisphosphoglycerate, the regulator of hemoglobin. In both cases, the bypass incurs an energy cost: ATP in glycolysis and GTP in the Krebs cycle. In glycolysis, 2,3-bisphosphoglycerate is a regulatory molecule and the energy cost is kept low by the fact that the flux through the bypass shunt is low. Ketone body oxidation can be a major pathway, but it leads to the generation of substantial amounts of energy, so its energy cost is negligible. Acetoacetyl-CoA is subsequently converted to two acetyl-CoA molecules, which then are oxidized by the Krebs cycle.

Many tissues can utilize ketone bodies for energy; heart is particularly active. In severe starvation, the brain can derive a considerable amount of energy by the oxidation of these compounds, utilizing more ketone bodies than glucose.

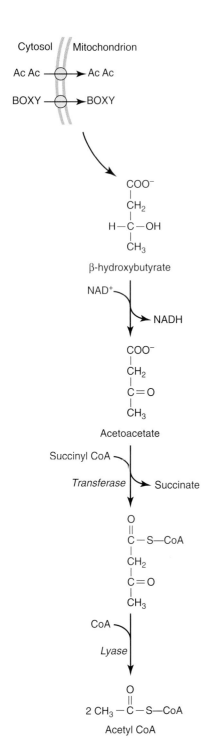

FIGURE 13.11 **Ketone Body Oxidation.**

## BOX 13.3: Acetone Breath

Acetone is the only volatile ketone body, so it is the only one that can be detected by smelling the breath. Under most conditions, no acetone appears to be present, but certain conditions of extreme carbohydrate deficiency, combined with an abundance of fatty acids, lead to the accumulation of acetoacetate, and subsequently acetone adds a distinctive sweet odor to the breath. Common conditions of this sign of ketoacidosis are acute alcoholic toxicity, severe starvation, diabetes mellitus, and diets very low in carbohydrates.

# 13.4 Fatty Acid Biosynthesis

Liver and adipose tissue can synthesize fatty acids from acetyl-CoA, which itself can be derived from carbohydrates or amino acids. Acetyl-CoA initially is formed in the mitochondria and must be exported to the cytosol for fatty acid synthesis. We can conceptually divide the pathway of fatty acids synthesis into three stages (FIGURE 13.12):

(**a**) Export of acetyl-CoA to the cytosol
(**b**) Carboxylation of acetyl-CoA to malonyl-CoA
(**c**) Sequential addition of two-carbon fragments to form palmitate

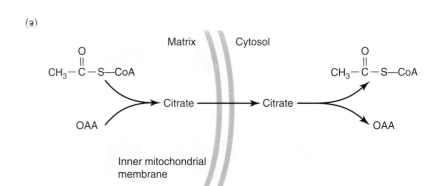

FIGURE 13.12 **General Steps in Fatty Acid Biosynthesis.** (A) Export of mitochondrial acetyl-CoA to the cytosol, carried by citrate. (B) Carboxylation of acetyl-CoA (C) The joining of one acetyl-CoA with the sequential input of seven malonyl-CoA molecules and the release of seven $CO_2$ molecules forms palmitate.

## Export of Acetyl-CoA to the Cytosol

The export of acetyl-CoA to the cytosol is achieved in an indirect way via the citrate shuttle (FIGURE 13.13). Essentially, acetyl-CoA produced in the mitochondria is carried by citrate into the cytosol. Mitochondrial citrate is formed from the condensation of acetyl-CoA with oxaloacetate in the citrate synthase reaction, encountered previously as part of the Krebs cycle. A selective transporter in the mitochondrial inner membrane enables export of citrate to the cytosol. Cytosolic citrate is a substrate for citrate lyase, and the products are acetyl-CoA and oxaloacetate. This reaction involves ATP cleavage and is distinct from the mitochondrial citrate synthase reaction. Phosphorylation of the 1-carboxyl group of citrate yields the acid phosphate (FIGURE 13.14a). This is then displaced by CoA-SH, leading to citryl-CoA (also an intermediate of citrate synthase). The steps leading to acetyl-CoA and oxaloacetate (Figure 13.14b) are familiar ones: acid–base catalysis and C–C bond cleavage.

While acetyl-CoA is destined for fatty acid synthesis, oxaloacetate, the other product of citrate lyase, continues through the citrate shuttle (Figure 13.13). First, a cytosolic isozyme of malate dehydrogenase catalyzes the conversation of oxaloacetate to malate, using NADH. Next, malate is oxidatively decarboxylated to pyruvate, catalyzed by malic enzyme. The electrons from malate are captured as NADPH,

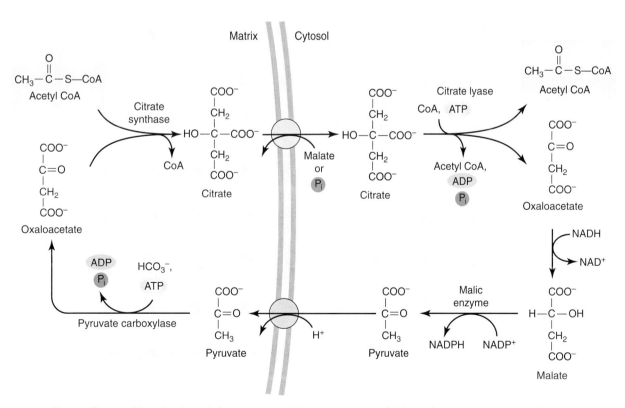

**FIGURE 13.13 Citrate Export Shuttle.** Acetyl-CoA reacts with oxaloacetate (OAA) to form citrate in the mitochondria. This citrate is then transported into the cytosol, where it regenerates acetyl-CoA via the citrate lyase reaction. The oxaloacetate product is reduced by NADH to malate, which is then decarboxylated at the malic enzyme reaction to pyruvate. The entry of pyruvate into the mitochondria and carboxylation to oxaloacetate completes the cycle, so that the net reaction is the moving of cytosolic acetyl-CoA to the cytosol and converting NADH to NADPH through the two cytosolic redox reactions.

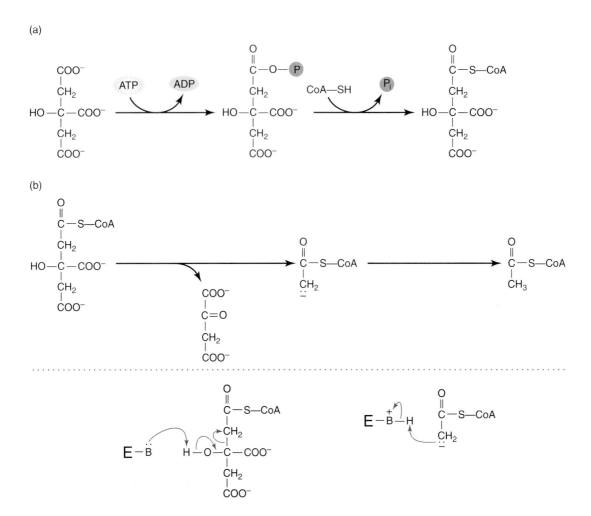

**FIGURE 13.14 Citrate Lyase Mechanism.** Citryl-CoA formation (a) is formed by the creation of an acid phosphate intermediate using ATP and displacing the phosphate with CoA-SH. The splitting portion (b) is initiated by proton abstraction and electron migration, producing the intermediate carbanion, which becomes protonated to acetyl-CoA.

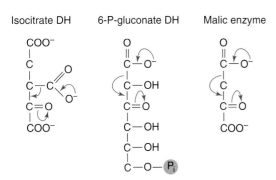

**FIGURE 13.15 Common Intermediates of Ketoacid Decarboxylation.** The movement of electrons to release $CO_2$ and subsequently to split a C–C bond is a theme common to isocitrate DH, 6-P-gluconate DH, and malic enzyme.

the reducing agent used in the latter steps of fatty acid synthesis. The mechanism of malic enzyme is the same as that of the other two oxidative decarboxylation reactions: isocitrate dehydrogenase (in the Krebs cycle) and 6-phosphogluconate dehydrogenase (in the pentose phosphate shunt). The three intermediates formed after oxidation of the alcohol and the electron rearrangements leading to decarboxylation are compared in **FIGURE 13.15**.

Once pyruvate has been formed, it can re-enter the mitochondria and reform oxaloacetate to complete the cycle. Just as we have seen in the Krebs cycle, the citrate shuttle accepts acetyl-CoA as a two-carbon input. Whereas the Krebs cycle produces two net $CO_2$ molecules per turn, the citrate shuttle outputs two carbons per turn in the form of one cytosolic acetyl-CoA molecule.

## Carboxylation of Acetyl-CoA to Malonyl-CoA

The carboxylation of acetyl-CoA, catalyzed by acetyl-CoA carboxylase, is the principal regulatory site of fatty acid biosynthesis. Malonyl-CoA

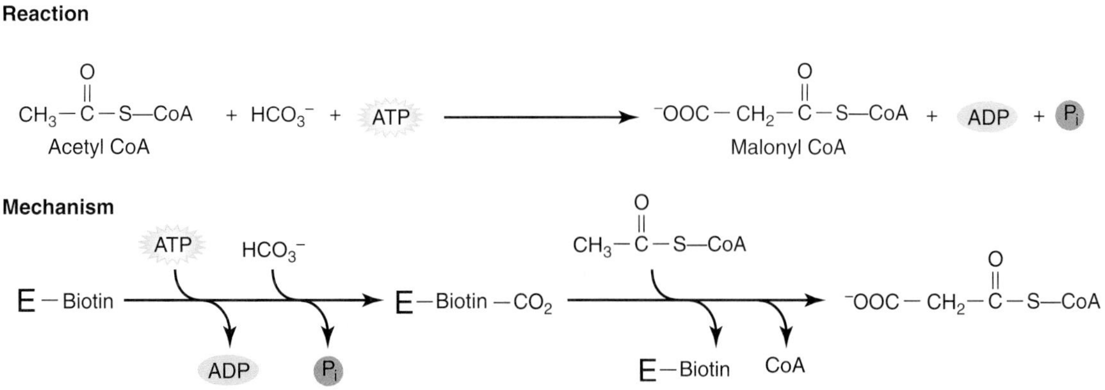

## Reaction

## Mechanism

## Regulation

Allosteric:  Stimulation — citrate
            Inhibition — fatty acyl CoA

Covalent:  Phosphorylation — AMP activated kinase
          and Protein Kinase A

FIGURE 13.16 **Acetyl-CoA Carboxylase**. The reaction is a carboxylation of acetyl-CoA, through a mechanism very similar to pyruvate carboxylase, that is, an intermediate carboxylation of a biotinylated enzyme. Both allosteric (citrate, fatty acyl-CoA) and phosphorylation (AMPK, PKA) control the rate of this enzyme.

formation is considered to be the "committed step" in the formation of fatty acids. The reaction itself (FIGURE 13.16) is another example of a biotin-dependent carboxylation, with strong similarities to the pyruvate carboxylase reaction. Just as with pyruvate carboxylase, ATP is used to activate bicarbonate to a carboxyphosphate that can be added to the biotin cofactor. This is then donated to the methyl carbon of acetyl-CoA to form malonyl-CoA at a separate site on the enzyme.

Acetyl-CoA carboxylase (Figure 13.16) is controlled by both allosteric regulation and covalent modification (phosphorylation and dephosphorylation). The positive regulator, citrate, can be viewed as a feed-forward activation, similar to that exerted by the product of fructose 1,6-bisphosphate on pyruvate kinase in glycolysis. The negative modulator, fatty acyl-CoA, decreases the activity of this enzyme, reflecting an increase in fatty acids and hence less need for *de novo* biosynthesis.

Acetyl-CoA carboxylase is subjected to phosphorylation in two ways. One involves cyclic AMP-dependent protein kinase (protein kinase A) and leads to an inactivation of the enzyme. This is an appropriate response to a starvation signal, with elevated cyclic AMP elicited in liver cells by the action of glucagon and in adipocytes by the action of epinephrine. This is opposed by insulin, which leads to a less phosphorylated, more active acetyl-CoA carboxylase. The other mode of phosphorylation is through the AMP-activated protein kinase. This enzyme is controlled by changes in the levels of AMP in cells and also causes phosphorylation and inhibition of acetyl-CoA carboxylase. AMP-activated kinase is a ubiquitous presence in cells, and an important regulator in many cells including liver, adipocyte, brain, and muscle. The regulatory circuit involving acetyl-CoA carboxylase in lipid metabolism is summarized in **BOX 13.4**.

## BOX 13.4: Regulatory Circuit of Lipid Metabolism

A number of regulatory circuits converge on the enzyme acetyl-CoA carboxylase (CBX), which in liver plays a central role in regulating the pathways of fatty acid synthesis, fatty acid oxidation, and triacylglycerol synthesis (FIGURE B13.4). Malonyl-CoA, the product of CBX, inhibits CAT I, the key enzyme required for entry of long-chain fatty acids into the mitochondria for β-oxidation. Fatty acyl-CoA, the substrate of CAT I, is itself an allosteric inhibitor of CBX. Other means of regulation of CBX are: allosteric stimulation by citrate, hormonal regulation (activation by insulin; inhibition by glucagon or epinephrine in adipocytes), and inhibition by 5'-adinosine monophosphate kinase (AMPK). The summation of cellular events can thus fine-tune the partition of carbon among the lipid pathways.

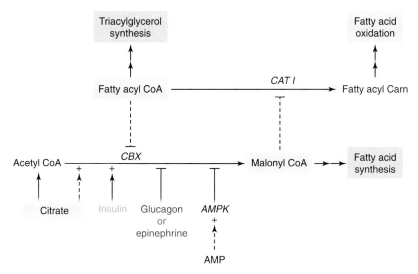

FIGURE B13.4 Pathways are shown in boxes, regulatory compounds in ovals, enzymes in italics, and hormones in blue. Allosteric regulation is indicated by dashed lines. Activation is indicated in green, and inhibition is in red.

### Sequential Addition of Two-Carbon Fragments to Form Palmitate

In mammals, the ensuing reactions of fatty acid biosynthesis are catalyzed by a protein complex called palmitate synthase. Like the pyruvate dehydrogenase complex of mitochondria, the individual reaction products remain attached to the complex. They are passed from one active site to the next until the final product, palmitate, is released. We will examine the individual reactions under descriptive headers: *loading, condensation, reduction, dehydration,* and *reduction.* These reactions are catalyzed by physically separate enzymes in bacteria.

### Loading

In preparation for the synthesis, the CoA-SH esters of both malonyl-S-CoA and acetyl-S-CoA are transferred to the thiol side chain of **acyl carrier protein (ACP-SH)**. The malonyl-ACP product is used for subsequent condensations, as detailed in the following step. As illustrated in FIGURE 13.17a, acetyl-S-ACP reacts with a thiol group

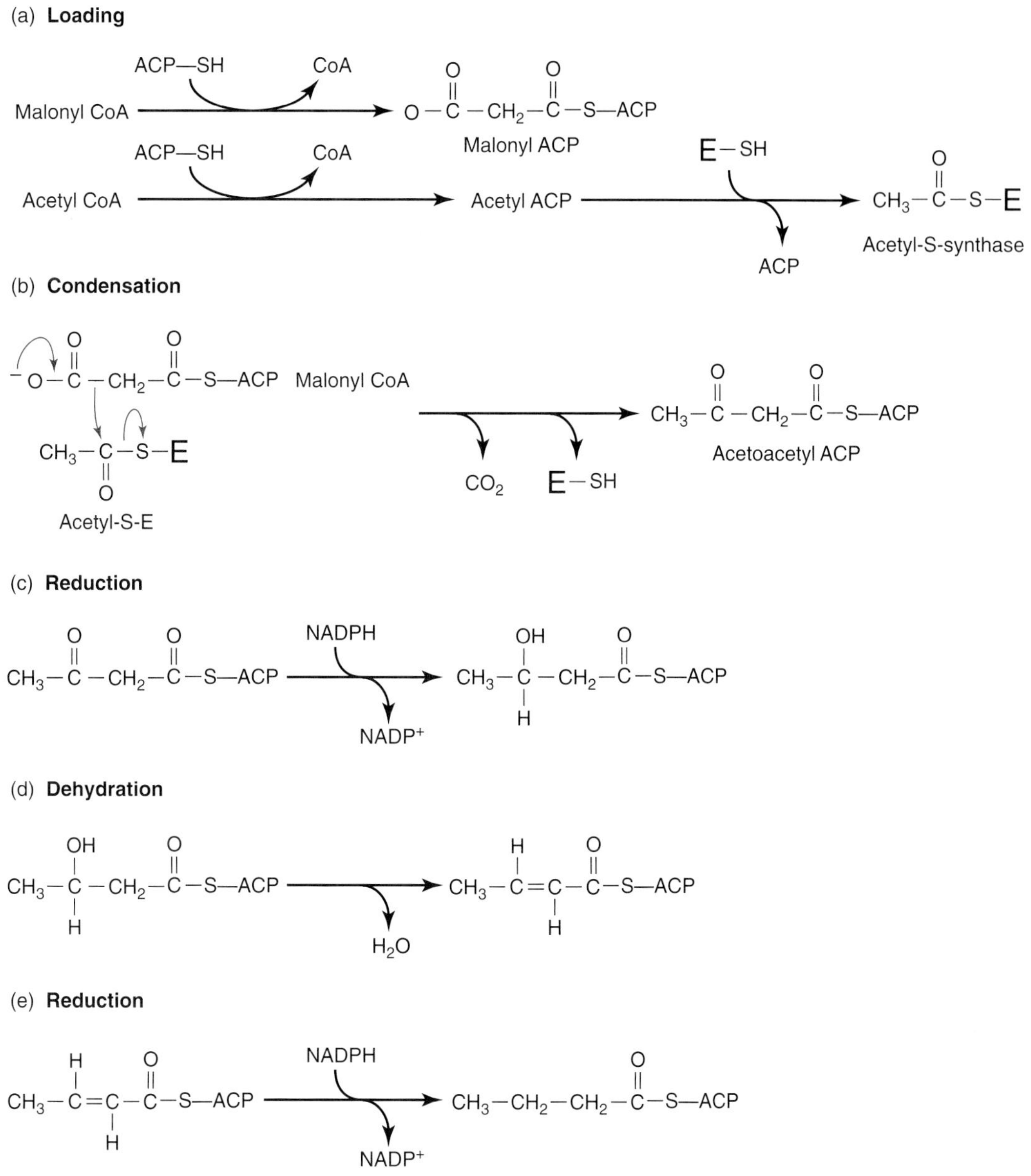

**FIGURE 13.17 Fatty Acid Synthase Steps.** (a) *Loading* reactions convert the CoA thioesters of acetyl-CoA and malonyl-CoA to thiol esters of acyl carrier protein (ACP). The acetyl-ACP then attaches to an –SH group of the synthase (E). (b) *Condensation* is the joining reaction between the attached acetyl group and the malonyl-CoA. The $CO_2$ from malonyl-ACP is released in the process. (c) The first *reduction* step is the NADPH-dependent conversion of the carbonyl to an alcohol. (d) In the *dehydration* step, water is removed to form the enoyl-CoA. (e) In the second *reduction* step, the double bond is saturated using electrons from NADPH.

(cysteine residue) of acetyl-S-synthase, releasing ACP-SH. This leaves the two-carbon acetyl moiety attached to the enzyme complex, which becomes the first two carbons incorporated into the fatty acid.

## Condensation

In the condensation step, a two-carbon fragment from malonyl-S-ACP is incorporated into the fatty acyl chain. As illustrated in Figure 13.17b, decarboxylation of the malonyl-S-ACP drives electron movement to the carbonyl carbon of the acetyl-S-synthase. The $CO_2$ released is the same carbon that was added in the acetyl-CoA carboxylation reaction.

Thus, no net $CO_2$ is incorporated in this pathway, as expected in animal cells. $CO_2$ release provides a thermodynamic drive for the reaction ($CO_2$ is a gas and, hence, is in a separate phase). The condensation product is β-keto-butyryl-ACP.

### Reduction

The reduction of the carbonyl (Figure 13.17c) is the first instance of a reaction utilizing NADPH reducing power. The standard chemical potential of the NADPH/NADP couple is identical to NADH/NAD$^+$, but in cells the NADPH/NADP$^+$ ratio is far more reduced. This allows NADPH to be used as a reductant for fatty acid synthesis in the same water space (cytosol) as the more oxidized NADH/NAD$^+$ redox pair that is involved in the distinct process of cellular energy transfer to ATP.

### Dehydration

The next step is the removal of the elements of water from the molecule to form a double bond (Figure 13.17d). This proceeds by the usual route of protonation, electron migration from the C–O bond to the double bond, and elimination. The molecule that remains is in the trans configuration, similar to the fumarase reaction of the Krebs cycle.

### Second Reduction

The double bond is reduced by hydride addition from the cofactor NADPH (Figure 13.17e). Thus, NADPH is used twice in each round to convert the carbonyl group to a methylene group. The product is butyryl-S-ACP.

In the next round, the sequence is repeated but with a slight variation: the four-carbon butyryl-S-ACP is used in place of acetyl-S-ACP (FIGURE 13.18). In subsequent cycles, the increasingly larger acyl-ACP is used until palmitoyl-S-ACP is formed. At this point, a thiolase specific for a 16-carbon chain catalyzes the hydrolysis of palmitoyl-S-ACP to form palmitate and ACP-SH (FIGURE 13.19).

Cells synthesize a variety of fatty acids starting from palmitate, using distinct enzymatic reactions, such as those for elongation and desaturation. These enzymes are embedded in the endoplasmic reticulum. They use acetyl-CoA for elongation and flavin-dependent (non-energy linked) oxidation to produce double bonds.

# 13.5 Triacylglycerol Formation

Most of the lipid in the body is stored as triacylglycerols. As indicated in FIGURE 13.20, the backbone for this class of molecules is glycerol-P, formed in the glycerophosphate DH reaction (BOX 13.5). The two hydroxyl groups of glycerol-P accept fatty acyl chains through successive acylation reactions with fatty acyl-CoA. Triacylglycerols are a group of molecules with varying fatty acyl chains that come either from the pathway outlined above (*de novo*), or from the dietary route outlined at the beginning of this chapter. The acylation reactions have a preference for incorporating unsaturated fatty acids into the 2-position of the glycerol-P, so that the resulting triacylglycerols are more unsaturated in this position.

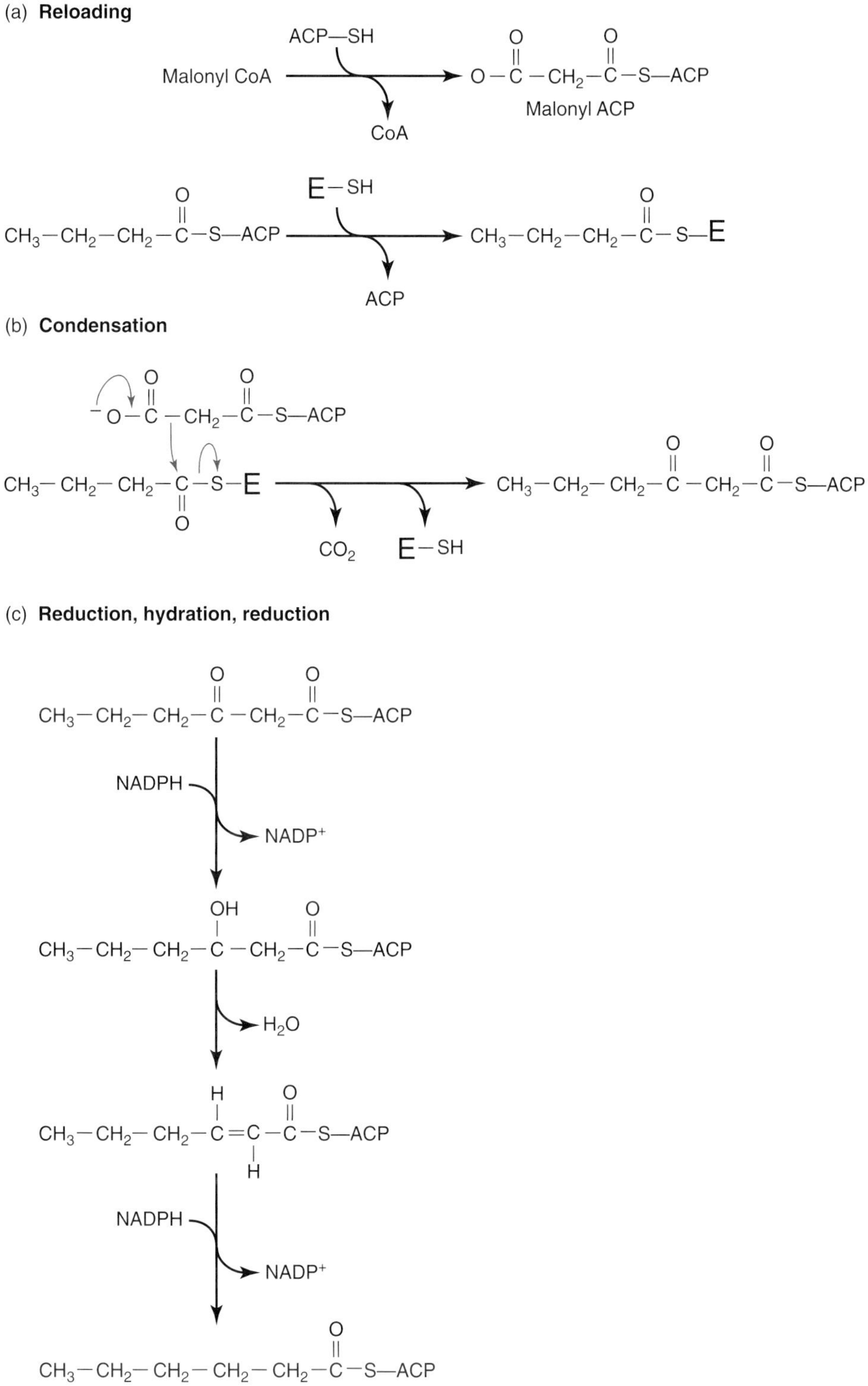

FIGURE 13.18 **Chain Elongation in Fatty Acid Synthesis.** The steps of fatty acid synthase are repeated with butyryl-CoA in place of acetyl-CoA for the second input of malonyl-CoA. Subsequent rounds of synthesis follow the pattern indicated here, until palmitoyl-ACP is formed.

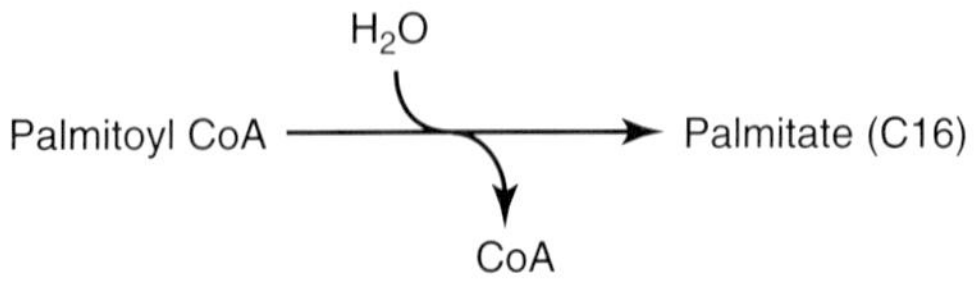

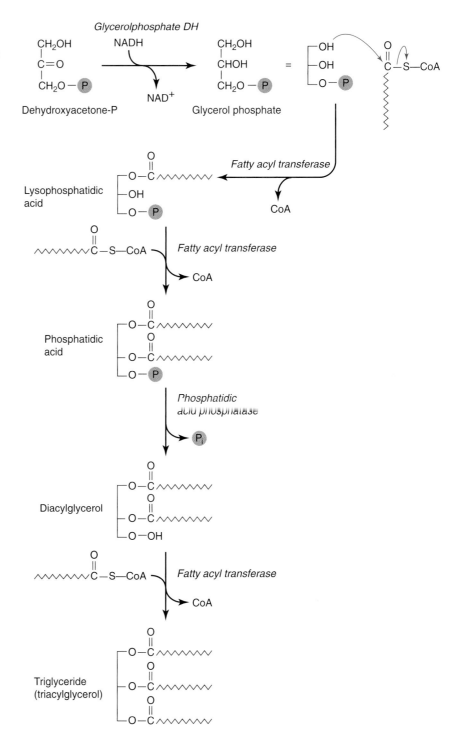

Palmitoyl CoA $\xrightarrow{\quad H_2O \quad}$ Palmitate (C16)

CoA

**FIGURE 13.19 Thioesterase.**

**FIGURE 13.20 Triglyceride Synthesis.** After forming the backbone of triacylglycerols by NADH reduction of dihydroxyacetone-P, two fatty acyl-CoA molecules are subsequently acylated to the hydroxyl groups of glycerol, forming phosphatidic acid. The latter is hydrolyzed to diacylglycerol, and a third fatty acyl-CoA is acylated to the backbone.

## BOX 13.5: Origins of the Triacylglycerol Backbone

Conversion of glucose to dihydroxyacetone-P is an established source of carbon for the glycerol backbone of triacylglycerols in mammals (**FIGURE B13.5**). However, an alternative route is **glyceroneogenesis**, the initial portion of the gluconeogenic pathway from pyruvate to dihydroxyacetone-P. This pathway requires the enzyme PEPCK, which is the key liver pathway for gluconeogenesis. PEPCK is found in adipose tissue, which, like liver, has the complete pathway for triacylglycerol synthesis. Note that adipose cells do not carry out gluconeogenesis themselves, because they do not have the enzymes fructose bisphosphatase or glucose phosphatase. The reason for the existence of glyceroneogenesis is not known. It is possible that it provides the metabolic flexibility to complete pathways within the fat cell that lead to Krebs cycle intermediates. As we've seen elsewhere, only acetyl-CoA can be oxidized by the cycle.

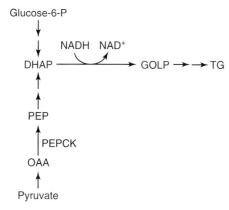

**FIGURE B13.5** Forming the backbone for triacylglycerols is glycerol P (GOLP), derived from glucose from the top part of glycolysis. GOLP can also arise from the bottom half of the gluconeogenic pathway as shown from pyruvate, involving the enzyme P-enolpyruvate carboxykinase (PEPCK).

With the two fatty acyl groups present in glycerophosphate, the resulting molecule is called a **phosphatidic acid**. Removal of the phosphate group is catalyzed by the enzyme phosphatidic acid phosphohydrolase, and the products of the reaction are inorganic phosphate and 1,2-diacylglycerol. Finally, fatty acyl transferase catalyzes the last acylation of a fatty acyl-CoA to complete the formation of the triacylglycerol.

# 13.6 Phospholipid Metabolism

The intermediates of triacylglycerol formation—diacylglycerol and PA—are also the precursors to phospholipid synthesis. As shown in FIGURE 13.21, each intermediate defines a distinct route. The *neutral phospholipid* pathway (e.g., phosphatidylcholine formation) originates with diacylglycerol, whereas the *acidic phospholipid* pathway (e.g., phosphatidylinositol formation) originates with phosphatidic acid.

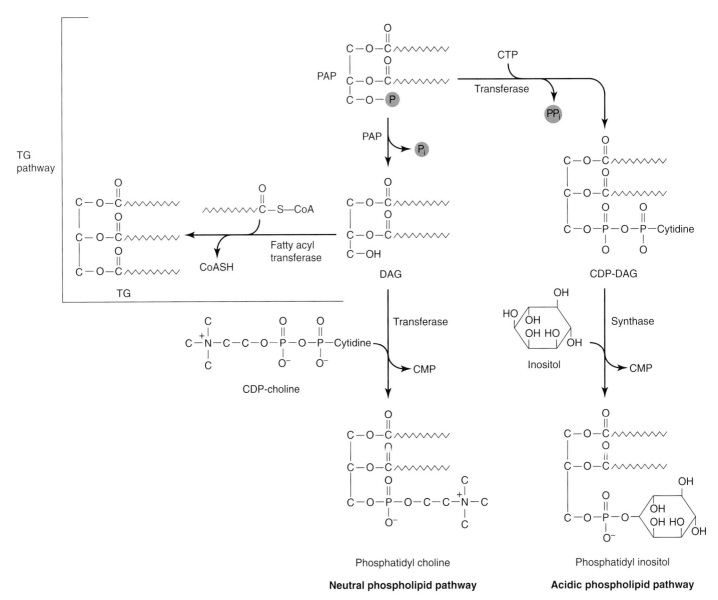

**FIGURE 13.21 Phospholipid Biosynthesis.** Two separate routes of phospholipid formation are indicated. For phosphatidyl choline (a neutral phospholipid), CDP-choline is preformed and reacts with diacylglycerol (DAG) to form the product. For phosphatidylinositol (an acidic phospholipid), a CDP adduct forms first with DAG, which is then displaced by the phospholipid head group, inositol, to form the product.

An example of the neutral phospholipid pathway is the formation of phosphatidylcholine, one of the most common phospholipids in membranes. This is formed by the transferase-catalyzed reaction of cytidine 5′-diphosphate–choline (CDP-choline) with diacylglycerol. Phosphatidylinositol, an example of the acidic phospholipid pathway, is formed in two steps from phosphatidic acid. First, a transferase catalyzes the condensation of cytidine triphosphate (CTP) with phosphatidic acid to form cytidine diphosphate-diacylglycerol (CDP-DAG). Subsequently, inositol reacts with CDP-DAG in a synthase reaction to produce the product, phosphatidylinositol. Both pathways produce cytidine monophosphate (CMP), but in neutral phospholipid

**CHAPTER 13** LIPID METABOLISM

synthesis, the head group (choline in this chase) first forms a CDP adduct, whereas in acidic phospholipid synthesis, the phospholipid itself forms a CDP adduct. Unlike triacylglycerol synthesis, which is limited mostly to liver and adipose cells, virtually all cells engage in membrane turnover; thus, most cells engage in phospholipid synthesis.

# 13.7 Cholesterol Metabolism

Cholesterol is an essential component of plasma membranes, a precursor to the steroid hormones and bile salts. Intermediates of the cholesterol biosynthesis pathway are precursors to other cellular products, including ubiquinone and lipid modifiers of proteins. The pathway for cholesterol synthesis, like fatty acid biosynthesis, requires cytosolic acetyl-CoA, and thus the citrate export cycle just described. Three acetyl-CoA molecules join to form HMG-CoA, using the same reactions as those for ketone body synthesis (Figure 13.9), except that the cholesterol biosynthesis enzymes are cytosolic isozymes. The next step, the reaction catalyzed by the enzyme HMG-CoA reductase, is rate limiting for this pathway.

As the control point of cholesterol synthesis, HMG-CoA reductase is a reasonable drug target for the treatment of hypercholesterolemia. The first clinically effective inhibitor of this enzyme was **lovastatin**, a natural product isolated from a fungus. As shown in FIGURE 13.22, the drug is a precursor to the acid form that bears a strong structural resemblance to the substrate, as expected from a competitive inhibitor. This drug, and similar ones that have been produced subsequently, are collectively known as **statins**.

FIGURE 13.23 shows a continuation of the cholesterol pathway up to the formation of the five-carbon intermediate isopentenyl pyrophosphate (isopentenyl-PP). After a reduction of HMG-CoA to mevalonate, a series of three phosphorylation reactions ensue. The last also involves a decarboxylation. Its mechanism should be familiar because it has an acid phosphate intermediate (like asparagine synthetase) and it removes water (like enolase).

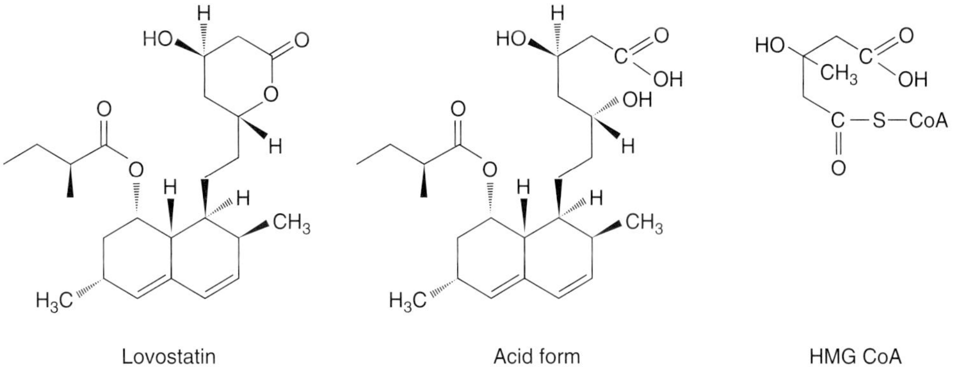

Lovostatin      Acid form      HMG CoA

FIGURE 13.22 **Lovastatin as an HMG-CoA Analog**. Lovastatin is a lactone. Its open-chain, acid form is produced in cells by an endogenous esterase, and is chemically similar to HMG-CoA.

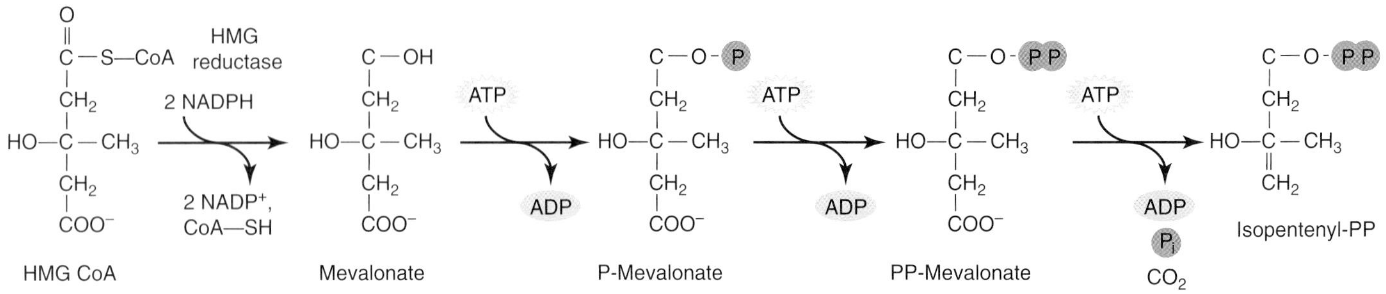

**FIGURE 13.23 Formation of Isopentenyl PP from HMG-CoA.** The first step, catalyzed by HMG-CoA reductase, uses two molecules of NADPH and forms mevalonate, rate limiting for cholesterol synthesis. Subsequent steps are phosphorylations and a decarboxylation, leading to isopentenyl-PP, the building block for cholesterol as well as other branched lipids such as ubiquinone.

**FIGURE 13.24** outlines the steps from isopentenyl-PP to cholesterol. First, an isomerase (reaction 1) catalyzes the formation of the dimethylallyl-PP. The reaction mechanism of the ensuing condensations proceed through carbocation intermediates. Typically, the pyrophosphate end of one precursor is called the *head* and the methylene end is called the *tail*, so that these joining reactions are either *head to tail* or *head to head*. The formation of geranyl pyrophosphate (reaction 2) is a head-to-tail reaction. Similarly, geranyl-PP reacts with isopentenyl-PP in another head-to-tail reaction to form the 15-carbon farnesyl-PP (reaction 3). Two farnesyl groups condense in the ensuing reaction (reaction 4) to produce the 30-carbon squalene in a head-to-head reaction. In this case, both pyrophosphate groups are released. The steps between squalene and cholesterol are detailed in the appendix.

Cholesterol can be incorporated directly into the plasma membrane. Alternatively, cholesterol can be esterified (often to the fatty acid oleate), and packaged into lipoprotein particles. The lipoproteins include very low density lipoproteins (VLDL) for export from the liver and chylomicrons for export from intestinal cells. Cholesterol is also converted into other products, albeit in lower amounts, such as the bile salt cholate (**FIGURE 13.25**). The hydroxylation steps of cholesterol and its derivatives, such as those evident in cholate, are catalyzed by the enzyme **cytochrome P450**. This is a mixed function oxidase, a family of enzymes that use $O_2$ as substrate, in which one of the oxygens is incorporated as a hydroxylation, and the other appears in the product $H_2O$. Several isoforms of this enzyme exist in liver and are also involved in the hydroxylation of **xenobiotics** (meaning foreign compounds such as drugs). The resulting hydroxylation aids their excretion from the body, both directly by increasing their solubility and indirectly by providing sites of glycosylation.

Cholate and other bile salts have all of their hydrophilic groups on one face of the molecule (below the plane as in the diagram in Figure 13.25). The molecule is thus an amphiphile able to form mixed micelles in the process of lipid digestion described at the beginning of this chapter. Most of the bile salts secreted into the intestine are reabsorbed back into the blood, removed by the liver, and exported back through the bile (the hepatobiliary circulation). However, a small

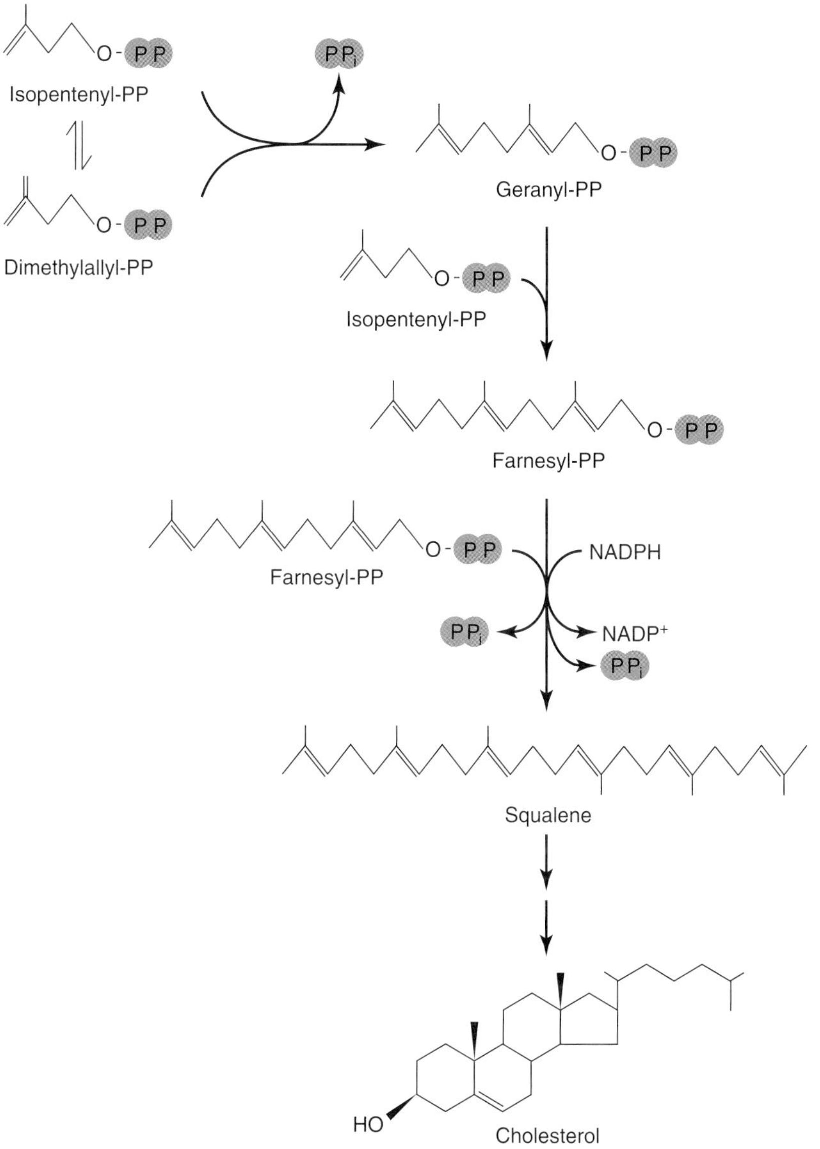

**FIGURE 13.24** **Isopentenyl PP to Cholesterol.** This abbreviated version of cholesterol formation shows the joining of the five-carbon building blocks to produce the 10-carbon geranyl-PP and the 15-carbon farnesyl-PP. The latter condenses with itself to form squalene, the last linear intermediate of cholesterol biosynthesis. The cyclization of squalene; several subsequent reactions lead to cholesterol.

fraction (<10%) escapes reabsorption and is excreted. This is the only route of cholesterol elimination from the body.

Figure 13.25 also presents other products of cholesterol, synthesized by other cells of the body. For example, Vitamin $D_3$ (dihydroxy-cholecalciferol) leads to increased formation of a transport protein that moves $Ca^{2+}$ into intestinal epithelia and subsequently into the bloodstream. Estradiol and testosterone are naturally-occurring sex hormones. The last entry in Figure 13.25 is one of the synthetic derivatives of testosterone, an **anabolic steroid**. First produced in 1935, these compounds were originally synthesized to mimic the ability of

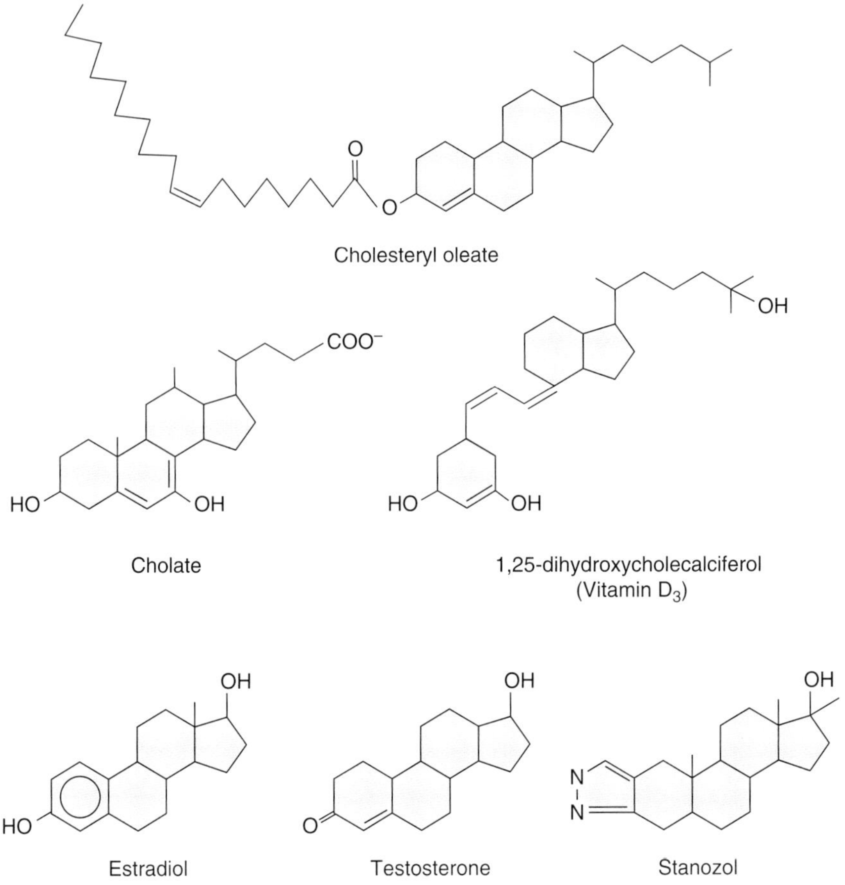

**Cholesteryl oleate**

**Cholate**

**1,25-dihydroxycholecalciferol**
(Vitamin D$_3$)

**Estradiol**        **Testosterone**        **Stanozol**

FIGURE 13.25 **Derivatives of Cholesterol.**

testosterone to reverse muscle wasting in certain disease states. Various modifications to the molecule allowed it to be taken orally, a problem in using testosterone itself. Currently they are better known as drugs of abuse in sports. While they strongly promote muscle growth, severe side effects are now well established, including enhanced tumor growth.

# 13.8 Other Lipids

The lipid species we have examined are those present in the largest amounts. Still, many minor routes of lipid metabolism have products with profound effects on cellular function. We consider here the eicosanoids, the sphingolipids, and some unusual fatty acids found in bacteria.

### Eicosanoids

The eicosanoids are a large class of relatively short-lived molecules (with half-lives as short as seconds) that are derived from the fatty acid arachidonate. Animal cells do not directly store this fatty acid, nor do they synthesize it, because they do not have desaturase enzymes that

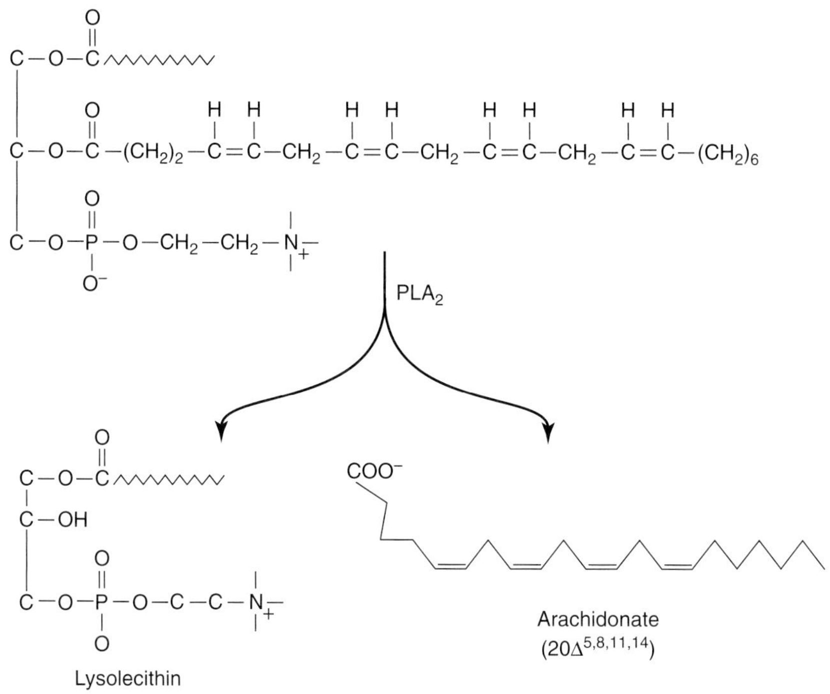

FIGURE 13.26 **Phospholipase A$_2$ Reaction.**

can introduce a double bond beyond the $\Delta^9$ position. Precursors of arachidonate, such as linoleic acid, are derived from plants in the diet, so they are called **essential fatty acids**. Once arachidonate is synthesized from them by elongation reactions, this fatty acid is incorporated into plasma membrane phospholipids in the 2-position. Production of arachidonate acid requires the action of phospholipase A$_2$ (FIGURE 13.26). Arachidonate is subsequently converted either to a class of cyclized molecules or to **leukotrienes**. The cyclized molecules include prostaglandins (such as that shown in FIGURE 13.27), which have local tissue effects, such as the stimulation of smooth muscle contraction. One

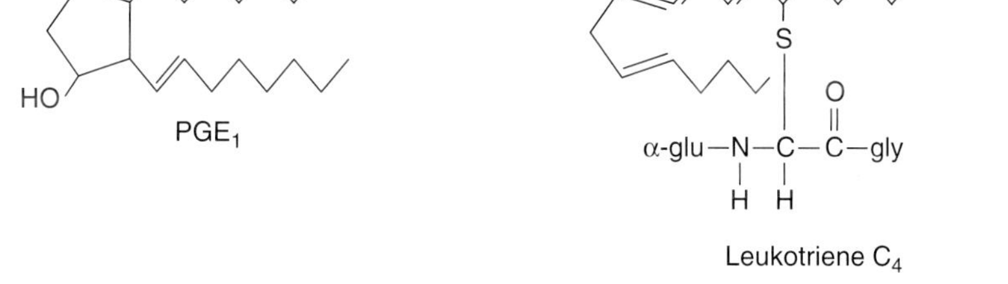

FIGURE 13.27 **Arachidonate Products.**

well-known example of a leukotriene (Figure 13.27) was previously known as the *slow reacting substance of anaphylaxis* due to its ability to cause contraction of smooth muscle around airway ducts in the lung.

The enzyme cyclooxygenase, the first step in the formation of cyclized arachidonate metabolites, is irreversibly inhibited by aspirin (acetylsalicylic acid). As a result, the products of this branch of the eicosanoid pathway are blocked, including the inflammation-producing prostaglandins.

Eicosanoids are involved in a multitude of physiological pathways, including muscle contraction, pain, and blood clotting. These molecules act by modulating ongoing hormonal signaling systems, such as the alteration of the activity state of G-proteins.

## Sphingolipids

Sphingolipids use serine rather than glycerol as a backbone. The key step is the first in the pathway, a reaction of serine with palmitoyl-CoA catalyzed by a transferase (FIGURE 13.28). The intermediate 3-ketosphinganine is converted after a few steps to ceramide. Ceramide is chemically similar to diacylglycerol; the two often copurify in chromatography. Accordingly, reactions similar to those of phospholipids produce the major sphingolipids shown in Figure 13.28: sphingomyelin (from phosphatidylcholine) and the cerebrosides (from UDP sugars such as UDP-galactose). The sphingolipids have a number of biological roles, including cell growth, migration, and the regulation of apoptosis. Several pathological states are associated with the inability to degrade these molecules, such as Neiman–Pick disease, which causes marked mental retardation. Sphingolipids make up a substantial portion of the lipid sheath (myelin) that wraps nerve fibers (axons), providing electrical insulation that enables rapid conduction of nerve impulses.

## Unusual Bacterial Fatty Acids

Two of the fatty acids found in bacterial lipids are illustrated in FIGURE 13.29. The first, an *iso*-C17:0 is a branched-chain fatty acid that uses carbon from branched-chain amino acids (e.g., isoleucine, leucine, or valine) as precursors. The other is a cyclopropane-containing fatty acid, which resembles the conformation of the *cis*-unsaturated fatty acids such as oleate. Both modifications alter membrane fluidity in response to the bacteria's environment. For example, bacteria synthesize cyclopropane-containing fatty acids when subjected to acidic media.

## 13.9 Overview of Lipid Metabolism in the Fed and Fasted States

In the fed state (FIGURE 13.30), dietary lipid is packaged into the lipoprotein chylomicron in the cells lining the intestine (epithelia). Ultimately, the lipid is stored in the adipocyte as a fat droplet occupying most of the cell volume. The other lipoprotein molecule, VLDL

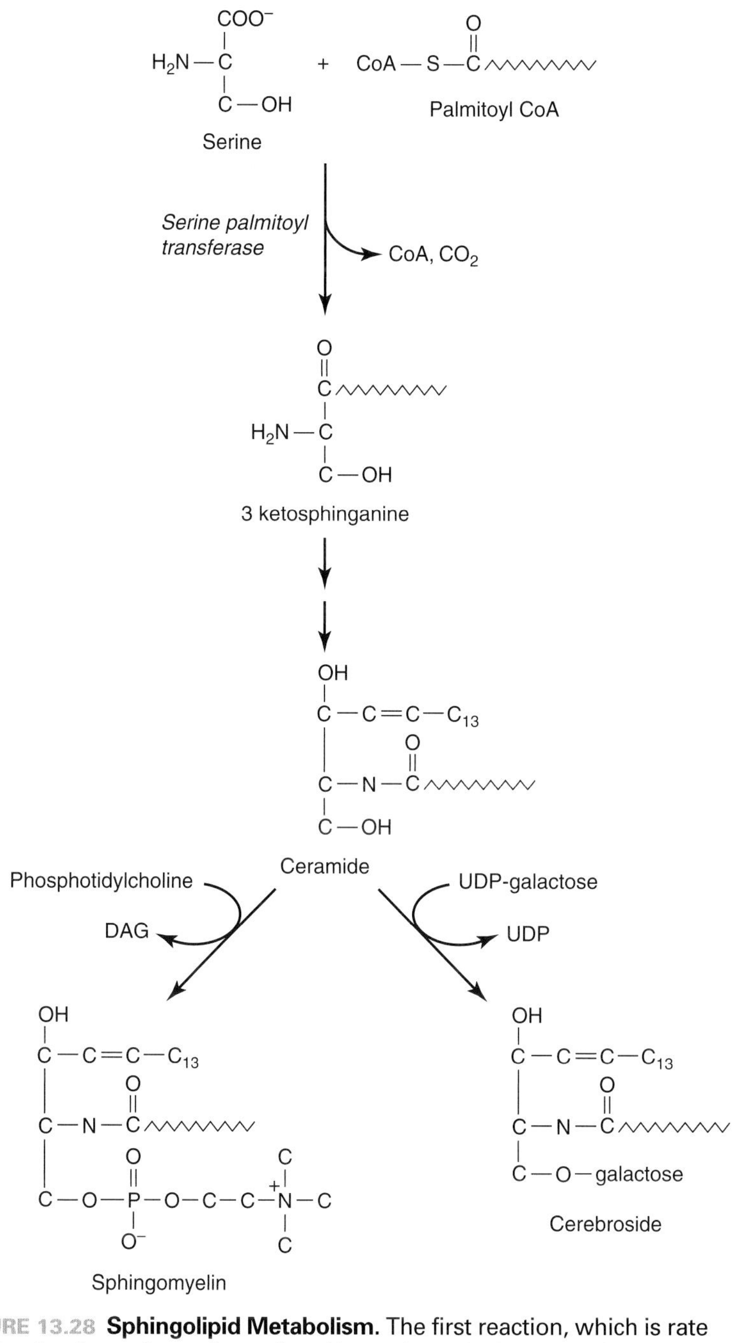

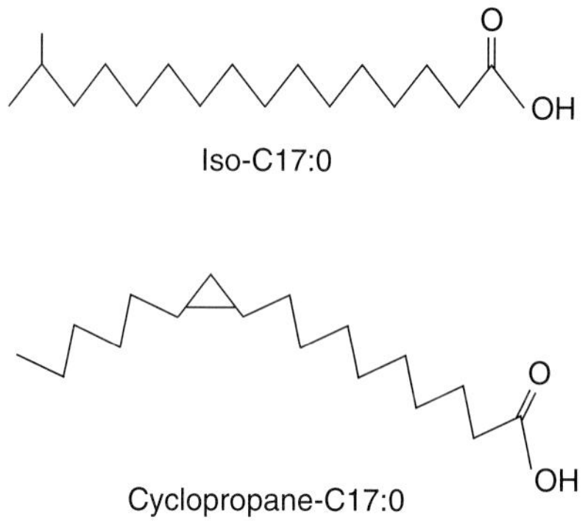

FIGURE 13.29 **Bacterial Fatty Acids.**

FIGURE 13.28 **Sphingolipid Metabolism.** The first reaction, which is rate limiting for the pathway, is catalyzed by serine palmitoyl transferase, leading to 3-ketosphinagnine, the backbone for further acylation. Desaturation and *N*-acylation leads to ceramine, which can either react with phosphatidyl choline to form sphingomyelin or with UDP-galactose to form a cerebroside. Due to the absence of an acyl group at the first position, and an amide at the second, these molecules are more stable membrane lipids than the major phospholipids.

is very similar in composition to chylomicrons, with approximately the same density (reflecting the high triglyceride content). VLDL is formed in the liver, as glucose is converted to fatty acids through fatty acid synthesis and the fatty acid is packaged with a protein coating (apolipoprotein). Thus, VLDL represents endogenous fat synthesis.

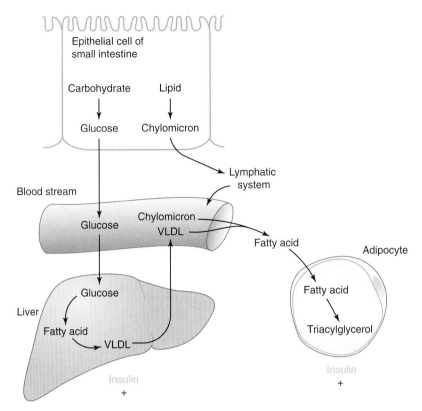

FIGURE 13.30 **Lipid Metabolism in the Fed State.** The intestine provides carbohydrate and lipid. Carbohydrates can be stored packaged into VLDL in the liver and can be transported to the adipocytes for storage as triacylglycerols. Dietary lipids can be packaged into chylomicrons and also taken into adipocytes for storage as triacylglycerols.

Like the triglyceride of chylomicrons, VLDL triglyceride is destined for the adipocyte triglyceride pool. The major hormone responsible for the shunting of lipid metabolism into these pathways is insulin, which stimulates fatty acid synthesis in liver and adipocytes, and favors triglyceride formation from fatty acids in adipocytes.

In the fasted state (FIGURE 13.31), the major metabolic events involve adipocytes, liver, and muscle. Fatty acids are produced by the adipocyte through triglyceride breakdown, which is stimulated by epinephrine. Liver takes up fatty acids, and either completely oxidizes them to $CO_2$ or partially oxidizes them to ketone bodies. Ketone bodies, as well as fatty acids, can be used by muscle for energy (shown in the figure as conversion to $CO_2$).

A focus on the key role of malonyl-CoA is outlined in FIGURE 13.32. Here we see that regulation of the acetyl-CoA carboxylase reaction, which exists in liver, adipose, and muscle, uses different factors in cells of those tissues. Malonyl-CoA, in the case of muscle and liver, is an important regulator of CAT I and therefore of fatty acid oxidation. However, malonyl-CoA is a significant precursor to fatty acids only in liver and adipose tissue. Finally, only the adipocyte provides fatty acids directly for use by other tissues through lipase-catalyzed cleavage, which is stimulated by epinephrine and inhibited by insulin.

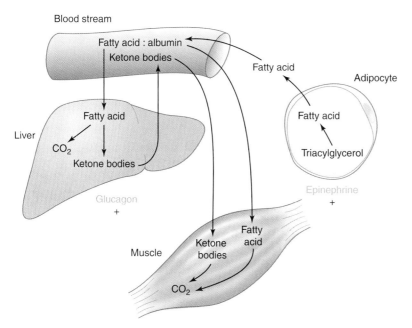

FIGURE 13.31 **Lipid Metabolism in the Fasted State.** In the fasted state, adipocytes catalyze the hydrolysis of triacylglycerols to fatty acids, hormonally stimulated by the presence of epinephrine. Blood fatty acids are oxidized in liver (a process stimulated by glucagon) completely to $CO_2$ and partially to ketone bodies. Muscle tissue can oxidize fatty acids and ketone bodies to $CO_2$.

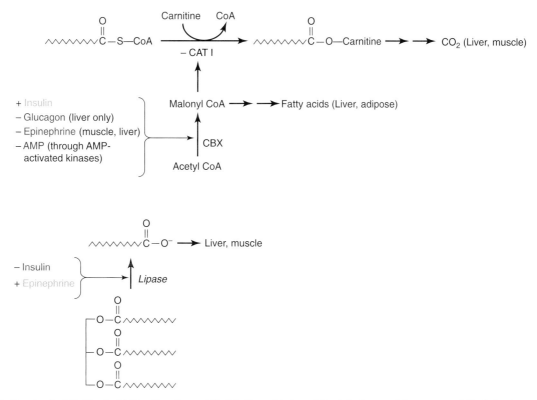

FIGURE 13.32 **Control of Fatty Acid Synthesis and Oxidation.** Malonyl-CoA is formed from acetyl-CoA in the acetyl-CoA carboxylase (CBX) reaction, where it can be further transformed into fatty acids in liver and adipose cells. Malonyl-CoA is also an allosteric inhibitor of CAT-I, which permits the entry of long-chain acyl-CoA into mitochondria for oxidation. Several factors influence CBX, including insulin, glucagon, epinephrine, and intracellular AMP concentration. In the adipocyte, epinephrine stimulates triglyceride hydrolysis, whereas insulin inhibits the process.

These differences reflect the distinct roles of the cell types: adipocytes synthesize, store, and release fat; muscle utilizes fat and ketone bodies for energy; and liver is involved in fat synthesis, VLDL secretion, fatty acid oxidation, and ketone body synthesis.

# 13.10 Integration of Lipid and Carbohydrate Metabolism

We have already observed several metabolic connections between lipid and carbohydrate metabolism. With the major pathways now in hand, we will consider three situations: the feeding–fasting transition, the resting–exercise transition, and the metabolic events in diabetes mellitus.

## Lipid and Carbohydrate Intersections in the Feeding–Fasting Transition

The fed state is characterized by storage routes for the two principal fuels of the body: glycogen for sugars and triglyceride for fatty acids. During feeding, elevated insulin concentrations drive these routes by the uptake of glucose, its conversion to glycogen as well as its conversion to fatty acids and thus triacylglycerols. Due to the ongoing fatty acid synthesis derived from glucose, malonyl-CoA levels are high while fatty acid oxidation and ketone body formation rates are minimal. Fatty acids are shunted into triacylglycerols and thus their blood levels are low.

In the fasted state, glucagon concentration becomes elevated, leading to liver glycogen breakdown to provide blood glucose, and fatty acid synthesis is diminished through the inhibition of acetyl-CoA carboxylase. Malonyl-CoA concentration is depressed, lifting the inhibition on fatty acid oxidation, which also increases ketone body formation. As the epinephrine concentration in blood is elevated by fasting, fat cells release more fatty acids to the blood, elevating their concentration and allowing more of them to be oxidized. Maintenance of the blood glucose concentration is essential to some tissues, including brain, and so gluconeogenesis becomes a critical pathway. Ultimately, the only substrate for that pathway in starvation is amino acids.

## Lipid and Carbohydrate Intersections in the Resting–Exercise Transition

During exercise, blood flow in the body is shunted away from most tissues except for muscles, and kept at a constant rate to supply the brain. Thus, we can expect that most of the metabolic changes involve those of the muscle. In anaerobic exercise, the muscle breaks down its glycogen store, stimulated by intracellular $Ca^{2+}$ as a result of the contractile process. This activation of glycogen phosphorylase was described; here we emphasize that lipid metabolism plays no role in the anaerobic state.

In aerobic exercise, a lipid contribution emerges, as fat cells under the influence of an increasing epinephrine concentration release fatty acids and they are oxidized by muscle mitochondria. Lowering of the malonyl-CoA concentration is now driven by an increased AMP kinase activity, in part through an increased AMP concentration. Muscle has little fatty acid synthesis, so the presence of acetyl-CoA carboxylase is largely for the purpose of malonyl-CoA formation for regulating fatty acid oxidation in these cells.

### Lipid and Carbohydrate Intersections in Diabetes Mellitus

Diabetes mellitus (commonly, just diabetes) is characterized by an elevated blood glucose concentration and is caused by a relative deficiency in insulin concentration. In **type 1 diabetes** (formerly known as juvenile or insulin-dependent diabetes), the pancreatic beta cells are impaired, resulting in insufficient release of insulin. In **type 2 diabetes** (formerly known as adult-onset or non-insulin–dependent diabetes), normally insulin-responsive tissues are **insulin-resistant**. This can lead to an abnormally elevated blood insulin level despite inadequate amounts of the hormone for action at target cells.

In both cases, the metabolic picture is similar, although in type 1 the symptoms are more severe. Due to the relative lack of insulin, the cells respond as they do in starvation: fatty acids are released from fat cells, ketone body levels are elevated, and gluconeogenesis is increased. Gluconeogenesis, combined with a depressed glucose uptake into muscle, leads to the elevated blood glucose concentration that is a signature of diabetes. Insulin therapy is essential for type 1 diabetes, and is sometimes used for the more commonly occurring type 2 diabetes, although type 2 is more commonly treated by drugs such as metformin. This drug inhibits liver gluconeogenesis and activates muscle glucose uptake through the activation of AMP kinase. Exercise leads to the stimulation of AMP kinase as well, which explains why increased exercise reduces the severity of diabetes, especially for type II patients.

While carbohydrates and lipids account for the bulk of the nutrients for mammals, there are critical roles for amino acids, such as that in nitrogen metabolism.

## Summary

Animals obtain lipids through their diets (exogenous lipid) or by de novo synthesis from carbohydrates and amino acids (endogenous lipid). The dietary route involves the formation of a mixed micelle in the intestine with bile salts, themselves cholesterol derivatives synthesized in the liver. Most of the dietary lipid appears in the blood as chylomicrons, a spherical lipoprotein particle with a core of triacylglycerols and cholesterol esters and a shell of protein (apolipoproteins) and phospholipids. Lipids are stored in the adipocytes, and they are released in times of need as free fatty acids, complexed to the protein albumin in the blood. Most fatty acids are oxidized to acetyl-CoA in three steps: activation to an acyl-CoA ester; transfer into the mitochondria; and β-oxidation to produce acetyl-CoA. The Krebs cycle enzymes catalyze the complete oxidation of acetyl-CoA to $CO_2$. A further possibility exists for liver, where acetyl-CoA can be converted to ketone bodies. Transport into the mitochondria, the middle step in the process of fatty acid oxidation, is catalyzed by carnitine acyl transferase I (CAT I) and allosterically inhibited by malonyl-CoA.

Malonyl-CoA, an intermediate of fatty acid biosynthesis, is formed from cytosolic acetyl-CoA. This acetyl-CoA is derived from mitochondrial citrate that is exported to the cytosol and then converted to acetyl-CoA and oxaloacetate. The carboxylation of acetyl-CoA to malonyl-CoA is catalyzed by acetyl-CoA carboxylase, the key regulatory enzyme of fatty acid biosynthesis. The carboxylase is inhibited by protein kinase A and AMP kinase. Following this reaction, the complex fatty acid synthase is responsible for seven sequential enzymatic reactions that lead to palmitate formation. In preparation, acetyl-CoA and malonyl-CoA esters are converted to thioesters of the protein ACP. Acetyl-ACP reacts with an –SH group of the fatty acid synthase. Next, malonyl-ACP decarboxylation is linked to its covalent attachment to the acetyl moiety to produce acetoacetyl-ACP. Reduction (using the cofactor NADPH) and dehydration steps lead to a butyryl-ACP intermediate. In the next cycle, the butyryl-ACP adds to the –SH group of fatty acid synthetase, the ensuing steps of reduction and dehydration occur; ultimately a 16-carbon fatty acyl-enzyme intermediate is formed that is released as palmitate. Ancillary reactions form other fatty acids of different chain lengths or with double bonds.

Ketone body synthesis, an exclusive pathway of the liver, represents a spillover pathway of excess acetyl-CoA that cannot be oxidized. The ketone bodies, acetoacetate and β-hydroxybutyrate, are utilized for energy by muscle and other tissues.

Triglyceride and phospholipid synthesis pathways involve fatty acyl-CoA addition to the hydroxyl groups of glycerol P, producing phosphatidic acid. Hydrolysis of the phosphate of phosphatidic acid forms diacylglycerol. The subsequent addition of a third fatty acyl-CoA produces a triacylglycerol. Phospholipids are produced by addition to either phosphatidic acid (acidic phospholipids) or diacylglycerol (neutral phospholipids) in reactions using CDP-activated precursors.

The first steps of the cholesterol pathway are the condensation of three acetyl-CoA groups to form hydroxymethyl-CoA (HMG-CoA), just like the ketone body pathway, except that cholesterol synthesis takes place in the cytosol. The next reaction of cholesterol synthesis is catalyzed by HMG-CoA reductase, the rate-limiting step of the process, and the site of inhibition of cholesterol-lowering drugs such as lovastatin. From this point, cholesterol biosynthesis proceeds through the formation of the five-carbon branched intermediate isopentenyl-PP. Several condensation reactions produce the 30-carbon branched lipid intermediate squalene; after cyclization and several more reactions, cholesterol is formed.

The major pathways of lipid metabolism—fatty acid oxidation, ketone body synthesis, and fatty acid synthesis—interact with carbohydrate metabolism in normal metabolic processes (such as fasting and exercise) as well as in disease states (such as diabetes). In general, when energy is needed for an extended period of time (e.g., longer than several minutes), lipid metabolism, largely through fatty acid oxidation, meets that need. Central to the control mechanisms are the hormones insulin, glucagon, and epinephrine, and the metabolic intermediate and CAT-I inhibitor, malonyl-CoA.

# Key Terms

acyl carrier protein (ACP-SH)
anabolic steroid
apolipoprotein
β-oxidation
bile
carnitine shuttle
chylomicron
chyme
cytochrome P450
emulsification
essential fatty acids (EFAs)
glyceroneogenesis
insulin-resistant
ketogenic
leukotriene
lovastatin
phosphatidic acid
statins
type I diabetes
type II diabetes
xenobiotics

# Review Questions

1. The system for metabolizing fatty acids in plants is similar to the one in animals. In the formation of erucic acid mentioned in this chapter (in *Ancillary Enzymes*), how many condensation steps are there in the fatty acid synthase reaction?

2. Desaturation reactions that produce double bonds following fatty acid synthesis are not linked to energy formation but still utilize oxygen. Why doesn't this compete with mitochondria and potentially impair ATP formation?

3. A separate class of phospholipids is derived from DHAP directly instead of GOLP, and forms a membrane lipid with an ether-linked fatty acid rather than an ester-linked one in the 1-position. What would you predict about the chemical stability of these ether phospholipids?

4. In the presence of fatty acids, lactate carbon is *not* oxidized for energy to support gluconeogenesis. Instead, lactate carbon is converted to glucose, in what is called the *glucose sparing* effect of fatty acids. How do you think fatty acids are able to do that?

5. In the fed state, glucose uptake in the liver decreases the oxidation of fatty acids. How do you think glucose is able to do that?

# References

1. Cooper, D. A.; Webb, D. R.; Peters, J. C. Evaluation of the potential for olestra to affect the availability of dietary phytochemicals. *J. Nutr.* 127:1699S–1709S, 1997.

2. Ikonen, E. Cellular cholesterol trafficking and compartmentalization. *Nat. Rev. Mol. Cell Biol.* 9:125–138, 2008.

3. Schanzer, W. Metabolism of anabolic androgenic steroids. *Clin. Chem.* 42:1001–1020, 1996.

4. Tobert, J. A. Lovastatin and beyond: The history of the HMG-CoA reductase inhibitors. *Nat. Rev. Drug Discov.* 2:517–526, 2003.

5. Vance, D. E.; Vance, J. E. *Biochemistry of Lipids, Lipoproteins and Membranes*; Elsevier: Amsterdam, 2008.
   Authoritative and detailed resource on lipid biochemistry.

6. Wymann, M. P.; Schneiter, R. Lipid signaling in disease. *Nat. Rev. Mol. Cell Biol.* 9:162–176, 2008.

7. Zhang, Y. M.; Rock, C. O. Membrane lipid homeostasis in bacteria. *Nat. Rev. Microbiol.* 6:222–233, 2008.

8. Alberts, A. W. Discovery, biochemistry and biology of lovastatin. *Am. J. Cardiol.* 62(15):J10–J15, 1988.

9. Kalhan, S. C.; Bugianesi, E.; McCullough, A. J.; Hanson, R. W.; Kelley, D. E. et al. Estimates of hepatic glyceroneogenesis in type 2 diabetes mellitus in humans. *Metab. Clin. Exp.* 57(3):305–312, 2008.

10. Klonoff, D. C. Replacements for *trans* fats—will there be an oil shortage? *J. Diabetes Sci. Tech.* 1(3):415–422, 2007.

11. Vaz, F. M.; Wanders, R. J. Carnitine biosynthesis in mammals. *Biochem. J.* 361(3):417–429, 2002.

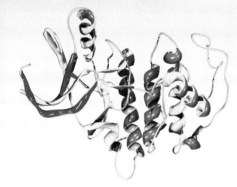

# 14

# Nitrogen Metabolism

Image © Dr. Mark J. Winter/Photo Researchers, Inc.

Take the simple glycerol molecule and nitrosylate the hydroxyl groups. The result is nitroglycerin (FIGURE 14.1), a compound that has both the explosive power to disintegrate boulders and the curative power to alleviate life-threatening blood vessel collapse. The explosion results from complete oxidation of the molecule to nitrogen gas ($N_2$). Nitroglycerin in the body is broken down to nitric oxide (NO), which relaxes arterial smooth muscles. Many nitrogen-containing compounds are more stable than oxygen-containing ones; for example, an amide is more stable than an ester. Most of the nitrogen in living species resides in the amino acids of proteins. Nitrogen metabolism revolves around the amino acids and ammonia, which are precursors to the nucleotides. We begin by considering the global nitrogen cycle.

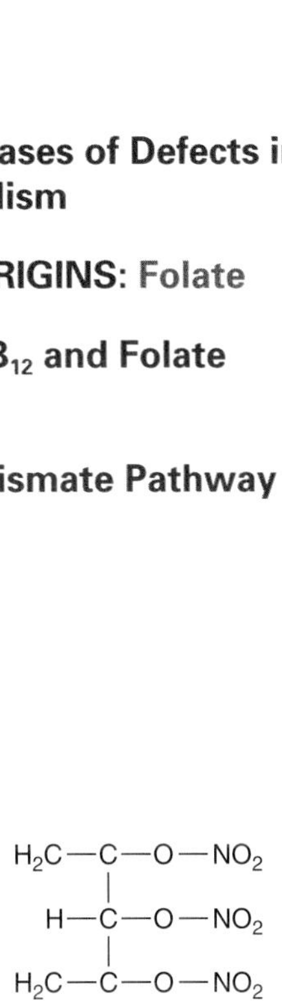

$$H_2C-C-O-NO_2$$
$$H-C-O-NO_2$$
$$H_2C-C-O-NO_2$$
Nitroglycerine

FIGURE 14.1 **Structure of Nitroglycerine.**

## 14.1 The Nitrogen Cycle

Like the carbon cycle in which $CO_2$ is incorporated into stable products and released back into the atmosphere, in the **nitrogen cycle** atmospheric $N_2$ forms stable products in organisms, some of which return it back to the atmosphere. $N_2$ is very stable and constitutes about 80% of the atmosphere. It takes the input of enormous amounts of energy before $N_2$ can be incorporated into and living systems.

FIGURE 14.2a summarizes the nitrogen cycle. The critical portion is the conversion of $N_2$ into ammonia ($NH_3$), a process called **nitrogen**

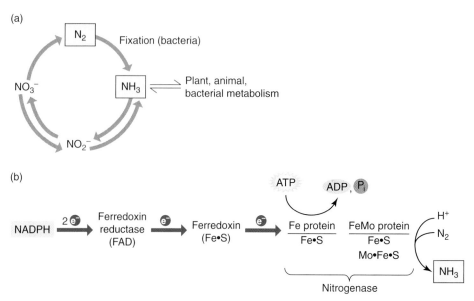

**FIGURE 14.2 Nitrogen Cycle.** (a) The global nitrogen cycle. The $NH_3$ branch is the entry point for mammals. (b) Nitrogenase accepts electrons from NADPH, utilizes ATP energy, and forms $NH_3$ from $N_2$.

**fixation.** This reaction can be accomplished biologically by only a few organisms, such as the *Azobacter* soil bacteria. About 40% of nitrogen fixation is accomplished synthetically using the Haber process, which combines $N_2$ with $H_2$ to form $NH_3$. The oxidation of $NH_3$ to nitrites and nitrates is accomplished by bacteria in a series of reactions known as **nitrification.** Completing the cycle, bacterial enzymes convert nitrates to $N_2$, called **denitrification.**

Many bacteria (and all plants) can also catalyze the reduction of nitrates and nitrites to $NH_3$ for subsequent use by all organisms, a process known as **assimilation.** Most of this chapter is concerned with reactions that begin with $NH_3$ and occur in animals. However, several energetic principles that we have already developed are evident in an examination of the process of bacterial nitrogen fixation, a key part of the nitrogen cycle.

An expansion of the process of nitrogen fixation is illustrated in Figure 14.2b. There is a flow of electrons, similar to an electron transport chain supplied from a nicotinamide nucleotide like that in the mitochondrial inner membrane (Chapter 10). Like the photosynthetic process (Chapter 11), redox states are boosted by an energy source. In the present case, the energy of ATP hydrolysis drives electron transfer. Electron pairs from NADPH reduce ferredoxin reductase, which donates electrons one at a time to the Fe–S centers of ferredoxin. Reduced ferredoxin, in turn, donates electrons to the two protein components of nitrogenase. The Fe-protein component of nitrogenase, which contains an Fe–S center, binds and hydrolyzes ATP, which facilitates electron transfer to the Fe—Mo protein of nitrogenase. Nitrogenase ultimately passes electrons to the a Mo–Fe–S center, which binds $N_2$ and reduces it in sequential steps to $NH_3$. A complication of nitrogenase is that both protein components are inactivated by $O_2$, which binds at the Fe–S centers and irreversibly inhibits the enzyme. Some bacteria avoid oxygen altogether because they are anaerobes; others keep oxygen at a level high enough to provide respiration but low enough to preserve nitrogenase. For example, *Rhizobium* bacteria grow in the root nodules of legumes, and the pink color of these nodules is due to the presence of **leghemoglobin**, which binds $O_2$, maintaining it at a low concentration.

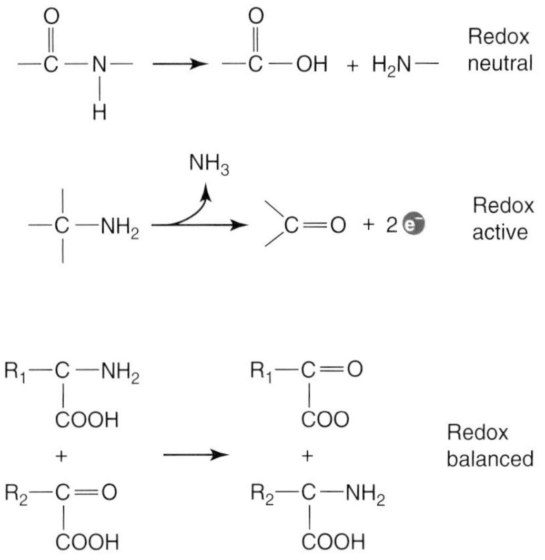

# 14.2 Reaction Types in NH₃ Assimilation

While we will consider a large number of metabolic reactions that ultimately begin with $NH_3$, we can better understand them by categorizing them as either metabolically irreversible or near-equilibrium reactions. In order to assist our thinking about reaction mechanisms, we can also divide the reactions into three separate groups based upon their redox changes (FIGURE 14.3).

## Redox Neutral

These are not redox reactions because the substrate and product atoms have the same oxidation numbers. An example is shown in Figure 14.3, the hydrolysis of an amide to a carboxylic acid plus ammonia.

## Redox Active

An overall redox reaction takes place in a redox active reaction, so these reactions require an exogenous electron acceptor. For example, conversion of an amine to a carboxyl group is an oxidation that produces two electrons.

## Redox Balanced

In a redox balanced reaction, one substrate is oxidized while the other is reduced. The overall change is two compensating redox reactions, with no net redox change. This is the case for the transaminases.

# 14.3 Metabolically Irreversible Nitrogen Exchange Reactions

One metabolically irreversible reaction is catalyzed by glutamine synthetase:

$$\text{Glutamate} + \text{ATP} + NH_3 \rightarrow \text{Glutamine} + \text{ADP} + P_i \qquad (14.1)$$

This reaction, in which an amide is formed from a carboxylic acid and ammonia, is redox neutral. A separate metabolically irreversible, redox neutral reaction, catalyzed by glutaminase, is the hydrolysis of the amide bond:

$$\text{Glutamine} \rightarrow \text{Glutamate} + NH_3 \qquad (14.2)$$

**FIGURE 14.3 Classifying Reactions in Nitrogen Metabolism by Redox Changes.** Redox-neutral reactions do not transfer electrons; redox-active reactions require an external mobile carrier for electrons; and redox-balanced reactions have an oxidized and reduced partner in both the substrate and the product, with no net electron transfer to external reactions.

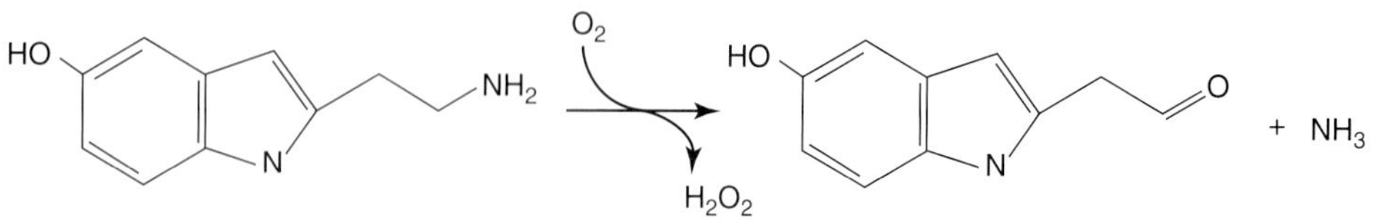

FIGURE 14.4 **Monoamine Oxidase.**

Combining the reactions in Equations 14.1 and 14.2 would result in net hydrolysis of ATP, similar to the substrate cycles of oxidative phosphorylation. However, cells usually do not contain substantial amounts of both of these enzymes in the same metabolic space, so these reactions constitute separate metabolic routes rather than a potential regulatory cycle. Only the release of ammonia, or the incorporation of ammonia takes place at one time.

Glutamine serves as a carrier of nitrogen between tissues. During amino acid breakdown in muscle, glutamine synthetase is active, and the muscle releases glutamine to the blood. Liver takes up this glutamine and converts it to glutamate and $NH_3$, via the glutaminase reaction. The nitrogen is converted to urea by the pathway we consider in Section 14.4.

A more specialized, metabolically-irreversible reaction is catalyzed by monoamine oxidase, which is responsible for the inactivation of the neurotransmitter serotonin (5-hydroxytryptamine) in FIGURE 14.4. This is a redox-active nitrogen exchange in which electrons are extracted from the monoamine to reduce $O_2$ to $H_2O_2$. Monoamine oxidase is embedded in the outer mitochondrial membrane, so that functionally it reacts with substrates in the cytosol and is not linked to energy capture. The enzyme contains a bound FAD cofactor, serially extracting electrons prior to their donation to $O_2$, with protons balancing the reaction. Most cells carry out one further catabolic step: the oxidation of the aldehyde product to the acid, an end product that exits the cells. Serotonin is a neurotransmitter responsible for digestive motility and mood changes; monoamine oxidase inhibitors were once widely used as antidepressants.

# 14.4 Near-Equilibrium Nitrogen Exchange Reactions

Glutamate dehydrogenase (DH) is the single most important near-equilibrium enzyme in nitrogen metabolism. Close behind is the class of reactions known as transaminases, intimately connected metabolically to glutamate DH.

## Glutamate DH

This reaction, found in the mitochondrial matrix, converts the Krebs cycle intermediate $\alpha$-ketoglutarate to glutamate:

$$\alpha\text{-Ketoglutarate} + NH_3 + NADH \rightleftharpoons Glutamate + NAD^+ \qquad (14.3)$$

The two steps involved in the direction of glutamate formation are shown in FIGURE 14.5. The first is the formation of the intermediate Schiff base. In the second, NADH reduces the double bond. One of the explanations for ammonia toxicity is the removal of α-ketoglutarate from the Krebs cycle, a result of the near-equilibrium nature of glutamate DH. Thus, an appreciable rise in ammonia levels could drive glutamate formation, limiting the concentration of α-ketoglutarate. This is particularly acute in brain cells, which strongly depend on a continuous oxidative metabolism.

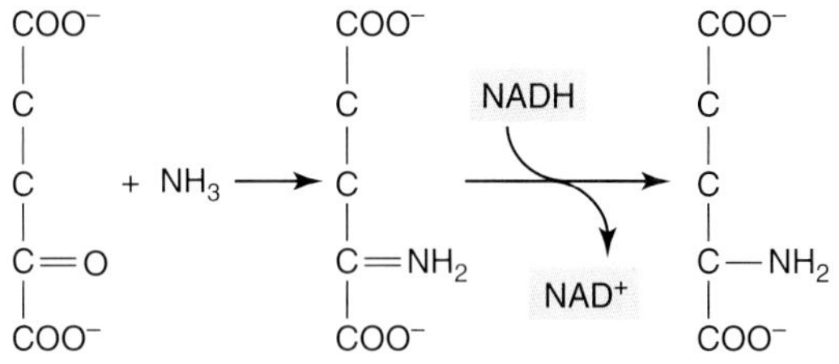

FIGURE 14.5 **Glutamate Dehydrogenase Mechanism.**

## Transaminases

Transaminases (aminotransferases) are enzymes that catalyze the interconversion of amino acids and keto acids. These are part of the malate/aspartate redox shuttle. The mechanism for aspartate aminotransferase is illustrated in FIGURE 14.6. The enzyme has the tightly bound cofactor pyridoxal-P, also used in glycogen phosphorylase. The first step of the aminotransferase reaction is Schiff base formation, which leads to an intermediate that is stabilized by hydrogen bonding to the dissociated hydroxyl group of pyridoxal phosphate. Next, electrons flow to the positively charged ring nitrogen (step 2), an electron sink. In step 3,

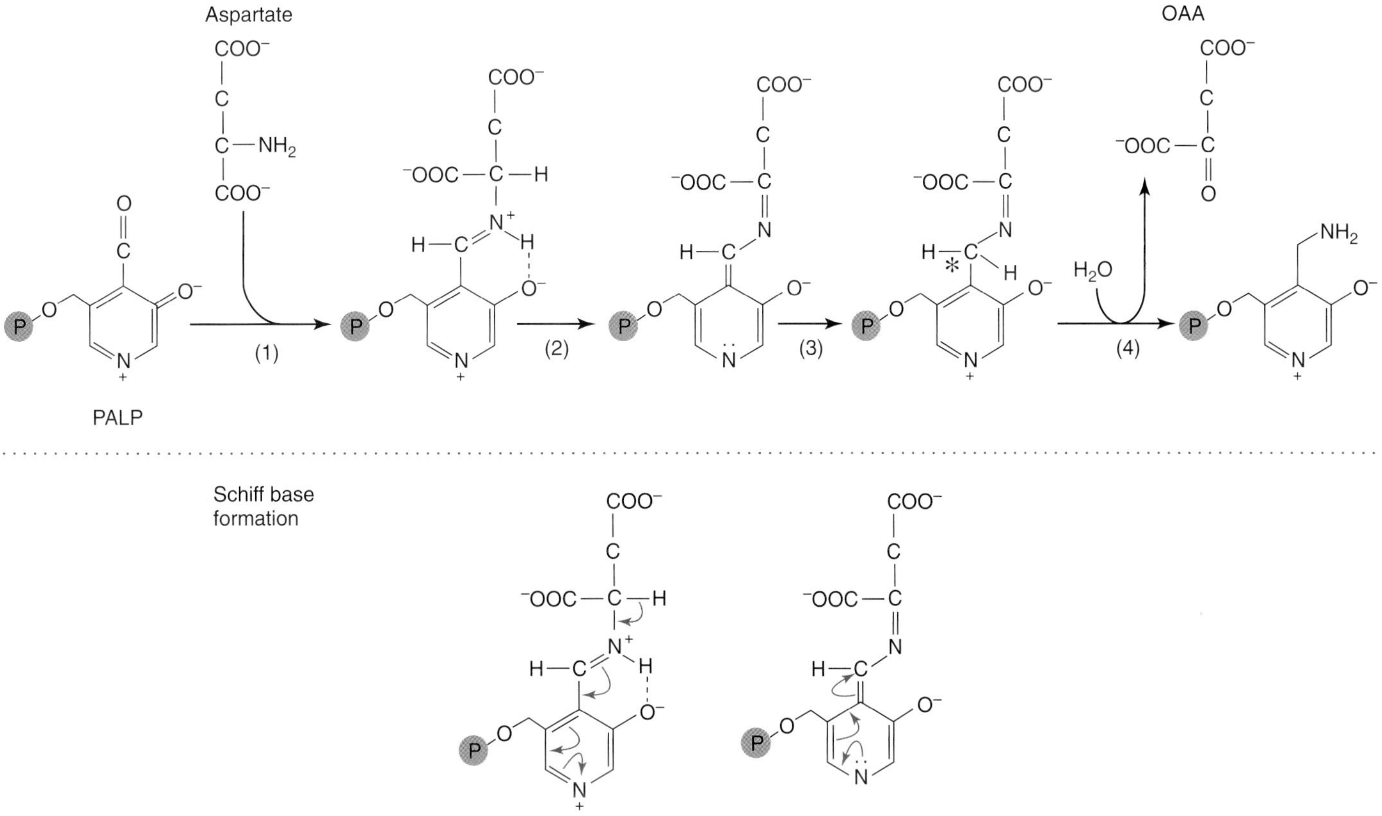

FIGURE 14.6 **Mechanism of Aspartate Transaminase.** In this representative transaminase, aspartate binds the cofactor PALP (pyridoxal phosphate), forming a Schiff base. After the electron rearrangements, the N atom is transferred to the bound PALP cofactor. This is the half-reaction. The remainder (not shown) is nearly a mirror image of this mechanism, in which α-ketoglutarate forms a Schiff base with the pyridoxamine form of the cofactor and is released as glutamate.

a partial reverse flow along the same path creates a new Schiff base, except the pyridoxal phosphate carbon (starred in Figure 14.6) is at the level of a methylene. Hydrolysis of the Schiff base (step 4) releases the ketoacid, oxaloacetate. This completes the half-reaction, leaving an enzyme-bound pyridoxamine. In a mirror image to the mechanism just outlined, α-ketoglutarate forms a Schiff base to the pyridoxamine, extracts the nitrogen, and ultimately is released as glutamate. Note that there is a redox transfer in each half-reaction, because the amine is more reduced (by two electrons) than the ketoacid. In the complete reaction, there is no net redox transfer (i.e., it is redox-balanced).

Because each half-reaction releases a product prior to the binding of the second substrate, there is an intermediate, separate stable form of the enzyme, with pyridoxamine-P bound. This is called a **ping-pong** mechanism because substrate binds and product releases, then another substrate binds and another product releases. By contrast, enzymes with multiple substrates that must all bind before product is released are called **sequential**.

Because transaminases catalyze exchange reactions, they do not remove or add nitrogen from substrates in a net sense. Instead, they funnel nitrogen into glutamate, a common amino acid that has broader metabolic connections. For example, the removal of nitrogen from alanine to form pyruvate plus $NH_3$ is the result of the following two reactions:

$$\text{Alanine} + \alpha \text{ Ketoglutarate} \rightarrow \text{Pyruvate} + \text{Glutamate} \qquad (14.4)$$

$$\text{Glutamate} + NAD^+ \rightarrow \alpha\text{-Ketoglutarate} + NADH + NH_3 \qquad (14.5)$$

An additional example of the combination of transaminase and glutamate DH reactions is the urea cycle, which we examine next.

# 14.5 The Urea Cycle

In humans, virtually all of the nitrogen is excreted as urea, which is formed from mixtures of amino acids and ammonia exclusively in the liver. The pathway for urea synthesis does not occur in isolation but is tied to gluconeogenesis. Moreover, for any mixture of ammonia and amino acids presented to the liver, half of the urea nitrogen comes from ammonia, and half comes from aspartate.

A portion of the pathway (FIGURE 14.7) emphasizes the transport of aspartate and citrulline from the mitochondria into the cytosol, and their condensation in the argininosuccinate synthase reaction. Viewed in this way, the requirement of urea synthesis for exactly the same amount of aspartate and citrulline (the latter representing the ammonia input) is simply a reflection of the stoichiometry of cytosolic argininosuccinate synthetase. In order to see how varying inputs of substrates are balanced, we need to examine what happens in just two limiting cases: when $NH_3$ in excess and when aspartate in excess.

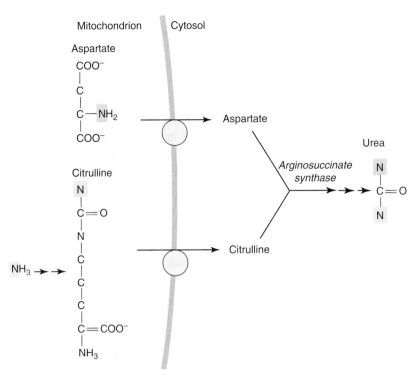

**FIGURE 14.7 Urea Cycle Segment: Stoichiometry of N Flows.** The pathway requires mitochondrial export of aspartate (Asp) and citrulline. The two condense in the cytosol, and the urea released has one N atom from aspartate and the other from citrulline. As indicated, the N atom of citrulline comes from $NH_3$.

## [$NH_3$] Exceeds [Aspartate]

With $NH_3$ in excess, glutamate DH runs in the direction of glutamate formation (**FIGURE 14.8a**). Subsequently, glutamate reacts with oxaloacetate through the aspartate aminotransferase reaction to produce the extra aspartate needed. Note that glutamate and $\alpha$-ketoglutarate are part of a small cycle in this process, and a *net* input of oxaloacetate occurs. The oxaloacetate itself is derived from pyruvate, through the pyruvate carboxylase reaction. The ultimate fate of this carbon is glucose formation.

## [Aspartate] Exceeds [$NH_3$]

With aspartate in excess, glutamate DH runs in the direction of $\alpha$-ketoglutarate formation (Figure 14.8b). Subsequently, $\alpha$-ketoglutarate reacts with aspartate through the aspartate aminotransferase reaction to regenerate glutamate to complete the small cycle just described in Section 14.5, but running in the opposite direction. The excess oxaloacetate produced ultimately is converted to glucose.

It is the ability of the near-equilibrium reactions to adjust to prevailing concentrations of substrates and to alter their reaction direction that makes it possible to exactly match the concentrations of aspartate and citrulline (representing ammonia) in the cytosol for urea formation. Like the nonoxidative pentose phosphate shunt reactions, the notion of near-equilibrium reactions provides a clear and simple

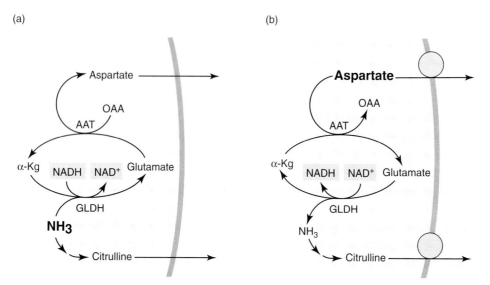

**FIGURE 14.8 Balancing N Flows for the Urea Cycle.** In order to match exactly the N from $NH_3$ and aspartate, the near-equilibrium reactions of glutamate DH (GLDH) and aspartate aminotransferase (AAT) run in different directions in a and b in order to balance the forms of nitrogen supplied to the liver.

understanding of how reaction flux can be diverted into separate routes depending only on their prevailing concentrations.

With this background, we can now consider the details of the cycle itself.

### Steps from $NH_3$ to Citrulline

The two reactions shown in **FIGURE 14.9** are required to synthesize citrulline from $NH_3$. First, $NH_3$, bicarbonate, and two ATP molecules react to produce carbamoyl P. This reaction, catalyzed by carbamoyl

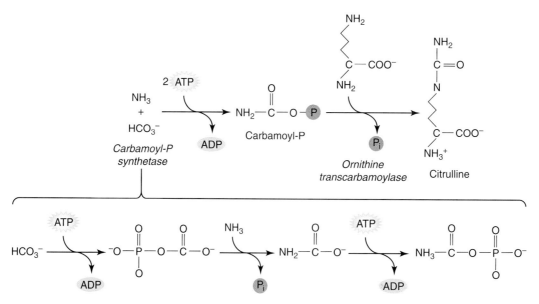

**FIGURE 14.9 Urea Cycle: From $NH_3$ to Citrulline.** The formation of carbamoyl P, the first step, is detailed in the inset. Next, ornithine transcarbamoylase catalyzes the transfer of the N from carbamoyl P to form citrulline. These are mitochondrial reactions.

P synthetase, resembles the reaction catalyzed by asparagine synthetase in that one ATP is used to form an acid phosphate that is displaced by ammonia. The sequence involving carbamoyl P synthetase is illustrated as an inset to Figure 14.9. A second ATP reacts to form carbamoyl P, the acid phosphate product. In the next reaction, catalyzed by ornithine transcarbamoylase, carbamoyl P is displaced by the amine of ornithine, a nonprotein amino acid, to form citrulline.

## Cytosolic Steps of the Urea Cycle

With the arrival of citrulline and aspartate in the cytosol, both of the nitrogen atoms for urea synthesis are represented. The condensation step, catalyzed by argininosuccinate synthetase, joins the molecules at the expense of two high-energy phosphates, the result of converting ATP to AMP and $PP_i$ (FIGURE 14.10). The mechanism is outlined in FIGURE 14.11. The adenosyl-P portion of ATP forms a phosphate ester with the **ureido** [NC(=O)N] group of citrulline. Next, the amine of aspartate displaces AMP, forming the product argininosuccinate.

Returning to the pathway of Figure 14.10, in argininosuccinate lyase reaction, hydrolysis leads to the products, arginine and fumarate. Fumarate is converted to malate through a reaction catalyzed by cytosolic fumarase, and malate is converted to glucose. Malate could enter the mitochondria, but cannot be oxidized by the Krebs cycle because it is an intermediate rather than a substrate for this pathway.

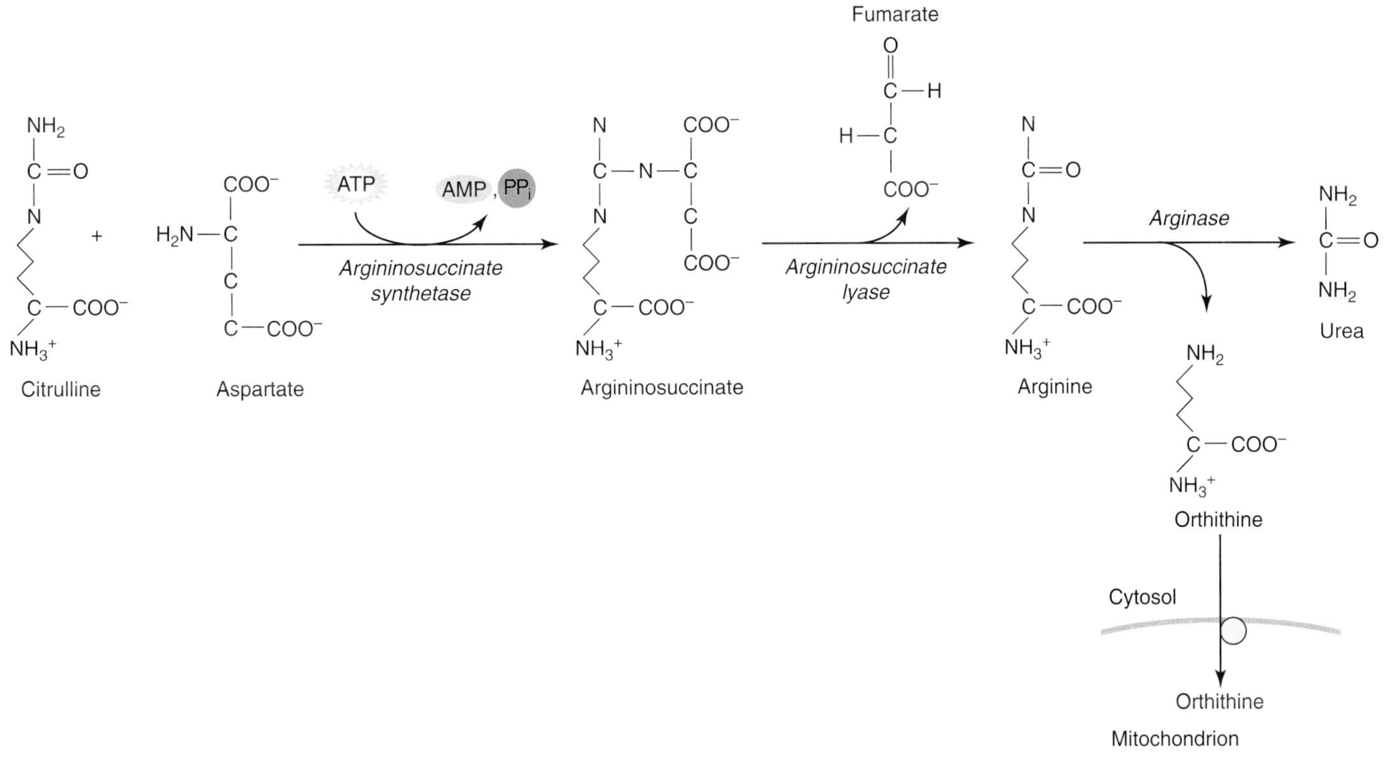

FIGURE 14.10 **Cytosolic Steps of the Urea Cycle.** The cytosolic portion of the urea cycle includes the condensation reaction between citrulline and aspartate, the subsequent formation of arginine, with release of fumarate, and the formation of urea, with ornithine, which is transported into the mitochondria to complete the cycle.

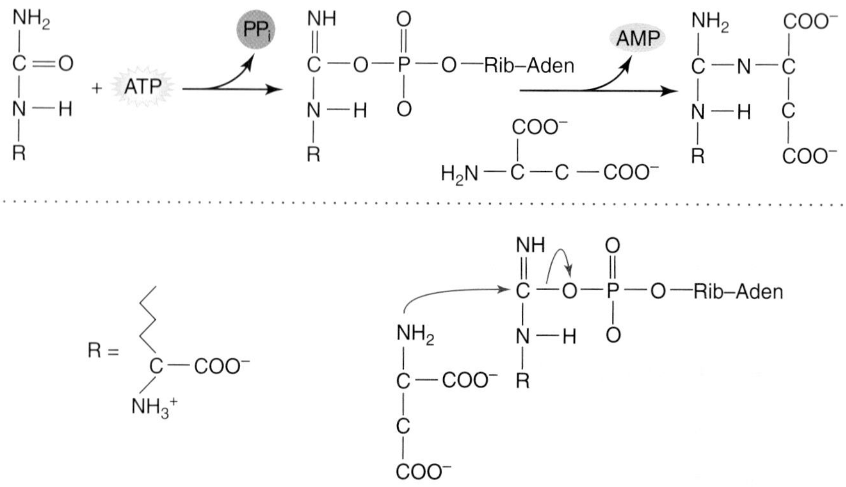

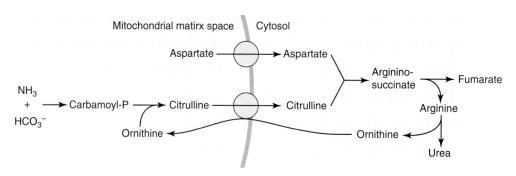

**FIGURE 14.11 Mechanism of Argininosuccinate Synthetase.** After formation of the acid phosphate with the ureido group of ornithine, the amine of aspartate displaces AMP to form the condensation product.

For the other product, arginine is hydrolyzed to urea and ornithine, a reaction catalyzed by arginase. The ornithine is transported into the mitochondria, completing the urea cycle.

## Overall Urea Cycle

FIGURE 14.12 provides an overview of the urea cycle. The mitochondrial reactions collect nitrogen, half in aspartate, and half in citrulline (representing ammonia). In the cytosol, two molecules exit the cycle: fumarate, destined for glucose synthesis, and urea, which is not further metabolized in humans. After it leaves the liver, urea enters the kidney and from there it is ultimately excreted from the body. While urea is metabolically inert in the kidney, it forms an osmotic gradient that assists in concentrating the urine, thus minimizing water loss.

Regulation of the urea cycle is exerted at three levels. First, near-equilibrium reactions control the delivery of nitrogen to urea from any mixture of amino acids and ammonia. Thus, as more nitrogen-containing molecules are provided to the liver, more urea is formed. Secondly, there is allosteric control exerted at carbamoyl P synthetase.

**FIGURE 14.12 Overall Urea Cycle.**

**CHAPTER 14** NITROGEN METABOLISM

This reaction is activated by the molecule N-**acetylglutamate**, and its levels correlate to states of increased urea synthesis (usually increased dietary protein). Finally, there is induction of all of the enzymes of the urea cycle. This genetic control is a long-term action that can raise the ultimate rate of urea formation by the liver. In fasting conditions, for example, protein breakdown is more extensive, and glucagon stimulates the liver to synthesize increased concentrations of all of the urea cycle enzymes in concert with the gluconeogenic enzymes.

# 14.6 Amino Acid Metabolism: Catabolism

For the 20 common amino acids, each with synthetic and catabolic pathways, there are 40 separate routes. The metabolic complexity is even greater, considering species differences, nonprotein amino acids, and the fact that the same organism may use multiple pathways. Accordingly, we present only partial amino acid metabolism pathways and general principles here; some further details are provided in the Appendix.

In amino acid metabolism, it is not always clear which steps can be strictly categorized as catabolic or anabolic. We will consider the removal of nitrogen and the disposal of carbon as catabolism. Catabolism can be subdivided into two groups: **ketogenic** and **glucogenic**. This means that, after nitrogen removal, the carbon skeleton is converted to acetyl-CoA and potentially ketone bodies (ketogenic), or to a precursor of glucose (glucogenic). Most amino acids are at least partially glucogenic; only a few are exclusively ketogenic. We begin by considering branched-chain amino acid catabolism, which illustrates routes of structurally-related amino acids that are ketogenic, glucogenic, and both.

## Branched-Chain Amino Acid Breakdown

The catabolism of leucine, isoleucine, and valine are similar (FIGURE 14.13). Each is first converted to the corresponding ketoacid by a transamination reaction, and then to a coenzyme A (CoA) thioester through the action of a dehydrogenase complex, which is analogous to the pyruvate dehydrogenase complex. Branched chain amino acid transaminases are particularly rich in muscle tissue, and the dehydrogenase complex (and subsequent steps) are particularly enriched in the liver. Oxidation steps following the formation of the CoA thioesters are distinct for each branched chain amino acid.

Ultimately, leucine forms acetyl-CoA, and is thus a ketogenic amino acid. Isoleucine forms acetyl-CoA and succinyl-CoA, and is thus partly ketogenic and partly glucogenic. Valine breakdown leads to succinyl-CoA and is thus glucogenic. A dramatic increase in blood concentrations of branched-chain amino acids and their ketoacid forms occurs in maple syrup urine disease, the result of a deficiency in the branched-chain dehydrogenase complex (**BOX 14.1**).

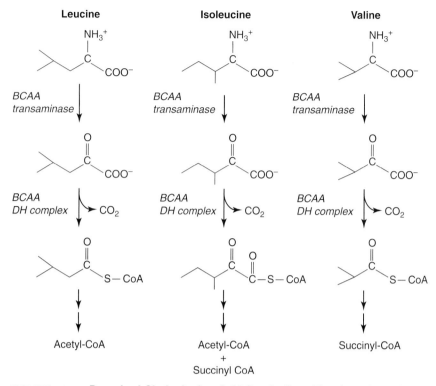

**FIGURE 14.13 Branched-Chain Amino Acid Catabolism.** The three branched-chain amino acids have a similar catabolic pathway: transamination, followed by a dehydrogenase complex that releases $CO_2$, and subsequent breakdown steps to either acetyl-CoA (Leu and Ile) or succinyl-CoA (Ile and Val).

## BOX 14.1: Clinical Cases of Defects in Amino Acid Metabolism

Inborn errors (*genetic deficiencies*) in amino acid metabolism are common, and they can have dramatic consequences. For example, a loss of phenylalanine hydroxylase results in an insufficiency of tyrosine and an accumulation of phenylpyruvate in the urine, producing a musty smell. This disease is called **phenylketonuria** (PKU). This was also one of the first enzyme deficiencies for which wide screening was implemented, and much of the damage from this disorder can be averted by a strict diet that is low in protein and includes supplemental tyrosine. Besides mental disorders (phenylalanine is a neurotransmitter precursor), a prominent feature is the near absence of skin pigmentation, because tyrosine is also a precursor to the dark pigment melanin. Another disorder of amino acid metabolism is **maple syrup urine** disease, which is due to an inability to oxidize branched-chain amino acids. Once again, the ketoacid products accumulate in the urine (causing a sweet, syrup-like smell). Like PKU, this disease also causes mental defects (as well as possible coma), and is treated by a strict dietary restriction of the branched-chain amino acids.

## Threonine

In mammals, the major pathway for the degradation of threonine is the conversion to 2-ketobutyrate, catalyzed by threonine dehydratase (FIGURE 14.14). The ensuing step is catalyzed by a dehydrogenase complex similar to the branched-chain dehydrogenases just introduced in the section above (as well as pyruvate dehydrogenase and 2-ketoacid dehydrogenase complexes). The resulting proprionyl-CoA is ultimately converted to succinyl-CoA, so threonine is glucogenic. While threonine is a straight-chain amino acid its degradation resembles those of the branched-chain amino acids.

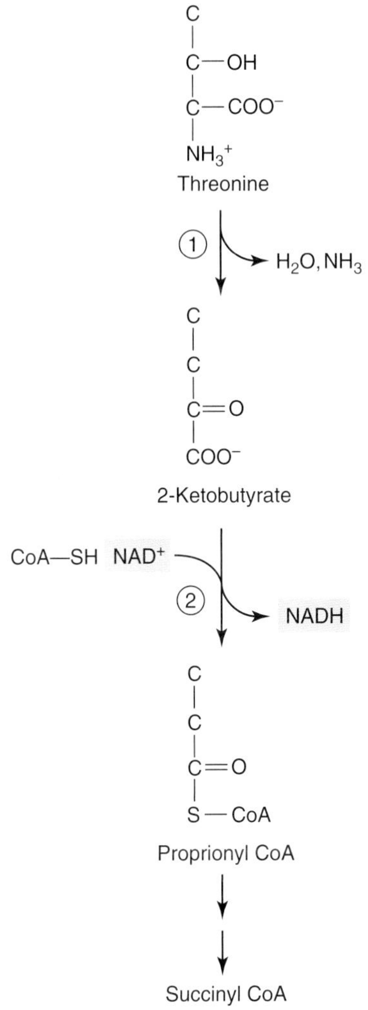

## Lysine

The initial stages of lysine degradation (FIGURE 14.15) are analogous to the first cytosolic reactions of the urea cycle, in that a condensation step and subsequent splitting results in the transfer of nitrogen between molecules. In the first step, lysine condenses with α-ketoglutarate, forming a Schiff base in a reaction catalyzed by saccharopine DH (NADP+ forming); the product is saccharopine. The next step, catalyzed by saccharopine DH (NADH forming) is an oxidative cleavage, forming the products glutamate and α-aminoadipate **semialdehyde**. (A semialdehyde is a molecule derived from a dicarboxylic acid, where one of the acid groups has become an aldehyde; in this case adipic acid is the parent dicarboxylic acid.) Subsequent steps convert the semialdehyde intermediate to acetyl-CoA and acetoacetate; thus, lysine is a purely ketogenic amino acid.

## Tryptophan

The major end products of tryptophan catabolism are alanine and acetoacetate, making this amino acid both glucogenic and ketogenic. A branch point in the pathway leads to the formation of NAD+. However, the amount of this redox cofactor synthesized in humans is insufficient for metabolic needs, so dietary input (as nicotinamide) is still required.

## Phenylalanine and Tyrosine Degradation

The initial stages of the breakdown of phenylalanine are illustrated in FIGURE 14.16. The first step is an aromatic ring hydroxylation, catalyzed by phenylalanine hydroxylase. In addition to the mobile cofactors indicated in the diagram, **biopterin** cofactors also participate in the reaction; two are illustrated in FIGURE 14.17. **Tetrahydrobiopterin** binds the enzyme along with the $O_2$ and phenylalanine, assisted by a nonheme iron at the active site. One oxygen atom is incorporated into the amino acid substrate to form the product tyrosine; the other is transiently attached to the biopterin. Further enzymatic steps remove the oxygen as water to form **dihydrobiopterin**, and reduce it to regenerate tetrahydrobiopterin. Transamination of tyrosine produces *p*-hydroxyphenylpyruvate, followed by several steps that lead

**FIGURE 14.14 Threonine Catabolism.** The deaminase step (1) is followed by a dehydrogenase complex (similar to the branched-chain enzymes). Ultimately, the pathway leads to succinyl-CoA, so that threonine is glycogenic.

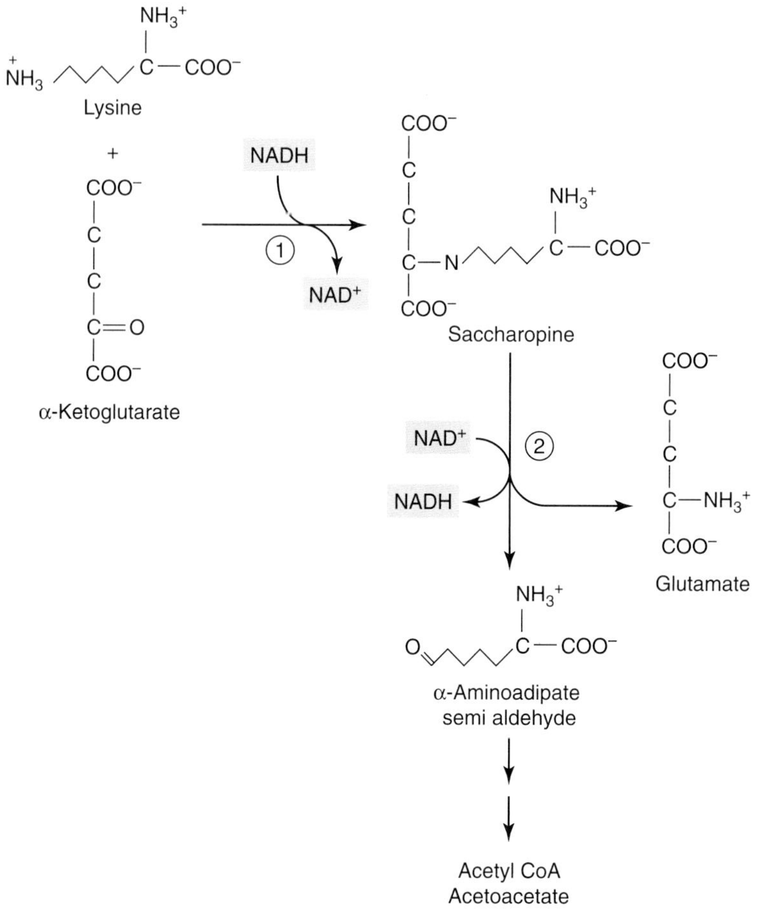

FIGURE 14.15 **Lysine Catabolism.** The first two steps, (1) condensation of lysine and α-ketoglutarate to saccharopine and (2) formation of the semialdehyde and glutamate, remove one nitrogen of the amino acid. Further steps produce only acetoacetate and acetyl-CoA; thus lysine is exclusively ketogenic.

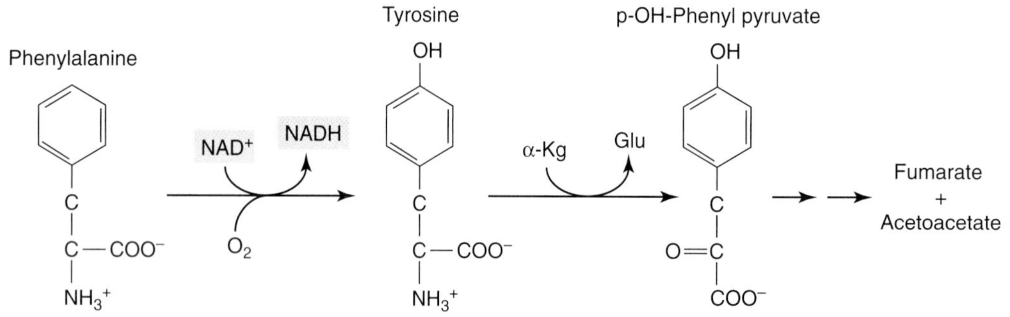

FIGURE 14.16 **Phenylalanine and Tyrosine Breakdown.**

to fumarate and acetoacetate. Thus, both phenylalanine and tyrosine are ketogenic as well as glucogenic amino acids.

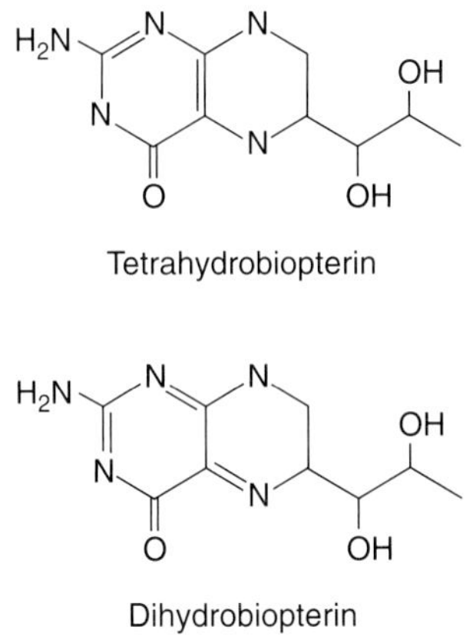

## Amino Acids Directly Connected to Major Metabolic Pathways

Several amino acids are degraded directly through a transaminase reaction to an intermediate of a major pathway, usually glycolysis or the Krebs cycle. For example, the reaction catalyzed by glutamate aminotransferase can convert glutamate to 2-ketoglutarate or aspartate to oxaloacetate. Alanine is converted in one step to pyruvate. These reactions are diagramed in FIGURE 14.18, which also illustrates the deamination reactions of asparagine and glutamine.

The degradation of arginine to ornithine that we examined in the urea cycle cannot be considered to be a true degradative pathway because the product ornithine is part of the cycle. Thus, while this amino acid is connected to a major metabolic pathway, its complete degradation is not. We consider arginine next, along with proline and histidine, two other amino acids that are ultimately converted to glutamate.

FIGURE 14.17 **Biopterin Structures.**

## Arginine, Proline, and Histidine Are All Converted to Glutamate

The degradation routes of arginine, proline, and histidine are shown in FIGURE 14.19. Arginine breakdown (Figure 14.19a) uses the urea

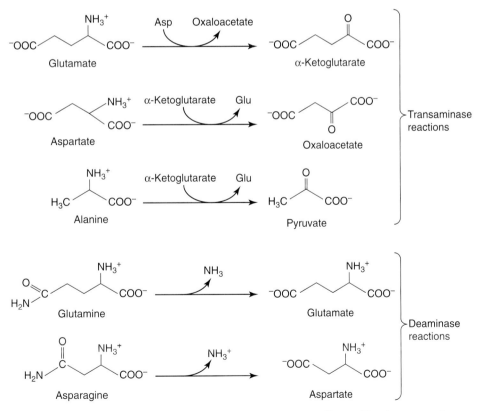

FIGURE 14.18 **Direct Degradation of Amino Acids to Major Pathway Intermediates.** Single-step pathways degrade glutamate, aspartate, alanine, glutamine, and asparagine.

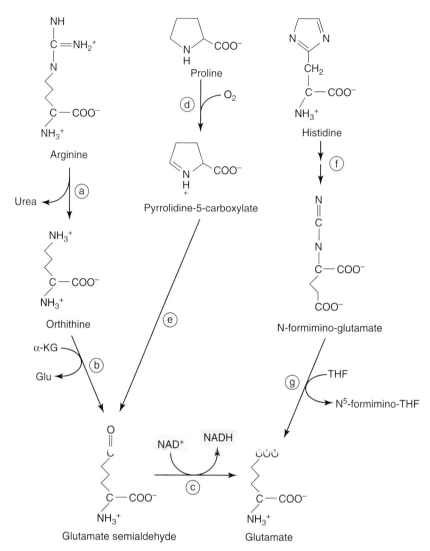

**FIGURE 14.19 Amino Acid Pathways Converging on Glutamate.** Arginine and proline converge at glutamate semialdehyde that forms glutamate. Histidine degradation, after transfer of a carbon to THF, forms glutamate. All three are thus glucogenic.

cycle enzyme arginase. The ornithine product is subsequently converted to glutamate semialdehyde through the action of a transaminase (Figure 14.19b) and to glutamate through a dehydrogenase reaction (Figure 14.19c).

Proline catabolism converges with the arginine route at glutamate semialdehyde after two reactions: an oxidation to pyrrolidine-5-carboxylate (Figure 14.19d) and hydrolysis of the ring (Figure 14.19e).

Histidine also is converted to glutamate via the abbreviated route shown in Figure 14.19f. The proximal intermediate to glutamate is *N*-formimino-glutamate. The formimino portion is transferred to a cofactor abbreviated THF in the figure, thus forming glutamate. **Tetrahydrofolate (THF)** is one of the key cofactors in what is called *one-carbon metabolism,* which is considered next in the context of the degradation of other amino acids that are more closely connected to this process.

**CHAPTER 14** NITROGEN METABOLISM

## One-Carbon (1C) Metabolism and Serine, Glycine, and Methionine Breakdown

The cofactor tetrahydrofolate just introduced in histidine catabolism is derived from **folic acid** (FIGURE 14.20a; **BOX 14.2**). Folic acid is a vitamin because it is not synthesized in humans. Once ingested into the body, folic acid is converted to tetrahydrofolate (Figure 14.20b), which has a reduced form of the pteridine, and a polyglutamate tail with two to four additional glutamate residues linked with amide bonds. The $N^5$ and $N^{10}$ positions (blue in the diagram) are the one-carbon attachment sites of the molecule (with the "active portion" of the molecule shown in Figure 14.20c). Either or both of these nitrogen atoms may be bound to a single carbon unit (or, in some cases, to a formimino group). In amino acid metabolism, the carrier molecule (i.e., THF) does not change its redox state; this happens only in one reaction of nucleotide metabolism (Section 14.8, A Unique Methylation to Form dTMP).

The catabolism of serine is illustrated in FIGURE 14.21. The first reaction, catalyzed by serine hydroxymethyltransferase, has an enzyme-bound pyridoxal phosphate group, which first binds the amine of serine. The substrate–pyridoxal phosphate intermediate is shown in the diagram. Note the similarity to the intermediate and subsequent electron flow in the transaminase reactions (Figure 14.6). In the case of the

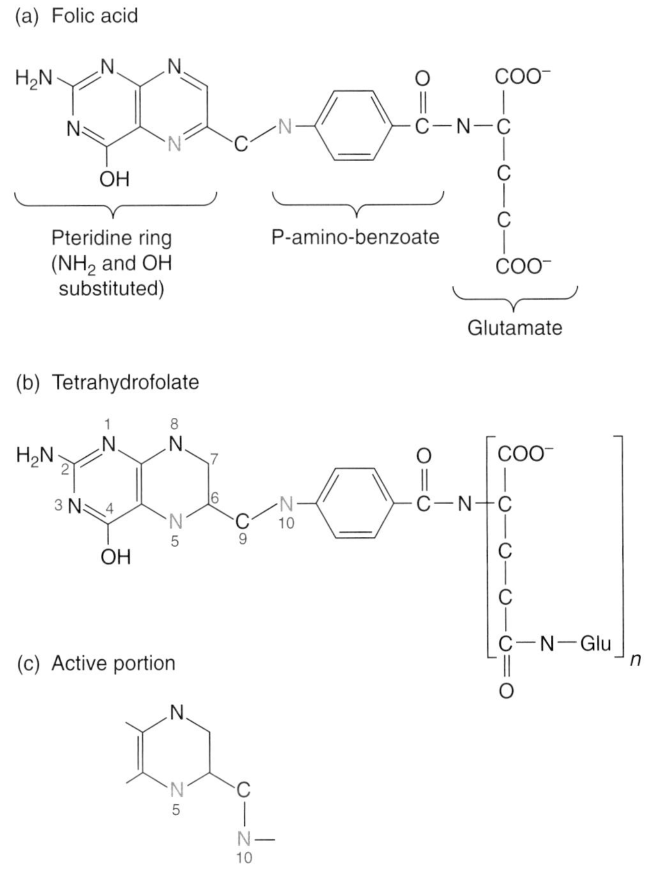

## BOX 14.2: WORD ORIGINS

### Folate

The isolation of a compound that was chemically a pteroylglutamate from leaves in 1941 was given the name **folic acid**, following a long tradition of assigning a name to a compound based on its biological origin (e.g., insulin for the "islands" of islet cells; formic acid from the ant [Latin: *formica*]). Folic acid is indeed found in leafy vegetables, such as spinach and collard greens, but also in others such as lentils. The use of *folio* is much older, a Latin word referring to a leaf of paper. Reference to leaves or *foliage* is still in common use. However, folic acid in great amounts is restricted to a small number of mostly dark-leaved vegetables.

(a) Folic acid

Pteridine ring
(NH₂ and OH
substituted)

P-amino-benzoate

Glutamate

(b) Tetrahydrofolate

(c) Active portion

**FIGURE 14.20 Folic Acid and THF.** (a) Folic acid is the vitamin and precursor to the cofactor, tetrahydrofolate (THF). (b) The cofactor THF. (c) The active portions in carbon transfer are highlighted.

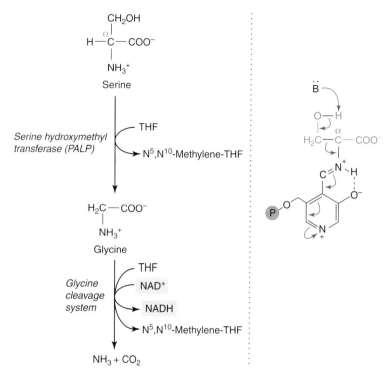

FIGURE 14.21 **Serine and Glycine Degradation**. Both enzymes of serine and glycine breakdown transfer carbons to THF. The intermediate of the transferase enzyme is a Schiff base bound to pyridoxal phosphate, with electron rearrangement similar to that of aminotransferases.

present enzyme, electron flow leads to splitting of a C–C bond as indicated, and the single carbon released becomes attached to THF. After rearrangement and hydrolysis, pyridoxal phosphate is regenerated, and glycine is released. The second reaction (Figure 14.21), catalyzed by the glycine cleavage system, has a mechanism similar to the pyruvate dehydrogenase complex, as well as to the branched-chain amino acid dehydrogenase complexes introduced in this section. As sketched in FIGURE 14.22, the enzyme complex has the same *swinging arm*, a lipoamide connected to a central protein (called H), which first reacts with the substrate glycine at the active site of the P protein (pyridoxal phosphate containing). The glycine-PALP complex is similar to that of the serine hydroxymethyltransferase reaction. In this case, $CO_2$ is released, and the remainder of the molecule becomes attached to the swinging arm as a sulfhydryl adduct. In the next step, the carbon fragment is transferred to THF, leaving the lipoamide in the reduced form. In the last reaction, reduced lipoamide formation is catalyzed by lipoamide dehydrogenase the same FAD-bound enzyme found in the other dehydrogenase complexes. In all cases, electrons are passed to the mobile cofactor $NAD^+$.

While there are other routes for serine oxidation, the one described here is the major pathway in mammals. Thus, in mammals, serine and glycine metabolism are strongly linked.

Methionine degradation is also connected to 1C metabolism. In an unusual addition reaction, methionine reacts with the adenosine moiety of ATP, forming an adduct called **S-adenosylmethionine** (**SAM**; FIGURE 14.23). The other products of this reaction, catalyzed

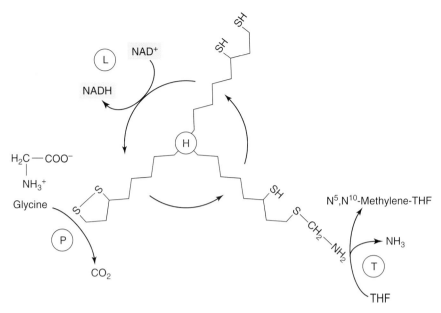

**FIGURE 14.22 Glycine Cleavage System.** The swinging arm mechanism with three separate enzyme activities in the complex resembles the dehydrogenase complexes for pyruvate and the branched chain amino acids.

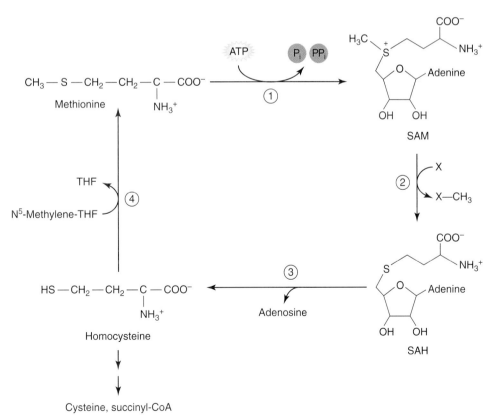

**FIGURE 14.23 Methionine Catabolism and the Methionine Cycle.** Abbreviations: SAM, S-adenosylmethionine; SAH, S-adenosylhomocysteine. The enzymes marked are 1) methionine adenosyltransferase, 2) methyltransferase, 3) SAHase, and 4) methionine synthase.

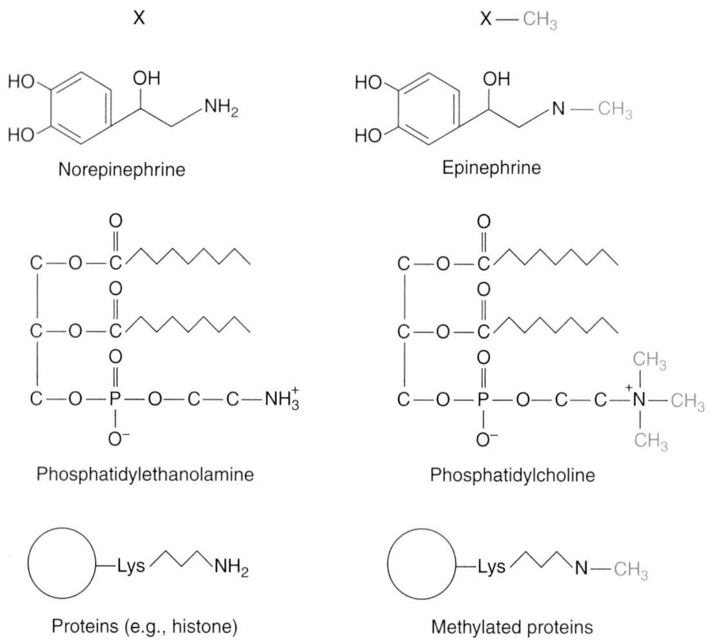

**FIGURE 14.24 Methylation Targets of SAM.** The methyl groups in the product are highlighted in blue.

by methionine adenosyltransferase, are $PP_i$ and $P_i$. SAM is a methyl donor for a variety of reactions, as indicated in the following step (where variable acceptors are indicated in the Figure as an X). After methyl donation, S-adenosylhomocysteine (SAH) is formed. Some methyl acceptors and products are shown in FIGURE 14.24. Methylated products (Figure 14.24) include the hormone epinephrine, the phospholipid phosphatidylcholine, and proteins such as histone, which are commonly methylated on lysine residues as shown. SAM also methylates the nucleic acids.

Reactions 1 to 4 of Figure 14.23 constitute the **methionine cycle**. This route connects the 1C metabolism of THF with the methyl donor SAM. The ultimate source of the carbon may arise from the "loaded" THF pool, much of which can arise from serine and glycine metabolism as just described. Various forms of THF can be interconverted. Reaction 4 of Figure 14.23, catalyzed by methionine synthase, has an additional bound cofactor **cobalamin**, or **Vitamin B$_{12}$**. This prosthetic group has a cobalt ion attached to a ring system very much like the iron ion in the heme group of hemoglobin and the cytochromes. The $Co^+$ of cobalamin directly binds the carbon involved in 1C transfers. Deficiencies in folate or Vitamin B$_{12}$ due to dietary insufficiency or poor absorption (such as in alcoholism) cause serious anemias (**BOX 14.3**).

# 14.7 Amino Acids: Anabolism

For purposes of organizing biosynthetic pathways, amino acids may be categorized as **essential** or **nonessential** (Table 14.1). The nonessential amino acids are those for which pathways exist in humans, whereas the

## BOX 14.3: Vitamin B$_{12}$ and Folate Deficiencies

The close relationship between THF and cobalamin (Vitamin B$_{12}$) is evident in the methionine cycle, but their association was noted many years before this pathway was established. Deficiencies of one or the other lead to anemias, a common condition in older patients. Moreover, any condition that alters absorption can also lead to a relative folate deficiency. B$_{12}$ absorption requires the presence of a protein synthesized in the stomach called **intrinsic factor** (involved in intestinal B$_{12}$ transport), and a lack of this protein leads to a relative deficiency even with normal dietary B$_{12}$. Cobalamin is not present in plants, so strict vegetarians run the risk of anemia. Extreme measures to counter severe obesity include surgery, which may involve removing part of the stomach and/or intestine, indirectly cause a deficiency in these vitamins. Finally, intake of certain drugs, such as glucophage for diabetes, is a cause of B$_{12}$ deficiency, as is the consumption of alcohol.

**TABLE 14.1 Essential and Nonessential Amino Acids**

| Essential | Nonessential |
|-----------|--------------|
| Cys | Ala |
| His | Arg |
| Iso | Asp |
| Leu | Glu |
| Lys | Asn |
| Met | Gln |
| Phe | Gly |
| Thr | Pro |
| Trp | Ser |
| Tyr | – |
| Val | – |

essential amino acids must be obtained from the diet. As omnivores, humans usually have no difficulty obtaining sufficient amino acids, mostly for protein synthesis, but also for other purposes as described in the following sections. However, vegetarians must consciously combine foods (e.g., rice and beans) to obtain essential amino acids because plant proteins do not contain the entire complement of amino acids.

## Nonessential Amino Acids

Many of the nonessential amino acids can be obtained by the reversal of near-equilibrium reactions that we have already considered, or by the conversion from other amino acids. For example, the synthesis of alanine, glutamate, and aspartate is accomplished immediately by near-equilibrium transamination reactions we have already considered (FIGURE 14.25). Glutamate is also formed as the product of the glutamate DH reaction. Also shown are asparagine and glutamine formation, catalyzed by the corresponding synthetase.

Proline and arginine are synthesized from the common precursor glutamate (FIGURE 14.26). After formation of glutamate γ-semialdehyde, the routes diverge. Nonenzymatic cyclization to pyrroline 5-carboxylate leads to proline, whereas transamination of the semialdehyde produces ornithine. The enzymes of the urea cycle are then employed to convert ornithine to arginine.

Serine and glycine are linked not only in their catabolic pathways (as in the prior section), but also in their biosynthesis. The amino acids are synthesized from the glycolytic intermediate 3-P-glycerate (FIGURE 14.27). First, 3-P-glycerate is reduced to 3-P-hydroxypyruvate, catalyzed by a dehydrogenase. Transamination produces P-serine, which is then dephosphorylated to serine. Serine is converted to glycine in a near-equilibrium reaction catalyzed by serine hydroxymethyltransferase. Thus, with the availability of excess glycine, this enzymatic step alone can produce serine.

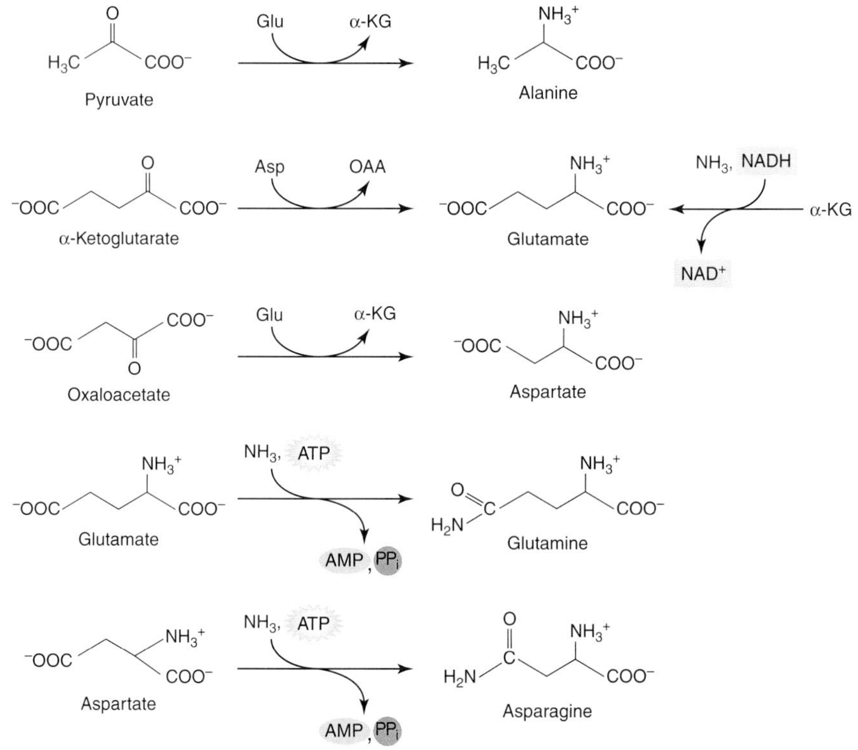

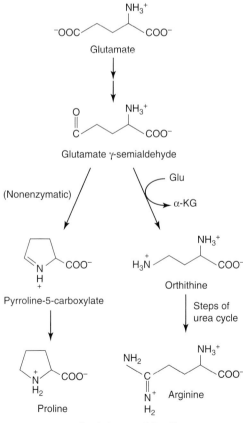

**FIGURE 14.25 Direct Formation of Amino Acids.** Single-step formation of alanine, glutamate, aspartate, glutamine, and asparagine.

## Essential Amino Acids

Roughly half of the amino acids required for protein synthesis must be obtained in the diet. Further details are presented and other pathways are described in the Appendix. Here we explore portions of a few of these pathways to demonstrate how they are connected to other metabolic routes. In some cases, at least a portion of the essential amino acid can be obtained from the metabolism of other amino acids. For example, in the degradation of methionine (Figure 14.23), we found one of its products was cysteine. In that same figure, the synthesis of methionine could be accomplished if we had a pathway for the formation of homocysteine, which arises from aspartate in bacteria. Similarly, some tyrosine can be synthesized from the hydroxylation of phenylalanine (Figure 14.16). However, the *de novo* synthesis of phenylalanine and the other aromatic amino acids requires a separate pathway that exists only in auxotropic organisms. We will consider them next in more detail as an example of more extensive essential amino acid biosynthesis.

## Aromatic Amino Acid Biosynthesis

The aromatic amino acids tryptophan, phenylalanine, and tyrosine are formed in the **chorismate** pathway, named for the common intermediate. Some of the steps are illustrated in FIGURE 14.28. The starting points are phosphoenolpyruvate (PEP) from glycolysis and

**FIGURE 14.26 Arginine and Proline Biosynthesis.** Glutamate, after forming the semialdehyde, branches to form the cyclic proline or the open-chain arginine. The formation of arginine uses enzymes of the urea cycle.

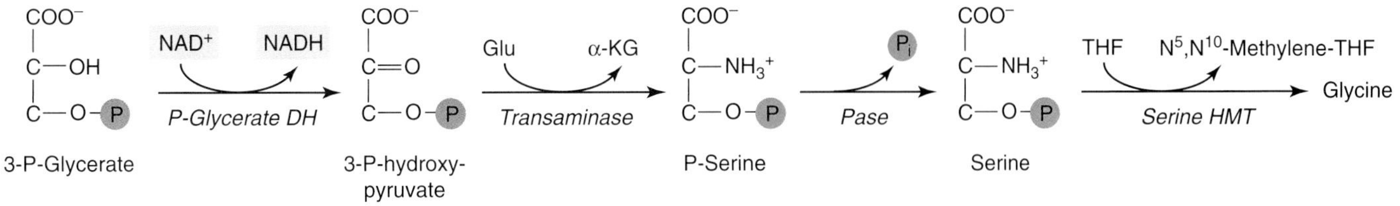

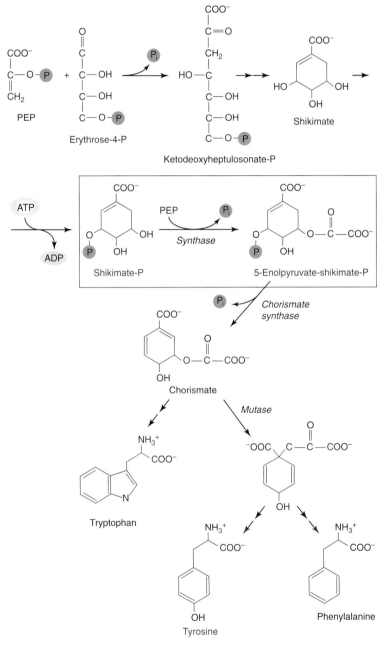

**FIGURE 14.27 Serine and Glycine Biosynthesis.** The glycolytic intermediate P-glycerate leads to serine formation in three steps. The serine is then converted to glycine, catalyzed by serine hydroxymethyltransferase (HMT).

**FIGURE 14.28 Aromatic Amino Acid Biosynthesis.** After a condensation reaction to form the heptulose analog, cyclization to shikimic acid and a phosphorylation step, a synthase and subsequent dephosphorylation produces chorismate, the precursor of the aromatic amino acids. The synthase reaction (boxed) is the site of inhibition of the herbicide glyphosate (Roundup; see Box 14.4).

## BOX 14.4: The Chorismate Pathway and Weed Control

The chorismate pathway can be selectively inhibited by a compound developed by Monsanto Chemical Company called glyphosate, popularly known as Roundup. The inhibited step is the synthase reaction shown in the boxed area of Figure 14.28, which is required for the synthesis of the three aromatic amino acids. Glyphosate inhibition is competitive with the substrate PEP; the structures are shown in FIGURE B14.4. Glyphosate is a **phosphonate** (note the P—C bond) and is broken down in the soil to the nontoxic products $P_i$, $NH_3$, and $HCO_3^-$. The specificity of glyphosate is remarkable. In fact, glyphosate appears to have no effect on other reactions that use PEP, including the one at the beginning of the chorismate pathway itself. Because the pathway does not exist in animals—the very reason the amino acids are essential—there is no human toxicity. Glyphosate kills virtually all plants, not just weeds. It can nonetheless be simultaneously used with crops genetically modified (and, conveniently enough, also supplied by Monsanto) with an altered synthase that does not bind glyphosate. Unfortunately, some weeds have evolved to become resistant to glyphosate.

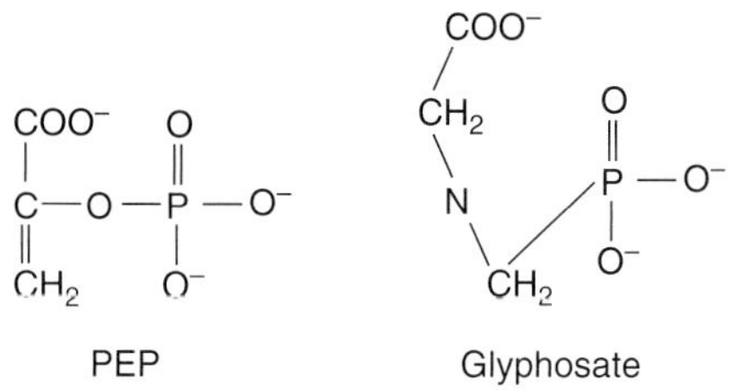

FIGURE B14.4. **Chorismate.** Structure of glyphosate compared to PEP.

erythrose-4-phosphate (E4P) from the pentose phosphate shunt. The seven-carbon condensation product is ultimately converted to shikimate, a ring structure that is the precursor to the aromatic rings in the product amino acids. After a phosphorylation step, a second PEP molecule condenses with the ring in a synthase-catalyzed step; this reaction is highlighted as an enclosed box in the figure. It is significant as the site of action of a well-known broad spectrum plant growth inhibitor, popularly known as Roundup (**BOX 14.4**).

# 14.8 Nucleotide Metabolism

Most of the nitrogen in organisms resides in the amino acids, largely within proteins. However, amino acids (along with $NH_3$) are extensively involved in the metabolism of other compounds. The products most central to the needs of the cell are the nucleotides, which are necessary as cofactors and for DNA and RNA synthesis. We examine here the synthesis of the pyrimidine and purine bases and aspects of the metabolism of these compounds.

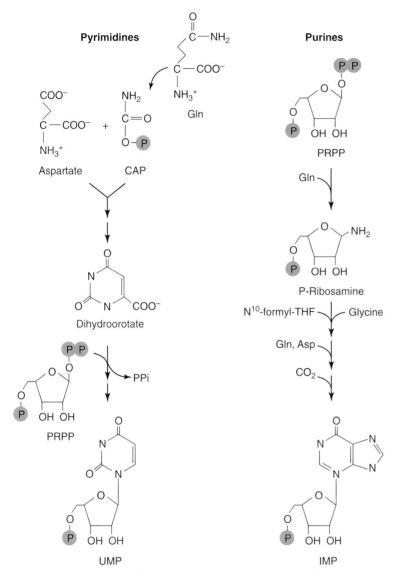

**FIGURE 14.29 Overview of Nucleotide Metabolism.** Pyrimidines (left side) form an aromatic ring to which ribose is added. Purines (right side) start with a ribose scaffold and, after several steps, lead to the finished nucleotide.

As an overview, consider the pathways outlined in FIGURE 14.29. In the left panel, pyrimidine bases are shown to originate from the amino acids aspartate and glutamine, forming the cyclic intermediate dihydroorotate. Dihydroorotate condenses with a ribose molecule that has a 5′-phosphate and a 1′-pyrophosphate, **phosphoribosylpyrophosphate (PRPP)**. The product of the pyrimidine pathway is uracil monophosphate (UMP). The purine pathway shown on the right panel of Figure 14.29 starts with the PRPP, and the pyrophosphate group is displaced by the nitrogen of glutamine, forming the intermediate P-ribosamine. The rest of the purine ring is built on that nitrogen, using inputs from 1C metabolism in the form of $N^{10}$-formyl-THF, nitrogen from glutamine, carbon from $CO_2$, and the incorporation of an entire glycine molecule. The product of the purine pathway is inosine monophosphate (IMP). We will first fill in the pyrimidine pathway, followed by aspects of the purine pathway; the complete route is shown in the Appendix.

## Pyrimidine Synthesis

The complete pathway for the formation of UMP is shown in FIGURE 14.30. The formation of carbamoyl-P (CAP) is catalyzed by carbamoyl-P synthetase II (CPS II). This reaction bears a strong similarity to CPS of the urea cycle, which is known as CPS I. For the CPS II reaction, glutamine is converted to ammonia on the enzyme surface, and the reaction then proceeds in the same way as CPS I. Whereas CPS I is localized exclusively in the mitochondrial matrix, CPS II is exclusively cytosolic. The condensation of CAP with aspartate, the

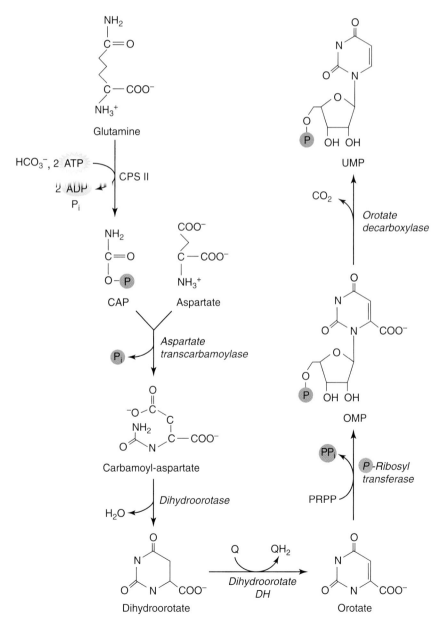

FIGURE 14.30 **UMP Biosynthesis.**

**CHAPTER 14** NITROGEN METABOLISM

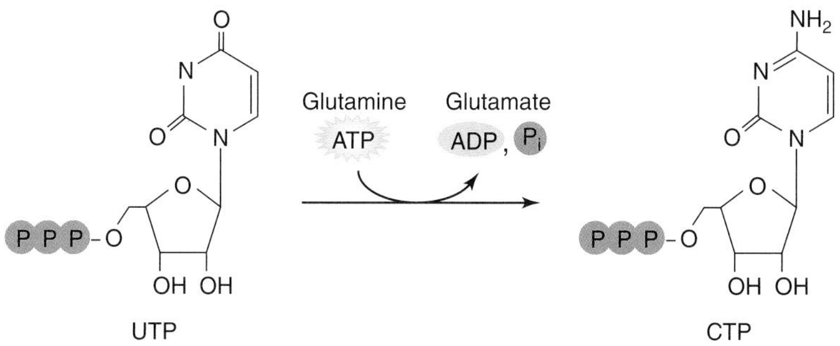

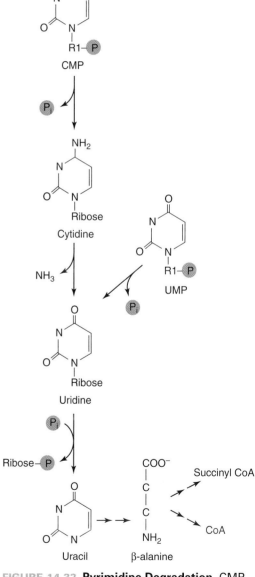

FIGURE 14.31 **CTP Synthetase Reaction.**

next step, is catalyzed by aspartate transcarbamoylase. Intramolecular dehydration of CAP, catalyzed by dihydroorotase, produces the cyclic compound dihydroorotate. Next, oxidation to orotate is catalyzed by dihydroorotase DH. This reaction is similar to glycerol phosphate oxidase because it is embedded in the mitochondrial inner membrane, its active site faces the cytosol, and electrons are donated to the mitochondrial respiratory chain via $QH_2$. In the next step, the ribose portion of PRPP attaches to orotate to produce orotidine 5′-monophosphate (OMP). Finally, orotate decarboxylase catalyzes the removal of $CO_2$ and the formation of UMP, the parent pyrimidine.

UMP is a substrate for nucleotide kinases, which use ATP as the donor to form first UDP and subsequently UTP. UTP is a substrate of the CTP synthetase reaction (**FIGURE 14.31**) in which the nitrogen of glutamine is incorporated into the product.

## Pyrimidine Degradation

The breakdown of pyrimidine nucleotides starting from CMP and UMP is illustrated in **FIGURE 14.32**. Dephosphorylation of CMP yields cytidine; deamination of this intermediate produces uridine. UMP directly forms uridine by dephosphorylation. Uridine phosphorylase, which catalyzes the next reaction, has a mechanism that is similar to glycogen phosphorylase. An inorganic phosphate ($P_i$) adds to an intermediate ribose carbocation at C1, producing uracil and ribose-1-P (R1P). A later intermediate of pyrimidine degradation is β-alanine, most of which is converted to succinyl-CoA and thus ultimately to glucose. However, some β-alanine can be used in the synthesis of CoA itself, so that pyrimidine degradation is connected to the formation of this cofactor. The other portion of CoA, pantothenic acid, must be obtained from the diet in humans, so this component of CoA is a vitamin.

## Purine Synthesis

Inosine monophosphate (IMP), the parent purine nucleotide, is shown in outline form in Figure 14.29. Some further details of purine biosynthesis are shown in **FIGURE 14.33**. The first step is the displacement of the pyrophosphate group of PRPP to form P-ribosamine. In contrast to pyrimidine synthesis, purine synthesis is built on a ribose phosphate scaffold, abbreviated in the Figure as R. The second

FIGURE 14.32 **Pyrimidine Degradation.** CMP and UMP breakdown converge at uridine, ultimately forming the intermediates CoA and succinyl-CoA.

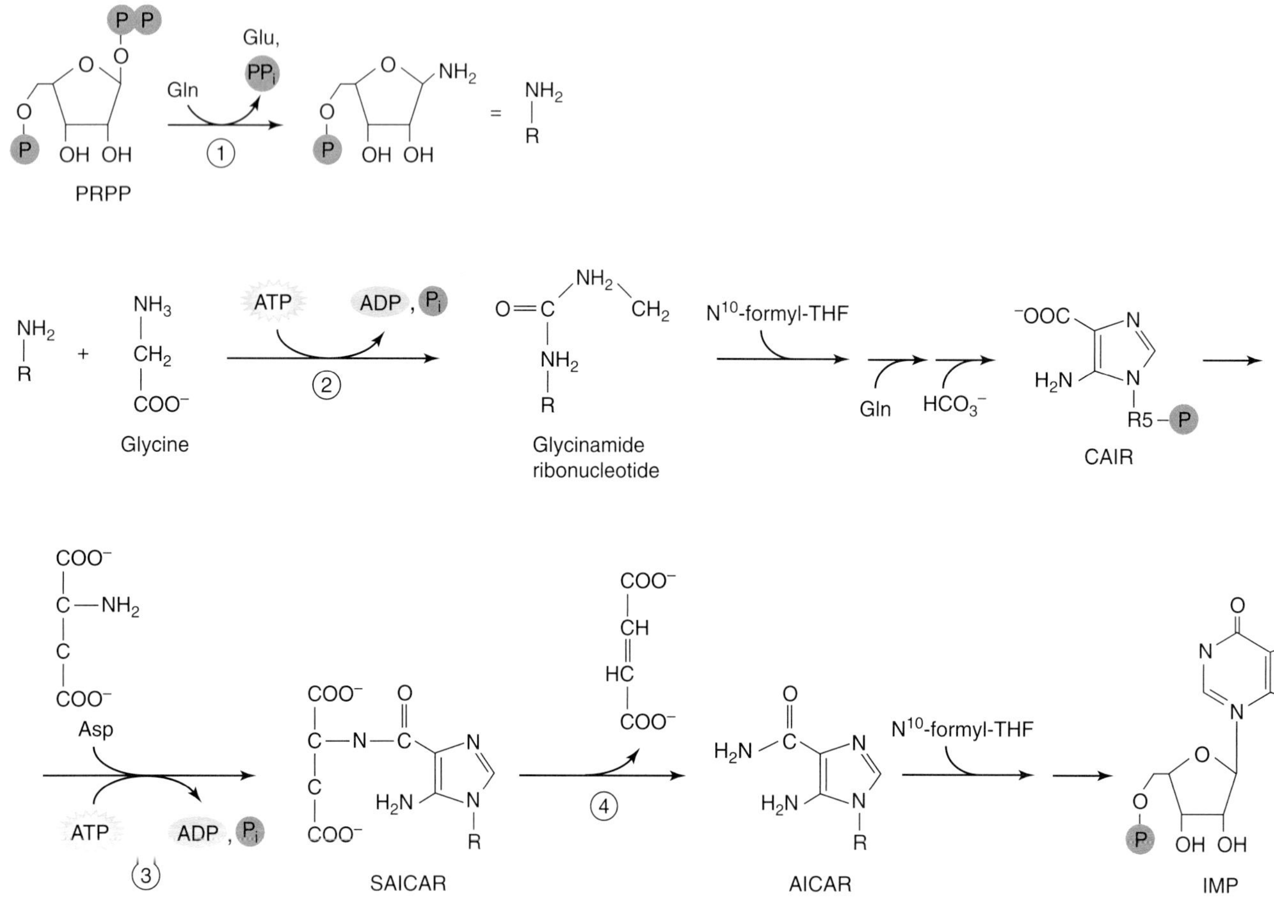

FIGURE 14.33 **IMP Biosynthesis.** The major steps in purine biosynthesis are catalyzed by 1) glutamine PRPP amidotransferase, 2) glycinamide ribose synthase, 3) SAICAR synthetase, and 4) adenylosuccinate lyase.

reaction is a condensation of the amine of ribose with the carboxyl group of glycine, catalyzed by a synthetase. The enzyme mechanism involves an acid phosphate intermediate, evident from the splitting of ATP to ADP and $P_i$ in the overall reaction. Several reactions follow that incorporate a formyl group (attached to the cofactor THF), the nitrogen of glutamine, and the carbon of $HCO_3^-$, leading to the first complete ring, carboxyaminoimidazole ribotide (CAIR). The next two reactions (3 and 4 of Figure 14.33) incorporate the nitrogen of aspartate in a manner similar to arginine formation in the urea cycle (Section 14.4). In step 3, aspartate condenses with the carboxyl group of CAIR, forming the succinylamide derivative of CAIR (SAICAR). In step 4, fumarate is released and the nitrogen atom is incorporated into the growing nucleotide, forming aminocarboxamide ribotide (AICAR). The last steps incorporate the last carbon from a second $N^{10}$-formyl-THF, and close the ring through a dehydration reaction, producing IMP.

Formation of GMP and AMP both involve the incorporation of nitrogen into the purine ring of IMP, using reactions similar to those we have just encountered in the biosynthesis of IMP itself. For GMP formation (FIGURE 14.34, upper branch), oxygen incorporation from water and an oxidation, catalyzed by a dehydrogenase, produces

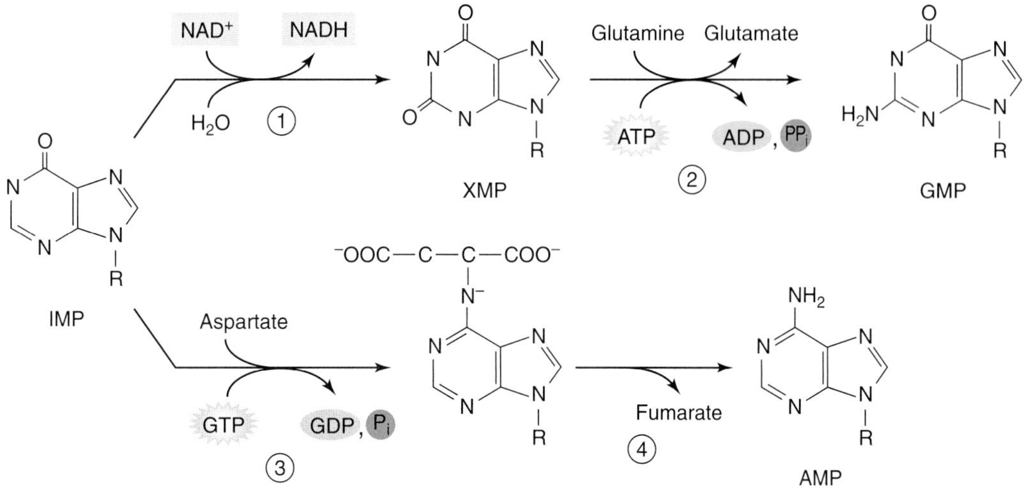

**FIGURE 14.34 GMP and AMP Formation from IMP.**

xanthine dehydrogenase (XMP). The incorporation of one nitrogen from glutamine, catalyzed by glutamine synthetase, produces GMP. AMP formation (Figure 14.34, lower branch) involves the incorporation of nitrogen from aspartate in two reactions. First (step 3) aspartate forms an adduct with IMP, then (step 4), fumarate is cleaved from the adduct, leaving the product, AMP.

As was the case with the pyrimidine nucleotides, an array of kinases catalyze the conversion of the pyridine monophosphates into di- and triphosphonucleotides. Furthermore, nucleoside diphosphate (NDP) kinase can catalyze the conversion of any of the diphosphates to triphosphates using ATP as the other substrate.

## Purine Degradation

The degradation of purines is a more active route than the pyrimidines, because these nucleotides are found in greater concentrations in cells and used in more metabolic routes. Adenine nucleotides are responsible for most phosphate-bond energy transfers, so it is fitting that they are present at 10-fold the concentration of the next major nucleotide: the guanine nucleotides.

The end product of purine nucleotide breakdown in mammalian systems is uric acid. The pathway is shown in FIGURE 14.35. AMP can be either deaminated, through the AMP deaminase reaction, or dephosphorylated, through the nucleosidase reaction. For example, muscle cells have mostly the deaminase pathway, whereas T-lymphocytes have mostly the nucleotidase pathway. Both of these routes converge on inosine, by a further dephosphorylation or deamination, as indicated in the figure. Next, the ribose portion is removed through a phosphorylase reaction of the type described above for pyrimidine catabolism (Figure 14.32); the product is hypoxanthine. Xanthine oxidase catalyzes the oxygenation of hypoxanthine to xanthine, as well as the further oxidation of xanthine to uric acid, which is excreted in humans. GMP is catabolized in

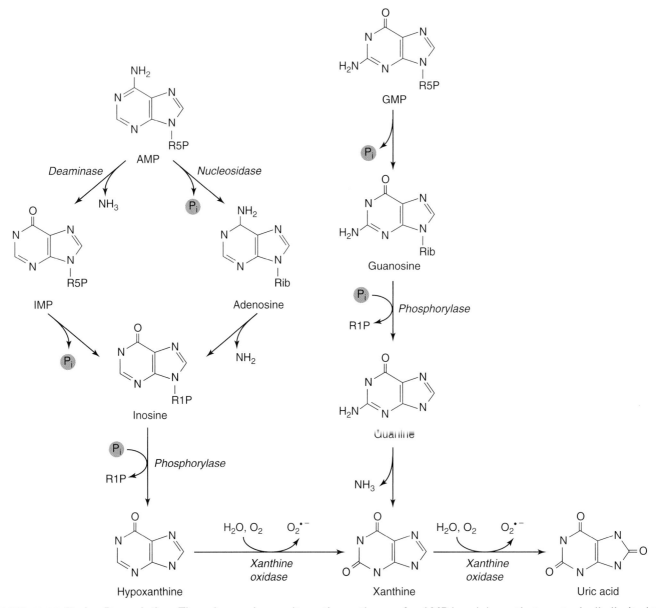

**FIGURE 14.35 Purine Degradation.** The scheme shows alternative pathways for AMP breakdown that are typically limited to separate cells. Purines ultimately form hypoxanthine and xanthine, which are converted by xanthine oxidase to uric acid, the excreted end product.

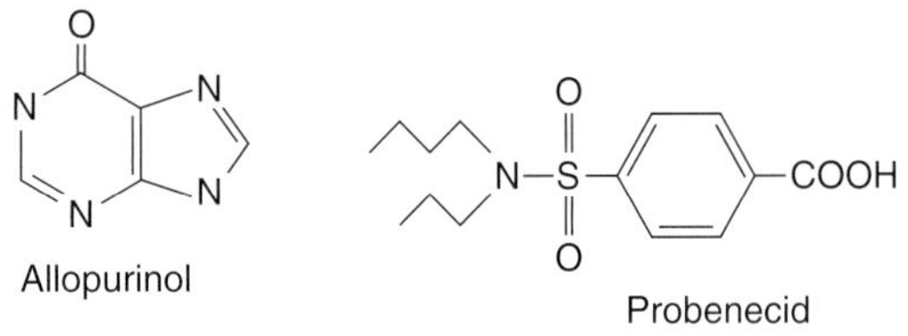

a similar way, through dephosphorylation, a phosphorylase reaction, and deamination to xanthine, which merges with the AMP degradation pathway.

Uric acid is poorly soluble in water, so when excessive amounts accumulate in humans, it can precipitate in joints, resulting in **gout.** Excess production is one cause, and this can be treated by inhibiting xanthine oxidase with **allopurinol.** A second treatment strategy is to block kidney reabsorption with **probenecid.** The structures of these compounds are shown in **FIGURE 14.36.** Uric acid is the principal nitrogen end product in birds. The white semisolid precipitate requires far less water, a weight-saving advantage for avian flight.

**FIGURE 14.36 Allopurinol and Probenecid: Gout Treatments.**

Allopurinol

Probenecid

## Salvage Reactions

Not all of the purine nucleotides that are broken down are excreted as uric acid. In many cases, they react with PRPP, using one of two enzymatic reactions to reform nucleotides. The first of these is adenosine P-ribosyl transferase:

$$Adenosine + PRPP \rightarrow AMP + PP_i \qquad (14.6)$$

The second, hypoxanthine-guanosine transferase, uses guanosine or hypoxanthine as substrate:

$$Guanosine + PRPP \rightarrow GMP + PP_i \qquad (14.7)$$

$$Hypoxanthine + PRPP \rightarrow IMP + PP_i \qquad (14.8)$$

The importance of the salvage pathway is evident in the pathology of Lesch–Nyhan syndrome, which stems from a deficiency in hypoxanthine-guanosine transferase. As a result, purine breakdown products accumulate, leading to gout. Other symptoms of the disease, such as mental defects, are less obviously related to the known enzyme deficiency.

Cells can take advantage of the salvage pathway in normal metabolism to bypass the far more extensive *de novo* synthesis pathway. This represents a significant energy saving for cells because purine turnover is extensive.

## Purine Nucleotide Regulation

Allosteric control by adenine and guanine nucleotides was demonstrated for isolated enzyme preparations in the late 1950s. Subsequent investigation focusing on microorganisms established long-term (genetic) control for several of the steps of the pathway. However, comparable control of *de novo* purine biosynthesis for animal species is not as well established. It is unlikely to be the same; for example, concentrations of ATP generally are unchanged during most metabolic states of mammalian cells. Hence, ATP is not itself a metabolic regulator. Free ADP and AMP concentrations do fluctuate, but on a far shorter time scale than that which would be appropriate for the control of nucleotide biosynthesis. Alterations in other nucleotide concentrations are far less well known.

The salvage pathways of purine biosynthesis are regulated in part by the concentration of PRPP. In birds, which have extensive flows through a slightly modified purine biosynthesis pathway in order to synthesize and excrete uric acid, increased synthesis of rate-limiting enzymes, in particular xanthine oxidase and PRPP amidotransferase, are important regulatory steps.

## Purine Nucleotide Cycle

A pathway specifically involving AMP is the **purine nucleotide cycle**. This involves just three reactions, one of which is the deamination

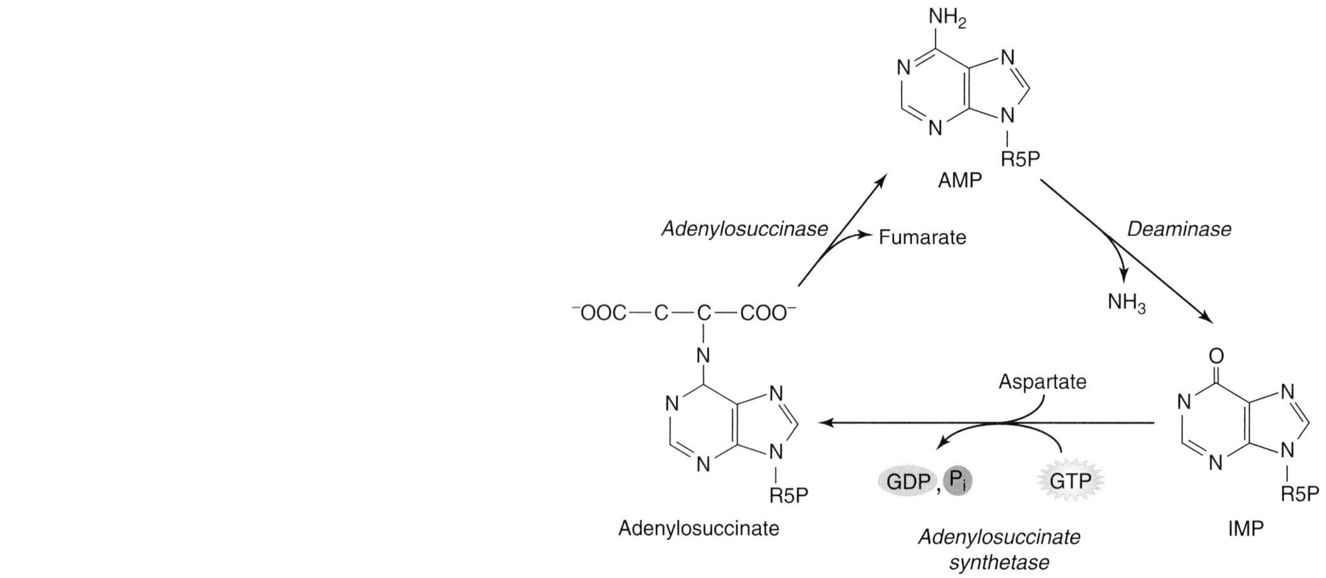

**FIGURE 14.37 Purine Nucleotide Cycle.** Cycling between these purine nucleotides converts aspartate to fumarate, which is used in muscle to maintain Krebs cycle intermediates from fumarate.

of AMP to IMP that we have already examined in the context of purine degradation. The cycle (FIGURE 14.37) regenerates AMP, using the nitrogen of aspartate, through a condensation reaction followed by a splitting reaction with release of fumarate. The enzymes of this pathway are found in particularly high concentration in muscle, which uses fumarate as an anaplerotic reaction (filling reaction) for the Krebs cycle. Muscle has little pyruvate carboxylase, which other cells use to convert pyruvate to oxaloacetate and maintain levels of Krebs cycle intermediates.

As shown in FIGURE 14.38, the aspartate condensation reaction followed by fumarate release is a common theme, present in the urea cycle, in purine biosynthesis, and in the purine nucleotide cycle.

## Deoxynucleotide Formation

Deoxynucleotides are formed from the corresponding nucleoside diphosphate (NDP) in the reaction catalyzed by ribonucleotide reductase (FIGURE 14.39). The reaction shown in Figure 14.39 is incomplete, because the enzyme is left in an oxidized form. Regeneration of the reduced form of ribonucleotide reductase (in order to complete the catalytic cycle for deoxynucleoside diphosphate [dNDP] formation) involves two other proteins in an electron-transfer chain, shown in FIGURE 14.40. First, **thioredoxin reductase** accepts a hydride from NADPH, transfers the electrons to a bound FAD, and then to the disulfide of **thioredoxin**, leaving that protein in the sulfhydryl form. Ribonucleotide reductase accepts electrons from thioredoxin. The action of thioredoxin reductase on thioredoxin is closely analogous to two other reactions in cells. One of these is the last enzyme of the dehydrogenase complexes (such as the pyruvate dehydrogenase complex), lipoamide dehydrogenase. The other is glutathione reductase

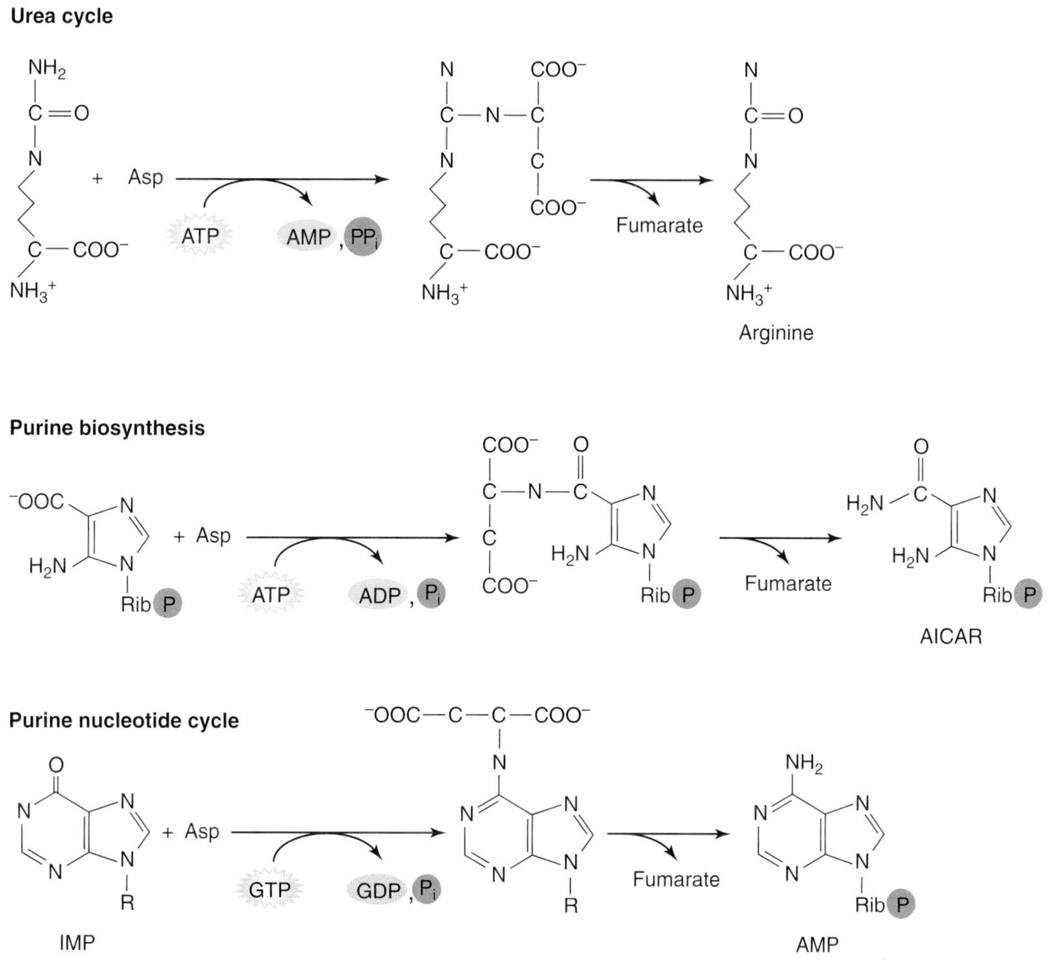

**Urea cycle**

Arginine

**Purine biosynthesis**

AICAR

**Purine nucleotide cycle**

IMP

AMP

**FIGURE 14.38 Three Instances of Aspartate N Incorporation.** The conversion of aspartate to fumarate in two steps is a metabolic route found in different pathways of nitrogen metabolism.

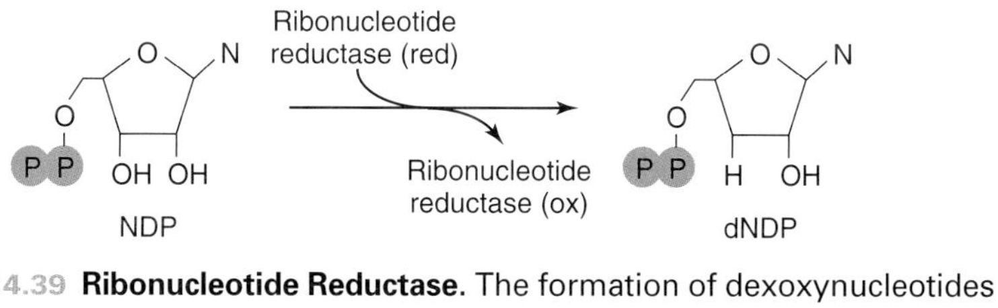

NDP

dNDP

**FIGURE 14.39 Ribonucleotide Reductase.** The formation of dexoxynucleotides leaves the enzyme in the oxidized form.

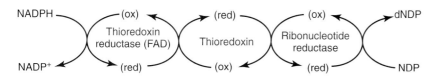

**FIGURE 14.40 Electron Flow Through Three Proteins for dNTP Formation.** The scheme for the restoration of the reduced form of ribonucleotide reductase.

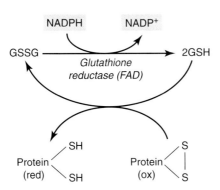

γ-glu-cys-gly
Glutathione (reduced) GSH

G—S—S—G

Glutathione (oxidized)

**FIGURE 14.41 Glutathione Reductase in Protein Sulfhydryl Reduction.** Glutathione (top panel, in reduced form) is utilized to maintain reduced protein sulfhydryl groups. The reducing equivalents are supplied by NADPH (bottom panel).

(FIGURE 14.41), which consumes NADPH in order to produce reduced **glutathione** from oxidized glutathione. Reduced glutathione reacts with a large number of intracellular proteins in order to maintain them in the sulfhydryl form as part of the antioxidant system in cells. All of these similar reactions have as their key the passing of electrons between an enzyme sulfhydryl, FAD, and a nicotinamide nucleotide (NAD+ or NADP+).

## A Unique Methylation to Form dTMP

Thymidine is required for the formation of DNA. It is formed through a methylation reaction that is unique in 1C metabolism. The methylation is catalyzed by thymidylate synthase, as illustrated in FIGURE 14.42 (top panel). The methyl donor is $N^5,N^{10}$-methylene-THF. Both the methyl group, as well as a pair of electrons, are extracted from this molecule to form the products, deoxythymidine 5′-monophosphate (dTMP) and dihydrofolate (DHF). In all other folate 1C reactions, the redox state of the folate itself is unchanged. The route of regeneration of THF is shown in the bottom panel of Figure 14.42, along with the key portions of the folate structures. The enzyme DHF reductase catalyzes the addition of a pair of electrons across the double bond containing the $N^5$ of DHF. Subsequently, THF can reacquire the methyl group through the serine hydroxymethyl transferase reaction as mentioned previously in the discussion of amino acid catabolism. The DHF reductase step can be selectively inhibited by compounds such as methotrexate (FIGURE 14.43), which thereby blocks DNA synthesis. This drug was one of the first used in cancer therapy, preferentially

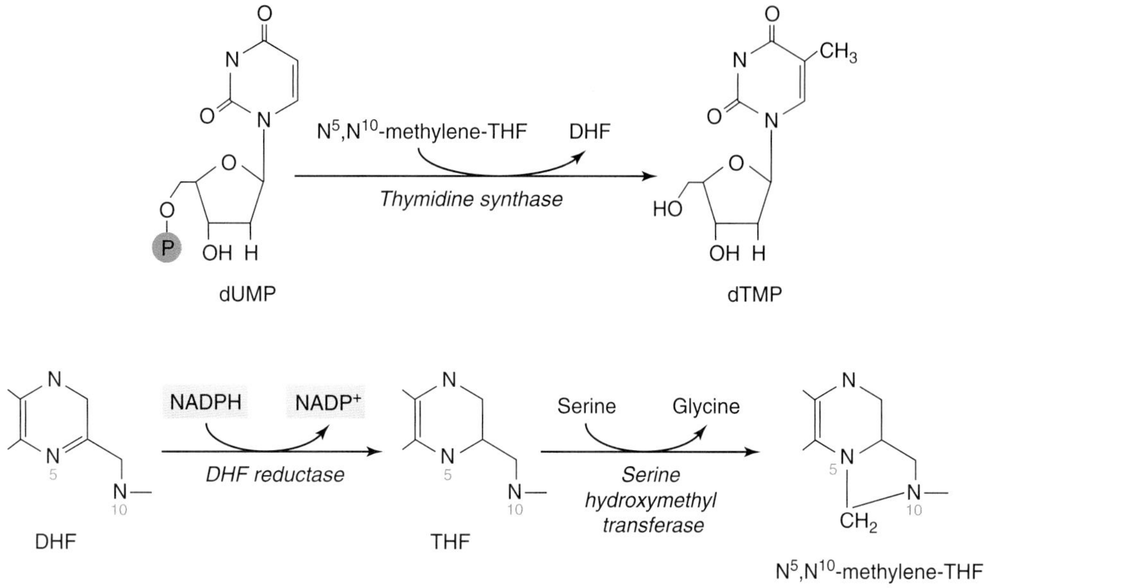

FIGURE 14.42 **dTMP Formation.** Formation of the DNA-specific nucleotide dTMP uses reductive methylation (top panel). Restoration of the methyl donor requires NADPH input and methylation of THF (bottom panel).

killing rapidly dividing cells. In addition, it is used in other cases of unusually rapid cell division, such as the inhibition of the excessive skin cell division in psoriasis.

With the two reduction states of folates, we can now appreciate the strong similarity between three bound cofactors: flavins, biopterins, and folates (FIGURE 14.44). The oxidized forms are on the left side of Figure 14.44. Both flavins and biopterins allow one- and two-electron

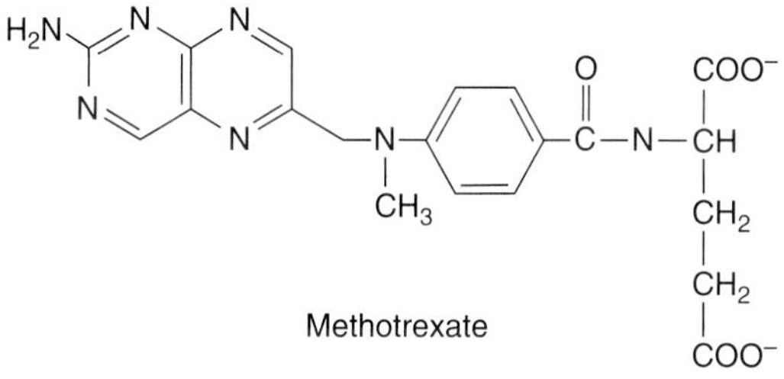

Methotrexate

FIGURE 14.43 **Methotrexate.**

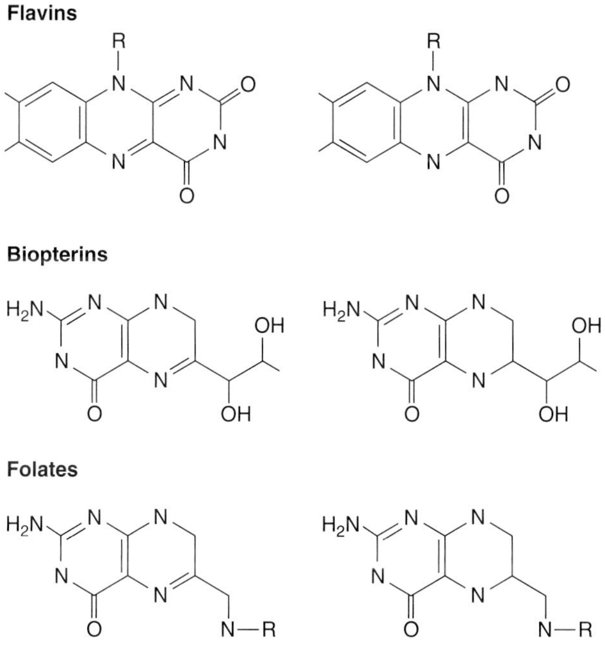

FIGURE 14.44 **Structural Similarities in Flavins, Biopterins, and Folates.**

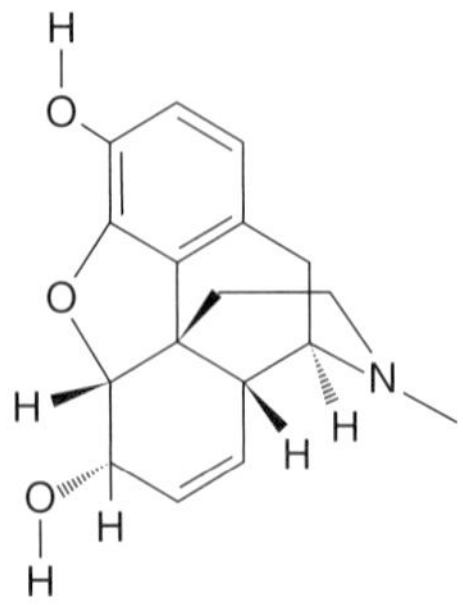

FIGURE 14.45 **Morphine**.

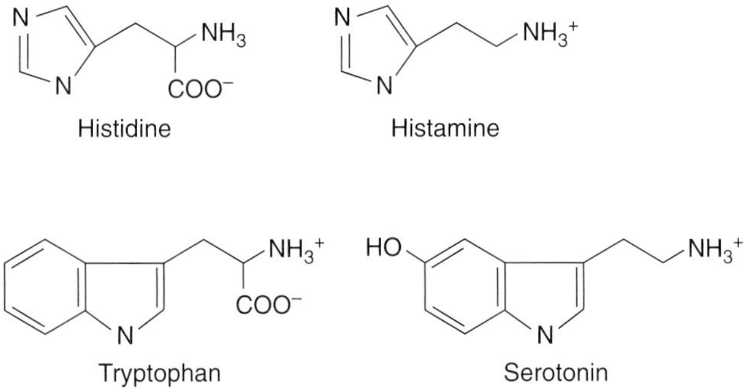

Histidine

Histamine

Tryptophan

Serotonin

FIGURE 14.46 **Signal Molecules from Amino Acids**.

transfers; they can react, for example, with metals or hydrides like NADH. The Figure underscores the strong chemical similarity in the pterin rings of biopterins and folates, indicative of similar origins.

# 14.9 Other Nitrogen Pathways

A large number of other routes exist in nitrogen metabolism. These are quantitatively more minor than those already described, and their biological distribution is more limited. In the plant and microbial world, the many unusual end products that contain nitrogen is vast; they are known collectively as the **alkaloids**. Their role in the organism that produces them is not always obvious. In many cases, they serve to discourage animals from eating them, because the compounds are typically bitter tasting. In some cases, alkaloids are important drugs, such as the analgesic morphine (FIGURE 14.45), isolated from the poppy flower.

A number of neurotransmitters and similar signal molecules are biosynthesized from amino acids (FIGURE 14.46). For example, decarboxylation of histidine produces histamine, which is released from white blood cells and is partly responsible for allergy symptoms. Moreover, hydroxylation and decarboxylation converts tryptophan into serotonin (5-hydroxytryptamine), the neurotransmitter introduced in Section 14.3.

Finally, we consider the production of nitric oxide (NO), which may be the chemically simplest cell modulator. The pathway for NO formation is the oxidation of arginine (FIGURE 14.47). The enzyme catalyzing this step is nitric oxide synthetase. Its reaction mechanism

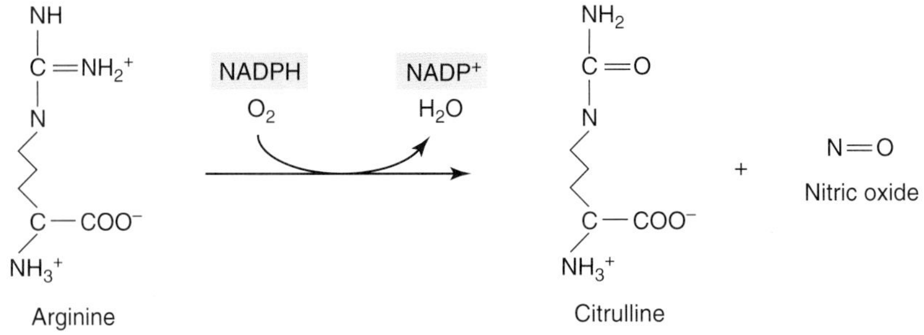

Arginine

Citrulline

Nitric oxide

FIGURE 14.47 **NO Synthesis**.

**CHAPTER 14** NITROGEN METABOLISM

is surprisingly complex, involving the bound cofactors tetrahydrobiopterin and flavin mononucleotide (FMN). In the body, NO synthase is found in the lining cells of the blood vessels (endothelia), and the target of NO is the smooth muscle, where it causes relaxation, and thus blood vessel dilation. Direct formation of NO can also be achieved by ingesting nitroglycerin, which nonenzymatically releases NO.

## Summary

Nitrogen has a global cycle, by which atmospheric $N_2$ gas is converted to $NH_3$ and then amino acids within organisms, to oxidized forms, and finally to $N_2$ once again. Most of biological nitrogen metabolism involves amino acids and $NH_3$. Two abundant near-equilibrium reactions, glutamate DH and transaminases, are prominent in the synthesis and degradation of amino acids as well as in nucleotide metabolism. The urea cycle is the mammalian pathway by which amino acids and $NH_3$ are converted into urea in the liver. This route is interconnected with gluconeogenesis, which is the fate of fumarate released from the urea cycle. Urea is excreted by mammals as the ultimate nitrogen end product. Breakdown of amino acids leads to either glucose formation (glucogenic), or acetyl-CoA (ketogenic). Some amino acids can form both products. While some amino acid degradation pathways are extensive (such as lysine), others are simple. For example, alanine is converted to pyruvate in a single transaminase step. The breakdown of serine, glycine, and methionine is linked to one-carbon (1C) metabolism, involving the transfer molecules THF (tetrahydrofolate) and SAM (S-adenosyl methionine). In turn, 1C transfer can contribute to the synthesis of other amino acids or other cellular structures. Anabolism of amino acids in humans can be divided into two groups: nonessential and essential. Nonessential amino acids can be synthesized from other precursors endogenously, whereas essential amino acids require ingestion of outside sources. For example, aspartate can be formed from oxaloacetate by a transaminase reaction, so aspartate is a nonessential amino acid. On the other hand, the aromatic amino acids, such as tryptophan, require dietary input. In the biosynthesis of pyrimidine nucleotides, a relatively short pathway leads to the construction of the pyrimidine base, and then addition of the ribose sugar to form UMP. The purine nucleotides are synthesized by a more elaborate pathway, but they begin with the ribose phosphate sugar and build the nucleotide on that scaffold. P-Ribose-PP also reacts with nucleosides to reform nucleotides in reactions collectively called the salvage pathway, saving the energetic cost of complete biosynthesis. The degradation of purines proceeds by deamination and dephosphorylation to an oxidized ring form, uric acid, which is excreted. The deoxynucleotides are formed from the corresponding nucleotide diphosphate, using a series of electron transfer reactions involving a disulfide protein exchange with thioredoxin and the enzyme thioredoxin reductase. The formation of dTMP is unique, requiring the donation of a methyl group from methylated-THF that produces an oxidized form, DHF (dihydrofolate). Regeneration of THF requires a reduction catalyzed by the enzyme dihydrofolate reductase. A large number of other, more specialized reactions occur in nitrogen formation, most of which are specific end products in plants or bacteria collectively known as alkaloids. One significant reaction in mammals is the conversion of arginine to nitric oxide, which causes the relaxation of smooth muscles in blood vessels.

## Key Terms

alkaloids
allopurinol
biopterin
cobalamin
denitrification
dihydrobiopterin
essential amino acids
folic acid
glucogenic
glutathione
gout
intrinsic factor
ketogenic

leghemoglobin
maple syrup urine
methionine cycle
N-acetylglutamate
nitrification
nitrogen cycle
nitrogen fixation
nonessential amino acids
phenylketonuria (PKU)
phosphonate
phosphoribosylpyro-
   phosphate (PRPP)
ping-pong (mechanism)

probenecid
purine nucleotide cycle
S-adenosylmethionine
semialdehyde
sequential (mechanism)
tetrahydrobiopterin
tetrahydrofolate
thioredoxin reductase
ureido
Vitamin $B_{12}$

## Review Questions

1. Many fish secrete ammonia as their nitrogen waste, but some secrete allantoin, which is a further degradation product of uric acid. The formation of allantoin, a more water-soluble molecule, is catalyzed by allantoinase. Describe how this enzyme might be used as a drug therapy in humans.

2. Ammonia toxicity in humans may result from liver damage, but the most severe effects are found in the brain. Describe the nitrogen metabolism pathways involved.

3. Explain how two key portions of the pyridoxal phosphate molecule—the carbonyl group and the phosphate—are used in transaminase reactions and in phosphorylase reactions.

4. How are the biopterin and flavin coenzymes similar?

5. SAM is sold commercially as "SAM-e" as a natural product that can supposedly treat liver disease and depression. Yet, no controlled studies have shown such benefits. Why would SAM appear to have these actions? Would ingested SAM enter cells?

6. How would urea synthesis from glycine differ from urea synthesis from alanine?

7. Why are branched chain amino acids considered metabolically as a unit?

8. After the formation of nitric oxide, what is the fate of the rest of the arginine molecule?

## References

1. Rolfes, R. J. Regulation of purine nucleotide biosynthesis: In yeast and beyond. *Biochem. Soc. Trans.* 34, 786–790, 2006.
2. Kim, Y. A.; King, M. T.; Teague, W. E. Jr.; Rufo, G. A. Jr.; Veech, R. L.; Passonneau, J. V. Regulation of the purine salvage pathway in rat liver. *Am. J. Physiol. Endocrinol. Metab.* 262, E344–E352, 1992.

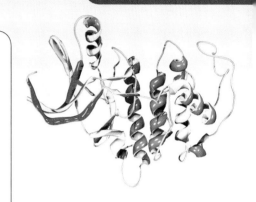

# 15

# Nucleic Acids

Image © Dr. Mark J. Winter/Photo Researchers, Inc.

I n 1952, James Watson and Francis Crick revealed the structure of DNA in a concise single-page paper in the journal *Nature*. It was the first announcement of the **double helix**, and showed how two complementary strands of DNA, noncovalently bound, spiraled around one another in pleasing symmetry. The landmark study concluded with the observation that the structure immediately suggests a means for DNA replication: as the strands separate, each could serve as a template for a new daughter strand.

A few years later, Crick proposed the *central dogma*

"DNA makes RNA makes protein,"

which succinctly summarizes this chapter. A slight revision to this pathway was required as it was later found that some viruses have RNA as their genetic material. The viral RNA is first converted to DNA in the host cell. After this step, the above pathway proceeds as shown.

We will examine some of the biochemical background within the framework of this central dogma, starting with the nucleic acids, DNA and RNA, the last biochemical polymers introduced in this book. The study of nucleic acid and protein metabolism is the essence of **molecular biology**, a term that blurs the lines between a biological and a chemical approach. In the present chapter, we will consider chemical features of the nucleic acids, the processes of duplication of DNA (**replication**), and the formation of RNA from DNA (**transcription**).

## 15.1 Strand Structures of the Nucleic Acids

The individual strands of DNA and RNA have a similar structure (FIGURE 15.1). The DNA polymer is linked by phosphodiester bonds between the deoxyribose moieties of the nucleotides. These bonds

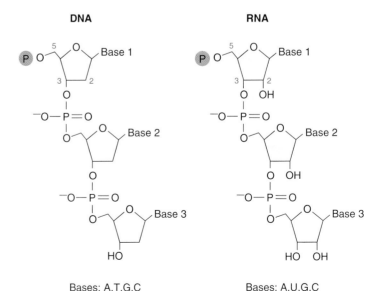

| DNA | RNA |
|---|---|
| Bases: A,T,G,C | Bases: A,U,G,C |

FIGURE 15.1 **Strand Structures of DNA and RNA.**

are the only covalent lineages in the polymer, and they also define a direction for the strand. The first subunit shown has a free phosphate group at the 5′ position; accordingly it is called the **5′ end**. The joining of the 3′ carbon of the first subunit to the 5′ OH of the second unit is repeated throughout the chain, which in the case of Figure 15.1 is only three nucleotides long. DNA polymers are much larger; in fact, DNA is the largest biological polymer (e.g., the human chromosome has 250 million base pairs). RNA, while considerably smaller—the average mRNA is five orders of magnitude less than DNA—shares a similar backbone structure, except that ribose sugars are present in place of deoxyribose sugars.

The other distinction between DNA and RNA strands is the identity of the bases, as indicated in Figure 15.1. DNA has four bases: adenine (A), guanine (G), cytosine (C), and thymine (T). RNA also has four bases: three are the same as those in DNA (A, G, and C ), but uracil (U) is uniquely present in RNA.

FIGURE 15.2 shows a detailed structure of a segment of RNA, with the nucleotide bases U and G shown. The nucleotides are the only unique portion of each unit of the chain. Accordingly, shorthand can be used to represent the nucleic acid strands. The position of the phosphate can only be 3′ or 5′. A lowercase *p* is written to the left to represent 5′, and to the right to represent 3′. Thus, the sequence in Figure 15.2 can be represented as,

> . . . pUpGp . . .

assuming the 5′ end of the chain was at the top. Intervening phosphates are usually omitted, so that a lengthy sequence such as,

> pCGAAUGCC

unambiguously identifies a strand of RNA with a phosphorylated cytosine at the 5′ end.

When DNA strands are written, it is possible to use shorthand that explicitly notes the deoxyribose nature of the bases:

> pdGdTdTdCdAdCdG

The abbreviation is usually the same as for the RNA, so that this strand would be written as:

> pGTTCACG

It is possible to confuse the strands if there are no thymine or uracil bases, but in context the representation is usually unambiguous. An alternative is to indicate the 5′ and 3′ ends explicitly with numbers:

> 5′-GTTCACG-3′

# 15.2 Structure of the Double Helix

DNA typically occurs as two strands, joined by noncovalent forces. FIGURE 15.3 shows the sugar–phosphate backbones of two separate strands in a ladder-like structure in which the "rungs" consist of a

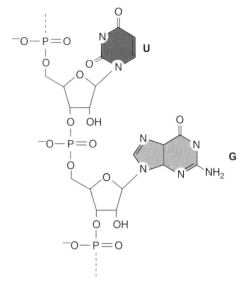

FIGURE 15.2 **RNA Segment.**

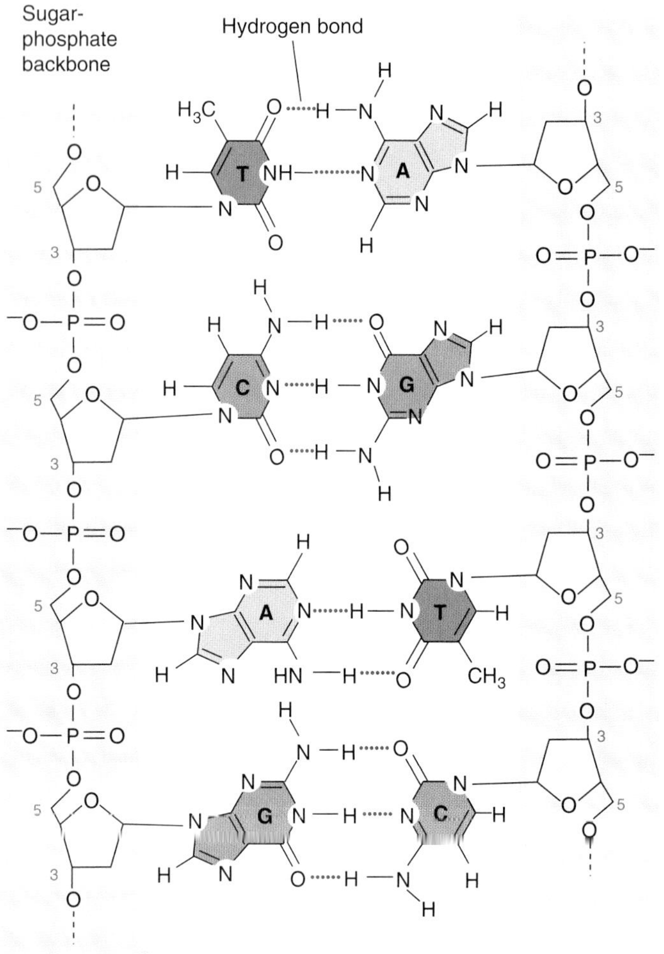

**FIGURE 15.3 Two-Dimensional Picture of DNA Base Pairing.**

nucleotide base contributed by each of the strands. Each rung has a purine and pyrimidine hydrogen bonded together, so that they have a uniform length. Invariably, a T always appears opposite an A, and a G always appears opposite a C. This explains the observation made prior to the discovery of the DNA structure that the total number of A bases equals that of T, and the total number of G bases equals that of C. These are known as *AT pairs* and *GC pairs*. Note that there are two hydrogen bonds in the AT pair and three in the GC pair. Because of this rule of **complementarity**, the same information is present in each half of the double helix. That is, once the sequence of one strand is known, the other can be readily predicted. It also predicts a mechanism for replication: if the two strands separate, then each can serve as a template for a new daughter strand.

There is further detail to the DNA structure. The base pairs in the center, due to their alignment and formation of hydrogen bonds, are planar and hydrophobic. In three dimensions, they stack on top of each other, as if they were flat rungs in a ladder, while the sugar–phosphate strands form the "sides" of the ladder, wrapped in a helix (FIGURE 15.4). The helix has features in common with the protein α-helix introduced previously in this text. The two are compared in FIGURE 15.5. In each case, there is an inner hydrophobic core that is internally hydrogen bonded and thus cannot form hydrogen bonds to water. The interiors of the protein α-helix and of the DNA duplex exhibit complete internal hydrogen bonding. The exterior of the α-helix can be hydrophilic or hydrophobic depending on the R groups of the individual amino acids. DNA, however, has a strongly polar, negatively charged exterior due to the dissociated phosphate groups that link the nucleotide subunits.

Helix formation in DNA is driven in large part by interactions between the "ladder rungs" (i.e., the AT or GC base pairs). Each hydrogen-bonded base pair rung experiences an attractive force between rungs above and below them. This attractive force is known as the **stacking interaction**. The base pairs are flat, aromatic, and hydrophobic. In order to explore the basis of the stacking interaction, we will turn to the benzene ring, which serves as a simpler model compound. FIGURE 15.6a depicts two benzene rings that illustrate the simplest possible stacking interaction. The same pair of benzene rings is shown in an edge-on view in Figure 15.6b, with the pi electron clouds explicitly drawn above and below each ring. Note that the edges of each ring have a relative positive charge. In Figure 15.6c, stability is achieved due to the offsetting of the rings, allowing positive and negative portions to interact.

Like the aromatic benzene rings, each rung in the DNA ladder is offset from the one below it. The three-dimensional structure forms as a spiral staircase. Hence, the stacking interaction directs the offsets of base pairs that give rise to the helical nature of the DNA double helix.

The positioning of the sugar residues of DNA is shown in FIGURE 15.7. Viewed from the top, the sugar residues are positioned unevenly, leading to a larger separation called the **major groove** and a smaller separation called the **minor groove**. The major and minor grooves provide one level of discrimination for binding: different proteins fit into the two grooves. Four distinct edges are pictured in Figure 15.7, resulting from differences between AT and GC base pairs and from differences between the major and minor grooves. Thus, a segment of DNA provides selectivity for proteins based upon the identity of the bases as well as the distinct binding of the protein to the major or minor groove.

FIGURE 15.4 **Three-Dimensional Form of DNA Helix with Base Stacking**.

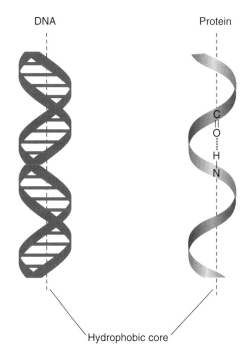

FIGURE 15.5 **Structural Similarities Between DNA and the α-Helix of Proteins**. The core in both cases is hydrophobic. DNA has a hydrophilic exterior, whereas the α-helix exterior may be either hydrophobic or hydrophilic, depending on the nature of the R-groups that extend to the outside of the structure.

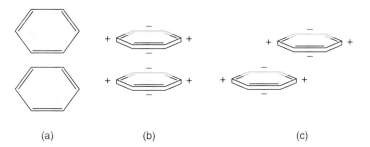

(a)          (b)          (c)

FIGURE 15.6 **Attractive Forces Behind Stacking Interactions**. (a) Two flat, aromatic bases are modeled by benzene rings. (b) Side view of the rings shows pi electron clouds above and below each ring (negatively charged), leaving positive charges at the edges. (c) Offsetting the rings occurs because of the attractive force of opposite charges.

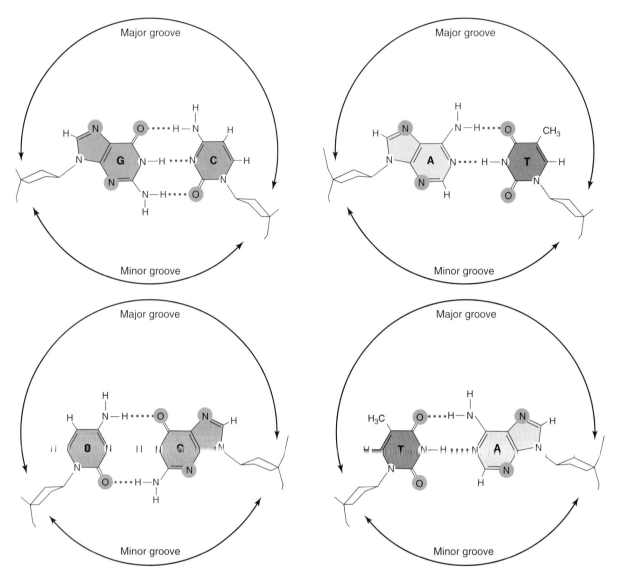

FIGURE 15.7 **Major and Minor Grooves**. Due to the way that the base pairs form and the sugars are attached to them, each flat "rung" of the DNA ladder has a larger gap or groove on one side (the *major groove*) than the other (the *minor groove*). As a result, the different base pairs offer different binding sites to proteins that bind DNA.

Unwinding of the DNA is required for its biological functions, such as replication and transcription. Intrinsically, the AT pair is split more easily than GC, which is reflected in **melting curve** experiments in which synthetic DNA strands containing only A and T are compared to synthetic strands containing only G and C (FIGURE 15.8). As the temperature is increased, the DNA strands separate, which is indicated by an increased absorption of ultraviolet light. The midpoint of the sharp increase in dissociation (called the **melting temperature**) is much lower for AT-rich strands than for GC-rich strands.

AT-rich regions exist at DNA replication origins, which suggests that the hydrogen-bonded bases in these segments are intrinsically easier to break. There are three hydrogen bonds in the GC pair and just two in the AT pair, but hydrogen bond energy differences are not the entire reason for the greater strength of the GC pairs. GC pairs

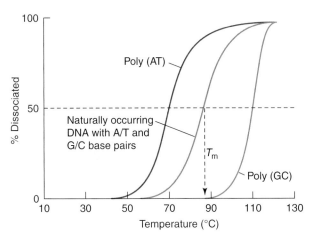

FIGURE 15.8 **Melting Curves for DNA**. Naturally occurring DNA has a midpoint in the melting curve ($T_m$) between poly-AT (low $T_m$) and poly-GC (high $T_m$). The results show experimentally that GC pairs are more stable than AT pairs.

also have greater stacking interaction energy. It is possible that, as a result of the increased hydrogen bonding, the GC pair is more planar, increasing the aromaticity of the dual ring system, accounting for the increase in the stacking interaction.

# 15.3 Supercoiling

Like twisting a coiled phone cord, the DNA helix can exist in **super-coils**. When the number of strand crossovers increases, the result is called a **positive supercoil**. Helices that have a decreased number of crossovers have a **negative supercoil**. Enzymes that catalyze changes in DNA supercoiling are called **topoisomerases**.

The topoisomerases are grouped into two categories, represented in bacteria by two enzymes, topoisomerase I and topoisomerase II. The first catalyzes hydrolysis of one of the phosphodiester bonds in one chain, moves a DNA strand through the opening, and then rejoins the chain. Hydrolysis of a phosphodiester bond joining the backbone sugars of nucleic acids causes a breakage, commonly called *cutting* (sometimes whimsically illustrated with miniature scissors). Here we use the term exclusively in italics as a shorthand. The mechanism of reaction is illustrated in FIGURE 15.9. A tyrosine hydroxyl group of the enzyme displaces a portion of the strand, forming an enzyme-bound intermediate. Subsequently, the 5′-OH group of ribose attacks the same phosphate group to rejoin the chain. Except for the change in topology (FIGURE 15.10), the product is the same as the substrate.

Topoisomerase II *cuts* both strands, allows the movement of a separate piece of double helix through, and then catalyzes resynthesis of the phosphodiester bond. In the course of this reaction, ATP is converted

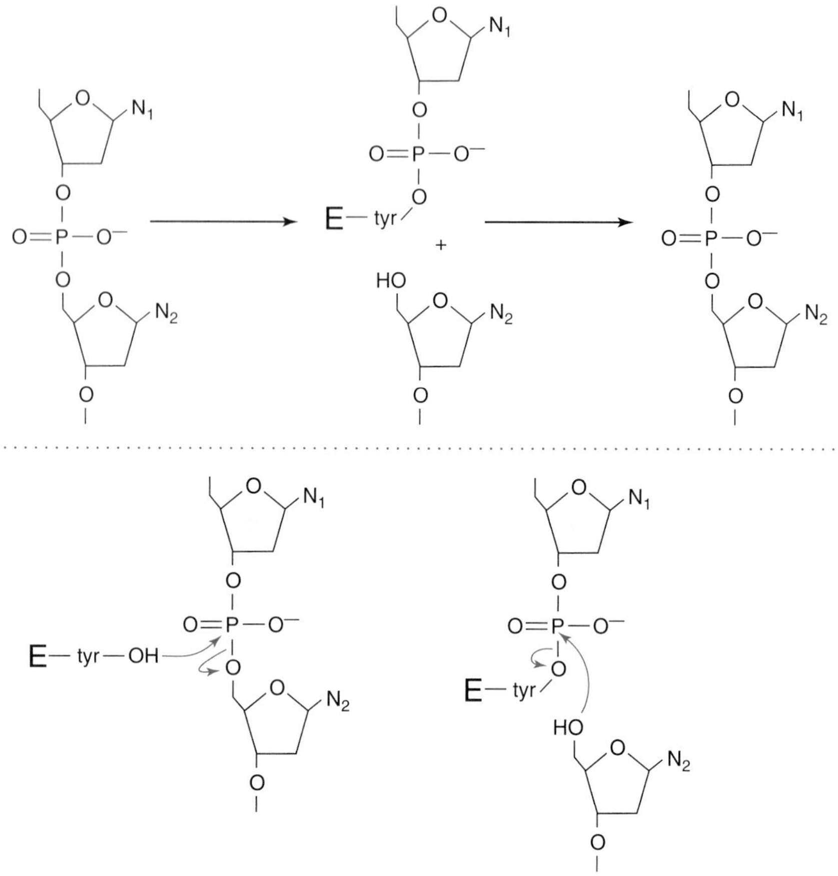

**FIGURE 15.9 Mechanism of Topoisomerase I.** The enzyme has an active tyrosine hydroxyl group that attacks the chain phosphate "backbone" and temporarily binds one portion. Displacement by the ribose OH of the other freed chain rejoins the DNA. The substrate and product are chemically identical but topologically distinct.

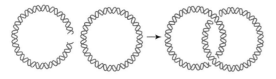

**FIGURE 15.10 Topology of the Topoisomerase I Reaction.**

to ADP and Pi, an indication that the process is energy dependent. The chemical mechanism for splitting and rejoining the phosphate groups is the same as topoisomerase I (Figure 15.9). Eukaryotic cells have similar enzyme activities, although more topoisomerase isoforms exist.

# 15.4 Histones

In eukaryotic cells, DNA strands bind basic proteins called **histones,** which exist in several isoforms. A cluster of four histone dimers roughly form a cylinder around which DNA is wrapped, and yet another histone isoform binds a portion of the free DNA between the histone clusters. The binding interaction is electrostatic: histones are basic proteins, attracted to the negative charges arising from the phosphates in the backbone of DNA. Proteins bearing a net positive charge are rare in cells, where the majority have net negative charges at the pH of the cytosol (about 7). In addition to histones, the other positively charged

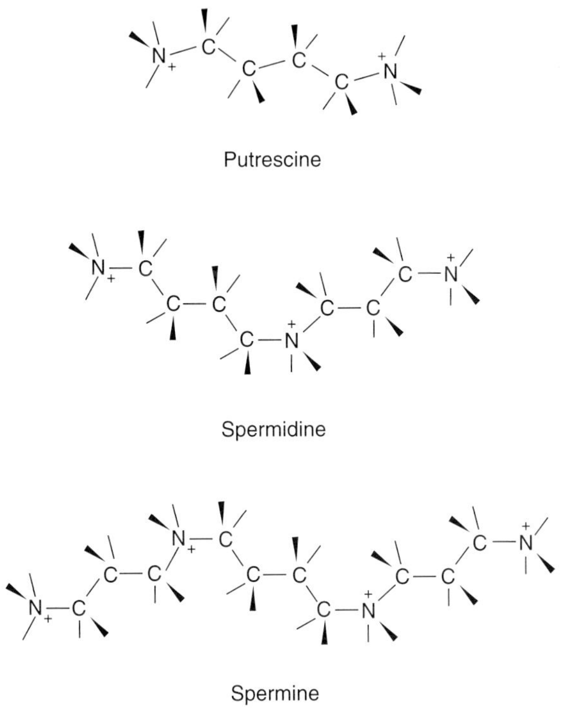

Putrescine

Spermidine

Spermine

FIGURE 15.11 **Polyamine Structures.**

molecules that bind DNA are small basic molecules, the **polyamines** (FIGURE 15.11). These molecules, found in both bacteria and eukaryotic cells, also bind DNA through electrostatic attraction.

# 15.5 Replication

The process of duplicating DNA—*replication*—involves opening the two strands of the molecule and then copying each. This is commonly divided into three stages: initiation, elongation, and termination. The enzyme that catalyzes the synthesis of the new bases is **DNA polymerase**, which specifically catalyzes addition of nucleoside triphosphates (NTPs, where N = A, G, C, or T) to a template starting from the 5′ end; hence, replication is said to proceed in the 5′→3′ direction. The earliest studies of replication involved *Escherichia coli* (*E. coli*), for which most of the mechanisms are best understood and are very similar to those in eukaryotes. We will note some of the distinctions in the course of our discussion.

## Initiation

*E. coli* DNA is circular in form, but in resting states (i.e., when replication or transcription is not occurring), it is supercoiled. As described in Section 15.3, topoisomerases catalyze the cleavage and rejoining of the phosphodiester bonds of the DNA backbone. A prominent topoisomerase in bacteria is known as DNA gyrase. Replication itself

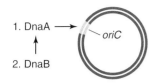

**FIGURE 15.12 Location and Sequence of Initiation Complex in *E. coli*.**

begins with the binding of the protein DnaA to a specific 245-base pair (bp) region of the *E. coli* chromosome called oriC, short for "origin of replication." This region is AT-rich, and thus intrinsically easier to unravel (recall that GC pairs are more tightly bound than AT pairs). A complex of multiple copies of DnaA bound to oriC serves as a binding site for another protein: the helicase enzyme DnaB (FIGURE 15.12). Helicases catalyze the separation of the DNA strands to present templates for replication. Two copies of the hexameric DnaB bind the DnaA:oriC complex and DnaB moves along a DNA strand. DnaB binds ATP, and the movement is driven by the transfer of bond energy. Upon displacement along the strand, the enzyme releases ADP and Pi. The movement of helicases along the surface of a DNA strand is an example of a molecular motor, like the $F_1F_o$ ATP synthetase complex.

## Replication Fork and the Replisome

Once the strands of DNA are separated, the binding of a **single-strand binding protein** maintains the open structure, thus making it possible to initiate replication. This protein is positively charged, and it selectively attaches to single strands through a central channel in its tetrameric structure.

With the two strands of DNA separated, ensuing reactions add nucleic acids to both single strands. The open "bubble" of separated strands is called the **replication fork** (FIGURE 15.13) and is the site of the key reactions of the process. Because the strands run in opposite directions, each is copied in a distinct manner. One chain, called the **leading strand**, is extended continuously in the 5′ to 3′ direction because its template starts with the 3′ end. The other chain, called the **lagging strand**, must be synthesized in the opposite direction in order to use the same DNA polymerase. As illustrated in FIGURE 15.14, the lagging strand is synthesized in short pieces, named **Okazaki fragments** after the investigator who first discovered them. The reason that the lagging strand must be formed in segments stems from the directionality restriction of nucleotide polymerases. While the leading strand always has a new template to act upon as the DNA opens up, the lagging strand grows in the direction opposite to the replication fork; of necessity, it must terminate and a new fragment must begin.

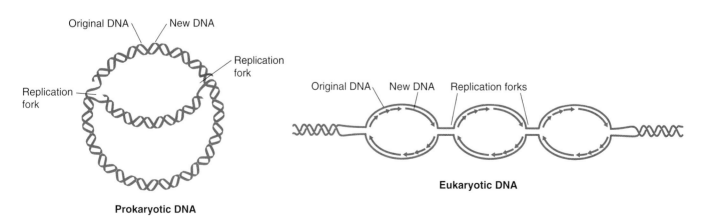

**Prokaryotic DNA**

**Eukaryotic DNA**

**FIGURE 15.13 Replication Fork in Prokaryotes and Eukaryotes.**

The synthesis of new DNA strands involves a large number of proteins bound together in a complex called the **replisome**. Both the leading and lagging strands are in contact with the same complex because the lagging strand loops back to engage the synthetic machinery (FIGURE 15.15). Accordingly, both strands are formed at the same rate.

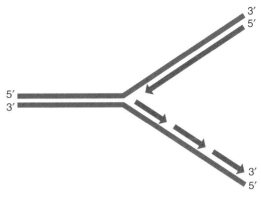

FIGURE 15.14 **Leading and Lagging Strands in DNA Replication**.

## Primer Formation

The first reactions that add bases to single-stranded DNA are catalyzed by an RNA polymerase, known as **primase**. The enzyme binds first to DNA gyrase and several other proteins, forming a complex called the **primosome**. The primase reaction product is a short chain of RNA, complementary to the DNA. The complementarity rules for DNA–RNA hybrids are similar to those for DNA–DNA hybrids, except that uracil substitutes for thymine in RNA. The reaction itself is similar to most synthetic reactions of nucleotides and essentially the same as other nucleophilic reactions involving phosphate groups. All nucleotides enter as triphosphates, and they are joined through a nucleophilic displacement mechanism. The first nucleotide retains the triphosphate at the 5′ end of the sugar, and its 3′-hydroxyl group attacks the alpha phosphate of the next incoming nucleotide triphosphate. This process is repeated until about 11 base pairs have formed, at which point primase disengages from the chain. Nucleotide polymerases generally engage the template for many rounds of synthesis before dissociating.

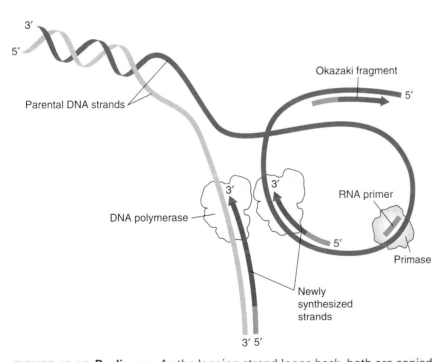

FIGURE 15.15 **Replisome**. As the lagging strand loops back, both are copied at the same rate. Several proteins (not shown) are required in this complex to unwind the DNA strands (topoisomerase), assist the DNA polymerase (a primosome complex, containing helicase and primase), and maintain portions of the lagging strand as a single chain (single strand binding protein).

This is called **processivity**, as opposed to the more common enzyme mechanisms in which a substrate is bound and released completely for each catalytic cycle.

The primer formation for the leading strand occurs just once in the replication of DNA but many times for the lagging strand. This is because the direction of synthesis runs opposite to the replication fork in the lagging strand, and new start sites are repeatedly made available as the duplex opens.

## Creating the Double Helix

Once the primers are established, the enzyme DNA polymerase can create new DNA strands. The reaction it catalyzes strongly resembles that of RNA polymerase: a nucleophilic attack by the 3′-hydroxyl group of one end can displace pyrophosphate and form a new phosphodiester bond, this time using incoming dNTPs as substrates. In bacteria, the major DNA polymerase isoform is DNA polymerase III (named for its order of discovery). The leading strand forms a long uninterrupted chain of bases, whereas the lagging strand once again only forms fragments, each about 1000 nucleotides, after which DNA polymerase dissociates, and a fresh section of lagging strand undergoes another cycle of priming with RNA and extension with DNA.

After the replication events described are complete, the lagging strand consists of short segments of RNA, followed by DNA, and then short interrupted single-strand segments. The enzyme DNA polymerase I catalyzes excision of the RNA as well as synthesis of new DNA to fill the gaps. The removal of RNA is catalyzed by **exonuclease**, a hydrolase activity. This leaves a gap in the DNA between the filled in segments (the result of polymerase I activity) and the beginning of another longer chain (the result of polymerase III activity). A separate reaction is required to form this "finishing touch," joining the ends in another phosphodiester bond. This is catalyzed by **DNA ligase**, which also uses ATP as a substrate.

## Distinctive Features of Eukaryotic Replication

The process of DNA duplication in eukaryotes is similar to that of bacteria, although more components are involved, and the details of the process remain less well understood. There are, for example, nine DNA polymerases. Because the length of the DNA is far greater in eukaryotes, it is not surprising to find that there are multiple initiation sites rather than just one per chromosome, as in prokaryotes. More binding proteins are involved in forming these replication complexes than in prokaryotes, and further controls exist to ensure that replication occurs only once in the cell cycle.

Perhaps the most striking distinction of eukaryotic DNA replication stems from the fact that its DNA is not only more extensive, but it is arranged linearly rather than in the circular form of prokaryotes. This produces a complication: the lagging strand cannot be completely replicated, because the exposed DNA reaches a free end (the 3′ end) and a new Okasaki fragment cannot be initiated due to insufficient length. This unreplicated segment thus leaves a single 3' overhang end, which requires a special system for its replication.

The sequence at these 3' ends (called **telomeres**; Latin, *telos*, meaning *end*) in vertebrate chromosomes consists of thousands of repeats of the sequence TTAGGG. The enzyme **telomerase** contains a complementary sequence as a cofactor, and accordingly binds and catalyzes the extension of the overhang. Upon completion, a small overhang remains, which is bound by a protein complex, a process called **capping**. This largely prevents degradation of the chromosomal telomeres, although there is some nuclease activity nonetheless, and shortening of chromosomes with age at the telomeres is a well-established phenomenon.

# 15.6 DNA Repair

Two major repair systems exist to ensure the fidelity of DNA reproduction. One is to correct errors caused by inserting the incorrect base during normal replication: this is called **mismatch repair**. The other is to correct damage that occurs to DNA after it has been replicated, an event strictly outside of the overall process of replication, but relevant for the proper functioning of DNA: this is called **excision repair**.

In bacteria, mismatch repair begins when specific binding proteins recognize an unpaired or mismatched DNA, causing this segment to form a loop. The proteins, in turn, contain a binding site for an endonuclease that catalyzes the hydrolysis of the looped strand. Strand recognition is achieved by the fact that the parental strands are methylated at the $N^6$ position of certain adenines: it takes time for new daughter strands to be methylated. Following removal of the bases including the mismatch, DNA polymerase I adds new bases, and the remaining gap in the phosphodiester backbone is joined by the action of DNA ligase. A similar set of binding proteins and enzymes exists for eukaryotic mismatch repair, although the process is less well understood. Nonetheless, it has been established that one form of colorectal cancer in humans is a result of inherited mutations in genes that code for proteins involved in mismatch repair.

Excision repair follows damage to existing DNA. One common form is a UV light-driven reaction that produces thymidine dimers when these bases are adjacent in DNA sequences (FIGURE 15.16). In this

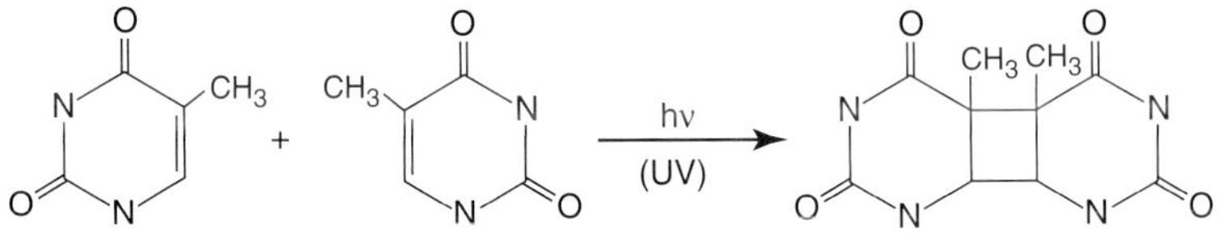

**FIGURE 15.16 Thymidine Dimer Formation.** Ultraviolet light catalyzes the joining of neighboring (vicinyl) thymidine bases, requiring a DNA repair process.

case, endonucleases—induced by ultraviolet radiation themselves—catalyze excision of the surrounding DNA segments. If not repaired, such damage can cause skin cancers.

Other types of DNA damage that require excision repair, such as that caused by other forms of radiation or pollutants, involve distinctive repair systems. These repair systems are in place not only for DNA fidelity during replication, but to ensure its integrity during transcription to form RNA.

# 15.7 Transcription

Transcription shares many features with replication. For example, producing RNA from DNA requires opening of the DNA duplex, a single-stranded DNA template, and polymerase activities that act in a 5′ to 3′ direction. In fact, we have already been introduced to one RNA polymerase, primase, which produces the RNA primer that was needed to begin DNA synthesis. RNA polymerases are distinctive because they do not require an existing polymer, and only one strand of DNA is copied. In transcription, only small portions of the genome are engaged at any given time, and they are copied into RNA at different rates. Moreover, RNA is far less stable than DNA.

Several distinct types of RNA are formed in transcription (FIGURE 15.17). Expression of cellular function is represented by messenger RNA (mRNA), which serves as the template for protein synthesis by the process of translation. The scaffolding for this process—the ribosomes—contains ribosomal RNA (rRNA). Carriers of amino acids and sequence recognition are the functions of transfer RNA (tRNA). A fourth class of small RNAs, such as the siRNA, also are involved in the control of cytosolic mRNA expression.

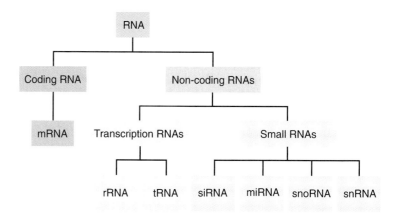

FIGURE 15.17 **RNA Families.** The mRNAs are a branch by themselves, because they code for proteins. The noncoding RNAs involved in transcription are ribosomal (rRNA) and transfer (tRNA). The small RNAs are regulatory: the small interfering RNAs (siRNA), the micro RNAs (miRNA; siRNAs are a class of these), the small nuclear RNAs (snRNA), and the small nuclear RNAs (snRNA).

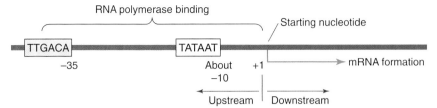

FIGURE 15.18 **DNA Segment Subject to Transcription in *E. coli*.** The first nucleotide copied into RNA is labeled +1; the RNA polymerase binds *upstream* of that site, as indicated. Two upstream regions are indicated that are widely found in prokaryotes, at −35 and at about −10 from the start site, which are involved in binding and positioning of the RNA polymerase.

The structure of a segment of DNA in prokaryotes is illustrated in FIGURE 15.18. The direction of transcription, as noted, is always from 5′ to 3′, but only one strand of DNA is a template for creating the RNA transcript. When the RNA polymerase first binds to DNA, it recognizes a site **upstream** of the first nucleotide in the RNA transcript. The sites in the DNA template are labeled as illustrated in Figure 15.18: nucleotide #1 is the first to be copied into RNA, and additional nucleotides in the RNA transcript are numbered sequentially. This is the positive direction, and is considered **downstream**. Those nucleotides prior to the start site are labeled sequentially as −1, −2, etc., and are *upstream*. Binding recognition is a result of distinctions in the outer portion of the DNA duplex structure that is known to be associated with specific sequences, called **promoters**. Typically, these regions are specific gene sequences. One that occurs frequently in bacteria is the TATAAT region (called a **TATA** box in eukaryotes), which serves as a binding site for the RNA polymerase.

## RNA Polymerase Binding to DNA

RNA polymerase binds selectively to one strand of DNA. In addition to RNA polymerase, many other proteins recognize specific single-stranded DNA sequences, often six to eight nucleotides long. This event may assist or hinder the binding of RNA polymerase itself.

While the binding interactions of RNA polymerase are dictated by a specific sequence, the initial binding of the enzyme to DNA is nonselective, and RNA polymerase both dissociates as well as scans along the surface until a selective binding site is encountered, locking the enzyme on the DNA.

## Transcription Events in *E. coli*

The well-studied *E. coli* serves to illustrate the basic events in the transcription of RNA: initiation, elongation, and termination. RNA polymerase (RNA Pol) binds upstream of the start site. In the DNA sequence of bacterial genes, a large number of similar sequences (called **consensus** sequences) exist in two upstream regions. One of these is the −35 region and the other is the −10 region (also called the **Pribnow** box, after its discoverer). These are the points of contact of RNA Pol and define the promoters. Thus, RNA Pol binds DNA with relatively high affinity and in close proximity to the start site.

The initiating event requires dissociation of the holoenzyme RNA Pol from DNA. The subunit structure of the holoenzyme is α₂ββ′σ, which binds relatively weakly to the DNA duplex and slides on its surface until it recognizes promoter elements. At this point, the RNA Pol binds tightly to the promoter elements, opens the DNA double helix (assisted by the fact that the consensus sequence promoters have a high AT content) and synthesizes an RNA transcript segment of about 10 nucleotides.

At this point, the σ subunit dissociates, producing the **core** complex, α₂ββ′, which binds less tightly to DNA, and **elongation** ensues. The dissociated σ subunit can bind to another core and form a separate holoenzyme to initiate another RNA transcript. As RNA polymerase proceeds, the unwound DNA creates superhelix segments around it, which are resolved by topoisomerases.

**Termination** results from the formation of a hairpin loop in the RNA transcript itself (FIGURE 15.19), as well as a **termination sequence**, also illustrated in Figure 15.19. Hairpin loops form from complementary base pairing in which a forward sequence is repeated in reverse; this is

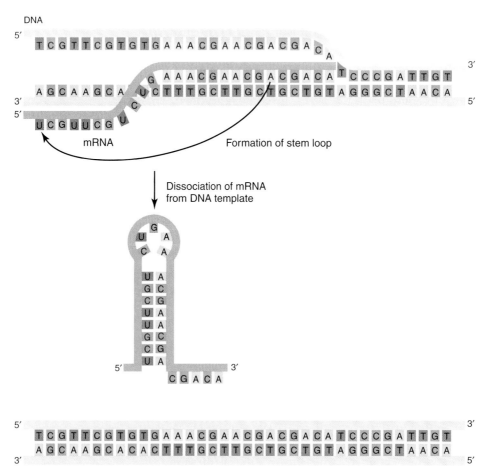

**FIGURE 15.19 Hairpin Formation Effects Transcription Termination.** As RNA polymerase proceeds, the enzyme is displaced as a hairpin loop is formed due to complementary base pairing.

known as a **palindrome** (**BOX 15.1**). At the termination sequence, the RNA Pol dissociates and the transcript is complete. Alternatively, some RNA sequences require the participation of separate proteins, most commonly the **rho protein**, which binds and disassembles the RNA transcript with concomitant ATPase activity.

Transcripts of RNA in *E. coli* produce multiple components. For the formation of rRNA and tRNA, a single transcript can produce several of these components in a single chain that are subsequently hydrolyzed (**FIGURE 15.20**). For the formation of mRNA, it is common to have several genes coded in a single transcript, which have related functional activity. The entire transcript, called an **operon**, thus represents the precursor to a set of proteins that are formed simultaneously. **FIGURE 15.21** shows the outline of the well-studied **lac operon**, which codes for a set of enzymes involved in bacterial uptake and metabolism of lactose.

## Eukaryotic Transcription

While the fundamental process of transcription is the same in both prokaryotes and eukaryotes, several differences exist because eukaryotes divide processes between organelles and among different cell types. One reflection of this is the existence of three RNA Pol enzymes (I, II, and III) in eukaryotes, which catalyze ribosomal (I), messenger (II), and transfer and small (III) RNA transcription. Of the three, RNA Pol I and III are most similar to the prokaryotic enzyme, in that both make a product that is subsequently cleaved into several smaller individual RNAs. For example, RNA Pol I catalyzes the transcription of a pre-rRNA, which is posttranscriptionally modified, and subsequently cleaved into 18S, 5.85S, and 28S rRNA. These rRNAs then assemble with several proteins to form ribosomes, the scaffold assembly for the synthesis of proteins. RNA Pol III catalyzes the synthesis of a precursor transcript that folds into loops that, upon hydrolysis, forms tRNA and is also a precursor for other small RNA transcripts, such as siRNA.

The most divergent form of eukaryotic transcription is that catalyzed by RNA Pol II, which leads to mRNA. Formation of the initiation complex by RNA Pol II is commonly the rate-limiting step for overall protein synthesis, although other important levels of control exist. Compared to prokaryotes, a greater number of proteins are involved

## BOX 15.1: WORD ORIGINS

### Palindrome

A palindrome is a sequence of letters that reads the same in the forward direction as the reverse. In the case of the RNA transcripts we have encountered, there are just four letters (G, C, A, and U), and the reverse sequence can meet the forward sequence in three dimensions, forming a double-stranded segment. This is often a loop (as the intervening sequence is typically not itself palindromic) called a **hairpin turn** as a result of proximal double-stranded binding. The function of the palindromic sequence is immediately clear, because it forms a control signal for protein–nucleic acid interaction.

The word itself is of Greek origin, a combination of two words, *palin,* meaning again, and *drom,* meaning run. In the English language, a common example is "Able was I ere I saw Elba." A related usage is in music, where the theme is run forward and then resolves itself in reverse. An example is Haydn's Palindrome Symphony (No. 47).

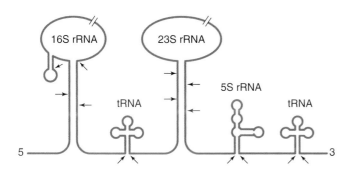

**FIGURE 15.20** **Formation of rRNA and tRNA from a Single Transcript in *E. coli.***

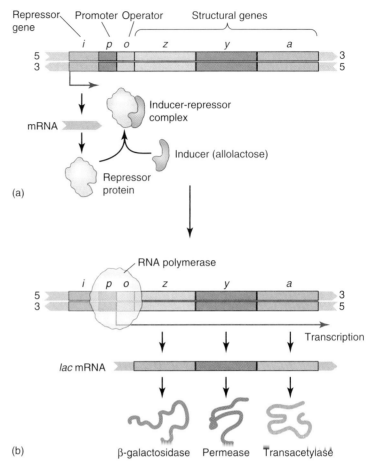

**FIGURE 15.21 The lac Operon.** The formation of three enzymes—galactosidase, permease, and transacetylase—required for the metabolism of galactose, are controlled by the presence of the sugar galactose, which upon entry is in equilibrium with allolactose. (a) Allolactose binds a protein repressor that can then bind to the operator region and prevent transcription of the genes for lactose catabolism. (b) A shortage of lactose decreases allolactose, dissociates the inducer-repressor complex, permitting RNA polymerase binding, followed by transcription of mRNA for all three proteins in sequence.

in the complex, and coregulators are common. For example, retinoids (FIGURE 15.22) are ligands for proteins that serve as co-activators that assemble along with other regulatory molecules to initiate specific transcription. A host of other proteins, including one catalyzing the acetylation of DNA assemble in the complex. Acetylation facilitates the opening of the DNA duplex and helps maintain the open complex form. A corresponding deacetylase enzyme returns the DNA structure to its inactive form.

Several processing reactions are unique to eukaryotic DNA formation. For example, shortly after initiation, the first base is modified to form a **cap** structure. That is, the end nucleotide reacts with a guanine nucleotide, and subsequently with S-adenosylmethionine, to produce the cap structure at the 5′ end of the transcript (FIGURE 15.23). Note that the cap nucleotide and the next nucleotide in the sequence are

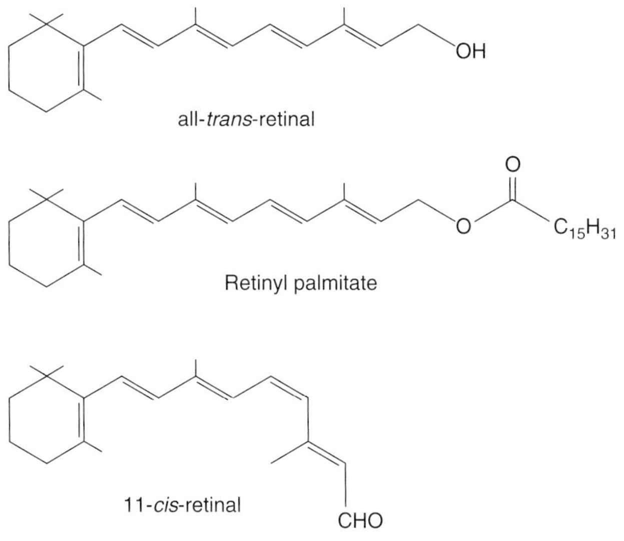

**FIGURE 15.22 Retinoids.** Common forms of retinoids in eukaryotes are retinol, retinoic acid esters (the common palmitate is shown), and retinal (the aldehyde form). The ester is cleaved in cells to form an active form as the free acid. The various oxidation states of retinoids are all regulators of cellular transcription, often as co-activators.

joined in a 5′→5′ linkage. At the 3′ end of mRNA, the enzyme Poly A polymerase catalyzes the addition of as many as 250 adenine residues, forming the **poly-A tail** of the transcript. The final processing step in the nucleus is the removal of intermediate portions of the transcript by splicing and rejoining to create the final mRNA. It is possible, in some cases, to produce distinct final products with differential splicing. The segments removed are known as **introns**, whereas the pieces

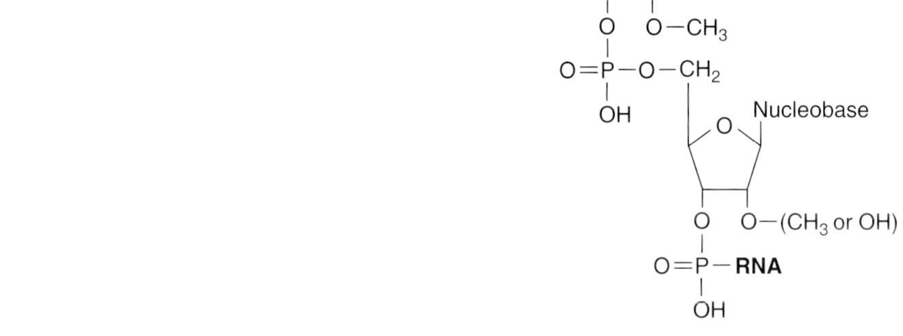

**FIGURE 15.23 RNA Cap.** The 5′ end of eukaryotic mRNA has a structure called a *cap,* utilizing the cofactor S-adenosylmethoinine and accounting for the methylated adenine at the terminus.

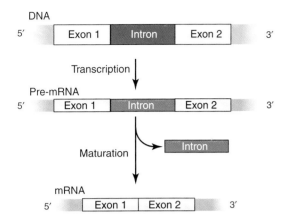

DNA
5′   Exon 1   Intron   Exon 2   3′

Transcription

Pre-mRNA
5′   Exon 1   Intron   Exon 2   3′

Maturation    → Intron

mRNA
5′   Exon 1   Exon 2   3′

FIGURE 15.24 **Exons and Introns**. The mRNA of eukaryotes contains both translatable RNA (exons) and non-translatable RNA (introns). The introns are excised and the exons are joined together in the nucleus prior to exiting as a mature RNA into the cytosol.

## BOX 15.2: Exome Sequencing

There are about 20,000 genes in the human genome, but this number represents only 1% of the total DNA. While the other segments have some known biological function, such as promoter and enhancer regions, it would be advantageous to study just the small fraction that was involved in the synthesis of proteins.

**Exome sequencing** is a technique for separating the exons from the remainder of the DNA and then sequencing just those nucleotides. The name is a nod to the original systematic study of genetics (genomics), followed by the same name applied to proteins (proteomics) and then to metabolism (metabolomics). As DNA is expressed in RNA, the RNA is first isolated and then converted to a complementary DNA (cDNA) library. Subsequently, cellular DNA is subjected to partial enzymatic hydrolysis of its phosphate esters and hybridized with cDNA (DNA copied in vitro from total cellular mRNA). Sequencing of just the variously sized hybrids comprises the *exome,* which, depending on the variant of the method used, represents the bulk of the exons.

Several rare diseases have already been related to their genes based solely on this technique, such as the identification of a defective gene for a chloride transporter in a kidney disorder. Another example is a loss of dihydro-orotate dehydrogenase that was discovered as the basis of Miller syndrome, a developmental disorder leading to abnormal facial and limb features. It is significant that the enzyme is part of the pyrimidine biosynthesis pathway. Known disorders in this pathway are distinct from Miller syndrome. This demonstrates the power of the technique in uncovering new meaning to metabolic pathways.

"spliced" together to form the final transcript are known as **exons** (FIGURE 15.24). Recently, techniques have been developed that attempt to directly sequence just the exons of genes (**BOX 15.2**). Finally, the finished mRNA exits through the nuclear pore to the cytosol, where translation occurs. Once in the cytosol, both cap and tail structures help orient mRNA for protein translation and protect the overall molecule from the action of cytosolic ribonuclease (RNase), which catalyzes the degradation of RNA into smaller pieces. With the production of RNA, the cell is poised to produce proteins, the subject of Chapter 16.

## Summary

The nucleic acids DNA and RNA direct cellular replication, transcription, and translation. DNA is a double-stranded polymer of nucleotides, linked in phosphate esters at the 3′ and 5′ positions of the deoxyribose. The bases present are adenine (A), thymine (T), cytidine (C), and guanine (G). The complete structure forms a "ladder" in which one side (chain) is arranged 5′ to 3′, the other 3′ to

5′, and the "rungs" are a pair of bases hydrogen bonded together, either an AT pair or a GC pair. RNA exists as a single polymer, with the same bases except that uracil (U) is present in place of thymine. The double helix is stabilized by stacking interactions, in which the pair of bases above and below are attracted as edge partial charges from one ring structure attracts the pi cloud charges from the one beneath it. DNA is normally kept in a supercoiled structure that requires topoisomerases to unravel the molecule for replication or transcription. In addition, DNA is stored wrapped around histones, basic proteins which interact with the negatively charged phosphate groups in the chain backbone. For replication, the DNA opens into two separate chains and allows hybridization of new nucleotides, in keeping with the AT and GC pairing rules. Each acts as a template, so two daughter strands are formed. The process in bacteria originates at an origin that is AT rich; these base pairs are held less tightly than GC pairs. Proteins are required to initiate, and continue the process of replication, always in the direction of 5′ to 3′. One of the chains is synthesized linearly (leading strand), but the other (lagging strand) is formed in small subunits because new template only becomes available upstream of the direction of replication. The actively replicating chain is held in a replisome, in which one of the strands is looped, which forces the rate of polymer formation for each chain to be the same. In eukaryotes, the process involves more proteins but is qualitatively different because these DNA molecules are linear rather than circular. Thus, sequences are formed at the ends called telomeres that require capping reactions to prevent shortening due to uneven overhang during replication of the DNA. Several processes that degrade DNA, such as oxidation or UV damage, can be repaired by mismatch repair, in which DNA is forced to form a loop. An endonuclease catalyzes hydrolysis of that segment. In excision repair, DNA segments are removed by endonucleases that recognize specific aberrant structures.

The transcription process forms RNA from a DNA template. One form of RNA ultimately codes for protein: mRNA. rRNA forms part of the protein synthesis scaffold in the cytosol (the ribosome), tRNA ferries amino acids to the translational complex, and the other small RNA species serves to regulate transcription and translation. Transcription requires binding of RNA polymerase to a nucleotide sequence that is upstream of the start site, and proceeds in a 5′ to 3′ direction. Regulation of transcription in bacteria occurs at the formation of the initiation complex. This is also true in eukaryotes, although the regulatory process is usually more involved, with multiple components that may need to be present to stimulate transcription. A further distinction of eukaryotes is the presence of introns within a finished mRNA; these represent sequences that must be excised and the remaining pieces of mRNA, the exons, spliced together. Prokaryotic mRNA is polycistronic, so that several proteins can be formed from a single pass by polymerase. Eukaryotic mRNA is monocistronic, and the mRNA is modified by methylation in the first nucleotide, the cap, and addition of multiple adenosyl residues to the 3′ end, the poly-A tail.

## Key Terms

cap
capping
complementarity
consensus
core
DNA ligase
DNA polymerase
double helix
elongation
5′ end
excision repair
exome sequencing
exons
exonuclease
hairpin turn
histones
introns
lac operon

lagging strand
leading strand
major groove
melting curve
melting temperature
minor groove
mismatch repair
molecular biology
negative supercoil
Okazaki fragments
operon
palindrome
polyamines
poly-A tail
positive supercoil
Pribnow
primase
primosome

processivity
promoters
replication
replication fork
replisome
rho protein
single-strand binding
   protein (SSBP)
stacking interaction
supercoils
TATA
telomerase
telomeres
termination
termination sequence.
topoisomerases
transcription

## Review Questions

1. Why do prokaryotes generally have no systems for direct control of protein synthesis?
2. One way in which antibody diversity is achieved is through the enzyme cytidine deaminase, which removes the ammonia from a cytidine residue. To what base would the product of this reaction bind?
3. Okasaki fragments must be joined by a ligase reaction. What would be the enzyme intermediate for this reaction?
4. Why are AT base pairs such as those present at the replication origins intrinsically weaker than GC base pairs?
5. The *lac* operon, a model for transcription regulation, produces a single mRNA that codes for three proteins. Does this mean the rate of formation of all three proteins must be the same?
6. What is the structural basis for the major and minor grooves in DNA?
7. Most of the time, DNA exists as a double strand. Enumerate the circumstances under which it is a single strand.

## References

Kaiser, J. Human genetics. Affordable "exomes" fill gaps in a catalog of rare diseases. *Science* 330:903, 2010.

Lewin, B. *Genes IX*. Jones and Bartlett Publishers, Sudbury, MA (2008).
   A broad treatment of nucleic acid biology.

Ng, S. B.; Buckingham, K. J.; Lee, C.; et al. Exome sequencing identifies the cause of Medellin disorder. *Nat, Genet.* 42:30–35, 2010.

Waters, M. L. Aromatic interactions in model systems. *Current Opinions in Chemical Biology* 6: 736–741 (2002).
   An explanation for base stacking based on through-space interactions of aromatic rings.

Watson, J. D., Baker, T. A., Bell, S. P., Gann, A., Levine, M., Losick, R. *Molecular Biology of the Gene.*, 6th Edition, Benjamin Cummings, San Francisco (2008).
   A classic text on molecular biology.

**CHAPTER 15** NUCLEIC ACIDS

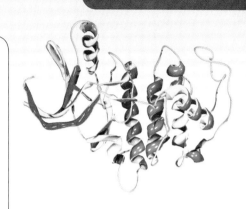

# Protein Synthesis and Degradation

# 16

## CHAPTER OUTLINE

Image © Dr. Mark J. Winter/Photo Researchers, Inc.

Our last topics are protein synthesis and protein degradation. Bacterial protein synthesis can be divided between initiation, elongation, and termination. While not fundamentally different, the eukaryotic process involves more regulatory proteins, a larger assembly complex (**ribosome**), and has more extensive mechanisms for its regulation. Bacteria synthesize proteins even before the mRNA is fully synthesized, and produce several proteins from a single mRNA. Hence, the regulation of protein translation separately from transcription is difficult to envision for these organisms. The major process for cellular protein degradation involves a molecular complex called the **proteasome**. Proteins are targeted for degradation at widely different rates. Just like small-molecule intermediates, proteins themselves exist in a steady state between synthesis and degradation, and the extent of their accumulation is largely determined by their rates of degradation. We begin our study of protein formation by examining the classes of molecules that are involved and the genetic code underlying protein translation.

## 16.1 Three Forms of RNA Are Employed in Protein Synthesis

Transfer (tRNA), ribosomal (rRNA), and messenger (mRNA) RNA are transcribed from a DNA template. In eukaryotes, these forms of RNA are exclusively synthesized in the nucleus and exit through nuclear pores into the cytoplasm. These three forms of RNA all interact during protein synthesis. We first consider their roles individually.

### tRNA

Transfer RNA is relatively small, having about 80 nucleotides in the mature molecule. Three views of a tRNA molecule are shown in FIGURE 16.1. The flat view (Figure 16.1a) shows extensive complementary base pairing, with three hairpin loops in a cloverleaf pattern. Several of the bases are modified, most commonly by methylation. Two key regions of the molecule are identified in each view: a three-base region called the **anticodon** site, and the 3′ end which terminates in CCA. The last ribose of the CCA end has a 5′ hydroxyl group that is esterified to the carboxyl group of an amino acid. This is called the **acceptor stem**. Figure 16.1b is a three-dimensional view of tRNA. The twisting shown in Figure 16.1b results from the same stacking interactions that cause spiraling in DNA. Finally, the abbreviated view in Figure 16.1c shows only the anticodon region and the acceptor stem.

The attachment of a tRNA to an amino acid is a reaction catalyzed by one of over 20 specific **amino-acyl-tRNA synthetases**. The generic reaction is:

$$\text{Amino acid} + \text{tRNA} + \text{ATP} \rightarrow \text{tRNA}^{\text{amino acid}} + \text{AMP} + \text{PP}_i \qquad (16.1)$$

Specificity in catalysis for one of the synthetases involves several regions of the tRNA molecule; in some cases, even the anticodon loop binds the enzyme. The reaction mechanism, shown in FIGURE 16.2, is

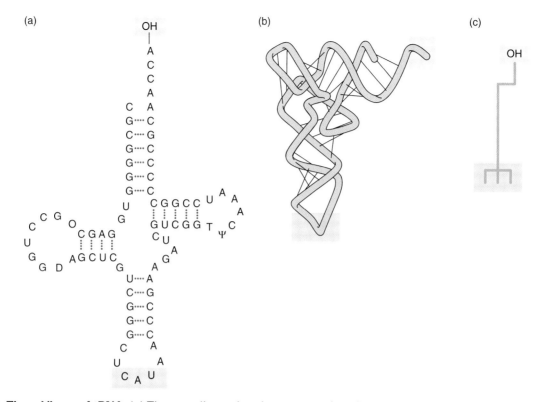

**FIGURE 16.1 Three Views of tRNA.** (a) The two-dimensional representation shows the extensive intrachain hydrogen bonding. (b) Twisting in three dimensions is the result of stacking interactions. (c) The essential view indicating the anticodon and acceptor sites.

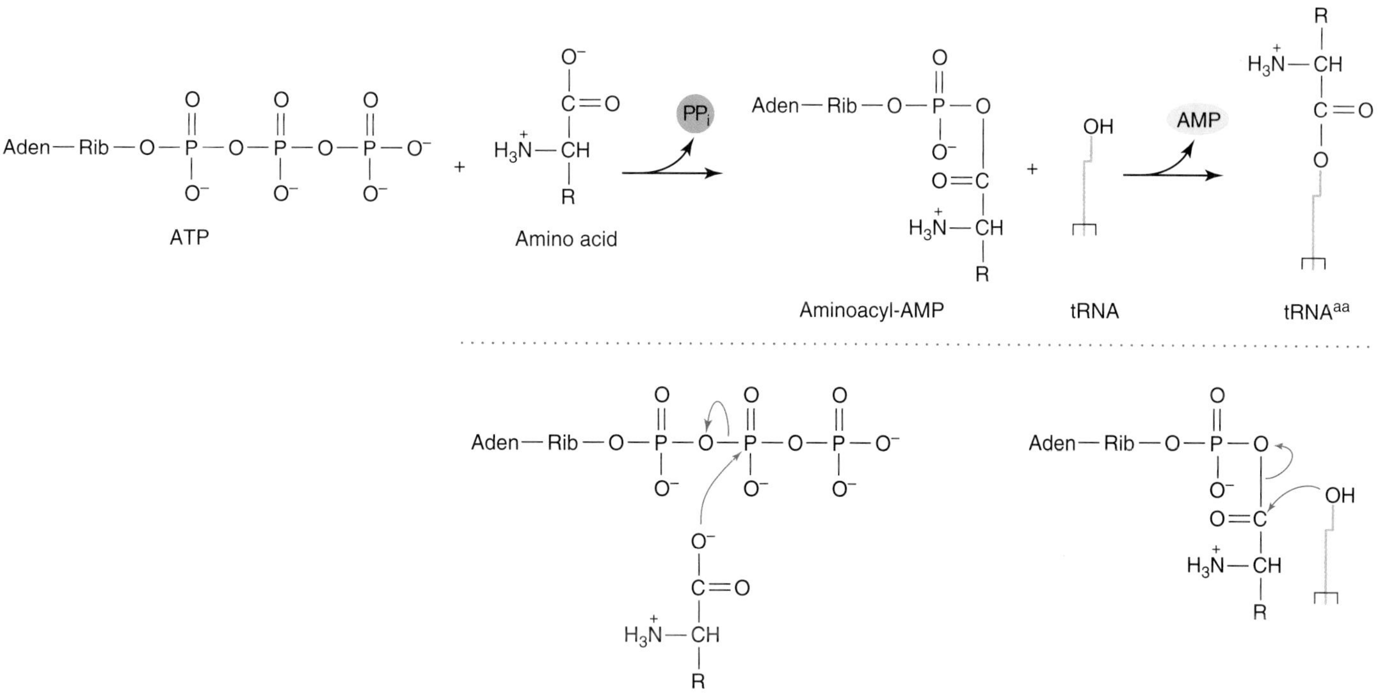

**FIGURE 16.2 Mechanism of tRNA Synthetase.** In the first step, an acid phosphate forms between the acid portion of the amino acid and the gamma phosphate of ATP. In the second step, bound AMP is displaced by the free hydroxyl group at the acceptor end of tRNA.

CHAPTER 16 PROTEIN SYNTHESIS AND DEGRADATION

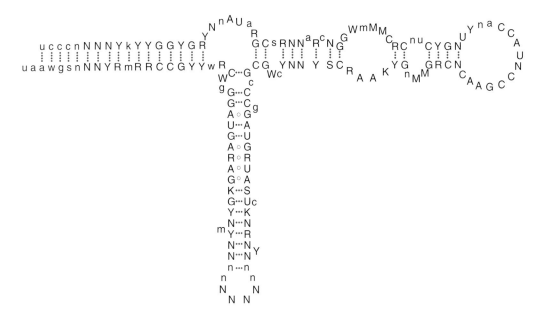

FIGURE 16.3 **Planar View of rRNA**. Like tRNA, the molecule has extensive intramolecular base pair bonding.

similar to that of asparagine synthetase. The enzyme first binds an amino acid and ATP, forming a carboxyphosphate adduct with the alpha-phosphate of ATP, displacing pyrophosphate. Subsequently, the CCA-OH group of the tRNA displaces AMP, attaching the amino acid. A tRNA with an amino acid attached is written as tRNA$^{\text{amino acid}}$ and is called a **charged tRNA**. This is the form that joins the complex responsible for protein synthesis.

Aside from the 20 amino acids, an additional tRNA is used for methionine. This molecule initiates all protein synthesis and is distinct from the tRNA$^{\text{Met}}$ species used for the interior portions of the protein. In bacteria, this special tRNA$^{\text{Met}}$ is formylated, using 10-formyl-THF as the precursor, to produce f-Met-tRNA.

## rRNA and the Ribosomes

A planar view of the rRNA structure is shown in FIGURE 16.3. Similar to tRNA, extensive base pairing in rRNA produces a more elaborate structure than the linear sequence would indicate. A complex of proteins and rRNA comprises the ribosome, the scaffold upon which proteins are synthesized. As indicated in FIGURE 16.4, the ribosomes can dissociate into two complexes. The prokaryotic ribosome is a 70S complex that decomposes into 30S and 50S complexes, whereas the eukaryotic 80S ribosome dissociates into 40S and 60S complexes. The unit of measure for these complexes is the **svedberg**. As discussed in **BOX 16.1**, it is based on the mobility of particles during ultracentrifugation.

Eukaryotic ribosomes are a more extensive complex of rRNA and protein. In both prokaryotes and eukaryotes, the quantity of RNA exceeds that of protein. The RNA is also used as a catalyst for the actual joining reaction that forms the peptide bond. Ribosomes that act as enzymes are called **ribozymes**.

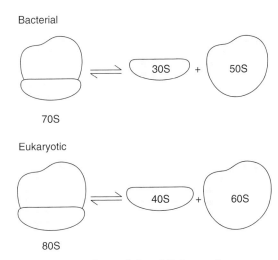

FIGURE 16.4 **Bacterial and Eukaryotic Ribosomes**.

The ribosome is the site of convergence of mRNA and tRNA. The three bases of the anticodon region in tRNA correspond to the **codon**, a triplet of bases in the mRNA bound to the ribosome. Each sequentially recognized codon corresponds to a specific amino acid. The matching of codons to their amino acids is known as the **genetic code**.

# 16.2 The Genetic Code

The genetic code was deciphered through experiments in which protein synthesis was replicated *in vitro*. The essential finding was that a mixture of ribosomes, tRNA molecules attached to amino acids, and mRNA can assemble together and form a protein chain. The key to the translational machinery is the tRNA molecule. The anticodon of tRNA binds a complementary triplet of bases in mRNA, and the amino acid on the opposing acceptor stem of tRNA forms a new peptide bond. Repeated many times, this forms the protein chain. The specific amino acid incorporated is determined by the anticodon of the tRNA.

Early experiments used synthetic mRNA as the input and the amino acid sequence of the resulting protein measured as the output. For example, a chain composed only of A (that is, a poly-A mRNA) produces polylysine, showing that AAA corresponds to Lys. More elaborate techniques along the same lines led to the elucidation of the entire three-base sequence code, shown in Table 16.1, which is fairly

## BOX 16.1: UNIT ORIGINS

### The Svedberg

A few molecular complexes, such as ribosomes and proteasome, are measured in svedberg units, abbreviated S. This is not an SI unit agreed upon by chemical commissions, but rather a curious hold-out from a technique used to measure mixtures of macromolecules, historically called **colloids**. The unit itself honors Theodor Svedberg (as with other eponymous units, such as the newton and the joule). Svedberg invented the ultracentrifuge, which is the device used to quantify sedimentation coefficients for colloids like ribosomes. The svedberg is a time unit: it is the ratio of velocity to acceleration in a centrifugal field. This is useful because the particle achieves terminal velocity when the centrifugal force is balanced by the viscosity of the media, just as the terminal velocity is achieved by a parachutist when the acceleration due to gravity matches the wind resistance. The value of 1 S is $10^{-13}$ seconds. It measures particles, although it is not just size- but also shape-dependent. Thus, the svedberg units are not additive, which explains why the bacterial 30S and 50S particles combine to make a 70S ribosome.

| TABLE 16.1 The Genetic Code[a] | | | | | |
|---|---|---|---|---|---|
| | **U** | **C** | **A** | **G** | |
| **U** | UUU Phe | UCA Ser | UAU Tyr | UGU Cys | **U** |
| | UUC Phe | UCC Ser | UAC Tyr | UGC Cys | **C** |
| | UUA Leu | UCA Ser | UAA **STOP** | UGA **STOP** | **A** |
| | UUG Leu | UCG Ser | UAG **STOP** | UGG Trp | **G** |
| **C** | CUU Leu | CCU Pro | CAU His | CGU Arg | **U** |
| | CUC Leu | CCC Pro | CAC His | CGC Arg | **C** |
| | CUA Leu | CCA Pro | CAA Gln | CGA Arg | **A** |
| | CUG Leu | CCG Pro | CAG Gln | CGG Arg | **G** |
| **A** | AUU Ile | ACU Thr | AAU Asn | AGU Ser | **U** |
| | AUC Ile | ACC Thr | AAC Asn | AGC Ser | **C** |
| | AUA Ile | ACA Thr | AAA Lys | AGA Arg | **A** |
| | AUG Met or **START** | ACG Thr | AAG Lys | AGG Arg | **G** |
| **G** | GUU Val | GCU Ala | GAU Asp | GGU Gly | **U** |
| | GUC Val | GCC Ala | GAC Asp | GGC Gly | **C** |
| | GUA Val | GCA Ala | GAA Glu | GGA Gly | **A** |
| | GUG Val | GCG Ala | GAG Glu | GGG Gly | **G** |

[a]The codon triplet is shown starting at the 5' end. The first nucleotide is in the left-hand column, the second in the top row, and the third along the right-hand column.

universal across species. The total number of possible amino acids that could be coded by using three different nucleotides, where each may be one of four possibilities, is $4 \times 4 \times 4 = 64$. Only 20 unique codons are necessary for amino acids, as well as a few more triplets to code for termination—the *termination codons*—so the genetic code is redundant. In fact, variation in the third position of the anticodon often results in the insertion of the same (that is, the correct) amino acid, which should be apparent from Table 16.1 (note, for example, that CGX, where X = U, A, C, or G, codes for Arg). This slight lapse of complete specificity, suggesting the first two bases are more critical for complementary base recognition, is known as the **wobble hypothesis.**

# 16.3 Steps in Protein Synthesis

The overall process of protein translation is commonly divided into initiation, elongation, and termination steps. Each requires a distinct set of complexes and regulatory features. We focus on bacterial protein synthesis, after which we consider key distinctions that exist for eukaryotes.

## Initiation

In bacteria, initiation begins with mRNA binding to the smaller, 30S ribosomal subunit (FIGURE 16.5). The positioning of the mRNA on this subunit is directed by a sequence of 4–6 nucleotides in the upstream region called the **Shine–Dalgarno** segment. This has a complementary sequence to an rRNA in the 30S complex, thus aligning the mRNA. Two binding proteins, initiation factor-1 (IF-1) and IF-3, attach to the complex. A further binding protein, IF-2, binds along with tRNA^f-Met. The f-Met-tRNA aligns with the start site (AUG) in the mRNA. The function of IF-1 is to prevent the premature joining of the 50S complex, whereas IF-3 prevents the addition of other tRNA molecules. IF-2 is a G-protein, so it is always bound to a guanine nucleotide, either GTP or GDP. In the active GTP-bound form, it enables the binding of tRNA^f-Met to the 30S ribosome. Once this assembly is complete, the 50S ribosome joins the complex and serves as a GTPase-activating enzyme, prompting the release of all of the initiation factors and resulting in the formation of the complete initiation complex.

As shown in Figure 16.5, the complex now has two sites within the full, 70S, ribosomal assembly: the **P site** (peptidyl), and the **A site** (acceptor); the f-Met-tRNA is in the P site. This is the form of the complete initiation complex, ready to undergo elongation.

## Elongation

The steps of elongation are illustrated in FIGURE 16.6. Starting with the completed initiation complex, a charged tRNA binds to the adjacent codon, positioning the two amino acids in close proximity, filling the

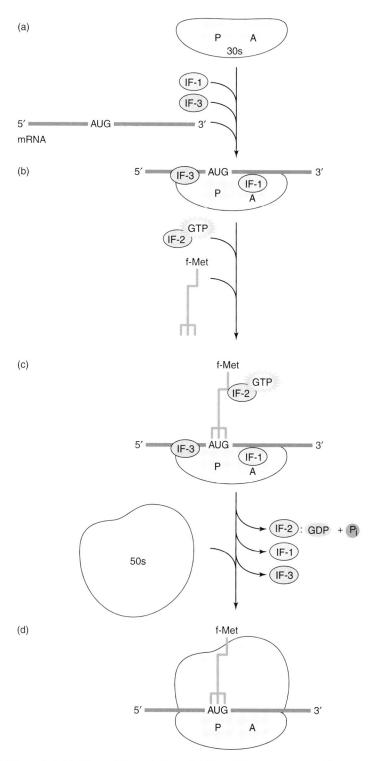

FIGURE 16.5 **Initiation of Translation.** (a) The 30S component binds initiation factors IF-1 and IF-3 and mRNA to form complex. IF-1 prevents incoming tRNA molecules from binding the A site. IF-3 prevents binding to the 30S subunit to prematurely form the complete complex. (b) IF-2 (a G-protein) and f-Met-tRNA bind at the initiation (AUG) site to form the complex (c). The 50S subunit binds to this complex, catalyzing GTPase activity at IF-2, releases all of the initiation factors, and forms the complete initiation complex (d).

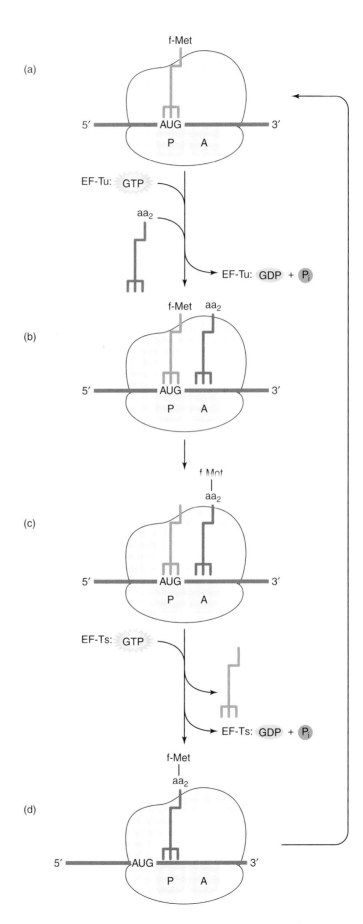

**FIGURE 16.6 Translation Elongation.** The initiated ribosome (a) binds the tRNA bearing the second amino acid, assisted by EF-Tu in the GTP form. Once the A site is bound to this tRNA (b), the EF-Tu:GDP (and P$_i$) is released. In the next step, f-Met-tRNA attaches to the second amino acid at the A site, catalyzed by the ribosome itself: this is the actual peptide bond synthesis (c). Finally, EF-Ts:GTP binds, shifting the mRNA over by three nucleotides, releasing the now free tRNA, and leaving the tRNA containing the short peptide in the P site (d). The cycle then resumes with the next incoming amino acyl-tRNA.

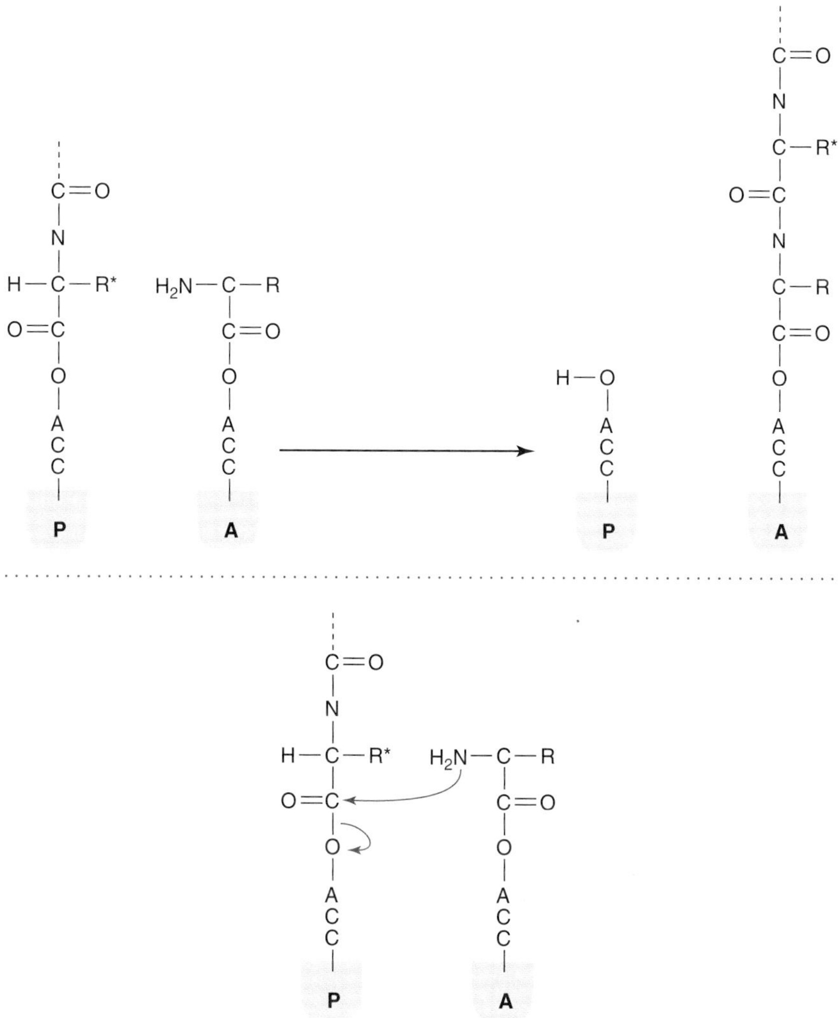

**FIGURE 16.7 Mechanism of Amino Acyl-tRNA Synthase.** The amine of the amino acyl group in the A site attacks the acyl group of the amino acyl group in the P site, leaving a free tRNA in the P site and a new peptide in the A site.

P and A sites. Binding of the tRNA to the A site is assisted by EF-Tu, an elongation factor G-protein that undergoes GTP hydrolysis upon binding, releasing the EF-Tu in its GDP-bound form. In the next step, the peptide bond forms, catalyzed by rRNA rather than a protein (**FIGURE 16.7** shows the mechanism). The amine from the amino acid at the A site displaces the esterified tRNA[f-Met] at the P site. Thus, the newly formed dipeptide becomes attached to the tRNA at the A site, leaving an empty tRNA at the P site. Next, binding of the elongation factor EF-Ts promotes the movement of mRNA so that the tRNA appears at the P site. This displaces the first tRNA from the ribosome, leaving the A site empty and a dipeptide bound to the tRNA in the P site. This complex enters into a second round of elongation, and the cycle is repeated until a triplet codon is encountered for which no tRNA exists; this is a termination codon.

While tRNA[f-Met] is used for the first amino acid, the internal Met residues of the protein use a distinct tRNA. Hence, only the tRNA[f-Met] can directly enter what becomes the P site, and all other amino acids enter at the A site. The rationale for using the formylated methionine is to prevent reaction of the amino end of the first amino acid with the ester of the second. Subsequent methionines cannot be formylated because they would be unable to elongate the protein chain.

## Termination

Upon encountering a termination codon, a separate protein (a release factor) binds the complex and catalyzes the hydrolysis of the ester bond linking the completed protein. The ribosomal assembly then dissociates into the two major subunits. Subsequently, the ribosome complex reforms and the entire cycle of translation is repeated at other sites. Because bacterial mRNA is **polycistronic**—that is, it contains sequences for several proteins in the same mRNA strand—this cycle can continue on a downstream segment of the same mRNA. Moreover, in bacteria the process of translation can begin well before transcription has ended.

## Distinctive Features of Eukaryotic Translation

While translation is generally very similar in eukaryotes and bacteria, a larger number of proteins are involved in eukaryotes, particularly during initiation. Additionally, the synthesis of mRNA (a nuclear event) takes place in a separate compartment from translation (a cytosolic event), and eukaryotic mRNA is monocistronic (it codes for just one protein). There are no Shine Dalgarno sequences in eukaryotic mRNA. Rather, the initial positioning of the mRNA relies upon several eukaryotic initiation factors—eIF proteins—as well as the 5′-cap structure and the 3′-poly-A tail of the mRNA. Thus, the initiation complex consists of eIF proteins and the beginning and end of the mRNA. The protein assisting the first incoming tRNA—which is a Met-tRNA, distinct from the internal Met-tRNA molecules—is also a G-protein, analogous to the one in bacteria. In eukaryotes it is eIF-2. Once bound, the proteins assist the tRNA in scanning along the mRNA for the first AUG codon sequence, which is typically the initiation point for translation. Similar to prokaryotes, hydrolysis of the bound GTP of eIF-2 leads to the dissociation of the proteins on the small ribosomal unit and the formation of the full ribosome with P and A sites now defined. Eukaryotic elongation factors—called eEF proteins—similarly assist the incorporation of subsequent amino acid charged tRNAs, and the formation of the protein continues until a termination codon is encountered in a similar way to the prokaryotic system.

## Regulation of Eukaryotic Translation

Because mRNA is translated even as it is being synthesized in prokaryotes, there is little opportunity for separate control of the translation process. This kind of regulation does occur in eukaryotes, however.

One example is the control of hemoglobin protein synthesis in reticulocytes by heme or iron. Reticulocytes have no internal compartments and no nucleus to produce new mRNA. A decrease in heme or iron concentration leads to the activation of a protein called *heme regulated inhibitor (HRI)*. HRI is itself a protein kinase that catalyzes the phosphorylation of eIF-2 (FIGURE 16.8). eIF2 is a G-protein. The form that is a substrate of HRI is the inactive, GDP-bound form. This species binds tightly to eIF2B, forming an inactive complex, and removes eIF2B from solution. Because eIF2B is the catalyst that activates eIF2 by displacing GDP in favor of GTP, the overall result of HRI is the inactivation of initiation, and hence protein synthesis. In this way, HRI diminishes hemoglobin synthesis under conditions of diminished heme or iron availability.

# 16.4 Posttranslational Modifications of Proteins

## Immediate Modifications

Immediate posttranslational modifications of proteins include the deformylation of the leading f-Met residue and the hydrolysis of one or more amino acids from the N terminus of the protein, which is catalyzed by aminopeptidase.

Protein folding, another immediate posttranslational modification, begins shortly after a stretch of amino acids emerges from the ribosome, and this process is assisted by a class of proteins called **chaperones**. Chaperones also catalyze the refolding of fully synthesized proteins. One common consequence of structural misalignment arises from proline residues that are nonenzymatically rearranged from the physiological *trans* configuration, in which neighboring side chains extend in opposing spatial directions, to the *cis* configuration. This occurs because proline has a relatively low energy barrier for this nonenzymatic transformation. The enzyme, *cis–trans* prolyl isomerase, lowers this barrier even further, which allows the ultimately more stable *trans* configuration to prevail, thus favoring normal protein folding (see the mechanism in FIGURE 16.9). Certain cyclic compounds, such as **cyclosporine**, have the unusual property of binding prolyl isomerases, inhibiting their activity. Surprisingly, the cyclosporine–prolyl

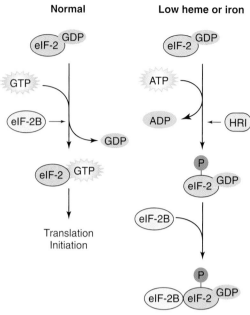

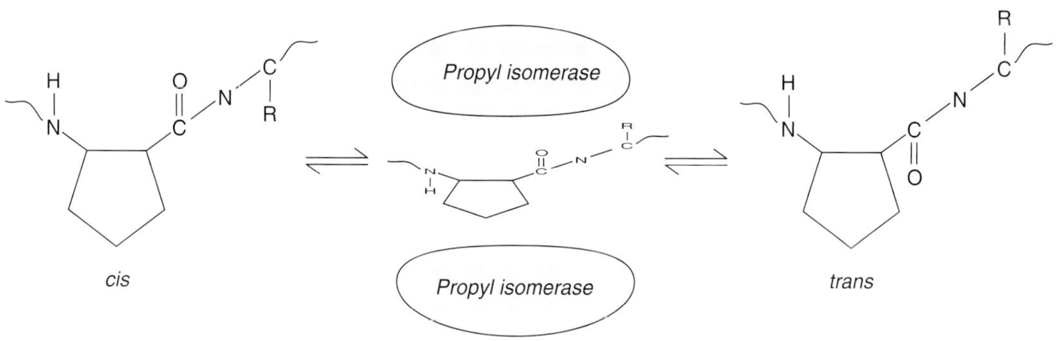

**FIGURE 16.8 Heme-Regulated Inhibitor (HRI) Control of Protein Translation.** In the normal case (left panel), eIF2 becomes activated by the G-activating protein eIF2B. However, when either heme or iron is in low concentration, HRI activity is elevated. HRI catalyzes the phosphorylation of eIF2:GDP, which tightly binds to eIF2B, forming an inactive complex.

**FIGURE 16.9 Mechanisms of *cis–trans* Prolyl Isomerase.** The enzyme provides a binding site that stabilizes the flat intermediate between the *cis* and *trans* forms. This increases the interconversion of the two forms. The *trans* form is ultimately more stable, so the interconversion leads to an increase in the *trans* form.

## (a) Cyclosporin

## (b) Tacrolimus (FK 506)

## (c) Rapamycin

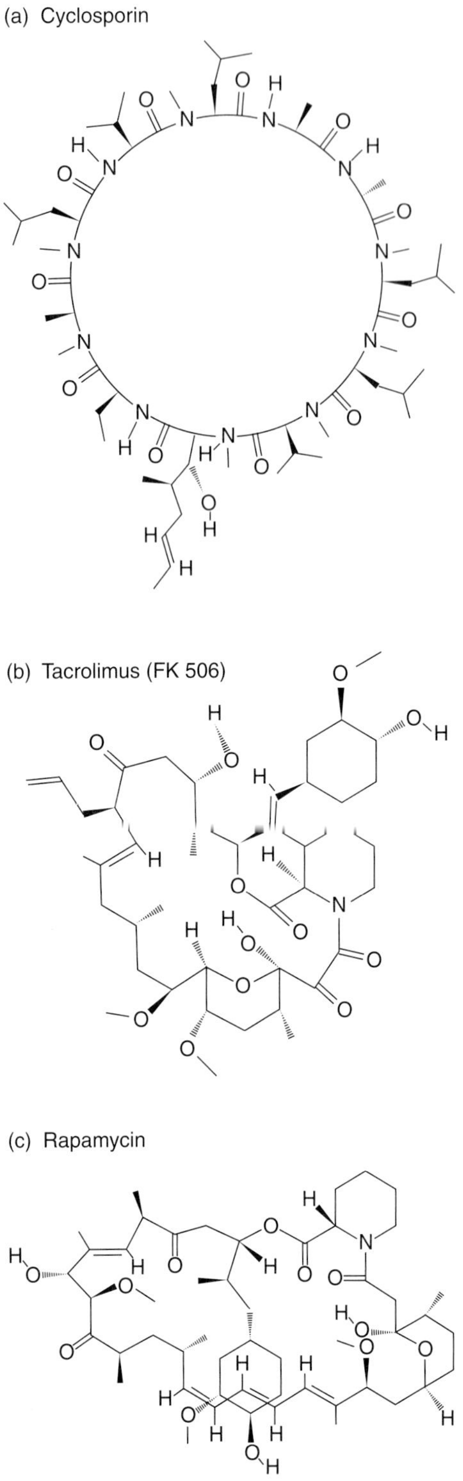

**FIGURE B16.2 Antibiotics that Bind Prolyl Isomerase.**

# BOX 16.2: Macromolecular Inhibitory Complexes

An unusual type of inhibition is exerted by certain cyclic compounds (**FIGURE B16.2**) that form a complex with the enzyme *cis–trans* prolyl isomerase. Tacrolimus and rapamycin both have a flattened prolyl residue that tightly binds the active site of the prolyl isomerase. However, because the residue is part of a small cyclic peptide, the enzyme cannot catalyze its reaction, thus forming a dead-end complex. Cyclosporin has no proline residue, but internal hydrogen bonding between the larger 11-amino-acid ring mimics this structure and similarly fits the prolyl isomerase. The cyclosporin-prolyl isomerase complex inhibits the protein phosphatase calcineurin. Calcineurin catalyzes the dephosphorylation of a transcription factor that prevents the formation of the T-cell–derived protein, interleukin-2 (IL-2). The absence of IL-2 prevents further cell division of the helper T cells, leading to immunosuppression. For this reason, the prolyl isomerases are sometimes called *immunophilins*. A second compound, FK506, binds a distinct prolyl isomerase, and the FK506–prolylisomerase complex also inhibits calcineurin, with an order of magnitude greater potency over cyclosporin. Rapamycin, on the other hand, binds a prolyl isomerase and inhibits its namesake target, mTOR. As discussed in the text, mTOR derives its name from rapamycin. The compound itself derives its name from its place of origin. It was discovered in a species of *Streptomyces* found on an island in the southern Pacific Ocean, 1500 miles west of Chile, known by outsiders as Easter Island, but by the natives as Rapa Nui.

isomerase complex itself serves as a selective inhibitor of an unrelated protein that leads to immunosuppression (**BOX 16.2**).

## Longer Term Modifications

Several modifications of proteins are classified as more long term. Some of these can be metabolically reversed (such as the protein phosphorylations discussed elsewhere in this text), which alter the activity of an enzyme, or acetylation, which alters the binding properties of histones and other proteins. Some modifications to side groups alter the functioning of proteins, such as the hydroxylation reactions that produce hydroxyproline residues in collagen. Most exported proteins have sugar residues covalently attached that are added in the Golgi apparatus, producing proteins that are more viscous and resist degradation. In some cases, proteins are modified and subsequent proteolysis yields a distinct product. For example, thyroxin formation involves the joining of tyrosine residues of the protein thyroglobulin, followed by iodination of the aromatic hydroxyls of these tyrosines. Proteolysis then releases the modified dipeptide thyroxin (tetraiododityrosine). In a similar way, creatine (the carrier of high-energy phosphates in nerve and muscle) and carnitine (the carrier of fatty acids into the mitochondria) are both formed by the modification of an amino acyl group of a protein, followed by the release of these molecules due to protein degradation. We next consider the process of protein degradation more broadly.

# 16.5 Protein Degradation

As we have just seen, degradation may be viewed as a modification of proteins. In some cases, proteins are only partially degraded. For example, the amino terminus of some proteins serves as a **signal sequence** to guide them into specific organelles, such as the endoplasmic reticulum or mitochondria. Once the protein has entered the new location, this portion is removed by proteolysis. Degradation may occur even as a protein is being synthesized, a strategy that eliminates misfolded proteins. However, usually proteins are completely degraded and the amino acids recycled, which we consider next.

## Extracellular Proteases

The earliest discovered proteases were those found in extracellular environments. For example, the proteolytic enzyme trypsin, found in the stomach, catalyzes the partial hydrolysis of proteins. Once the stomach contents enter the intestine, several proteases secreted from the pancreas further digest the proteins and protein fragments to produce amino acids and some dipeptides. The mechanisms of the proteases have been well studied, particularly a major group known as the serine proteases. These employ a **triad**: three amino acids positioned by hydrogen bonds in the active site (FIGURE 16.10a). The key catalytic components are the amino acids His and Ser, as indicated in Figure 16.10b. The mechanism of action involves the series of nucleophilic attacks illustrated in FIGURE 16.11. A related class of proteases uses cysteine (rather than serine) as a nucleophile; still others use a metal to activate the amide for attack by water. Some selectivity exists in these proteases, directed by the functional groups near the amide that is to be split. For example, chymotrypsin is relatively selective for aromatic residues on the N side of the peptide bond. Using the known selectivities of the proteases, early protein sequencing studies were able to map peptides and eventually the amino acid sequence of proteins.

Another important class of extracellular proteases is involved in the clotting reaction involving a series of enzymes that catalyzes the sequential peptide bond hydrolysis and ultimately the creation of a protein plug that prevents the leakage of blood vessels. Because each step catalyzes many copies of the product, the sequence is an amplification that rapidly produces large amounts of clotting material. A common form of hemophilia results from the genetic loss of one of these enzymes, called factor VIII.

## Intracellular Proteases

Organelles are degraded and recycled by fusing them with the lysosome, which has an acidic environment, produced by a proton pump analogous to the $F_1F_o$ATP synthetase of mitochondria and chloroplasts. A group of proteases with an optimal pH in the acidic range catalyze the degradation of the protein component of these organelles.

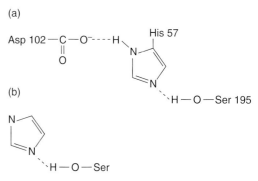

**FIGURE 16.10 Serine Protease Triad.** (a) The nucleophile responsible for the hydrolysis of peptides is a serine that is hydrogen bonded to a histidine, which is locked into position by an aspartate residue, thus forming the triad. (b) The essential catalytic unit consists of the serine and the imidazole of the histidine.

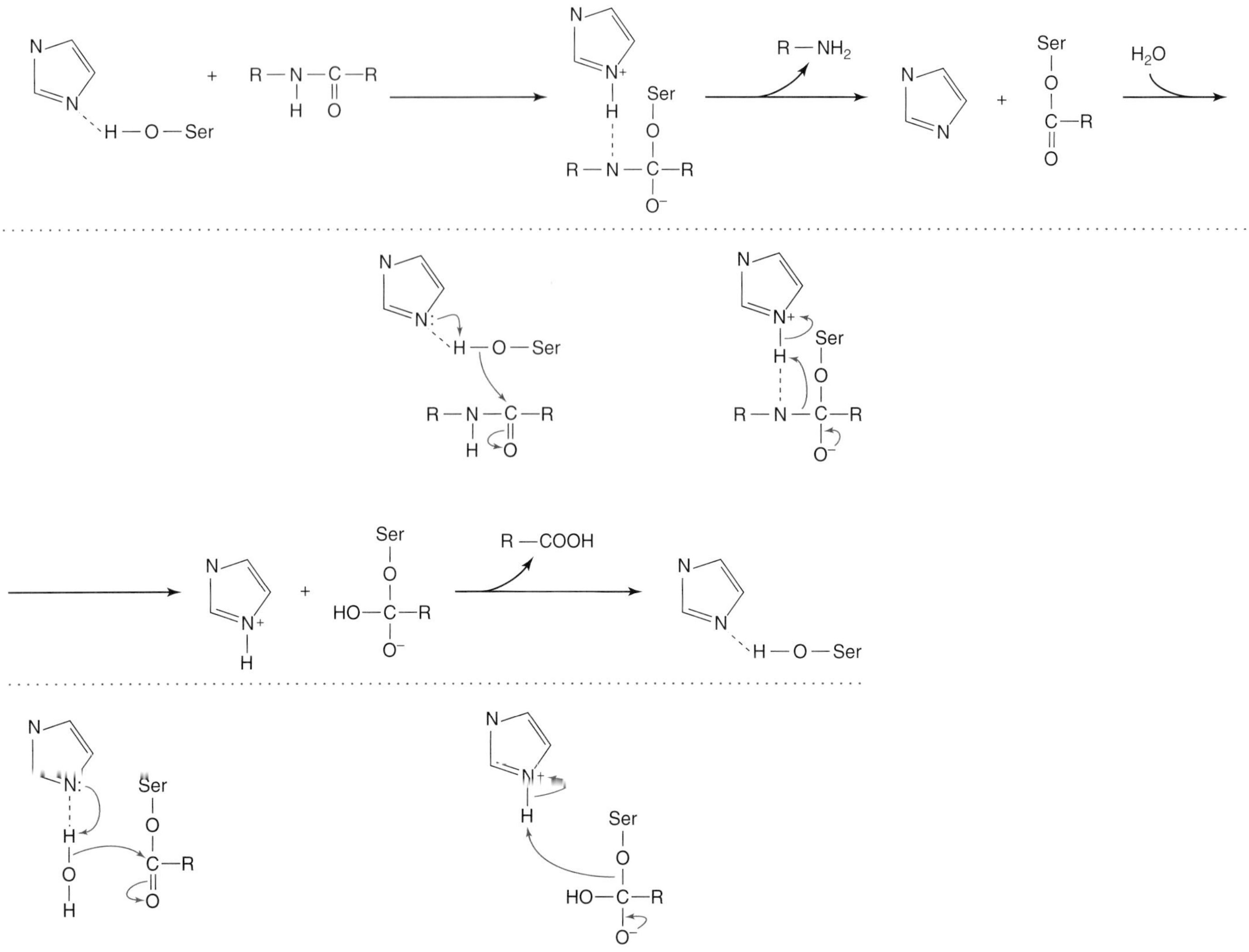

**FIGURE 16.11 Mechanism of Serine Protease.** The imidazole enhances the hydroxyl group of serine for nucleophilic attack and then provides an electron sink for subsequent electron rearrangement that leads to the cleavage of the peptide bond. Next, the imidazole assists in both the hydration and hydrolysis of the enzyme-bound carboxyl group of the remaining protein.

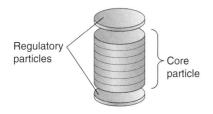

**FIGURE 16.12 The Proteasome.** The core particle is the cylinder, and the hydrolysis of proteins occurs in its interior. Regulatory particles are protein complexes at both ends that cap the cylinder.

Self-degradation (**autophagy**) and vesicular uptake are examples of cellular activity that involve lysosome fusion.

Proteolytic degradation of cytosolic proteins involves a separate particle, the proteasome (**FIGURE 16.12**). This structure is a complex of proteins that creates an interior environment separated from the cytosol in which proteins can be degraded once they are threaded into it. Unfolding depends upon a posttranslational modification called **ubiquitination**, in which molecules of **ubiquitin** (a small 76-amino-acid protein) are covalently attached to proteins. Proteins that unfold become substrates for attachment to ubiquitin (UbiQ) by the reaction outlined in Equation 16.2:

$$UbiQ + ATP + E\text{-}Lys\text{-}NH_2$$
$$\rightarrow E\text{-}Lys\text{-}N\text{-}C(O)\text{-}UbiQ + AMP + PP_i \quad (16.2)$$

**CHAPTER 16** PROTEIN SYNTHESIS AND DEGRADATION

## TABLE 16.2 Protein Half-Lives[a]

| Protein | Tissue | Half-Life |
|---|---|---|
| Ornithine decarboxylase | Rat liver | 11 min |
| RNA polymerase I | Rat liver | 1.3 h |
| Phosphoenolpyruvate carboxykinase | Rat liver | 6 h |
| Hexokinase | Rat liver | 1 day |
| Acetyl-CoA carboxylase | Rat liver | 2 days |
| Arginase | Rat liver | 5 days |
| Myosin (heavy chain) | Rat heart | 6 days |
| Myosin (light chain) | Rat heart | 9 days |
| Tropomyosin | Rabbit skeletal muscle | 25 days |
| Myosin | Rabbit skeletal muscle | 30 days |
| Actin | Rabbit skeletal muscle | >50 days |

[a]Source: Kay, J. Intracellular protein degradation. *Biochem. Soc. Trans.* 6:789–797, 1978.

At least four copies of ubiquitin must be added to a protein to allow it to thread itself into the cavity of a proteasome.

While many different proteins are degraded by the proteasome, they do so at distinctive rates. As a result, the degradation half-lives for different proteins vary over a wide range, as shown in Table 16.2. The shortest known half-life is that of ornithine decarboxylase, which catalyzes the formation of putrescine in the polyamine pathway (FIGURE 16.13). Polyamines bind the negatively-charged phosphate residues of DNA. An enzyme with a relatively short half-life is hepatic PEP carboxykinase, which catalyzes the rate-limiting step of gluconeogenesis; its half-life is a few hours. Other proteins have half-lives that can range up to many days or weeks.

The variation in the degradation rates is particularly important when we consider that proteins themselves are in a steady state, just like intermediary metabolites. To appreciate its significance for regulation, we conclude with an examination of a regulatory pathway for protein synthesis.

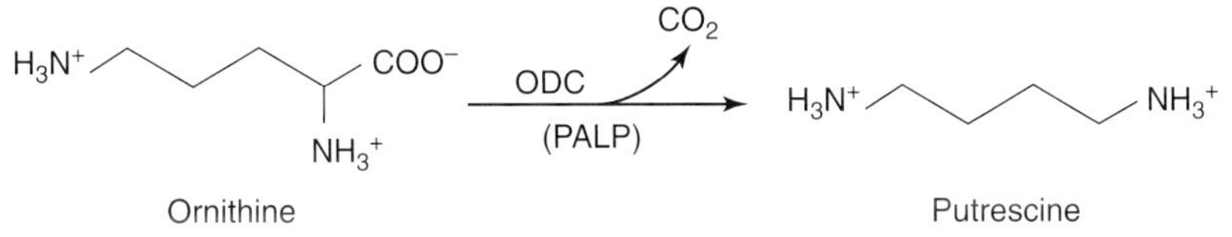

FIGURE 16.13 **Ornithine Decarboxylase.** The reaction uses a bound pyridoxal phosphate group (PALP) that attaches to the alpha-amine of ornithine, leading to the release of the alpha-carboxyl group and the formation of putrescine, one of the polyamines that serves to neutralize the negative charge of DNA.

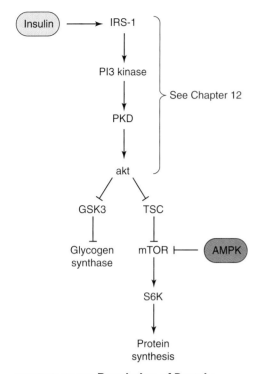

**FIGURE 16.14 Regulation of Protein Synthesis at mTOR.** Insulin, via steps outlined in Chapter 12, leads to the activation of the kinase akt, which phosphorylates and inactivates the G-protein TSC. This lifts the inhibition that TSC exerts on mTOR, leading to the activation (*phosphorylation*) of S6K, a kinase that leads to increased protein synthesis. Activation of AMPK (through increased concentrations of AMP), on the other hand, leads to the phosphorylation and inactivation of mTOR, and the suppression of protein synthesis. Shown for reference is another action of akt: namely, the phosphorylation and inhibition of GSK3, which lifts the inhibition of glycogen synthase.

## 16.6 The mTOR Pathway in the Control of Protein Synthesis

The protein mTOR is a kinase that is central to the control of protein synthesis. It can be regulated in both a positive and negative sense: insulin activates it, whereas AMP kinase inhibits it (FIGURE 16.14).

mTOR, an abbreviation for *mammalian target of rapamycin* (Box 16.2), was discovered after rapamycin itself. Insulin (and other tyrosine kinase mediated growth factors) leads to the activation of the kinase akt. In turn, akt catalyzes the phosphorylation of the G-protein TSC (*tuberous sclerosis complex*, named for a tumor that develops when the protein is dysfunctional). This relieves the inhibition of mTOR, similar to how the inhibition of GSK3 relieves the inhibition of glycogen synthase, another insulin-related function (Chapter 12). AMP kinase is activated under circumstances when the concentration of AMP increases, such as when increased ATP utilization that is not matched by ATP supply leads to increased ADP concentration. ATP is then regenerated through the near-equilibrium adenylate kinase reaction:

$$2 \text{ ADP} \rightarrow \text{ATP} + \text{AMP} \qquad (16.3)$$

When this happens, the concentration of AMP increases, activating AMP kinase and inactivating mTOR.

mTOR catalyzes the phosphorylation of several targets including those controlling protein synthesis. One such target is a binding protein that activates initiation upon phosphorylation by mTOR. Another is S6 kinase (S6K), a kinase present in the small subunit of the ribosome (the S stands for *small ribosomal subunit*). Phosphorylation of S6K promotes elongation in protein synthesis. The net effect of mTOR activation is the stimulation of cellular protein synthesis.

Insulin stimulates protein translation, a reflection of the growth-promoting properties that it shares with other cellular growth factors. AMP kinase, by suppressing protein translation, antagonizes growth promotion, and allows energy conservation by the cell. While we have seen that several high-energy phosphate groups are directly consumed during protein synthesis, the mechanism of energy sparing is the diminished formation of the enzymes that catalyze the pathways that use energy.

Due to the variation in rates of protein degradation, inhibition of the overall rate of protein synthesis does not affect all proteins equally. Those with long half-lives will be little affected by alterations in protein synthesis rates, because the amount accumulated will take a very long time to be affected. Thus, while this means of regulation appears to be completely nonselective, in fact it produces differential effects that can largely be read from a table of half-lives such as Table 16.2.

Finally, all forms of regulation in cells and organisms can occur together. We may formally divide regulation into *short term* and *long term* for experimental or conceptual convenience. In fact, while translation can be considered another long-term effect, most cell regulation occurs at the transcriptional level. mRNA turnover, being relatively rapid, allows the cell to produce precisely the amount of mRNA needed to drive protein synthesis, based on its needs and sensitive to the control elements upstream of the gene in the DNA. This is tempered by protein translation short term control mechanisms and, most fundamentally, the rate of ATP utilization in the cell cytosol. All regulatory features move us from one steady state to the next.

## Summary

Protein synthesis requires the convergence of three types of RNA: ribosomal, messenger, and transfer. The ribosome is the scaffold composed of a small and a large subunit, each of which are complexes of proteins and rRNA. The small subunit forms first, binding a charged (amino acid bound) tRNA, mRNA, and proteins serving as initiation factors. With the addition of the large ribosomal subunit, the protein factors are released and elongation ensues. Elongation factors assist the binding of charged tRNA to the ribosome and the peptide bond is formed. Termination occurs at a codon for which no corresponding tRNA exists, and the completed protein is released. Formation of each charged tRNA requires a distinct tRNA synthetase enzyme. Recognition of the tRNA on the mRNA at the ribosome is due to a triplet anticodon sequence in tRNA base pairing with a three-nucleotide codon in mRNA. The correspondence of the codon to the amino acid is the genetic code, which at a glance provides the protein sequence for a given mRNA sequence. Chaperone proteins catalyze the folding of newly synthesized proteins into their active configuration. Proteins are also subject to various posttranslational modifications that alter their properties, such as phosphorylation and methylation. The process of translation is subject to regulation in eukaryotes. For example, the heme-regulated inhibitor is a kinase that phosphorylates a eukaryotic initiation factor, thus attenuating protein synthesis. mTOR is a kinase that stimulates protein synthesis. mTOR is stimulated by insulin and inhibited by AMP kinase. Proteins are constantly degraded in cells, in a process that occurs at distinct rates for different proteins. A major route of protein degradation depends on the covalent incorporation of a small protein called ubiquitin, followed by breakdown in a protein complex, the proteasome. Proteins are constantly synthesized and broken down; that is, they exist at a steady state within cells.

## Key Terms

acceptor site
acceptor stem
amino-acyl-tRNA
  synthetases
anticodon
autophagy
chaperones
charged tRNA

codon
colloids
cyclosporine
genetic code
peptidyl site
polycistronic
proteasome
ribosome

ribozymes
Shine–Dalgarno
  signal sequence
svedberg
triad
ubiquitin
ubiquitination wobble
  hypothesis

## Review Questions

1. Tetracycline, an antibiotic used against bacterial infections, binds to the A site in bacterial protein synthesis. Which phase of protein synthesis does this compound specifically affect?

2. Several G proteins are involved in protein synthesis. How is the action of these proteins similar to G proteins involved in hormonal regulation?

3. Why are there no examples of regulation of prokaryotic translation?

4. Leucine, one of the branched chain amino acids, is a potent activator of mTOR. Why is this signaling appropriate?

5. While there is less control over proteolysis than protein synthesis, there are disease states associated with dysfunctional proteolysis. One example is Alzheimer's disease, in which neural cells accumulate massive quantities of a protein called amyloid. Which specific proteolysis system is likely responsible for this accumulation? If this is the primary cause of the disease, what might be a possible avenue of treatment?

6. A few proteins are found to have amino acids other than the 20 we have considered (formally called the "canonical" 20 amino acids) that are in fact incorporated translationally (i.e., not arising by modification after the protein is synthesized). One example is seleocysteine, found, for example, in glutathione peroxidase. Would this amino acid have its own amino acid synthetase? Its own codon? Does this make seleocysteine the "21st amino acid"?

## References

1. Chen, J-J.; London, I. M. Regulation of protein synthesis by heme-regulated eIF-2a kinase. *Trends Biochem. Sci.* 20: 105–108, 1995.

2. Komander, D.; Rape, M. The ubiquitin code. *Annu. Rev. Biochem.* 81:31. 1–31.27, 2012.
   Ubiquitin in proteolysis and beyond.

3. Watson, J. D.; Baker, T. A.; Bell, S.; Gann, A.; Levine, M.; Losick, R. *Molecular Biology of the Gene,* 6th Edition; Benjamin Cummings: San Francisco, 2008.
   Contains a more expansive but still general review of protein synthesis.

4. Zoncu, R.; Efeyan, A.; Sabatini, D. M. mTOR: from growth signal integration to cancer, diabetes and ageing. *Nature Revs.: Molec. Cell Biol.* 12:21–35, 2011.
   A somewhat more expansive but still general treatment of protein synthesis.

5. Kapp, L. D.; Lorsch, J. R. The molecular mechanics of eukaryotic translation. *Annu. Rev. Biochem.* 73:657–704, 2004.

LIBRARY, UNIVERSITY O  CHESTER

# Appendix

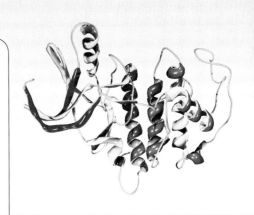

Image © Dr. Mark J. Winter/Photo Researchers, Inc.

# 1. Mathematical Ideas

## Vectors

Many common entities have only a *magnitude*; speed, weight, and density are examples. Others require both a magnitude and a *direction* to capture their essence: these are *vectors*. For example, a car's speed can be represented as 30 mph, but a description of its velocity needs more, such as 30 mph north.

Our first encounter with a vector in this text is in the consideration of polarity. The magnitude of this vector depends on the relative attraction for the binding electrons between the two nuclei. The direction is from the lesser to the more electronegative atom. When we add these vectors (to determine molecular polarity), we must take into account both magnitude and direction. If we consider a simple two-dimensional addition, we can accomplish it graphically (**FIGURE A.1**). The addition is performed by reconstructing the vectors A and B so that their origins intersect, and then constructing the sum as shown in the figure. Note the direction and magnitude of the sum C is (in general) distinct from the sum of the magnitudes of A or B. If A and B are exactly 180 degrees apart and of equal magnitude, the vector sum is zero; this is the case of the net polarity of $CO_2$.

A second important case of vectors is more conceptual: the vectorial flow of protons in mitochondria, bacteria, and chloroplasts establishes the *proton motive force*, the basis of energy production. In this case, we did not consider vector mathematics explicitly; only the directional traversal of the membrane is essential.

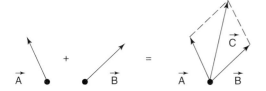

**FIGURE A.1 Vector Addition.** Slide the two vectors in the plane of the page, maintaining their direction and magnitude, until their origins coincide. Construct parallel segments (dashed lines) to each line and complete the parallelogram. The sum is the arrow from the common origin to the intersection of the constructed (dashed) lines.

## Logarithms

The hierarchy of operations in mathematics can be written as:

    I. Addition and its inverse, *Subtraction*
    II. Multiplication and its inverse, *Division*
    III. Exponentiation and its inverse, *Logarithms*

Level I is the familiar process of addition. Even if this was the only operation available, we could still perform operations that usually employ levels II and III, but it would be tedious. A step up from addition, multiplication effectively tallies objects by grouping them into multiples. For example, rather than add 300 individual pencils, we can arrange them into groups of 10; finding 30 of these means 30 multiples of 10.

For the next level, consider bacterial growth: each cell divides, each of those cells divide, and so forth. This is exponential growth. For three events, rather than multiplying a number by itself repeatedly, say $2 \times 2 \times 2$, we can abbreviate this by writing $2^3$. The superscripted 3 here tells us how many times we have multiplied 2 by itself. The 2 is called the *base* and the 3 is the *exponent* or *power*. This is expressed as *two to the third power*. The interesting property of exponents is that we can add the exponents themselves rather than just carrying out the whole operation. We thus could say:

$$2^3 * 2^4 = 2^7 \tag{A.1}$$

which is easily verified by multiplying out both sides.

Now consider the inverse function of exponentiation: the logarithm (or just *log*). In our example of exponents, the base 2 raised to the power 3 produces the answer 8. Then the log of 8 *to the base 2* is 3.

We are usually interested in just a few bases: 2, 10, and e. The latter is the base of the natural logarithms. The rationale for e is somewhat far afield of this treatment; it is easily understood from the operations of calculus (the derivative of the function $e^x$ is equal to the very same function, $e^x$; thus, e is the base of the most important recursive function). Base 2 is commonly used in measuring population growth. We use base 10 in pH discussions; in fact, the pH is itself a base 10 logarithm of hydrogen ion concentration (actually its reciprocal), and useful because the concentration of hydrogen ions is very low in solution. Thus, we can represent the concentration of hydrogen ions of $10^{-7}$ as simply a log form; we say it is a pH of 7.

As logarithms can be viewed as *picking off the exponents*, and because exponents as we have just seen are additive, this also must be true of logarithms. Hence the rule:

$$\log a * b = \log a + \log b \qquad (A.2)$$

which is used in the derivation of the Henderson–Hasselbalch equation.

Two important points must be noted about logarithms. First, there are no units. This means that the phrase *pH units* has no meaning. The hydrogen ion concentration has units, but once we take its logarithm, we are extracting the exponent, producing a purely mathematical number. Secondly, the value of a change in pH is not a constant difference. It depends on the range. In the study of chemiosmosis, we commonly encounter expressions that employ $\Delta pH$ as part of the electrochemical potential. However, this is an inexact quantity; clearly the same $\Delta pH$ starting from say pH 8 must be an order of magnitude greater than one starting from pH 9. Thus, logarithmic transformation can complicate understanding as it makes exponential changes appear linear and eliminates the units of measure.

## Geometric Mean

An uncommon method of taking an average, this arises prominently in the calculation of isoelectric points in amino acids. The definition of pI for an amino acid with no additional dissociable group (i.e., an extra acid or base) is:

$$pI = (pK_1 + pK_2)/2 \qquad (A.3)$$

While this seems like a straightforward arithmetic mean, it isn't. We are dealing with logarithmic expressions, so arithmetic means do not apply. We need what is known as the *geometric mean*, where the values are multiplied before dividing by two, rather than being added. This makes sense when we realize that this must be the case for finding an average value for equilibrium constants, because these are always related by multiplication in cases of connected equilibria with polyprotic acids like inorganic phosphate.

The rationale for using the geometric mean is that combining equilibrium constants for sequential reactions involves multiplicative rather than additive operations. This goes back to the original derivation of an equilibrium constant for an equation: the products *multiplied* together divided by substrates *multiplied* together. It should be expected that a combination of two reactions in series is also summarized by an overall equilibrium constant that is the multiple of each individual constant. Thus, in forming the isoelectric point, what appears to be an arithmetic average is masked by the fact that what is being added are log values.

# 2. Chemical Fundamentals

Even after taking two years of chemistry, students entering biochemistry are often unclear or uncertain about some elementary chemical ideas. In part, this is due to the enormous amount of information presented, and the difficulty in knowing what is essential and what is not.

Another factor that hinders understanding starts at the beginning of a modern chemical education: presenting the periodic table as based on C having a relative weight of 12. This is technically true, but only of real importance to the Bureau of Weights and Measures. It has no fundamental significance.

In fact, we should think of the periodic table as based on hydrogen, the lightest element, assigned the relative weight of 1. All other elements are multiples of H. Experimentally, C is 12 times the mass of H, and so C has a molecular weight of 12. The *mole* (abbreviated mol) is the number of atoms in 1 gram (g) of H: that number is $6 \times 10^{23}$ (*Avogadro's number*). Thus, 12 g of C or 16 g of O contain the same number of atoms as 1 g of H.

A subtlety of Avogadro's number is that its units are themselves variable. For example, it can be applied to a number of atoms or molecules. Usually, we consider a mol as an Avogadro number of an entity. Take the connected collection of atoms that is glucose, $C_6H_{12}O_6$. This entity has a molecular weight of 180 g. This means that 180 g of glucose is equal to 1 mol of glucose. In what we can call *chemical arithmetic*, a reaction can be considered to follow:

$$1 \text{ mol of A} + 1 \text{ mol of B} \rightarrow 1 \text{ mol of C} \tag{A.4}$$

for a joining reaction. This adds up in the usual mathematical manner only when we take into account the actual weights of each term. Otherwise, abstracting the numerical part of (Eq. A.4) without units would produce: $1 + 1 = 1$!

Protons and electrons have an opposite charge, and attract; this notion is a primitive fact of chemistry (that is, we don't know why it is true since charge itself is a mystery). It is the origin of all electrical phenomenon. The neutrons have no charge but co-exist in the nucleus for the purpose of atomic stability. This has no bearing on the chemistry that we consider, but it is interesting to note that without neutrons, protons could not possibly exist so close together. The mutual electrostatic repulsion would cause them to fly apart. Protons and neutrons are attracted to one another by another force that acts at very short range and overcomes the proton–proton electrostatic repulsion. The approximate balance of these forces explains why most elements have a similar number of protons and neutrons, making the atomic mass roughly double the atomic number.

The electron arrangement outside the nucleus is all-important, and is a balance between two electrostatic influences. From the standpoint of the atom, there is the attraction between proton and electron, which keeps the electrons within the *orbit* of the nucleus. Broadly speaking, there are two types of electrons: those that we call *valence electrons*, and the rest. Only the valence electrons are considered to interact with other atoms, as they are the furthest from the nucleus and thus influenced the least. There are other subtleties. For example, the outer electrons also are shielded by the inner electrons from the charge of the proton, and thus experience a reduced electrostatic attraction. The non-valence orbitals are important in ions such as $Fe^{2+}$, which form coordinate bonds to ligands like heme; the inner orbitals of $Fe^{2+}$ allow it to have six binding positions.

Electrons do not actually orbit the nucleus like the planets around the sun. Despite this, the volume of space that is occupied by the electrons is called an *orbital*. The actual shape of orbitals (such as the first one appearing in the text for water) is calculated from probability considerations based on quantum mechanism, developed in the early 20th century. While there is some complexity to quantum chemistry, the results are extremely simple: the orbitals are *probability clouds* in which there is a high probability of finding particular electrons. Electrons associated with atoms have a regularity that matches the types of orbitals that exist within them. In order, they are called s, p, d, f, etc. The atoms we are concerned with usually have electrons only in the first few—s and p—and only occasionally the d orbitals. Each individual orbital can have up to two electrons in it. The s orbital is spherical and can contain two electrons. There are a set of three p orbitals, each with two electrons, so that p orbitals can hold up to six electrons. The d orbitals are a set of five, each with two electrons, so that these can accommodate up to ten electrons. Each succeeding atom has just one more proton than the preceding one, and is also accompanied by one more electron. Thus, for hydrogen, there is just one proton and one electron; the electron occupies an s-orbital, which is spherical. Carbon has six electrons, two of which are in an s-orbital, and the other four in p-orbitals.

A molecule is a collection of atoms that are covalently linked. This covalent linkage means that orbitals from two atoms overlap and, thus, electrons are shared between nuclei. This joining of atoms creates molecular orbitals. Similar to the atomic orbitals just considered, these orbitals also can hold only a maximum of two electrons. For example, the bond between carbon and hydrogen atom in methane, $CH_4$, is an overlap (merge) between the s-orbital of H and one of the p-orbitals of C. All four C–H bonds in methane are identical. This merger is called a *hybridization of orbitals*, and this particular one is called the $sp^3$ orbital, indicating the amount of atomic s and atomic p qualities that contribute. While this is obviously a rough view of electronic bonding, it is useful in understanding chemical behavior. Because the electrons in each C–H orbital repel the electrons in the other C–H orbitals, they are arranged in space so that they are the maximal distance apart. This produces the tetrahedral structure of the methane molecule as well as most other single-bonded hydrocarbon molecules.

# 3. Derivation of the Henderson-Hasselbalch Equation

We start with the acid dissociation equilibrium equation,

$$HA \rightleftharpoons H^+ + A^- \tag{A.5}$$

which has the dissociation constant,

$$K = [H^+][A^-]/[HA] \tag{A.6}$$

The Henderson-Hasselbalch equation is the transformation of this equation into logarithmic form. Taking the log (base ten) of both sides:

$$\log K = \log [H^+] + \log [A^-]/[HA] \tag{A.7}$$

Rearranging,

$$-\log [H^+] = -\log K + \log [A^-]/[HA] \tag{A.8}$$

we already know that pH = −log [H+]. A similar definition applies to K:

$$pK = -\log K \tag{A.9}$$

then we can replace both terms in the rearranged equation:

$$pH = pK + \log[A^-]/[HA] \tag{A.10}$$

which is the Henderson-Hasselbalch equation. The variables are:

a) pH, which is the negative logarithm of the hydrogen ion concentration
b) pK, which is the negative logarithm of K, the equilibrium constant
c) [A⁻], the concentration of the anion form of the acid
d) [HA], the concentration of the protonated form of the acid

Key properties of the variables are summarized below.

- All variables are expressed in log form; thus, changes observed should be considered to be larger than they "feel." For example, a change in pH from 3 to 2 is much greater than a change in pH from 11 to 10, by 8 orders of magnitude. Thus, knowing just a "pH change" is not meaningful unless you know the starting and ending pH values.

- Often it is useful to consider the ratio of [A⁻]/[HA] together. In this case, the ratio can be considered in three general ways: where [A⁻] = [HA], when [A⁻] is much greater, or when [HA] is much greater.
- If [A⁻] = [HA] then pH = pK. This is because the log of 1 is zero; it can alternatively be appreciated simply from the nontransformed equation (the equilibrium expression itself), which shows that K = [H⁺]. This is reasonable; they are in the same units. In a titration curve of pH versus added base, the relatively flat region in the center, where pH changes little with increasing base, pK is the center of this *buffering region*. The corresponding value of pH at this point is the experimental means for determining the pK. At this point, [A⁻] = [HA]; there are ample amounts of both forms to react with either excess acid or base, so changes in pH are minimized.
- When [A⁻] is in excess, the ratio is greater than one, and the pH will be that much greater than the pK. An excess of the salt form that is of [A⁻] always has this effect, and we can turn the argument around and say that when pH is greater than pK, then [A⁻] is dominant.
- When [HA] is in excess, the roles are reversed. Now, there is more acid form, and this means that the ratio of [A⁻]/[HA] will be less than one, *the log of the* ratio less than zero, and thus that the pH will be lower than the pK. If this argument is reversed, we can say that whenever the pH is lower than the pK, the HA is in excess.

The Henderson-Hasselbalch equation can be used for bases as well as acids. Starting with the equation for acid dissociation,

$$HA \rightleftarrows H^+ + A^- \tag{A.11}$$

HA is an acid, and A⁻ is defined as the conjugate base. Writing the association equation for a base,

$$B + H^+ \rightleftarrows BH^+ \tag{A.12}$$

where B is the base, BH⁺ is the conjugate acid. If we reverse the equation (the new dissociation constant will thus be the reciprocal of the one for the above equation):

$$BH^+ \rightleftarrows B + H^+ \tag{A.13}$$

We now have an acid dissociation equation again, albeit a conjugate acid. Using the same derivation as before,

$$pH = pK + \log[B]/[BH^+] \tag{A.14}$$

Thus, we arrive at an equation that can be put in the same form as the originally derived Henderson-Hasselbalch equation. In order to use the equation in general for acids or bases, we need only note that the denominator term will always have the more protonated form.

# 4. Derivation of the General Free Energy Equation

Changes in enthalpy, entropy, and free energy at standard state are indicated with a superscript degree symbol (°) appended to their abbreviation: $\Delta H°$, $\Delta S°$, and $\Delta G°$, respectively. The standard free energy change ($\Delta G°$) is of particular interest, and can be related to the equilibrium constant ($K_{eq}$). To show this relationship, we begin with the following equation:

$$\Delta G = \Delta G^0 + RT \ln [C][D]/[A][B] \tag{A.15}$$

is an equation of broad nature, although it does require a specific chemical reaction, for example:

$$A + B \rightleftarrows C + D \tag{A.16}$$

Essentially, the equation stems from the free energy equations for each reaction component, such as

$$G_A = G_A^0 + RT \ln[A] \tag{A.17}$$

The logarithmic relationship between free energy and concentration of the component is the result of a derivation that uses the ideal gas law. The logarithmic term arises from an integration of a reciprocal pressure, which is then related to concentration. Perhaps surprisingly, the resulting equations have wide applicability to situations which are not ideal, and to liquids as well as gases. Using equations similar to (Eq. A.17) for the other components of the reaction (Eq. A.16), and the linear free energy relationship:

$$G_A + G_B = G_C + G_D \tag{A.18}$$

The final equation (Eq. A.16) results from collecting terms.

# 5. Lipids

## Fatty Acids

Table A.1 shows a more extensive listing of fatty acids, along with their structures and melting points. The latter show a regular increase with chain length for the unsaturated fatty acids. Table A.1 also shows that double bonds profoundly decrease the melting points.

## Phospholipids

A listing of phospholipids along with their structures is shown in FIGURE A.2.

## TABLE A.1 Fatty Acids

| Identity | Common Name | Systematic Name | Structure | Melting Point |
|---|---|---|---|---|
| **Saturated** | | | | |
| 4:0 | butyric | butanoic | $CH_3(CH_2)_2CO_2H$ | −5.5 |
| 6:0 | caproic | hexanoic | $CH_3(CH_2)_4CO_2H$ | −4 |
| 8:0 | caprylic | octanoic | $CH_3(CH_2)_6CO_2H$ | 16 |
| 10:0 | capric | decanoic | $CH_3(CH_2)_8CO_2H$ | 31 |
| 12:0 | lauric | dodecanoic | $CH_3(CH_2)_{10}COOH$ | 44 |
| 14:0 | myristic | tetradecanoic | $CH_3(CH_2)_{12}COOH$ | 52 |
| 14:0 | palmitic | hexadecanoic | $CH_3(CH_2)_{14}COOH$ | 63 |
| 18:0 | stearic | octadecanoic | $CH_3(CH_2)_{16}COOH$ | 70 |
| **Unsaturated** | | | | |
| 16:1($\Delta^9$) | palmitoleic | hexadecanoic | $CH_3(CH_2)_5CH=CH(CH_2)_7CO_2H$ | 0 |
| 18:1($\Delta^9$) | oleic | octadecanoic | $CH_3(CH_2)_7CH=CH(CH_2)_7COOH$ | 16 |
| 18:2($\Delta^{9,12}$) | linoleic | octadecadienoic | $CH_3(CH_2)_4(CH=CHCH_2)_2(CH_2)_6COOH$ | −5 |
| 18:3($\Delta^{9,12,15}$) | linolenic | octadecatrienoic | $CH_3CH_2(CH=CHCH_2)_3(CH_2)_6COOH$ | −11 |
| 20:3($\Delta^{5,8,11,14}$) | arachidonic | eicosatetraenoic | $CH_3(CH_2)_4(CH=CHCH_2)_4(CH_2)_2COOH$ | −50 |

# 6. Aspects of Enzyme Kinetics

## Derivation of the Michaelis–Menten Equation

Considering our simple mechanism:

$$E + S \underset{k_{-1}}{\overset{k_1}{\rightleftharpoons}} ES \overset{k_2}{\rightarrow} E + P \qquad (A.19)$$

This is composed of three separate reactions, shown with their corresponding rate equations:

$$E + S \rightarrow ES \qquad rate(20) = k_1 * [E][S] \qquad (A.20)$$

$$ES \rightarrow E + P \qquad rate(21) = k_2 * [ES] \qquad (A.21)$$

$$ES \rightarrow E + S \qquad rate(22) = k_{-1} * [ES] \qquad (A.22)$$

Exact solutions of the rate equations would require differential calculus. However, the solution can be reduced to algebra by using the steady-state assumption:

$$\text{Rate of ES formation} = \text{Rate of ES destruction} \qquad (A.23)$$

This is true when ES reaches a constant concentration, which is commonly the case in enzymatic reactions. We can recast Equation A.23 by substituting

> Rate(20) for the rate of ES formation, and

> Rate(21) + rate(22) for the rate of ES destruction.

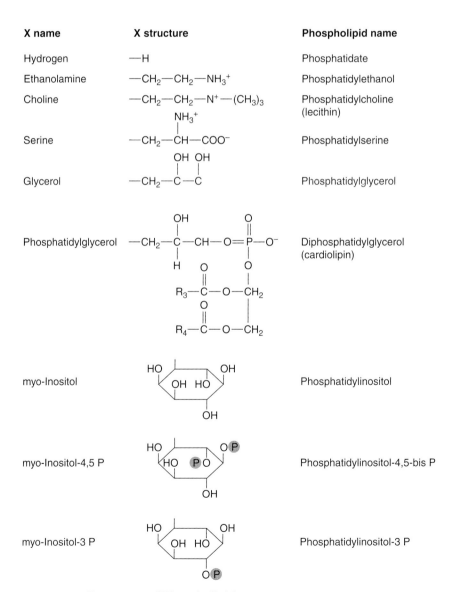

General structure:
R = fatty acyl groups

| X name | X structure | Phospholipid name |
|---|---|---|
| Hydrogen | —H | Phosphatidate |
| Ethanolamine | —$CH_2$—$CH_2$—$NH_3^+$ | Phosphatidylethanol |
| Choline | —$CH_2$—$CH_2$—$N^+$—$(CH_3)_3$ | Phosphatidylcholine (lecithin) |
| Serine | —$CH_2$—$CH$—$COO^-$ | Phosphatidylserine |
| Glycerol | —$CH_2$—$C$—$C$ | Phosphatidylglycerol |
| Phosphatidylglycerol | | Diphosphatidylglycerol (cardiolipin) |
| myo-Inositol | | Phosphatidylinositol |
| myo-Inositol-4,5 P | | Phosphatidylinositol-4,5-bis P |
| myo-Inositol-3 P | | Phosphatidylinositol-3 P |

FIGURE A.2 Structures of Phospholipids.

APPENDIX

Making this substitution, we can recast Equation A.23 as:

$$k_1 * [E][S] = k_2 * [ES] + k_{-1} * [ES] \tag{A.24}$$

and factoring and rearranging constants to one side and concentrations to the other,

$$[E][S]/[ES] = (k_2 + k_{-1})/k_1 \tag{A.25}$$

A new constant for the constant terms on the right is defined as the $K_m$:

$$K_m = (k_2 + k_{-1})/k_1 \tag{A.26}$$

And thus,

$$[E][S]/[ES] = K_m \tag{A.27}$$

We need two more assumptions in order to produce an equation with easily measurable quantities, that is $[E]_{tot}$ (total enzyme) and $[S]$. The first of these is the conservation equation:

$$[E]_{tot} = \text{sum of all enzyme forms} = [E] + [ES] \tag{A.28}$$

In the present case, there are just two forms in which the enzyme can exist. The second assumption is that the initial velocity of the enzyme under steady-state conditions is the rate at which product is formed. In the mechanism we are considering (Eq. A.19), it can be seen by inspection that formation of P arises from ES, which gives us the rate:

$$v_i = \text{rate}(21) = k_2 * [ES] \tag{A.29}$$

Applying the assumptions, we can first use the conservation Equation A.28 to eliminate the variable $[E]$ from (Eq. A.27), and produce an expression for $[ES]$ that we then substitute into (Eq. A.29). Thus, we have first of all:

$$([E]_{tot} - [ES])[S]/[ES] = K_m \tag{A.30}$$

by substitution of (Eq. A.28) into (Eq. A.27). Rearranging (Eq. A.30) algebraically, we solve for $[ES]$:

$$[ES] = [E]_{tot}[S]/(K_m + [S]) \tag{A.31}$$

Substituting this into (Eq. A.29):

$$v_i = k_2 * [E]_{tot}[S]/K_m + [S] \tag{A.32}$$

Finally, we define a new term for the first portion of the right hand side of (Eq. A.32):

$$V_{max} = k_2 * [E]_{tot} \tag{A.33}$$

which is justified by the fact that initial velocity under any conditions is,

$$v_i = k_2 * [ES] \tag{A.34}$$

When $[E]_{tot}$ replaces $[ES]$ in this equation, the velocity is constant, and has reached the maximum. Substituting (Eq. A.34) into (Eq. A.33) leads to the equation in final form:

$$v_i = V_{max}[S]/(K_m + [S])$$  (A.35)

## Derivation of Equations for Reversible Enzyme Inhibition

Reversible inhibitors add one or more equilibria to the enzyme forms of the enzyme mechanism. Each of the three common cases is considered in turn.

### Competitive Inhibition

By definition, the inhibitor I binds to free enzyme E at equilibrium, defined in the direction of dissociation,

$$EI \rightleftarrows E + I$$  (A.36)

$$K_{is} = ([E] * [I])/[EI]$$  (A.37)

The presence of the new enzyme form EI requires a new conservation equation:

$$[E]_t = [E] + [ES] + [EI]$$  (A.38)

Because EI is a dead end complex, this reduces the concentration of enzyme present in the other forms, and thus must decrease the initial velocity. The derivation is similar to the uninhibited case, except that the presence of the extra enzyme form means that equations (Eq. A.36) and (A.37) are used in order to eliminate all enzyme terms from the equation apart from ES; otherwise the derivation proceeds as before. Rearranging Equation A.36,

$$[EI] = ([E] + [I])/K_{is}$$  (A.39)

Substituting (Eq. 38) into (Eq. 37),

$$[E]_t = [E] + [ES] + ([E] + [I])/K_{is}$$  (A.40)

Rearranging and solving for [ES] as before, with the same definition of $V_{max}$ produces the final equation,

$$v_i = V_{max}[S]/\{K_m(1 + [I]/K_i) + [S]\}$$  (A.41)

### Uncompetitive (Anticompetitive) Inhibition

The derivation for the uncompetitive case is similar to that of competitive inhibition. In this case, the extra equilibrium involves the ES complex:

$$ESI \rightleftarrows ES + I$$  (A.42)

$$K_{ii} = ([ES] * [I])/[ESI]$$  (A.43)

The conservation equation then becomes,

$$[E]_t = [E] + [ES] + [ESI] \qquad (A.44)$$

The derivation proceeds as before: these new equations are used to first eliminate the [E] term and then solve for [ES]. The final equation governing uncompetitive inhibition is:

$$v_i = V_{max}[S] / \{K_m + [S](1 + [I]/K_{ii})\} \qquad (A.45)$$

## Noncompetitive (Mixed) Inhibition

In the final case, we have a mixture of competitive and uncompetitive. Both equilibria already introduced (Equations A.35 and A.41) occur, as both E and ES reversibly bind I. The conservation equation is,

$$[E]_t = [E] + [ES] + [EI] + [ESI] \qquad (A.46)$$

and the resulting equation for noncompetitive inhibition is:

$$v_i = V_{max}[S] / \{K_m(1 + [I]/K_{is}) + [S](1 + [I]/K_{ii})\} \qquad (A.47)$$

## The Problem with Double Reciprocal Plots

It is common to represent enzyme inhibition by showing double reciprocal plots, as the pattern of each type is distinctive (**FIGURE A.3**). Once the reciprocal is taken of the equations developed in the section on Reversible Enzyme Inhibition above, plotting $1/v_i$ versus $1/[S]$ at various concentrations of inhibitor [I] produces a pattern of lines that intersect at the $1/v_i$ axis for the competitive case, parallel lines for the uncompetitive case, and lines that intersect to the left of the $1/v_i$ axis for the mixed inhibition case.

However, the interpretation that these patterns *define* enzyme inhibition, while widespread, is incorrect. They are used as data presentation in kinetic experiments and can guide interpretation, but

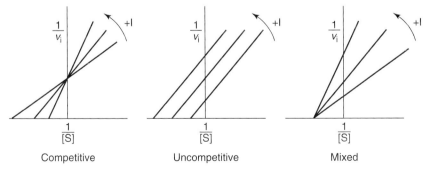

Competitive          Uncompetitive          Mixed

**FIGURE A.3 Double Reciprocal Plots.** Three patterns of double reciprocal forms for reversible enzyme inhibition are indicative of: lines intersecting at the y-axis ($1/v_i$ axis), competitive; parallel lines, uncompetitive; lines intersecting to the left of the y-axis ($1/v_i$ axis), mixed or noncompetitive.

the patterns themselves are not mechanistic. Rather, the direct plots, as shown in the main text, clearly show inhibition that is evident at low [S] (competitive), high [S] (uncompetitive), or both (mixed). The appeal of the double reciprocal plots are that they are:

a. linear
b. distinct patterns
c. traditional

Linearity was critically important prior to the advent of computers. Today, kinetic data are derived from fitting of data to the direct, nonlinear data. Even the originators Lineweaver and Burk pointed out in 1934 that using the linear transforms to extract data was error-prone: the worst data (smallest numbers) get the most emphasis in the reciprocal.

Beyond this, it is not always obvious when lines are really parallel or not, or when they are exactly intersecting on the $1/v_i$ axis. A simpler means is to just compare values of $K_{is}$ and $K_{ii}$ to specify the relative contributions of inhibition at low and high substrate. The modern use of double reciprocals is to replot data strictly for presentation purposes after it has been calculated and evaluated analytically.

## The Lipid–Water Interface

Enzymes with lipid substrates are often themselves water soluble, but act at the lipid–water interface. Because most of our thinking about enzyme action assumes that all reaction components are in the same aqueous phase, a departure from this viewpoint is required to understand interfacial catalysis.

The concept of interfacial catalysis is already implicit in the operation of membrane transporters, which take a substance from one aqueous phase, pass it through the membrane, and deposit it into a separate aqueous phase. However, this situation requires no new kinetic treatment; transport kinetics are indistinguishable from enzyme kinetics.

With lipid enzymes that act at an interface, the situation is different. We will consider the example of phospholipase $A_2$, the enzyme most commonly used for analysis of the general problem of lipid–water interface. A major characteristic of this enzyme is burst kinetics. **FIGURE A.4** compares the kinetics of hydrolysis of phospholipase $A_2$ with an esterase enzyme, such as acetylcholine esterase. Notice the unusual shape of the phospholipase catalyzed reaction: rather than a monotonic increase, there is a small rise in activity that nearly levels off, and then at a point marked CMC, an abrupt rise in velocity until a saturation point is reached.

This curve can be explained in the following way. The first region, below CMC, represents Michaelis-Menten type kinetics in which individual substrate molecules interact with the enzyme and it becomes nearly saturated in the typical way. However, once the lipid molecules reach the point where a cluster is formed—the critical micelle concentration (CMC)—the reaction shifts to essentially a new, secondary enzymatic activity. This is because the substrate has changed. Rather

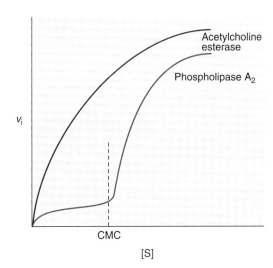

**FIGURE A.4** **Kinetics of Enzymes at an Interface**. Compared to a water soluble enzyme like acetylcholine esterase, the kinetic curve for the interfacial acting enzyme phospholipase $A_2$ has two regions. Below the critical micelle concentration (CMC), the enzyme resembles a normally saturated reaction. When CMC is reached, velocity rises sharply with substrate concentration as a new surface is created for the enzyme to bind to; this appears as *burst kinetics.*

than being an individual lipid molecule, it is now in the form of a micelle. Now the enzyme binds a new surface and has far greater activity. Thus, the figure can be viewed as two separate, sequential saturation curves.

# 7. Vitamins and Cofactors

Vitamins are human dietary essentials that mostly serve as enzyme cofactors. These are listed in Table A.2. Note that the majority are named as a B vitamin with a subscript, largely in the order of their discovery. Often in the popular literature (or sales brochures for vitamins) this grouping is designated as the *B complex*, but this is strictly a clerical grouping. There is no chemical association or even relationship among the B vitamins, apart from the fact that they are all water soluble. Table A.2 also includes three entries for cofactors that are not derived from vitamins and can be synthesized by humans: ubiquinone, lipoate, and menaquinone. Often these are included in commercial dietary supplements. The last compound is menaquinone, also known as Vitamin $K_2$, produced by resident colon bacteria. $K_2$ provides about half of the daily requirement for Vitamin K (the carboxylation cofactor), the other half being provided in the diet as Vitamin $K_1$. Corresponding structures are displayed in FIGURES A.5 through A.8.

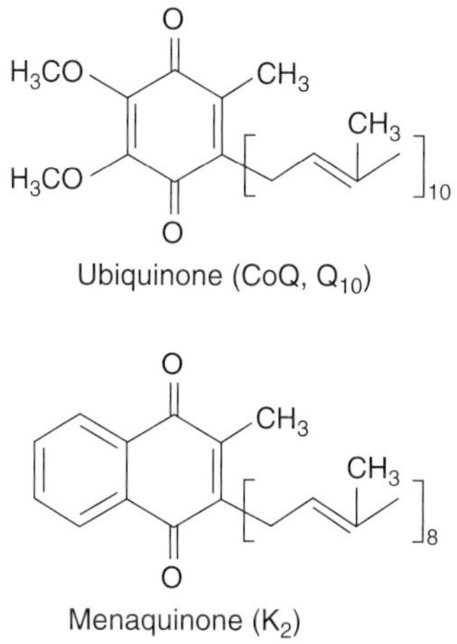

Ubiquinone (CoQ, $Q_{10}$)

Menaquinone ($K_2$)

FIGURE A.5 **Quinones.**

| TABLE A.2 Vitamins and Cofactors | | | | |
|---|---|---|---|---|
| **Vitamin** | | **Cofactor** | **Type** | **Usage Example** |
| Ascorbate | C | Ascorbate | Mobile | Collagen synthesis, maturation |
| Biotin | $B_7$ | Biotin | Bound | Pyruvate carboxylase |
| Carotene | A | Retinal, retinol | Bound | Rhodopsin (visual cycle) |
| Cholecalciferol | D | Vitamin $D_3$ | Mobile | Steroyl hydroxylase |
| Cobalamin | $B_{12}$ | Cobalamin | Bound | Methylmalonyl mutase |
| Folate | $B_9$ | Dihydrofolate, Tetrahydrofolate | Bound | Serine hydroxymethyl transferase |
| Niacin | $B_3$ | $NAD^+$, $NADP^+$ | Mobile | Lactate dehydrogenase |
| Pantothenate | $B_5$ | Coenzyme A | Mobile | Carnitine acyltransferase |
| Phylloquinone | $K_1$ | Phylloquinone | Mobile | Gamma-glutamyl carboxylase |
| Pyridoxine | $B_6$ | Pyridoxal P | Bound | Transaminase |
| Riboflavin | $B_2$ | FAD, FMN | Bound | Succinate dehydrogenase |
| Thiamine | $B_1$ | Thiamine PP | Bound | Pyruvate dehydrogenase complex |
| Tocopherol | E | Tocopherol | — | Unknown |
| — | — | Ubiquinone (Q) | Mobile | Succinate dehydrogenase |
| — | — | Lipoate | Bound | Pyruvate dehydrogenase complex |
| — | — | Menaquinone ($K_2$) | Mobile | Gamma-glutamyl carboxylase |

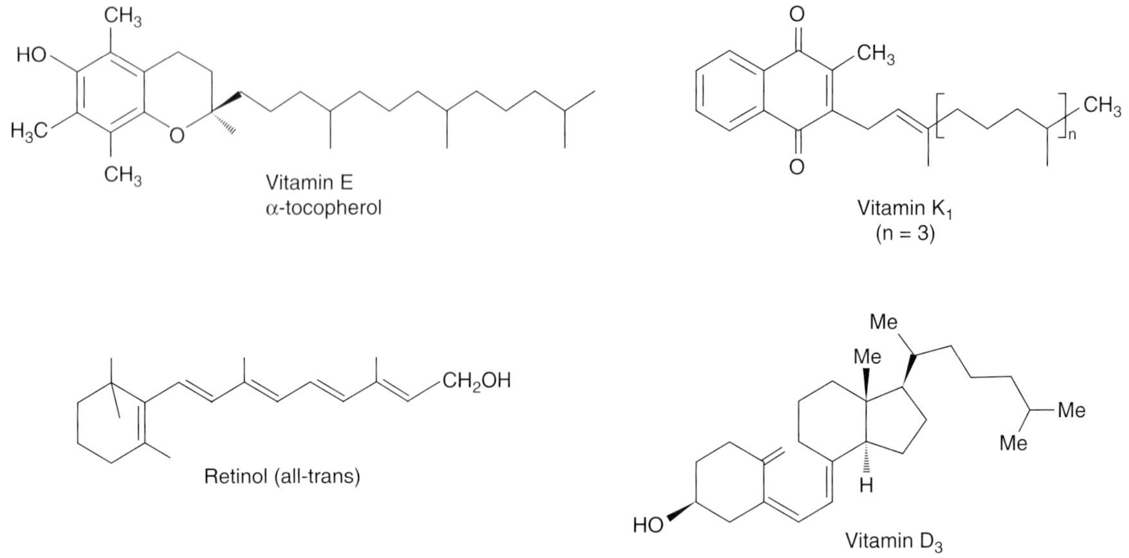

**FIGURE A.6 Fat Soluble Vitamins.**

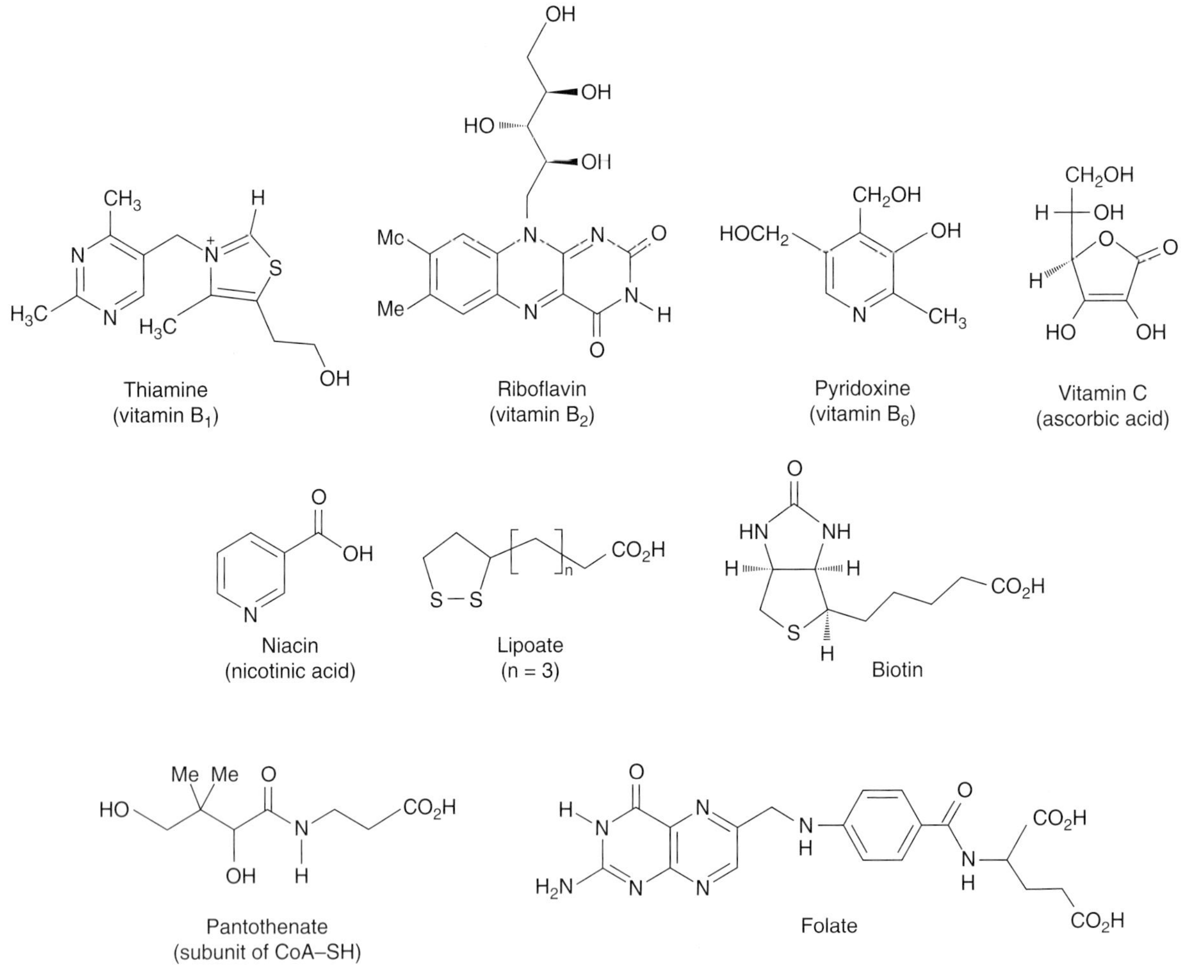

Vitamin E
α-tocopherol

Vitamin K₁
(n = 3)

Retinol (all-trans)

Vitamin D₃

**FIGURE A.7 Water Soluble Vitamins.**

Thiamine
(vitamin B₁)

Riboflavin
(vitamin B₂)

Pyridoxine
(vitamin B₆)

Vitamin C
(ascorbic acid)

Niacin
(nicotinic acid)

Lipoate
(n = 3)

Biotin

Pantothenate
(subunit of CoA–SH)

Folate

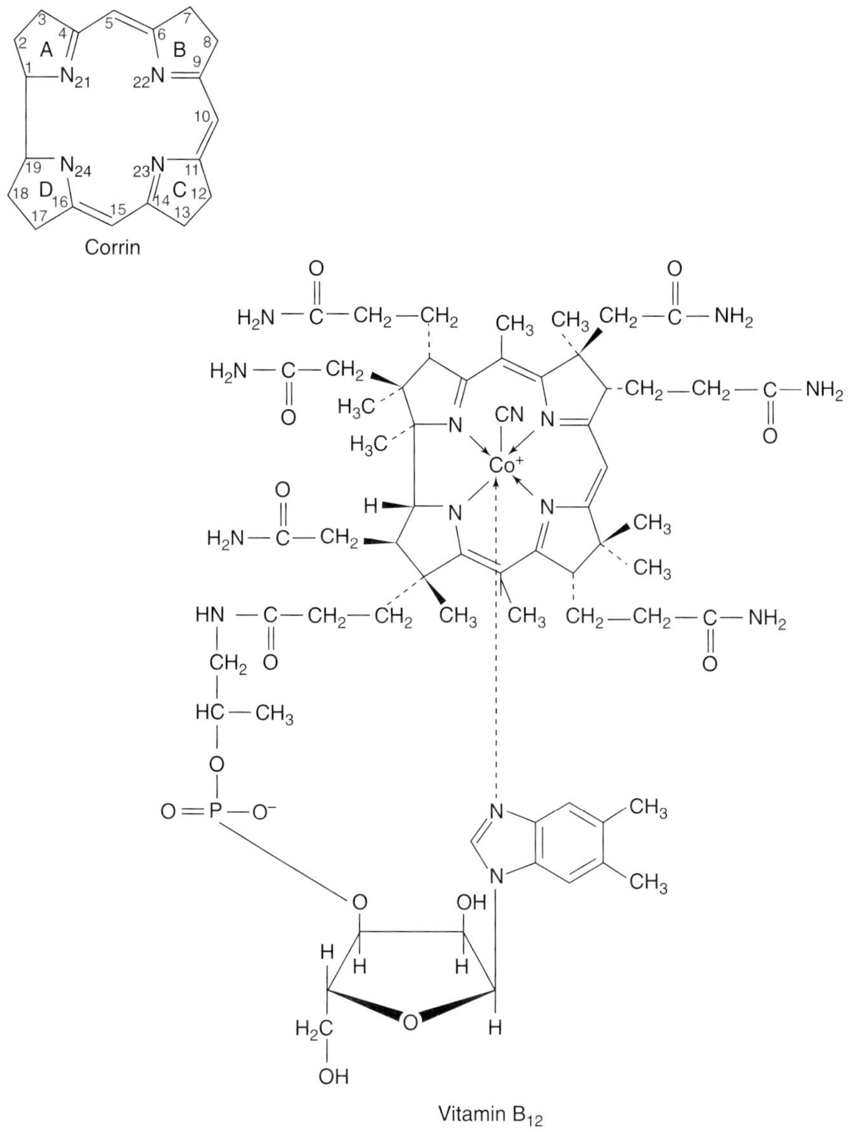

Corrin

Vitamin B$_{12}$

FIGURE A.8  **Vitamin B$_{12}$.**

# 8. The RS System of Stereochemistry

An unambiguous method for assignment of stereochemical centers is the RS system, which is more generally applicable than the DL system described in the text. While essential for naming certain compounds, it is less widely used in introductory treatments, or in specific research areas of biochemistry such as sugars and amino acids.

Consider for example the molecule isocitrate. It is not readily compared to the reference compound D-glyceraldehyde that is required to assign the stereoisomers as either D or L. Clearly, a distinct method is required that does not require a reference compound.

For the RS system, we start with a chiral center as in FIGURE A.9, where the substituents are a, b, c, and d. We place these groups attached

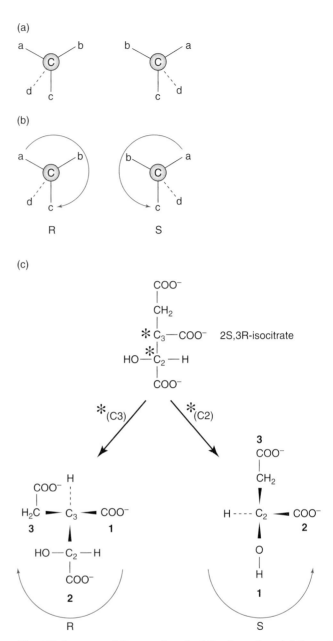

FIGURE A.9 **The RS System of Stereochemical Designation.** (a) Enantiomers of a molecule with a chiral carbon and four distinct substituents a, b, c, and d. (b) Positioning from highest [a] to lowest [d] molecular weight, with the lowest attached species placed behind the plane of the paper, tracing in the clockwise direction is the R configuration; tracing in the counterclockwise direction is the S configuration. (c) Stereochemistry of 2R, 4S isocitrate.

to the chiral carbon in the order of their molecular weights, so that a > b > c > d. The tetrahedron is arranged so that the three largest substituents project above the plane of the page, and the smallest behind it. The two possible arrangements (enantiomers) are shown in Figure A.9a. We next imagine that the three projecting substituents (a, b, and c) are connected by drawing an arc connecting them from a to b to c (Figure A.9b). If the arc is clockwise as in the figure on the left, the molecule is defined as R (Latin, *rectus*). If the arc is counter-clockwise, as in the figure on the right, the molecule is defined as S (Latin, *sinister*).

There are a few more rules needed to assign some chiral compounds. First, it is possible that groups directly attached to the chiral compound are the same. In this case, we move further down the substituent in order to assess the weight of the next substituent. Secondly, it is possible that a substituent has multiple bonds. In this case, the atom across the double bond is counted twice (double bond) or three times (triple bond).

As an example, consider the stereochemical assignment of the isomer of isocitrate, an intermediate of the Krebs cycle. Two chiral centers (at C2 and C3) exist, and their orientation is illustrated in Figure A.9c. Thus, the full designation for this compound is 2R,3S isocitrate.

Like the DL system, there is no relationship between the optical activity of a molecule and its RS assignment. Finally, the two systems also are similar in that they are formal assignments that allow us merely to classify and name a specific molecule.

# 9. Reactive Oxygen Species

It is established that mitochondria produce reactive oxygen species (ROS), which are relatively unstable forms of molecular oxygen, such as superoxide ($O_2^{\cdot-}$). Virtually all mitochondrial ROS production arises from Complex I, although there remains some debate about the exact mechanism. Despite the existence of several Fe/S clusters within Complex I, none of them seem to be able to react directly with molecular oxygen. This is due to their presence deep inside the hydrophobic interior of the complex, which suggests a reason for the complex itself: shielding the intermediates from alternative reactions. We have previously noted this feature of protein complexes, such as the pyruvate dehydrogenase complex: side reactions are excluded.

The reaction with oxygen can occur at the binding site of NADH, or with that of ubiquinone. It is known that the semiquinone form of FMN can react with oxygen. While electrons from succinate can flow backwards *in vitro,* and even larger amounts of ROS are formed in this manner, such reversed flow is unlikely *in vivo.*

The reaction that produces superoxide at the cytosolic face of the mitochondria catalyzes the reaction:

$$Q^{\cdot-} + O_2 \rightarrow Q + O_2^{\cdot-} \tag{A.48}$$

However, there are two separate species of $Q^{\cdot-}$ and only the one on the cytosolic face reacts with $O_2$. The cause of the distinction is the membrane potential. Due to the charge difference across the membrane, the reduction potential of cytosolic-facing $Q^{\cdot-}$ is greater than the matrix-facing $Q^{\cdot-}$ and so superoxide is generated from $Q^{\cdot-}$ only in the cytosolic compartment.

The greater the membrane potential, the greater the reduction potential of the cytosolic $Q^{\cdot-}$ and the more $O_2^{\cdot-}$ is produced. The other site of $O_2^{\cdot-}$ formation is at Complex I. While details are unknown, the superoxide is produced in the matrix space, and insensitive to changes in membrane potential.

$O_2^{\cdot-}$ is broken down in slightly different ways in the two spaces. In the cytosol, $H_2O_2$ is formed in a reaction catalyzed by superoxide dismutase. Subsequently, the $H_2O_2$ is converted to water, catalyzed by peroxidase, an enzyme largely confined to its own vesicle, the peroxisome. In the mitochondrial matrix, the first step is similar: a mitochondrial superoxide dismutase catalyzes $H_2O_2$ formation. The next step, however, is catalyzed by glutathione peroxidase:

$$GSH + H_2O_2 \rightarrow GSSG + H_2O \qquad (A.49)$$

Glutathione is a cellular thiol, a tripeptide of glutamate–cysteine–glycine, in which glutamate uses its side-chain carboxyl in amide bond. The GSH designation emphasizes the central –SH group; when these are combined in the reduced form, the oxidized glutathione is composed of two molecules, written as GSSG to emphasize the disulfide bond.

Regeneration of the reduced form, GSH, requires reducing power in the form of NADPH, and the enzyme glutathione reductase, which catalyzes the reaction,

$$GSSG + NADPH \rightarrow GSH + NADP^+ \qquad (A.50)$$

The hydride for NADPH is derived from NADH, through the mitochondrial energy-linked transhydrogenase:

$$NADH + NADP^+ \rightarrow NAD^+ + NADPH \qquad (A.51)$$

There is also a glutathione reductase system in the cytosol, reducing GSSG to GSH in part for the purpose of maintaining cytosolic enzymes in the largely–SH form.

ROS are a current medical concern as strong correlative evidence suggests they are a key factor the aging process, as well as many pathologies. However, the origin of most ROS stems from cells of the immune system in mammals, and mitochondrial ROS physiologically may be of little importance for most other cells. Still, tissues with very active oxidative metabolism and many mitochondria, such as heart and skeletal muscle, may contribute to overall ROS production in humans.

# 10. Enzyme Mechanisms

## Aconitase

An alternative substrate for citrate synthase is fluoroacetyl-CoA, produced from the poison fluoroacetate ($CH_2FCOO^-$) through the thiokinase reaction (FIGURE A.10). The mechanism is analogous to

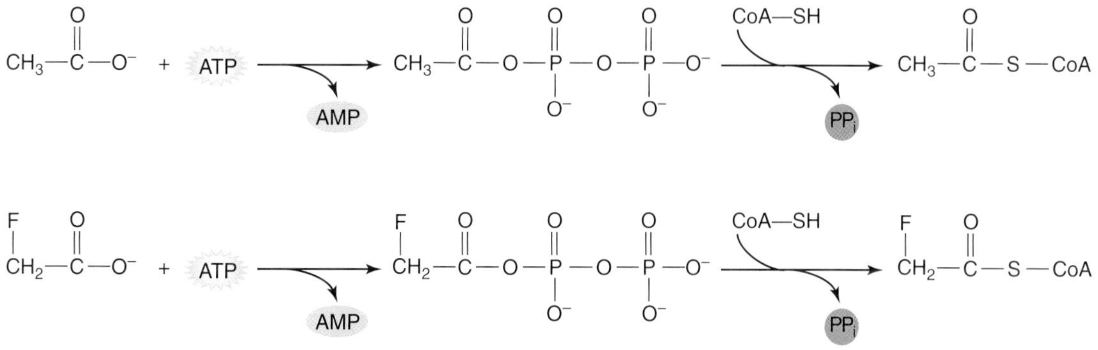

**FIGURE A.10 Thiokinase Reactions with Acetate and Fluoroacetate.**

asparagine synthetase, in that the first step is a nucleophilic attack of the carboxylate on the β-phosphate of ATP, producing AMP as a product and a carboxy-pyrophosphate intermediate. The other step is a nucleophilic displacement by the sulfur of CoA to the carbonyl, releasing pyrophosphate and producing acetyl-CoA. The fluoride substitution does not prevent formation of the CoA ester or the condensation reaction to fluorocitrate. Aconitase, the next Krebs cycle reaction, can react with fluorocitrate citrate as well. Examination of this mechanism in parallel to reaction with citrate as substrate helps elucidate the mechanism of aconitase, the toxicity of fluoroacetate, and the unusual stereochemistry of the Krebs cycle.

The orientation of citrate in the first enzyme substrate complex is 180° rotated ("flipped") compared to the orientation of *cis*-aconitate in the second enzyme-substrate complex.

In the diagram, the original acetyl-CoA carbons (or fluoroacetyl-CoA carbons) are labeled with asterisks. The F atom at carbon 2 serves as a "built-in label."

FIGURE A.11 presents the mechanism of aconitase with citrate as substrate (left side) and fluorocitrate as substrate (right side). The hydrogen transferred to the enzyme and back is indicated in blue. Formation of the *cis*-aconitate intermediate is the same in each case (A). The substrate then flips its orientation to the enzyme—by means currently unknown—for both intermediates (B). Finally, addition of the hydroxyl group (C) to one side of the double bond, leads to abstraction of the proton from the enzyme to the other side, producing isocitrate (left panel). In the fluoride-containing case, the reaction is different. Here, rearrangement of the electrons from the double bond leads to splitting of the C—F bond, because F⁻ is a good leaving group. This produces a molecule with a double bond and an alcohol that stays bound to the enzyme. The free enzyme is not regenerated; the protonated form is locked in. Thus, aconitase is irreversibly inhibited and the Krebs cycle is blocked.

The path for fluorinated citrate supports the mechanism for citrate itself and shows that a new stereochemical center can be formed from an achiral molecule.

## Prenyltransferase

This enzyme is responsible for the condensations of 5 and 10 carbon branched chain lipids. In the example of FIGURE A.12, dimethylallyl-PP

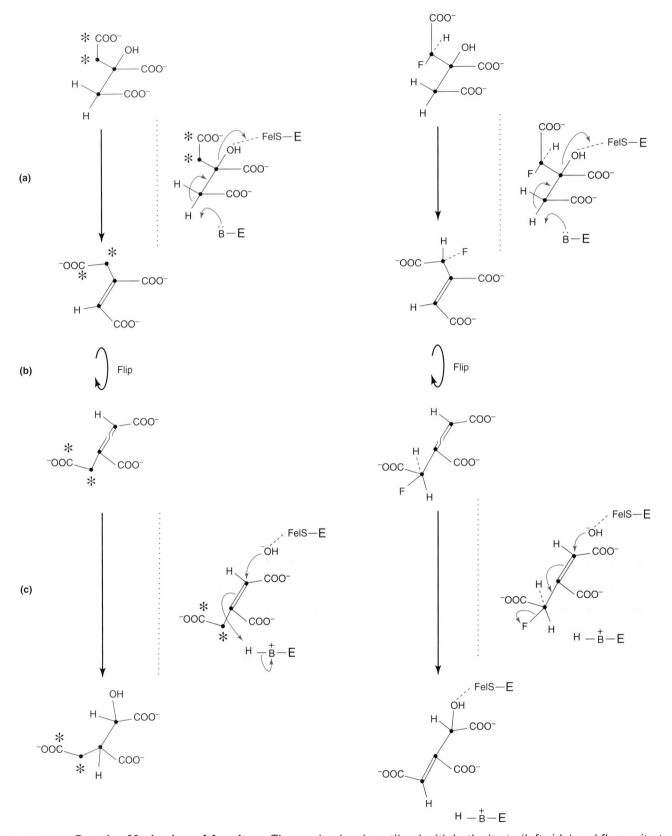

**FIGURE A.11 Reaction Mechanism of Aconitase.** The mechanism is outlined with both citrate (left side) and fluorocitrate (right side). Both involve proton abstraction with rearrangement (a) and a flipping of the substrate on the enzyme face (b). In (c), the OH⁻ attack leads to the isocitrate product (left), but to an enzyme-bound dead-end complex with the fluorinated intermediate.

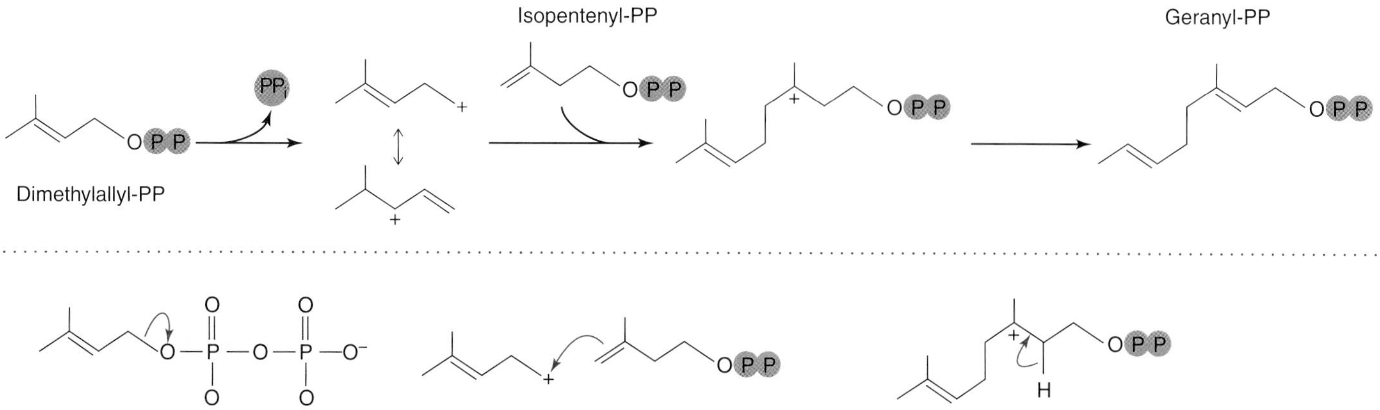

**FIGURE A.12 Mechanism of Prenyltransferase.** Loss of pyrophosphate is stabilized by the resonance forms of the intermediate cation. The latter reacts with isopentenyl-PP to form a 15C cation intermediate that attracts the electrons of a neighboring C—H bond, forming the geranyl-PP product.

reacts with isopentenyl-PP to form the 10C product geranyl-PP. In the first step of the mechanism, the electrons from the bond to the pyrophosphate migrate to the oxygen, releasing $PP_i$. This is possible because of the resonance-stabilized carbocation product; both resonance forms are shown. Next, the pi electrons of the isopentenyl-PP attach to the carbocation to form the 10C carbocation adduct. Migration of electrons from the C—H bond creates the second double bond of the product, geranyl-PP.

The carbocation intermediate is not a common one in metabolic reactions, although it appears prominently in glycogen metabolism in both of the key reactions: glycogen synthetase and glycogen phosphorylase.

# 11. Metabolic Pathways

## Glyoxylate Cycle

The glyoxylate cycle is a means of converting acetyl-CoA to glucose, a pathway that does not occur in animals but does in some bacteria, yeast, and plants. The bacterial route is shown in FIGURE A.13, as a bypass of the Krebs cycle.

Two enzymes unique to the glyoxylate pathway are boxed in the figure: isocitrate lyase, which catalyzes hydrolysis of isocitrate to glyoxylate and succinate; and malate synthase, which catalyzes condensation of acetyl-CoA with glyoxylate to form malate. Note the parallels of these enzymes to citrate lyase and citrate synthase, respectively.

When the two enzymes are induced in bacteria, they enable a net conversion of acetyl-CoA to glucose. One acetyl-CoA condenses with oxaloacetate to form citrate; a second acetyl-CoA condenses with glyoxylate to form malate. This malate can then be converted to glucose.

An example of the medical significance of this pathway is the case of the tuberculosis bacteria, which survive metabolic isolation resulting from phagocytosis in the lysosome by induction of the glyoxylate cycle. This greatly improves bacterial metabolic flexibility, as now glucose can be formed from acetyl-CoA. This enables the bacteria to

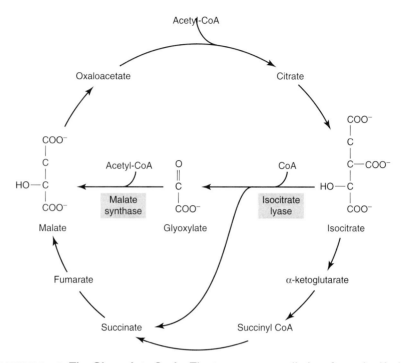

**FIGURE A.13 The Glyoxylate Cycle.** The two enzymes distinct from the Krebs cycle are in boxes; together, isocitrate lyase and malate synthase bypass the Krebs cycle and produce net four-carbon intermediates rather than $CO_2$.

resist the innate immune system. The glyoxylate cycle is also present in yeast and plants; however, the pathway is more complex in these organisms as the two specific glyoxylate cycle enzymes exist exclusively in a separate organelle, the glyoxysome. Thus, in these organisms the pathway spans the glyoxysome, the cytosol, and the mitochondria.

## Odd Chain Number Fatty Acids

When an odd number chain fatty acid undergoes beta oxidation, the final product is proprionyl-CoA rather than acetyl-CoA. As indicated in FIGURE A.14, proprionyl CoA can be converted to methylmalonyl CoA through a carboxylation reaction catalyzed by proprionyl-CoA carboxylase. This enzyme, which uses ATP and contains a bound biotin, has the same reaction mechanism as pyruvate carboxylase. The carboxylation product shown in Figure A.14 is the S isomer, which is then converted to the R isomer utilizing a racemase enzyme. Finally, a complex reaction utilizing Vitamin $B_{12}$ (cobalamin) converts R-methylmalonyl-CoA to succinyl-CoA. The reaction is a carbon migration that involves an intermediate carbon-cobalt bond. The cobalt ion is chelated in turn to the $B_{12}$ cofactor, which is tightly attached to the mutase enzyme.

In this manner, the small amount of dietary odd carbon chain number fatty acids can be converted to succinyl-CoA and (mostly) acetyl-CoA. The succinyl-CoA product can be converted to glucose if the oxidation occurs in the liver. Otherwise, it is possible, using reactions of the Krebs cycle to convert this first to malate, and then to pyruvate through malic enzyme; the pyruvate can then be completely oxidized through the pyruvate DH complex and the Krebs cycle.

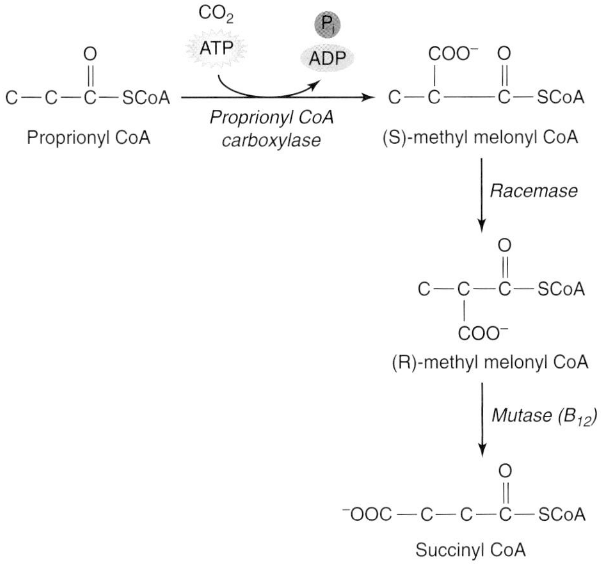

FIGURE A.14 **Key Steps in Odd Chain Fatty Acid Oxidation.**

## Later Steps in Cholesterol Synthesis

From the formation of the 30C squalene to the 27C cholesterol are over 20 enzymatic steps. The first few are the cyclization of the open chain squalene to the first four-ring form lanosterol (FIGURE A.15). This is accomplished in two enzymatic steps, with the electron rearrangements leading to ring cyclization and incorporation of oxygen into the molecule indicated in the figure. Electrons rearrange in a concerted fashion to close four rings, creating a fused ring system consisting of three six-membered and one five-membered form that persists throughout cholesterol synthesis as well as in modified sterols including the steroid hormones and the bile salts.

Most of the remaining 19 enzymatic reactions are transformations of this ring structure; a few of the intermediates are illustrated in FIGURE A.16. From lanosterol, the ring is slightly modified, for example

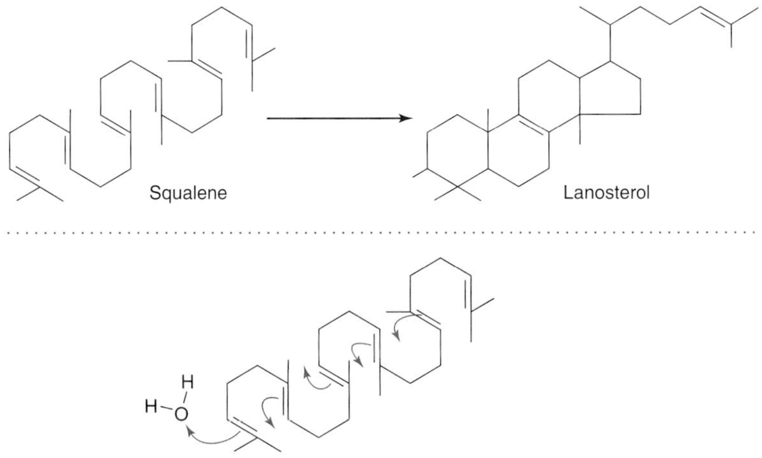

FIGURE A.15 **Cyclization of Squalene to Lanosterol.** Fused rings are formed simultaneously through concerted electron rearrangement.

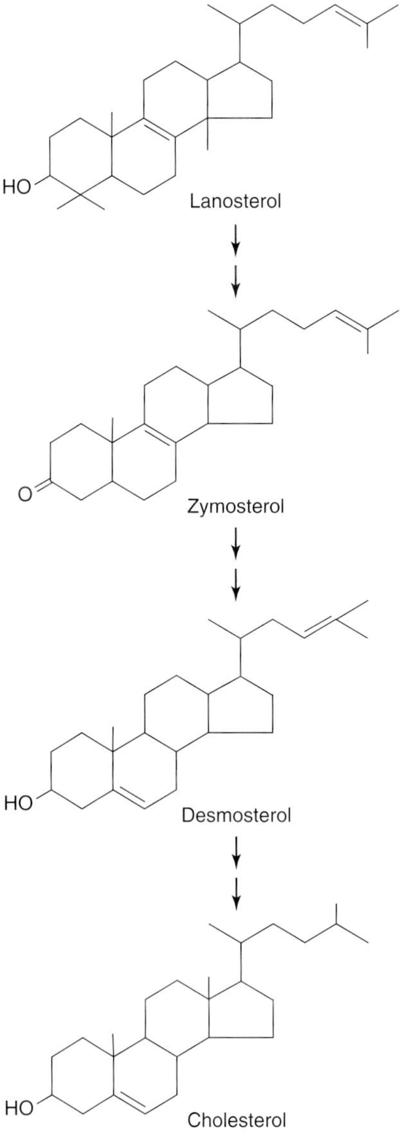

FIGURE A.16 **Final Steps in Cholesterol Synthesis from Lanosterol.**

| TABLE A.3 Enzymes of Purine Metabolism | |
|---|---|
| Step | Enzyme |
| a | Aminophosphoribosyltransferase |
| b | GAR synthetase |
| c | RAG transformylase |
| d | FGAM synthase |
| e | AIR synthetase |
| f | AIR carboxylase |
| g | SAICAR synthetase |
| h | adenylosuccinate lyase |
| i | AIRCAR transformylase |
| j | IMP cyclohydrolase |

in the intermediate formation of a ketone and decarboxylation to remove three methyl groups to form zymosterol, the formation of the correct ring double bond in desmosterol, and the reduction of the chain on the D ring (the 5-membered ring) in the product cholesterol. Nearly all of the steps in this pathway use NADPH as a reductant or co-reductant.

## The Purine Biosynthesis Pathway

The pathway from PRPP to IMP is shown in FIGURE A.17. Note that the steps (see Table A.3) already shown in Chapter 14 are indicated in numbers in circles. These are repeated here; the complete pathway is shown.

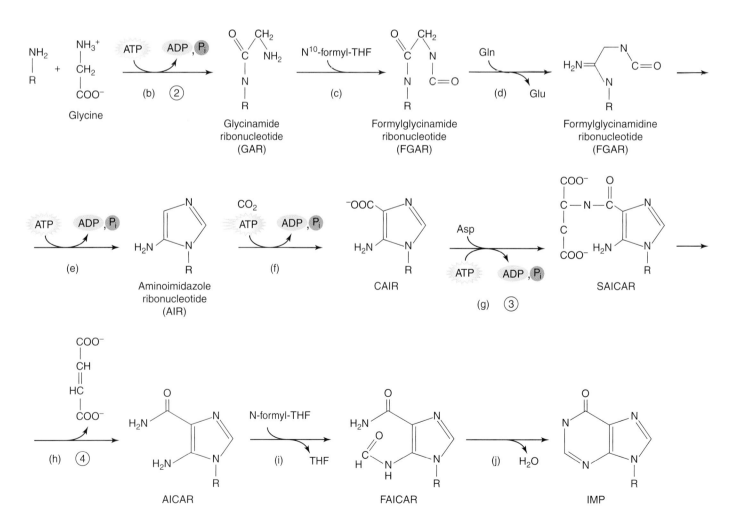

FIGURE A.17 Detailed Pathway for Purine Biosynthesis.

# References

1. Andreadis, E. A.; Katsanou, P. M.; Georgiopoulos, D. X.; et al. The effect of metformin on the incidence of type 2 diabetes mellitus and cardiovascular disease risk factors in overweight and obese subjects. The Carmos Study. *Exp. Clin. Endocrinol. Diabetes* 117:175–180, 2009.
2. Lineweaver, H.; Burk, D. The determination of enzyme dissociation constants. *J. Am. Chem. Soc.* 56:658–666, 1934.
3. Lorenz, M. C.; Fink, G. R. Life and death in a macrophage: Role of the glyoxylate cycle in virulence. *Euk. Cell* 1:657–662, 2002.
4. Maor, E. *The Story of a Number.* Princeton University Press: Princeton, NJ, 1994.

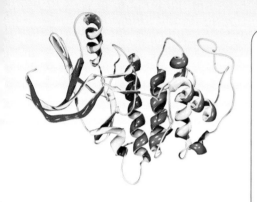

# Glossary

## A

**α-carbon** — Carbon adjacent to a carboxyl group; often referred to as the carbon bearing a carboxyl group and amino group in an amino acid.

**acceptor site** — During protein synthesis, the location of the ribosomal binding site for aminoacyl-transfer RNA (also termed *A site*).

**acceptor stem** — Located at the opposite end of the anticodon on a transfer RNA, the 3′ end with a CCA sequence, where the appropriate amino acid bonds with the terminal end adenine (A) residue.

**active site** —A specific region of an enzyme where substrates bind and reaction catalysis takes place.

**acyl carrier protein (ACP–SH)** — A small protein containing a free –SH from a cysteine residue that forms thioester intermediates in fatty acid biosynthesis.

**affinity chromatography** — A form of chromatography in which the stationary phase contains molecules that selectively combine with components of the mobile phase.

**agar** — A mixture of agarose and agaropectin with a gelatinous consistency that can support growth of microorganisms for *in vitro* analysis.

**agaropectin** — A polysaccharide similar to agarose, except that it has branched sugars and sulfuric acid esters to some of its hydroxyl groups.

**agarose** — A polysaccharide found in certain seaweeds, it has alternating galactosyl and anhydro-L-glucosyl subunits.

**α-hydrogen** — The hydrogen attached to the α-carbon of an amino acid.

**aldoses** — Sugars containing an aldehyde (at carbon 1).

**alkaloids** — A class of naturally-occurring nitrogenous organic compounds found in plants, some fungi, and animals. Examples: caffeine, morphine, nicotine, quinine, and strychnine.

**allopurinol** — A synthetic drug (a hypoxanthine analog) used to treat gout through its inhibition of xanthine oxidase and thus uric acid formation.

**alpha (form)** — The hydroxyl group of an anomeric carbon that is on the same side of the ring as the hydroxyl group that determines the D-form of most monosaccharides.

**amide** — A bond formed (formally by dehydration) between an amine and a carboxyl group.

**amino acids** — The building blocks of proteins and intermediates in metabolism, all contain an amine group and a carboxyl group attached to the α-carbon.

**amino-acyl-tRNA synthetases** — Enzymes that selectively recognize and bind an amino acid to a matching transfer RNA.

**amino end** — The end of a protein or peptide that terminates in a free α-amine (also, the *N end*).

**amphipathic** — Having both polar and nonpolar parts within a single molecule. Example: fatty acids.

**anabolic steroid** — A synthetic hormone that promotes protein synthesis and cell growth, sometimes used by athletes to increase muscle size and strength.

**annihilation** — A mechanism of formation of the mitochondrial proton gradient in which protons are removed from the matrix space through their reaction with oxygen to form water.

**anomeric carbon** — In a sugar ring, the carbon bearing the hydroxyl group that was a carbonyl in the open chain form.

**antennae** — In photosynthesis, a large pool of chlorophyll molecules that absorb light.

**anticodon** — Located on one end of transfer RNA, a sequence of three adjacent nucleotides. Functions during the translation phase of protein synthesis to bind to the complementary coding triplet of nucleotides in messenger RNA (see *codon*).

**anticompetitive inhibition** — Occurs when an enzyme inhibitor only binds to enzyme substrate (ES) complex (see *uncompetitive inhibition*).

**antiparallel** — In a β-sheet of an amino acid, when the strands run in opposite directions.

**aquaporins** — Protein channels selective for water transport.

**apoprotein** — The protein portion of a larger complex, formed with a prosthetic group.

**assay** — Quantitative or qualitative procedure to analyze a substance.

**assimilation** — The incorporation of nutrients into living tissue.

**autophagy** — Lysosomal digestion of a cell's own cytoplasmic material.

**auxotropic** — Organisms requiring carbon sources other than $CO_2$.

## B

**:B** — Enzyme-bound base.

**β-barrel** — Molecular structure of an amino acid that is a closed β-sheet.

**beta (form)** — The hydroxyl group of an anomeric carbon that is on the opposite side of the ring as the hydroxyl group that determines the D-form of most monosaccharides.

**β-oxidation** — The process of oxidative degradation of saturated fatty acids, where two-carbon units are sequentially removed from the carboxyl end of the fatty acid chain.

**bile** — A mixture of bile salts, phospholipids, and cholesterol that is secreted by the liver, stored in the gallbladder, and released into the duodenum, where it forms emulsions and ultimately mixed micelles with dietary lipids to facilitate their hydrolysis.

**bile salts** — Covalent combination of a bile salt with a base in an amide linkage, such as cholate with glycine (glycocholate).

**binding change** — A mechanism for complex V ($F_1F_o$ATPase), proposed by Paul Boyer, that a protein segment of the complex moves among the three adenine nucleotide binding sites, driven by proton entry.

**biopterin** — A pterine (heterocycle) redox cofactor that can transfer single electrons as well as electron pairs.

**biotin** — A carboxylation cofactor in which an intermediate $CO_2$ is transiently attached to a nitrogen atom of this molecule.

**Bohr effect** — The diminished binding of oxygen to hemoglobin in response to an increase concentration of protons.

**branch points** — Location on a polymer chain in which a chain diverges to form a new polymer end. This usually refers to polysaccharides, which have multiple hydroxyl groups, in which the branch points are often $\alpha1\rightarrow6$ linkages.

**buffering region** — The region of a titration curve in which addition of acid or base has minimal effects on the pH; visually, a flat region in the center of the curve.

**bundle sheath** — Inner plant cells that converts the four-carbon malate to pyruvate and $CO_2$; the high concentrations of the latter prevent $O_2$ from reaction at RuBisCo.

## C

**C3 plants** — Plants that directly incorporate $CO_2$ into P-glycerate (the C3) rather than intermediate conversion to a C4 for the purpose of localizing $CO_2$.

**C4 plants** — Plants with specialized pathways to protect against RuBisCo using $O_2$ as an alternative substrate to $CO_2$. The C4 refers to a four-carbon intermediate that is synthesized from PEP, carrying $CO_2$ as the fourth carbon and releasing it in a separate reaction.

**calmodulin** — A calcium-binding protein found in all nucleated cells that mediates a variety of cellular responses to calcium.

**Calvin cycle** — Cyclic pathway in photosynthetic organisms that serves to convert three $CO_2$ molecules to 3-P-glycerate.

**CAM plants** — Crassulacean *acid metabolism*. These plants have a modification to prevent $O_2$ from reaction at RuBisCo. Pores open at night to admit $CO_2$.

**cap** — A 7-methylguanosine in 5′-5′ triphosphate linkage to the first nucleotide of the mRNA; this is added posttranscriptionally and is not encoded in the DNA.

**capping** — At the terminal point of a chromosome (see *telomere*), multiple pathways and at least four molecular structures protect the ends from degradation and fusion, thus ensuring chromosome stability and preventing it from being recognized as damaged DNA.

**carbon cycle** — Global reactions of $CO_2$ between auxotropes (such as animals) and photosynthetic organisms (such as plants).

**carbon fixation** — Net incorporation of $CO_2$ into sugars and ultimately other carbon compounds in cells.

**carboxyl end** — The end of a protein or peptide that terminates in a free $\alpha$-carboxyl group (also, the *C end*).

**carnitine shuttle** — A pathway for transport of fatty acyl-CoA molecules into mitochondria that involves intermediate conversion of the thioester into a carnitine ester.

**catalytic rate constant** — Rate constant for the species directly leading to product in an enzyme mechanism. In our simple enzyme mechanism, that species is ES and the catalytic rate constant is $k_2$.

**cell theory** — A concept that cells are the fundamental unit of living systems, arising from other cells.

**cellulose** — A linear polymer of glucose that has $\beta1\rightarrow4$ linkages and a very regular three-dimensional solution structure; an insoluble polysaccharide that comprises plant cell walls, vegetable tissue fibers (e.g., cotton), and wood; the most abundant macromolecule on earth.

**change in state** — Thermodynamic variables that are defined by the difference between one condition and another undergo a change in state if the change is independent of the path taken between the two.

**chaperones** — A protein that assists folding of another protein into its active conformation, either during the synthesis of the protein or afterward.

**charged tRNA** — A tRNA with an amino acid attached (tRNA[amino acid]).

**chemiosmotic hypothesis** — The theory that protons are the essential intermediate between oxidation of substances (with attendant electron transfer) and the phosphorylation of ADP to form ATP.

**chiral** — An asymmetric carbon center in a molecule; Latin for *handedness*. A carbon atom to which four distinct groups are attached is a chiral center.

**chloroplast** — Plant cellular organelle in which photosynthesis occurs.

**chorismate** — An intermediate in the biosynthesis of aromatic amino acids. The common pathway to aromatic amino acids bifurcates at chorismate: one leads to phenylalanine and tyrosine; the other leads to tryptophan.

**chromatography** — A group of analytic techniques for purification of materials that works by different molecular affinities within the mixture for a mobile phase and a stationary phase.

**chylomicron** — A lipoprotein particle synthesized in intestinal epithelial of the small intestine that is transported by the lymphatic system to the blood and subsequently various tissues in the body, where triglyceride components are released through lipoprotein lipase activity.

**chyme** — Liquefied digest following physical processing in the stomach that exits through the pylorus to the duodenum for digestion.

**citric acid cycle** — A cyclic series of reactions that catalyzes the complete oxidation of acetyl-CoA to $CO_2$. (see *Krebs cycle, tricarboxylic acid cycle*).

**classical thermodynamics** — Thermodynamic principles developed in the 19th century based upon postulates (called "laws of thermodynamics") that have never been contradicted. The first law is energy conservation; the second postulates an increase in entropy (roughly, disorder) in the universe.

**cobalamin** — (see *Vitamin B_{12}*).

**codon** — A triplet of bases in the messenger RNA that, when bound to the ribosome, bind to the three bases of the anticodon region in transfer RNA (see *anticodon*).

**colloids** — Early name for macromolecules.

**collision theory** — A concept of reaction kinetics that the rate of a reaction is proportional to the number of collisions.

**competitive inhibition** — A form of reversible inhibition in which the inhibitor binds exclusively to the free enzyme form.

**complementarity** — The selective binding of nucleotides within a DNA or RNA chain to another nucleic acid that follows the rule that A binds T (or U, if the nucleic acid is RNA), and G binds C.

**complete pathway** — A metabolic pathway that is balanced, so that all mobile cofactors are regenerated.

**concerted** — Electron movements within a chemical mechanism that occur at the same time; a single-step reaction.

**condensed phases** — A form of matter in which the molecules (or atoms) are incompressible, such as a solid or liquid.

**conjugated** — Electron clouds of pi bonds of at least two unsaturated centers are able to interact, enabling delocalization and stabilization. Commonly applies to double bonds separated by a single methylene moiety.

**consensus (sequences)** — Similar sequences between strands of DNA in different species that have a common function. The term can also be applied to RNA or protein sequences.

**cooperativity** — A model to explain deviations from the normal Michaelis-Menten curve for enzyme activity, or the S-shaped curve for binding oxygen to hemoglobin. The concept usually involves conformational changes of individual subunits in a multi-subunit protein that subsequently alter the affinity of other subunits for substrate or ligand.

**coordination number** — The number of directed molecular bonds to electron-rich atoms.

**core** — After RNA is synthesized in transcription, the $\sigma$ of the holoenzyme $\alpha_2\beta\beta'\sigma$ dissociates, thereby producing the core complex, $\alpha_2\beta\beta'$.

**Cori cycle** — An inter-organ metabolic pathway in which glucose forms lactate in the muscle (glycolysis), and the lactate is converted to glucose in the liver (gluconeogenesis).

**coupled** — In oxidative phosphorylation, it refers to the oxidation of substrates, usually through the intermediacy of NADH, to $O_2$, tied to the production of ATP.

**coupling** — (see *coupled*).

**critical micelle concentration (CMC)** — The concentration of fatty acids (or other lipid molecules) added to a water solution that enables them to form a micelle, so that further addition will not increase the free concentration. The concept is also applied to molecules that form non-micelle aggregates such as certain phospholipids.

**cyclosporine** — A cyclic peptide (1200 Da) that causes inactivation of the protein phosphatase calcineurin and suppresses formation of the cytokine IL-2 from T cells, leading to immunosuppression. The drug is used to prevent organ rejection in transplant patients.

**cytochrome P450** — Any of a group of enzymes catalyzing mixed-function oxidation, leading to hydroxylation of xenobiotics, including drugs, which renders them more water soluble and enables their excretion from the body.

**cytosol** — The major water space within a cell.

## D

**dark reactions** — In plant photosynthesis, reactions in which $CO_2$ is converted to organic compounds, largely sugars.

**dehydrogenases** — A category of enzymes that catalyzes the exchange of hydrides between substrates and nicotinamide nucleotides ($NAD^+$ and $NADP^+$).

**denitrification** — Process of removing nitrogen or nitrogen groups from a compound or reducing nitrites and nitrates to $N_2$.

**diester** — A double ester; for example, a phosphate group esterified both to choline and to glycerol in phosphatidyl choline.

**diffusion** — Random motion of molecules that results in their redistribution from regions of high to low concentration.

**dihydrobiopterin** — A redox cofactor with electron transfer properties similar to flavins.

**dinuclear copper center** — Redox transferring copper metal ion complex that occurs in cytochrome oxidase.

**disaccharide** — Two sugars linked by a glycosidic bond.

**DNA ligase** — An enzyme that catalyzes diester formation between adjacent nucleotides in double-stranded DNA.

**DNA polymerase** — Enzymes active in the replication and repair of DNA that catalyze the synthesis of a new DNA strand from an existing DNA template.

**domain** — A protein structure that may have separate secondary structural elements that have a specific function (e.g., NADH binding).

**double helix** — The two strands of DNA polymers linked together by hydrogen bonds.

**double-reciprocal plot** — (see *Lineweaver-Burk plot*).

**down regulation** — A mechanism that decreases the number of receptors on a cell surface target, making the cell less sensitive to excess hormones or other ligands by degrading or internalizing them.

**downstream** — Direction of a nucleic acid polymer that is further away from a given point (often the start site) for transcription.

## E

**EF hand** — A 40-amino-acid–residue domain containing a helix-loop-helix design that binds $Ca^{2+}$ ions with high affinity (micromolar concentrations).

**eicosanoids** — A group of endogenous lipid signaling molecules synthesized from arachidonic acid.

**electrogenic** — An exchange process for molecules across a biological membrane that is not charge balanced.

**electronegativity** — The ability of an atomic nucleus to attract electrons in a bond to itself.

**electron paramagnetic resonance** — A type of spectroscopy in which there is resonant absorption of radiation from chemical substances with unpaired electrons. A spectrographic technique that identifies and measures free radicals.

**electron sink** — An electron deficient portion of a molecule that attracts electrons in a reaction mechanism.

**electron transport chain** — A connected series of redox reactions that exist in the mitochondrial membrane that accept electrons from NADH and donate them to $O_2$.

**electrophoresis** — A form of chromatography in which the mobile phase is driven by an electric field imposed by electrodes; the stationary phase is a gel composed of a polymer, such as polyacrylamide (see *chromatography*).

**elongation** — The process of extending a polymer, such as an RNA chain or a protein.

**emulsification** — The dispersion of insoluble lipids (e.g., triglycerides) into water solutions.

**enantiomers** — Molecules whose structures are non-superimposable mirror images.

**5′ end** — One end of a linear polynucleotide strand at which the 5′-hydroxyl group of the terminal nucleoside residue is not joined to another nucleotide, i.e., the free end.

**endergonic** — A positive change in free energy. An endergonic reaction will only be possible in the reverse direction.

**endoplasmic reticulum** — A membrane network of tubules forming an enclosed space within the cell that is active in lipid transport, vesicular transport, protein export, and $Ca^{2+}$ release to the cytosol.

**energy** — A nonmaterial driving force for reactions, classified as potential (stored) or kinetic (expressed). Different forms of energy include electrical, gravitational, heat, and mechanical.

**energy coupling** — The pairing of a reaction with positive free energy to one with even greater negative free energy. This process is often applied to partial reactions that make up a whole; in any case, it is a formalism, since every reaction in a metabolic pathway must have a negative free energy change.

**enthalpy** — A heat change for a system at constant pressure; a defined energy change for thermodynamic systems. The enthalpy change takes into account largely bond energies in reactions.

**enthalpy driven** — A chemical reaction in which the free energy change is dominated by the enthalpy term ($\Delta H$).

**entropy** — An expenditure of energy that is unavailable to apply to work. The increased entropy is due to an increased dispersion of a system. For example, when two bonded atoms are split apart, the reaction entropy increases.

**entropy driven** — A chemical change in which the free energy change is dominated by the entropy term ($\Delta S$).

**enzyme (E)** — A biological catalyst that lowers the activation energy for a reaction without affecting the equilibrium position and can do so with minimal changes in temperature.

**enzyme activity** — Measure of an enzyme function in units of moles of substrate converted to product per unit time.

**enzyme complex** — A group of enzymes that catalyzes a metabolic sequence without releasing intermediates.

**enzyme conservation** — A statement that the total amount of enzyme ($E_{tot}$) is constant for the course of a reaction and can be expressed as the sum of all enzyme forms.

**epimerase** — A type of enzyme that catalyzes the rearrangement of hydroxyl groups in a sugar.

**epimers** — Sugars that differ at chiral centers other than the D or L positions.

**epinephrine** — A catecholamine hormone released by the adrenal glands and some neurotransmitters of the central nervous system in response to stress.

**equilibrium** — A state that exists when the forward and reverse rates of a reaction are equal. In chemistry, it is when the products and reactants are in a constant ratio.

**ES** — Intermediate enzyme–substrate complex.

**essential amino acids** — Any of the amino acids not synthesized by the body—and so must come from the diet—that are required for the formation of proteins.

**essential fatty acids (EFAs)** — A group of unsaturated fatty acids essential to human health that cannot be manufactured in the body, so they come from oils and fats found in plants and animals. Examples: omega-3 and omega-6 fatty acids.

**evolution** — The fundamental thesis of biology explaining how species arise: from previous species. The principle applies also to macromolecules, such as DNA and protein, which can be mapped from their earliest origins and indicate their relationship to other, similar forms.

**excision repair** — A process to correct DNA damaged after replication by removing the faulty segment and, through DNA synthesis, replacing it with the correct segment replicated from a template of an undamaged DNA strand.

**exciton transfer** — A form of resonance energy transfer that occurs between chlorophyll molecules in photosynthesis.

**exergonic** — A reaction for which $\Delta G < 0$, where $G$ is free energy. Every reaction in a functional metabolic pathway is exergonic.

**exome sequencing** — A laboratory technique for separating the exons from the remainder of the DNA and then sequencing just those nucleotides.

**exons** — A segment of a gene that contains information used in coding for protein synthesis. Genetic information within genes is discontinuous, split among the exons that encode for messenger RNA and absent from the DNA sequences in between (see *introns*). Genetic splicing, catalyzed by enzymes, results in the final version of mRNA, which contains only genetic information from the exons.

**exonuclease** — An enzyme capable of detaching the terminal nucleotide from a nucleic acid chain; any of a group of enzymes that catalyze the hydrolysis of single nucleotides from the end of a DNA or RNA chain.

## F

**fatty acids** — Molecules with two parts: a long hydrocarbon segment (the tail), and a smaller region that typically consists of a carboxyl group (head). The hydrocarbon tail consists mostly of chemically unreactive methylene groups with a variable number of double bonds.

**feed-forward activation** — A modulator (stimulation or inhibition) of an enzyme that arises from a reaction that occurs earlier in the same metabolic pathway.

**fermentation** — An early name for yeast glycolysis, which produces ethanol and $CO_2$.

**Fischer projection** — A two-dimensional representation of the three-dimensional structure of an organic molecule.

**flavonoids** — A large group of water-soluble plant pigments that are polyphenols that have reputed antiviral, antioxidant, and anti-inflammatory health benefits.

**fluid phases** — Forms of matter that assume the shape of their container. These include both gases and liquids.

**folic acid** — A water-soluble vitamin (classified as a B vitamin) that is a precursor to tetrahydrofolate (THF) and dihydrofolate (DHF), cofactors in one-carbon metabolism.

**fractional saturation** — Proportion of binding of a ligand to a macromolecule, such as oxygen binding to myoglobin or hemoglobin.

**free energy** — A thermodynamic term that describes the overall directional change of a reaction, combining the essence of the first law (energy changes as enthalpy) and the second law (entropy).

**fructose bisphosphatase** — Hydrolase enzyme converting the bisphosphate of fructose into fructose-6-P.

## G

**galactose** — A sugar that is the 4-epimer of glucose; the other half (with glucose) of the disaccharide lactose.

**galactose catabolism** — Breakdown of galactose, converging with the glycolytic pathway at glucose-6-P.

**gas liquid chromatography** — An analytical technique where the sample and carrier fluid are converted into the gas phase (the mobile phase); the stationary phase is packed into a long, thin column.

**genetic code** — The rules for matching of codons to their amino acids. In the polynucleotide chain, the pattern of nucleotides governing the sequence of amino acids that make up each protein synthesized in the cell.

**glucagon** — Hormone secreted by the α-cells of the pancreas during fasting conditions, as blood glucose concentrations are decreased. Glucagon acts on the liver to increase glucose release into the bloodstream.

**glucogenic** — Giving rise to or producing glucose.

**gluconeogenesis** — A process by which glucose is made in the liver from noncarbohydrate precursors (e.g., lactate, amino acids, and glycerol).

**glucose-6-phosphatase** — An enzyme that catalyzes the hydrolytic dephosphorylation of glucose-6-phosphate, which is the principal route for hepatic gluconeogenesis, thus allowing glucose stored in the liver to enter the blood.

**glutathione** — A peptide composed of glycine, cysteine, and glutamic acid. Reduced glutathione (due to the cysteinyl –SH group) reacts with intracellular proteins in order to maintain their sulfhydryl groups in the reduced form, part of the antioxidant system in cells.

**glycerol phosphate shuttle** — A mechanism for the transport of electrons from cytosolic NADH to mitochondrial carriers of the oxidative phosphorylation pathway.

**glyceroneogenesis** — A metabolic pathway occurring in adipose tissues and liver in mammals that can convert pyruvate to the glycerol of triacylglycerol (the major storage form of fat).

**glycogen** — A branched polysaccharide that is the major form of carbohydrate storage in animal cells.

**glycogen branching enzyme** — In glycogen synthesis, the enzyme that catalyzes the transfer of a linear ($\alpha 1 \rightarrow 4$) segment of the molecule to a hydroxyl group at the 6 position to create a branch.

**glycogen phosphorylase** — In glycogenolysis, the major enzyme leading to a release of glucose-1-phosphate from glycogen.

**glycogen phosphorylase kinase** — An enzyme that catalyzes the conversion of phosphorylase b to phosphorylase a (the phosphorylated, active form); activated by $Ca^{2+}$ and cyclic AMP dependent protein kinase.

**glycogen storage diseases** — A set of inborn genetic errors of metabolism deleterious to glycogen metabolism.

**glycogen synthase** — An enzyme that catalyzes the transfer of a glucosyl unit of uridine diphosphate glucose (UDP-glucose) to glycogen.

**glycogen synthase kinase 3 (GSK3)** — A cytosolic protein kinase that catalyzes the phosphorylation of serine residues in glycogen synthase and inactivates it.

**glycogen synthesis** — Pathway of formation of glycogen from individual glucosyl subunits.

**glycogenin** — A protein at the core of the glycogen particle, linked to the glucose subunit through a tyrosine residue.

**glycogenolysis** — Degradative pathway converting glycogen to glucosyl subunits (glucose-1-P).

**glycosidic bond** — The molecular link between sugars, chemically an acetal or a ketal.

**gout** — A metabolic arthritis resulting from overproduction of uric acid (see *allopurinol*).

**G protein** — Signaling proteins that are either bound to GTP (active form) or GDP (inactive form). On the inner surface of plasma membranes, G proteins transmit some hormone signals to their effectors such as in the production of cyclic AMP. Within cells, G proteins act as intermediate control systems such as those regulating protein synthesis.

**grana** — Overlapping stacks of membranes within the chloroplast organelle.

# H

**hairpin turn** — In RNA transcripts that have a palindromic sequence, the formation of a double-stranded segment functions as a control signal for protein–nucleic acid interactions.

**half-cell** — One of two solutions of electrolytes with an electrode enabling separate oxidation and reduction in a galvanic cell.

**half-reaction** — The oxidation or reduction portion of a redox reaction, either a formal representation, or a physical one within a half-cell.

**Haworth projection** — A perspective representation of the molecular ring structure of sugars, using pentagons, hexagons, and shading to depict bonds in three dimensions.

**heat** — A form of energy that transfers among molecular particles through the kinetic force of those particles, always in the direction of high temperature to low temperature.

**helix breakers** — Amino acids that disturb the ability of a protein segment to form an alpha helix, either through restricted bonding (like proline), too much rotational freedom (glycine), or neighboring charge interactions (like multiple proximal glutamate residues).

**heme** — An iron binding cofactor made up of linked imidazoles. The iron may be redox active, as in cytochromes, or at fixed oxidation state, as in hemoglobin.

**hemiacetals** — The condensation product of an alcohol and an aldehyde. The hydroxyl group is used to form glycosides in sugars.

**hemiketals** — The condensation product of an alcohol and a ketone. The hydroxyl group is used to form glycosides in sugars.

**high-energy electrons** — The electrons carried in redox transfer cofactors such as NADH that ultimately lead to energy generation.

**high-energy molecules** — Molecules containing localized electrons that can be redistributed and used as an energy source in a further reaction, such as NADH and ATP.

**high-energy phosphate** — A phosphoryl group that is exchanged between molecules, usually involving at some point the mobile cofactor intermediate, ATP.

**high pressure liquid chromatography (HPLC)** — An analytical technique using many small beads and a correspondingly large surface area for the stationary phase. High pressure is needed to drive flow of the mobile phase (see *chromatography*).

**histones** — Small basic proteins that bind the negatively charged backbone of DNA and contribute to the condensation and packing of DNA.

**holistic** — A view that focuses on system properties rather than its parts.

**hydrated** — Attached to water molecules.

**hydride** — A compound of hydrogen with another more electropositive element or group.

**hydride ion** — The anion of hydrogen (H:⁻).

**hydrogen bond** — An interaction involving a hydrogen atom attached to a N or O atom and another N or O atom. The bond strength is intermediate between van der Waals forces and covalent bonds.

**hydrolases** — A category of enzymes that are actually a type of transferase (itself a separate category of enzymes) in which water is a substrate.

**hydrophobic effect** — A clustering of nonpolar molecules driven by more favorable water-water interactions.

## I

**inhibitor (I)** — A substance that decreases the rate of an enzymatic reaction.

**insulin** — A hormone secreted from the β cells of the pancreas into the bloodstream in response to an increase in blood glucose. Insulin leads to a decrease in blood glucose concentrations by glucose uptake and metabolism in cells.

**insulin receptor substrate 1 (IRS-1)** — A cytosolic regulatory protein that is phosphorylated by the insulin receptor and serves as a binding site for other regulators in the cell.

**insulin-resistant** — A condition in which cells respond poorly to the presence of insulin, found almost universally in obese individuals, and a precursor state to diabetes.

**initial velocity** ($v_i$) — The first (linear) portion of a reaction progress curve, either disappearance of substrate, or appearance of product. In theory, the slope this curve as time approaches zero is defined as the initial velocity; in practice, the first 5% to 10% of the progress curve is approximately a straight line.

**inner mitochondrial membrane** — One of two membranes in the mitochondrion, this membrane is infolded, providing a greater membrane area for the components of oxidative phosphorylation and transport proteins (see *outer mitochondrial membrane, matrix space*).

**inorganic phosphate** — An isolated phosphate group that is a weak acid and can combine with an alcohol to form an ester.

**intermediary metabolism** — Connected reaction sequences in cells that represent the major pathways, common to cells in many biological organisms.

**internal energy** — The energy of the system under study, defined as the sum of heat and work.

**intrinsic factor** — A protein found in cells of the stomach lining (epithelia), secreted into the stomach, and used later in the small intestine in the uptake of dietary vitamin B₁₂.

**introns** — A segment of a gene situated between exons that does not function in coding for protein synthesis. After transcription of a gene to messenger RNA, the introns are removed, and the exons are spliced together by enzymes before translation and assembly of amino acids into proteins (see *exons*).

**ionization of water** — The dissociation of water into hydrogen ions (H⁺) and hydroxide ions (OH⁻).

**ionophore** — A molecule that binds an ion and transports it across a membrane.

**isolated (double bond)** — A type of double bond in which the pi orbitals do not overlap with those of other double bonds (also termed, *nonconjugated*).

**isoelectric point (pI)** — The pH at which a molecule has a net zero charge.

**isoforms** — Distinct proteins that function identically; isomeric form of the same protein, with different amino acid sequences but with the same activity.

**isomerases** — A class of enzymes that catalyze rearrangement reactions, such as the interconversion of ketoses and aldoses.

**isozymes** — Enzymes with different structures that catalyze the same reaction.

## K

**ketogenic** — Amino acids that are catabolized to acetyl-CoA and thus potentially to ketone bodies.

**ketoses** — Sugars in which the carbonyl group is a ketone, located at carbon two.

**Krebs cycle** — One of the major metabolic pathways in cellular metabolism. A cyclic pathway that converts acetyl CoA to $CO_2$ and extracts energy intermediate molecules NADH, $QH_2$, and GTP (see *citric acid cycle, tricarboxylic acid cycle*).

# L

**lac operon** — A grouping of genes and regulatory products in bacteria that controls the synthesis of three enzymes that enable utilization of lactose as a carbon source.

**lactose intolerance** — A common genetic deficiency of lactase (an enzyme that catalyzes conversion of lactose to galactose and glucose) that impairs the digestion of dairy products. The partial bacterial oxidation of lactose in the large intestine is responsible for the symptoms of bloating and gas production.

**lagging strand** — In DNA replication, the strand in which the nascent strand is synthesized in discontinuous segments because all transcription must be in the 5′ to 3′ direction and both DNA strands (oppositely arranged) must be replicated (see *Okazaki fragments, leading strand*).

**leading strand** — In DNA replication, the strand that is synthesized continuously during replication (see *lagging strand*).

**lecithin** — A common name for phosphatidylcholine.

**leghemoglobin** — A monomeric heme containing protein structurally similar to myoglobin, found in the root nodules of leguminous plants (such as soybeans) that are host to nitrogen-fixing bacteria. Oxygen inhibits nitrogenase; leghemoglobin maintains a low oxygen concentration that can still supply bacterial respiration.

**leukotrienes** — A group of biologically active compounds derived from arachidonic acid that are responsible for allergic and inflammatory reactions.

**light reactions** — Steps in photosynthesis in which light energy is utilized to produce the high energy intermediate compounds ATP and NADPH.

**Lineweaver–Burk plot** — Double reciprocal transform plot of the Michaelis-Menten equation, which linearizes the resulting graph (also termed *double-reciprocal plot*).

**lipids** — Relatively unreactive hydrocarbons such as oils and fats; they are entirely or mostly water insoluble. Functions include energy storage and membrane formation.

**liposomes** — A spherical particle formed by a lipid bilayer enclosing an aqueous compartment; used to convey viruses, drugs, and other substances.

**loop** — Describes the shape of a type of secondary structure of protein formed with the backbone atoms and their hydrogen bonds that link other secondary structures together (also termed *random coil*).

**loop mechanism** — A mechanism for creating the proton gradient in the Q cycle, involving the movement of oxidized and reduced Q moving between the faces of the mitochondrial membrane (see *Q cycle*).

**lovastatin** — Inhibitor of hydroxymethylglutaryl CoA reductase and thus cholesterol synthesis; it is a natural product isolated from a fungus used to treat hypercholesterolemia.

**lumen** — Most interior water space of the chloroplast, bounded by the thylakoid membrane. Also a general anatomical term for an extracellular space interior to the body, such as the lumen of the intestines.

**lyase** — A category of enzymes that catalyze the splitting of a substrate into pieces; lyase reactions do not involve water as a substrate.

# M

**macromolecules** — A large molecular polymer such as a protein that exhibits new properties that distinguish it from small molecules, such as domain structures.

**major groove** — One of two grooves along the DNA double helix, this has the larger gap between the sugar base structures when viewed from the top of the helix structure (see *minor groove, double helix*).

**malate/aspartate shuttle** — A pathway for importing NADH from the cytosol to the mitochondria, using reduction of oxalacetate to malate in the cytosol, malate to oxalacetate in the mitochondrial matrix, and interconversion between oxalacetate and aspartate; neither NADH nor oxalacetate can cross the mitochondrial membrane.

**maple syrup urine disease** — A genetic disease caused by the accumulation of three amino acids—leucine, isoleucine, and valine—due to their inability to be oxidized through the branched chain ketoacid dehydrogenase complex. Named for the distinct burned sugar odor of urine (also termed *branched chain ketoaciduria*).

**mass action ratio** — The ratio of the reaction products multiplied together to the substrates multiplied together for a reaction at any given point. When the forward and reverse reaction rates are the same, the mass action ratio is equal to the equilibrium constant.

**mass spectrometry** — A system for identifying molecules by fragmenting them into ions in the gas phase and then separating them in an electric field by mass.

**matrix space** — The water space enclosed by the mitochondrial inner membrane.

**melting curve** — A plot of strand separation of DNA with increasing temperature.

**melting temperature** — The midpoint of the DNA melting curve.

**membrane fluidity** — The relative ability of individual molecules to move within the bilayer of a cell.

**metabolically irreversible** — Metabolic reactions within cells that are greatly displaced from their equilibrium positions and never run in the reverse direction under cellular conditions. These reactions are the sites of metabolic control, through allosteric or covalent modification.

**methionine cycle** — Metabolic pathway of one carbon metabolism in which methionine is converted to S-adenosyl methionine and a methyl group is donated to an acceptor.

**methylcelluloses** — A biologically inert polymer formed in the laboratory by creating methyl esters to some of the hydroxyl groups of cellulose; their partial solubility in water can be controlled by the extent of methylation.

**micelles** — A form taken by lipid amphiphiles in water solution in which the polar portions of the molecules face the aqueous exterior and the nonpolar portion forms an interior core excluded from water.

**minor groove** — One of two groove structures on the DNA double helix, this has the smaller gap between the sugar base structures when viewed from the top of the helix structure (see *major groove, double helix*).

**mismatch repair** — A system within the cell during replication that corrects errors in DNA by detecting and replacing bases in the DNA that are wrongly paired; an enzyme catalyzes removal of the mismatched segments, and then DNA polymerases fills the gaps in the strands with the appropriate bases.

**mitochondria** — Organelle evolutionarily derived from bacteria, responsible for most of the energy generation of the cell through oxidative phosphorylation. Mitochondria contain portions of several metabolic pathways, such as urea synthesis and gluconeogenesis.

**mixed inhibition** — A type of reversible enzyme inhibition in which the inhibitor binds both the free enzyme and the enzyme–substrate complex (also *noncompetitive inhibition*).

**mobile cofactors** — Part of an enzymatic reaction that can dissociate from the enzyme at the conclusion of the reaction and participate in a reaction in a separate pathway in the cell. Examples are NADH, $QH_2$, and cytochrome *c*. By contrast, bound cofactors never leave the enzyme during the catalytic cycles.

**molecular biology** — A broad term encompassing genetics, biochemistry, and growth processes of the cell.

**monomers** — Individual protein chains (subunits).

**motif** — (see *domain*).

## N

**N-acetylglutamate** — A compound produced from acetyl coenzyme A and glutamate; an allosteric activator of carbamoyl phosphate synthetase in the urea cycle.

**NAD binding domain** — The site of binding of this mobile cofactor, common to a large number of enzymes.

**near-equilibrium** — A reaction in cells that has a mass action ratio close to the equilibrium constant, and can run in the reverse direction in a separate pathway. Such reactions are controlled by substrate and product concentrations and are not regulated by external means, i.e., by allosterism or covalent modification.

**negative cooperativity** — Cooperativity in which successive ligand molecules appear to bind with decreasing affinity.

**negative supercoil** — Occurs when the double helix strand of DNA is twisted in a counterclockwise direction (the opposite direction as turns in the helix), which either unwinds or twists the entire structure (see *supercoils, double helix*).

**nitrification** — A process where the oxidation of ammonia ($NH_3$) to nitrites and nitrates is accomplished by bacteria in a series of reactions.

**nitrogen cycle** — A global nitrogen cycle by which atmospheric nitrogen gas ($N_2$) is fixed into nitrogen oxides and ammonia, which can be incorporated into biological systems, and then converted back to $N_2$.

**nitrogen fixation** — The incorporation of atmospheric nitrogen, catalyzed by nitrogenase into ammonia by various bacteria; the process whereby certain microorganisms, such as *rhizobia*, convert atmospheric nitrogen into compounds that plants and other organisms can assimilate.

**non-cooperative** — A binding process in which ligands bind independently to monomers of a multimeric protein.

**nonessential amino acids** — Amino acids that can be synthesized by the organism and, thus, do not need to be present in the diet (see *essential amino acids*).

**nonpolar lipids** — Lipids that are not amphiphiles. Example: triacylglycerols.

**nonreducing end** — The sugar residue end of a polysaccharide that has no free anomeric hydroxyl group.

**nonreducing sugars** — Any saccharide that has no free anomeric carbon atoms. Examples: sucrose and trehalose.

**N side** — Negatively charged side of an energy producing vesicle (chloroplast, bacteria, or mitochondria) produced when hydrogen ions leave this space.

**nuclear magnetic resonance (NMR) spectroscopy** — A method for the study of molecular structure in which a magnetic field is imposed and the absorption of electromagnetic radiation at specific frequencies is measured.

**nuclear membrane** — The double-layered membrane surrounding the nucleus of a cell.

**nucleotides** — Modified sugar molecules that contain a nitrogenous base and at least one phosphate ester; the most common are AMP, ADP, and ATP, key energy transfer molecules widely used in metabolic reactions in cells. Additionally, nucleotides form the polymers DNA and RNA.

## O

**oil** — The liquid form of lipids.

**Okazaki fragments** — During DNA replication, the relatively short fragment of DNA synthesized on the lagging strand; later these fragments will be covalently connected into a continuous strand.

**oligosaccharides** — A carbohydrate whose molecules are composed of a relatively small number of monosaccharides linked via glycosidic bonds.

**operon** — A group of genes or a segment of DNA that functions as a single transcription unit, comprised of an operator, promoter, and one or more structural genes that are transcribed into one polycistronic mRNA.

**osmosis** — The ability of water to move across a semipermeable membrane due to osmotic differences across the membrane.

**osmotic pressure** — The applied pressure needed to equalize the fluid levels across a semipermeable membrane that exhibits osmotic imbalance.

**outer mitochondrial membrane** — One of two membranes in the mitochondrion; porous to molecules with a molecular weight up to about 10,000 (e.g., ions, ATP, ADP, and nutrient molecules) (see *inner mitochondrial membrane*).

**oxidoreductases** — A category of enzymes in which the substrate changes the oxidation state during the process of going to product; there must be a mobile cofactor that removes or adds those electrons, often $NAD^+$.

## P

**P** — Phosphoryl group, often referred to as a phosphate group, such as in the carbohydrate derivative glucose-6-phosphate (glucose-6-P).

**palindrome** — A self-complementary nucleic acid sequence. In general, any sequence of letters that reads the same in the forward direction as in the reverse.

**parallel** — In a β-sheet of a protein, when the strands run in the same direction.

**partially ionic bond** — A description of a covalent bond that emphasizes its partial ionic character due to electronegativity differences of the atoms on either side of the bond.

**Pasteur effect** — A 19th-century observation by Louis Pasteur that glucose utilization is decreased in the presence of oxygen in yeast metabolism.

**path independent** — A change from one state to the next that is independent of the route that is taken. This describes several changes in thermodynamics such as the change in internal energy, enthalpy, and free energy.

**pathway completion** — In a metabolic pathway, all mobile cofactors must be balanced so that the route can be traversed again by further pathway substrates.

**pathway flux** — The overall rate of a pathway that is also the rate of every individual reaction in the pathway under a given steady state condition (see *pathway substrate, pathway product*).

**pathway product** — The final product of a series of reactions that defines a pathway in cells.

**pathway substrate** — The initial product of a series of reactions that defines a pathway in cells.

**pentose phosphate shunt** — Pathway branching from glycolysis at glucose-6-P that produces ribose carbon for nucleotides and NADPH.

**peptides** — Short chains (roughly 25) of amino acids that have sufficient flexibility in solution so that a fixed higher order structure does not exist.

**peptide bond** — An amide linkage found between the amino acids of proteins.

**phenylketonuria (PKU)** — A genetic disorder in humans caused by an inability for the body to metabolize phenylalanine to tyrosine.

**pheophytins** — Part of the reaction center in the light reactions of photosynthesis, these molecules are structurally similar to chlorophyll molecules except that the chelate ring has no metal inside it (see *reaction center*).

**phosphatidic acid** — A phospholipid with a single fatty acyl chain attached to a glycerol phosphate backbone.

**phosphatidylinositol phosphate 3-kinase (PI3K)** — An enzyme that catalyzes the incorporation of a phosphoryl group into the 3-position of the inositol ring.

**phosphodiesterase** — An enzyme that cleaves phosphodiesters into a free hydroxyl group and a phosphomonoester.

**phosphoenolpyruvate carboxykinase (PEPCK)** — The rate-limiting step of gluconeogenesis under most conditions, this enzyme converts oxaloacetate to carbon dioxide and phosphoenolpyruvate.

**phospholipid-dependent kinase (PDK)** — A regulatory enzyme that phosphorylates serine and threonine residues in response to protein kinase B and membrane phospholipids.

**phosphonate** — A phosphate in which one of the P—O bonds is substituted by a P—C bond.

**phosphoribosylpyrophosphate (PRPP)** — This multiply phosphorylated ribose is a precursor in the synthesis of nucleotides.

**phosphotyrosine binding domain (PTB)** — Domains of 100–150 residue modules that commonly bind the Asn-Pro-X-Tyr motifs where the tyrosine residue is phosphorylated. Found in proteins involved in regulatory cascades.

**photosynthesis** — Pathway for utilization of light energy to form high energy intermediates (the light reactions); usually also refers to the ultimate synthesis of carbon compounds such as carbohydrates as well (the dark reactions).

**photosynthetic** — Type of organism that can utilize light to synthesize carbon compounds.

**photosystem (PS)** — Complex of proteins involved in photosynthetic energy transfer that accept light input and boost electron energy of other molecules.

**photosystem I (PS I)** — In green plant photosynthesis, contains iron-sulfur proteins and transfers electrons from plastocyanin to a [2Fe-2S] ferredoxin.

**photosystem II (PS II)** — In green plant photosynthesis, contains an iron center and transfers electrons from the oxygen evolving complex to plastoquinone.

**photon** — Minimal unit of light energy defined from the equation $E = h\nu$ where E is energy, h is the Planck constant, and $\nu$ is the frequency of light.

**ping-pong (mechanism)** — An enzyme mechanism in which one substrate binds and the product is released before a second substrate binds. This mechanism occurs in transaminase reactions.

**plasmodesmata** — A bridge of cytoplasm that spans adjacent plant cells to permit communication and transport between individual cells. Gap junctions are the functional equivalents in animal cells.

**polar bond** — A bond between two atoms that have large differences in electronegativity.

**polarimeter** — A device that detects chiral molecules and differentiates the two forms by determining the rotation of the plane of polarized light as it passes through a solution.

**polar lipids** — Lipid molecules that are amphipathic, with one portion of the molecule interacting with water. These include cell membrane lipids.

**polar molecule** — Describes a molecule with a net polarity due to the overall vector sum of its bonds.

**polyamines** — Molecules with multiple amines, bearing a net positive charge and interacting with DNA through electrostatic attraction. Examples: spermidine, spermine.

**poly-A tail** — A structure of up to 250 adenylate residues located on the 3′ end of eukaryotic mRNAs used in translation and protecting the mRNA against nuclease action.

**polycistronic** — Bacterial mRNA that contains sequences for several different proteins in the same messenger RNA strand.

**polymers** — Large molecules composed of many repeating subunits.

**polysaccharides** — A carbohydrate polymer composed of monosaccharide residues bound together via glycosidic bonds.

**polyunsaturated** — Fatty acids with more than one double bond.

**porins** — Protein molecules lining the outer membrane of the mitochondria, which enable passage of molecules up to 10,000 Da.

**positive cooperativity** — Cooperativity in which successive ligand molecules appear to bind with increasing affinity.

**positive supercoil** — Occurs when the double helix strand of DNA is twisted in a clockwise direction (the same direction as the turns in the helix), which tightly coils the structure (see *supercoils, double helix*).

**postprandial state** — A period immediately after ingestion of a meal.

**Pribnow (box)** — A highly conserved sequence element located upstream from the start site of transcription, to which the sigma subunit of RNA polymerase binds; found in prokaryotes such as *E. coli* and phage genes.

**primary structure** — The sequence of amino acids in a linear protein chain.

**primase** — An RNA polymerase that catalyzes single-stranded DNA in the first biochemical reactions to add bases to the strand.

**primosome** — During DNA replication, the protein complex containing primase and helicase that initiates the RNA primers on the lagging DNA strand.

**probenecid** — A drug used to treat gout by increasing its excretion by the kidneys (see *gout*).

**processivity** — The movement and catalysis of a molecule on a substrate surface without dissociation between catalytic cycles.

**prochiral center** — A carbon center having two identical substituents that can be discriminated by binding to an enzyme surface.

**promoters** — A site in a DNA molecule at which RNA polymerase and transcription factors bind to initiate transcription of messenger RNA.

**prosthetic group** — A small molecule that is bound to a protein to assist its function. These molecules remain bound to the protein surface as it functions, and may be considered an extension of the protein itself.

**protease** — An enzyme that hydrolyzes peptide bonds in proteins.

**proteasome** — A cytosolic protein complex in which proteins enter and are degraded to small peptides.

**protein** — Linear chains of amino acids that form a specific three-dimensional structure (thus distinguishing them from smaller peptides that are usually on the order of 25 amino acids or less). Proteins are the most diverse biomolecules, including virtually all of the enzymes, the biological catalysts.

**protein kinase C (PKC)** — A protein kinase that catalyzes the phosphorylation of serine or threonine amino acid residues. The "C" in the name derives from an early finding that a calcium-stimulated protease activated the enzyme; however, this proteolysis is not the mechanism for its rapid activation in cells. An increase in the lipid mediator diacylglycerol is the activator of PKC.

**PS I** — Photosystem I of chloroplasts (see *photosystem*).

**PS II** — Photosystem II of chloroplasts (see *photosystem*).

**P side** — Positively charged side of an energy producing vesicle (chloroplast, bacteria, or mitochondria) produced when hydrogen ions enter this space.

**purine nucleotide cycle** — A metabolic cycle between AMP and IMP that serves to convert aspartate into fumarate as an anapleurotic (refilling) reaction for the Krebs cycle.

**pyridoxal phosphate** — A cofactor derived from vitamin $B_6$ that is the prosthetic group of glycogen phosphorylase and amino acid transaminases.

**pyrrole** — A five-member heterocycle containing one nitrogen atom and four carbon atoms.

**pyruvate carboxylase** — An enzyme that catalyzes the carboxylation of pyruvate to oxalacetate, important in lactate gluconeogenesis, fatty acid synthesis, and as an anapleurotic reaction for the Krebs cycle.

**pyruvate dehydrogenase complex** — A noncovalent association of enzymes that converts pyruvate to acetyl CoA.

**pyruvate transporter** — In the inner mitochondrial membrane, transports pyruvate into the mitochondrial matrix space.

# Q

**Q cycle** — The mechanism for complex III production of the proton gradient across mitochondria in which oxidized and reduced forms of ubiquone (Q, $QH_2$) move between the membrane faces and ferry protons across the membrane.

**quaternary structure** — A structural level for proteins that have more than one protein chain.

# R

**racemase** — An enzyme that catalyzes the inversion around an asymmetric carbon atom.

**radical anion** — A molecule containing both a free radical and a negatively charged ion.

**raffinose** — A sugar found in foods such as cabbage and beans that is poorly digested due to the presence of the galactose-$\alpha1\rightarrow6$-glucose bond in the first two sugar moieties of the trisaccharide; its bacterial degradation in the large intestine causes gas.

**random coil** — Indeterminate in structure, it links the other secondary structures together in proteins.

**ras oncogene** — A family of genes that undergo muta-tion in the oncogene resulting from an inactive GTPase, which leads to a state of constant activa-tion and hyperactive cell growth, replication, and eventually cancer.

**rate constant ($k$)** — The proportionality between veloc-ity and substrate concentration for a reaction.

**reaction center** — In the context of photosynthesis, a special pair of chlorophyll molecules that accept energy transferred from other chlorophyll molecules.

**receptor tyrosine kinases** — A class of membrane receptors that have kinase activity selective for tyro-sine residues.

**redox** — Oxidation-reduction reactions involving an electron transfer.

**reducing end** — The end of a polysaccharide that has an equilibrium form with an open-chain aldose or ketose.

**reducing equivalents** — Energy rich electrons that can be transferred from cofactors and utilized for energy extraction or reduction of compounds, such as in the hydride in NADH or NADPH.

**reducing sugar** — Any saccharide that has at least one free anomeric carbon (i.e., an anomeric carbon *not* involved in a glycosidic bond). A small amount of this open-chain form exists in solution, so its carbonyl group can reduce certain metal ions in diagnostic tests.

**reductases** — Enzymes that catalyze the reduction of their substrate.

**reductionist** — A "bottom up" perspective; in bio-chemical terms, focusing on specific reactions or chem-ical species or decomposing a more complex system into its elements in order to understand the whole.

**reduction potential** — A measure of the capacity of a redox half-reaction to donate electrons.

**replication** — The process by which DNA can be used as a template to form daughter DNA molecules for cell division.

**replication fork** — The section of a DNA molecule at which the two strands are just separated; here, both daughter molecules are being synthesized.

**replisome** — In DNA replication, a protein complex containing two DNA polymerase molecules for syn-thesizing the leading and lagging ends of the DNA strands.

**rho protein factor** — A protein involved in prokary-otic transcription termination, which binds and disassembles the RNA transcript with concomitant ATPase activity.

**resonance** — In conjugated double bonds, where elec-trons can spread out over more than two nuclei.

**respiration** — Energy linked process in cells that uti-lizes oxygen to enable electron flow that ultimately leads to the formation of ATP.

**rhodopsin** — Molecule formed from the covalent link-ing of retinal with the protein opsin, responsible for light absorption in the visual cycle of mammals.

**ribosome** — A cytosolic complex of proteins and ribo-somal RNA (rRNA) that binds mRNA and tRNA and functions in the synthesis of proteins.

**ribozymes** — An RNA segment able to catalyze the cleavage and formation of covalent bonds in specific sites of RNA strands; ribosomes are a nonprotein type of enzyme.

**rotor** — The moving (spinning) part of a motor mecha-nism such as that in $F_1F_o$ATP synthase.

## S

**S-adenosylmethionine (SAM)** — A condensation prod-uct of ATP and methionine that functions as a one-carbon donor.

**saturated** — Fatty acids containing no double bonds.

**saturation** — A high and unchanging enzyme velocity achieved at high substrate concentration.

**secondary metabolism** — Pathways that produce spe-cialized metabolic products (secondary metabolites) that are not essential for the physiologic function of the organism.

**secondary metabolites** — Molecules that are end prod-ucts of pathways specific to one or a few organisms and, thus, are less universal than primary metabolic routes. Secondary metabolites protect the organisms from being consumed by predators.

**secondary structure** — A regular, repeating structural element of an amino acid that is formed when a segment of the molecular chain folds in three dimensions.

**semialdehyde** — A molecule derived from a dicarbox-ylic acid, where one of the acid groups has become an aldehyde.

**semipermeable** — A membrane that permits the trans-port of molecules but acts as a barrier to others; partially permeable.

**sequential (mechanism)** — An enzyme mechanism with multiple substrates in which all must bind to the enzyme in turn before any product is released.

**Shine–Dalgarno (sequence)** — Located on a prokary-otic mRNA molecule upstream of the translational start site, the short (3–10), purine-rich sequence of nucleotides that binds ribosomal RNA, thereby aligning the ribosome on the initiation codon on the messenger RNA. Proposed by John Shine and Lynn Dalgarno in 1975.

**shunt pathway** — A pathway that branches from another pathway and terminates at some other point in that pathway. Example: the 2,3-bisphosphoglycerate shunt of glycolysis.

**shuttles** — A pathway for transferring portions of a compound across membranes. Examples: the malate/aspartate shuttle for reducing equivalents; the citrate shuttle for acetyl-CoA.

**single-strand binding protein (SSBP)** — Proteins binding to separated DNA strands, such as those at the replication fork of DNA that enable replication (see *replication fork*).

**signal sequence** — During protein translation, a sequence of amino acid residues at the amino terminus recognized by the signal recognition particle, which permits transport of the nascent protein to its new destination in an organelle. Once it enters into the new location, this set of amino acids is removed by proteolysis.

**soft nucleophile** — An electron rich center that is less localized and more easily polarized.

**species hierarchies** — The classification of plants and animals into classes, orders, families, etc.

**spontaneous** — A term that has diffuse meanings in thermodynamics; often a substitute for exergonic, but at other times used to mean nonenzymatic. It is not a well-defined thermodynamic term.

**spontaneous generation** — The erroneous hypothesis that living organisms arise from nothingness.

**src** — Sarcoma virus protein (*src* is the sarcoma virus gene), it is the parent of a family of protein kinases that lead to phosphorylation of tyrosine residues, called src kinases.

**Src homology 2 (SH2)** — A protein-building domain in the src family of cytoplasmic tyrosine kinases, which can bind phosphotyrosine-containing proteins. First found in the src (sarcoma virus protein), from which the name derives (Src homology).

**stacking interaction** — An attractive force between each hydrogen-bonded base pair (AT and GC, the "rungs" of the double helix "ladder") above and below them along the DNA double helix (see *double helix*).

**state** — A thermodynamic condition in which certain variables, such as temperature, pressure, volume, and number of moles, is fixed.

**state variables** — Variables that determine the state of a system, including temperature ($T$), pressure ($P$), volume ($V$), and number of moles ($n$).

**statins** — Inhibitors of hydroxyl-methyl-CoA reductase, the rate-limiting step of cholesterol biosynthesis, these drugs are used to lower blood cholesterol concentration particularly in low density lipoproteins (LDLs).

**statistical thermodynamics** — A derivation of the thermodynamic principles that uses the laws of large numbers, calculating the behavior of atoms and molecules on the basis of statistical laws.

**stator** — The stationary part of a motor mechanism (see *rotor*).

**steady state** — A condition in which the intermediates of a process, such as a metabolic pathway, are constant with time, while the pathway substrate decreases and pathway product increases.

**steady-state assumption** — In enzyme kinetics, the assumption that the intermediates of a reaction mechanism are constant with time (see *steady state*).

**steady-state constant** ($K_m$) — A ratio of rate constants for an enzymatic reaction, the constant is also the concentration of substrate corresponding to the half-maximal initial velocity for an enzyme.

**stroma** — Water space in the chloroplast between the inner membrane and the thylakoid membrane.

**strong acid** — An acid that completely dissociates in water, forming $H^+$ and an anion.

**strong base** — A base that dissociates completely in water, forming $OH^-$ ion (lowering $H^+$ ion concentration) and a cation.

**subcutaneous fat** — Less saturated lipid-containing adipocytes found just under the skin.

**substrate (S)** — Reactants for enzyme reactions or pathways.

**substrate cycle** — Reactions that involve the same pathway substrates and products, but in opposing directions. Together, the reactions catalyze a net hydrolysis of a high energy phosphate intermediate such as ATP. These cycles enable more precise metabolic control (also termed *futile cycle*).

**substrate-level phosphorylation** — Reaction(s) that form a high energy phosphate intermediate such as ATP by direct chemical means, such as phosphoglycerate kinase or succinyl CoA thiokinase.

**sugar acid** — A monosaccharide derivative in which the carbonyl group is oxidized to an acid, such as glyceric acid.

**supercoils** — Circular or closed loops of DNA that occur when DNA is twisted around its own axis, which changes the number of turns in the double axis (see *double helix*).

**supersecondary structure** — In protein molecules, the combinations of alpha-helices and beta-structures connected through loops that form folding patterns stabilized through the same types of linkages as in the tertiary level (also, *motif*).

**surroundings** — Everything in the universe apart from the system under study.

**svedberg** — A unit of measure equal to $10^{-13}$ second used as a measure of size of large macromolecules and particles. The term refers to the material's sedimentation during ultracentrifugation.

**symmetrical intermediate** — Within some enzyme mechanisms that are rearrangement reactions, such as isomerases, addition of protons and/or electrons lead to an intermediate that displays symmetry. The second half of the mechanism will then be a mirror image of the first.

**synthase** — Enzymes that catalyze joining reactions that do not involve a high energy intermediate molecule like ATP. Synonym: lyase (named for the reverse direction).

**synthetases** — See *ligases*.

**system** — That portion of the universe under study.

# T

**TATA** — Located in the promoter region of most eukaryotic genes, a consensus sequence (5′-TATAAAA-3′) at ~25 nucleotides upstream of the site of the initiation of transcription.

**telomerase** — An enzyme that forms, maintains, and repairs the ends of chromosomes (see *telomeres*); regulates cell proliferation in humans (see *consensus*).

**telomeres** — The repetitive nucleotide segments located at the ending segments of chromosomes; these protect the chromosome from deterioration.

**termination** — Occurs when the RNA reaches a specific sequence (terminator) and the transcription complex dissociates from the DNA, thereby releasing the RNA polymerase to begin a new initiation event (see *transcription, termination sequence*).

**termination sequence** — In RNA transcription, a polymerase-specific DNA binding protein found close to the end of a coding sequence; rho-independent sequences are inverted and form a hairpin back onto the sequence; rho-dependent terminators rapidly unwind the DNA-RNA hybrid formed during transcription, thereby freeing the newly synthesized RNA (see *transcription*).

**tertiary structure** — The entire three-dimensional arrangement, or conformation, of a protein chain in three dimensions.

**tetrahydrobiopterin** — Redox cofactor involved in the hydroxylation of phenylalanine to tyrosine.

**tetrahydrofolate** — Cofactor involved in one-carbon metabolism.

**thioredoxin reductase** — Enzyme catalyzing the reduction of thioredoxin, which is involved in forming deoxynucleotides. The mechanism of thioredoxin reductase is the same as that of glutathione reductase.

**through-space effect** — An electrostatic interaction within a molecule that involves interaction between attached groups, such as the carboxyl and amine groups of an amino acid.

**thylakoid membrane** — Most interior membrane of the chloroplast; separates the inner, lumen space, and the stroma.

**titration curve** — A plot of pH versus the equivalents of a strong base (or strong acid) in a solution of a weak acid (or weak base).

**topoisomerases** — A class of enzymes that catalyze changes in double-stranded DNA by transiently cutting one or both strands of the helix; in supercoiling the twisting results in a shortened molecule (see *supercoils, double helix*).

**transaldolase** — Transferase enzyme that exchanges carbon fragments between substrates using the same mechanism as aldolase.

**transcription** — The synthesis of ribonucleic acid (RNA) from a deoxyribonucleic acid (DNA) template.

**transferases** — A category of enzymes that catalyze reactions that move a piece of one substrate on to another.

**transhydrogenase** — Enzyme that catalyzes movement of electrons between NADH and NADPH.

**transketolase** — Transferase enzyme that uses thiamine as a bound cofactor to exchange carbon fragments between substrates.

**translation** — A cellular process in which messenger RNA (mRNA) provides a template for protein synthesis to form new protein.

**translational** — At the level of protein synthesis.

**trehalose** — A disaccharide (glucose $\alpha1-\alpha1$ glucose) found in bacteria, fungi, plants, and insects.

**triad** — Three amino acids—histidine, serine, and aspartic acid—positioned close to one another at the center of the active site of the enzyme that are essential to the amide hydrolysis activity of these proteases.

**tricarboxylic acid cycle** — Another name for the Krebs cycle or the citric acid cycle (see *Krebs cycle, citric acid cycle*).

**triose** — A three-carbon monosaccharide that is the smallest sugar.

**triosephosphate isomerase (TIM) domain** — A domain that is donut-shaped, with an inner portion composed of a β-sheet and an outer portion of α-helices (see *domain*).

**triosephosphate isomerase** — A glycolytic enzyme that catalyzes the interconversion of two isomeric three-carbon phosphorylated sugars.

**type I diabetes** — A disease in which blood glucose concentrations are above normal due to little or no insulin production by the pancreas (formerly known as *juvenile* or *insulin-dependent diabetes*).

**type II diabetes** — A disease in which blood glucose concentrations are above normal due to tissue insensitivity to insulin (formerly known as *adult-onset* or *noninsulin–dependent diabetes*).

# U

**ubiquitin** — A small protein that can become attached to cytosolic proteins and serve as a signal for degradation in the proteasome (see *proteasome, ubiquitination*).

**ubiquitination** — The process of attachment of ubiquitin to a protein (see *ubiquitin*).

**uncompetitive inhibition** — A reversible inhibitor that binds exclusively to the enzyme-substrate complex (also termed *anticompetitive inhibition*).

**uncoupling** — The dissociation of ATP formation (phosphorylation) from mitochondrial electron transfer. Uncouplers accelerate electron transport but diminish ATP formation (see *coupling*).

**universe** — All that exists; in thermodynamics, the sum of system and surroundings.

**unsaturated** — Fatty acids containing double bonds. Examples: arachidonic acid, oleic acid.

**upstream** — In transcription, the sequence before (see *downstream*).

**ureido** — The $NH_2$–CO–NH– group, present in citrulline and urea.

# V

**vectorial** — A process that involves both a quantity and a direction, such as the movement of protons across the mitochondrial membrane during oxidative phosphorylation.

**visceral fat** — Adipose deposits located in the organs, especially of the abdomen (also termed *organ fat, intra-abdominal fat*).

**vitalism** — The principle (erroneous) that living systems do not obey the same chemical principles as inert materials.

**vitalist** — One who believes (erroneously) that living systems do not obey the same chemical principles as inert materials.

**vitamins** — Precursors to enzyme cofactors, classified as fat or water soluble molecules needed in small amounts.

**vitamin $B_{12}$** — Vitamin involved in methyl transfer reactions, having a central cobalt ion; also known as cobalamin.

# W

**weak acid** — An acid that only partially disassociates in water. Examples: acetic acid, oleic acid.

**weak base** — A base that only partially dissociates in water. Examples: ammonia, diethylamine.

**wobble hypothesis** — During the process of transferring genetic code, a possible explanation of how a specific transfer RNA molecule can translate different codons in a messenger RNA template to the same amino acid. The third base of the transfer RNA anticodon does not have strict base pairing rules.

**work** — An energy of motion (displacement), the product of force and distance.

# X

**X-ray crystallography** — A method for determining the structure of a crystal (including proteins and nucleic acids) by subjecting them to x-ray radiation, measuring the scattering pattern that results, and mathematically constructing a three-dimensional image from that data.

**xenobiotic** — Any substance (often, a drug) that is introduced into an organism that is foreign (xeno) to that organism.

# Z

**zwitterions** — Molecule having no overall charge containing balanced negatively charged and positively charged parts (German word *zwitter*, meaning a hybrid of two forms).

# Index

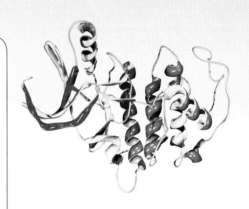

*Note:* Italics *f*, *t*, or *b* indicates the term is found in a figure, table, or box, respectively, on that page.

from protein digestion, 69
related ketoacids, 162–163, 162f
R groups, 62–63, 62f, 63t, 64f–65f
secondary amine, 65f
signal molecules from, 316, 316f
stereochemistry, 62, 62f
thioether, 65f
zwitterions, 67
Amino-acyl-tRNA synthase mechanism, in protein synthesis, 349, 349f
Amino-acyl-tRNA synthetases, 342, 343f, 344
Amino end (N end), 68
Ammonia (NH₃), 22
Ammonia (NH₃) assimilation, 282–283, 282f
Ammonium ion (NH₄₊), 25
AMP (adenosine monophosphate)
    in adenylate kinase mechanism, 118, 118f
    measurement of, 114–115
    structure, 55, 57f
Amphipathic fatty acids, 33
Amphipathic molecules, lipid-water interactions of, 37–38, 37f
AMP kinase (AMPK), regulation of glycogen metabolism, 228–229
Amylopectin, 54
Amylose
    nonreducing end, 52, 52f
    reducing end, 52, 52f
    in starch, 54
Anabolic steroid, 269–270, 270f
Angiotensin II, 68–69, 69f
Animals
    carbon fixation in, 233b
    fats from, 31b
Annihilation, 178
Anomeric carbon, 47, 59
Antennae system of chlorophyll, light absorption and, 196–197, 197f
Antibiotics, prolyl isomerase binding, 352b
Anticodon, 342
Anticompetitive inhibition, 92, 93–94, 94f, 96b, 97f
Antimycin A, 170, 173f
Antiparallel β-sheets, 71, 72f
Apolipoprotein, 247
Aquaporins, 39
Arachidonate products, 271, 271f
Arachidonic acid, 31t
Arginase, half-life, 355t
Arginine, 64f
    degradation routes, 295–296, 296f
    properties, 63t
    synthesis, 301, 302f
Argininosuccinate synthetase, mechanism, 289–290, 290f
Arsenate poisoning, 143, 143f, 144f
Ascorbate (ascorbic acid; vitamin C), 373t
A site (acceptor site), 343f, 346, 347f
Asparagine, 63t, 64f
Asparagine synthetase, energy coupling mechanism, 115–116f
Aspartate, structure, 64f
Aspartate condensation reaction, 312, 313f
Aspartate transaminase, mechanism, 285–286, 285f
Aspartic acid, 63t
Aspirin, 98
Assays, 78, 86
Assimilation
    definition of, 282
    reaction types in, 282f, 283

ATP (adenosine triphosphate)
    energy coupling with, 115–119, 115t, 116f–118f
    formation, electron flow and, 8
    in glycolysis, 138
    hydrolysis reaction, 113–114, 114f, 114t
    as mobile cofactor, 120–121, 121f
    structure, 55, 57f
    synthesis, 166
        free energy change of, 172
        in mitochondria, 185
        in oxidative phosphorylation, 178–179, 179f, 180f
AT pairs, 322
ATP energy equivalents, estimating, 160b
Atpenins, 173b
ATP synthetase, 116
Autophagy, 354
Auxotropic organisms, 192
Avogadro's number, 362

**B**

Bacterial fatty acids, 272, 273f
Baker's yeast, 140
β-Barrel, 72
Bases
    definition of, 22
    strong, 22
    weak, 22
B complex vitamins, 139b, 373, 373t
Beer's law, spectrophotometer and, 86
Benzene, boiling point, 17t
Beta (form), 47
β-Hydroxybutrate, 254
β-Oxidation, 251–253, 252f
β–Sheet, 80
    antiparallel, 71
    parallel, 71
Bicarbonate buffering, 26
Bile, 247
Bile acids, 268–269
Bile salts, 38
Binding change, 179
Biochemical standard state, 113b
Biochemistry, origins of, 2
Biological systems, 10, 10f
Biopterins
    in phenylalanine degradation, 293
    structure, 295f, 315–316, 315f
Biotin (vitamin B₇)
    in gluconeogenesis, 230, 230f
    structure, 374f
    usage, 373t
2,3,-Bis-P-glycerate, 142, 142f
Bohr effect, 77, 78f
Boiling points, of liquids, 17, 17t
Bonds
    covalent, 15–16
    definition of, 16b
    hydrogen, 15–16, 15f
    ionic vs. covalent, 16b
    vibrational energy of, 21
Branched-chain amino acids. See Amino acids, branched-chain
Branched polysaccharides, 53–54, 54f. See also Glycogen
Branch points, 53–54, 54f, 218
Brown adipose tissue, 184b
Buchner, Eduard, 125b
Buffering regions
    definition of, 26
    in derivation of Henderson-Hasselbalch equation, 365
    in glycine titration curve, 64, 66f

Bundle sheath, 208–209
Butadiene, resonance in, 33
Butyric acid, 30–31, 31t

**C**

CAIR (carboxyaminoimidazole ribotide), 308
Calcineurin, 352b
Calmodulin, 73
Calories, 106b
Calvin cycle, 195–196
    carbon fixation to glyceraldehyde-P reactions, 203–204, 204f
    ribulose bisphosphate carboxylase reaction, 202–203, 202f, 203f
    RuBisCo formation from GAP, 204–205, 204f–207f, 207–208
    stoichiometry, 204, 204f
cAMP, 224
CAM plants, 210
Cap, 336–337, 337f
CAP (carbamoyl-P), formation, 306, 306f
Capping, 331
Carbamoyl-P (CAP), formation, 306, 306f
Carbamoyl-P synthetase I (CPS I), 306–3076f
Carbamoyl-P synthetase II (CPS II), 306–3076f
Carbohydrates
    definition of, 44
    derivatives
        simple modifications, 54–55, 55f, 56f
        substitutions, 55–57, 56f–58f
    metabolic pathways, 215f
    monosaccharides, 44–45, 44t, 45f
    polysaccharides, 51–54, 52f, 54f
    terminology, 44
    weight, compared to lipids, 33
Carbon, electronegativity, 14
Carbon cycle, 192
Carbon dioxide (CO₂)
    C3 plants and, 208, 208f
    C4 plants and, 208–209, 209f
    generation, in CAM plants, 210
    incorporation into carbohydrates. See Carbon fixation
    lack of polarity in, 15
    transformation to glyceraldehyde-P, 203–204, 204f
Carbon dioxide cycle, global, 8, 8f
Carbon fixation
    in animals, 233b, 233f
    in C3 plants, 208, 208f
    in C4 plants, 208–209, 209f
    dark reactions of photosynthesis and, 195–196
    definition of, 193
    transformation of CO₂ to glyceraldehyde-P, 203–204, 204f
Carbon tetrachloride, boiling point, 17t
Carboxyaminoimidazole ribotide (CAIR), 308
Carboxyl end (C end), 68
Carboxyl group, in amino acids, 62, 62f
Carboxylic acid head group, dissociation and resonance, 33, 33f
Carnitine, 352
Carnitine acyl transferase I (CAT I), 250–251, 251f
Carnitine shuttle, 250–251, 251f
Carotene (vitamin A), 373t
Catalysis, equilibrium balance and, 83, 83f
Catalytic inhibition, 94b
Catalytic rate constant, 87
CAT I (carnitine acyl transferase I), 250–251, 251f
CDP-choline (cytidine 5' diphosphate choline), 266, 266f

Methylamine, 67
Methylcelluloses, 53
Methylene residues, lack of polarity in, 18, 18f
Methyl hydrogens, 154–155, 154f
MgATP, 114, 114f
Micelles
  bilayer formation, 38, 39f
  in lipid-water interactions, 37–38, 37f
Michaelis-Menten equation
  curve, midpoint of, 90, 90f
  derivation of, 88–89, 367, 369–370
  Double-reciprocal plot (Lineweaver-Burk plot), 96–97, 96f, 97f
  enzyme behavior and, 89–90, 89f
  $K_m$ (steady-state constant), 89, 90–91, 90f
  modified by inhibition, 92–93, 93f
Minor groove, 323, 324f
Mismatch repair, 331
Mitochondria
  control of, 185
  cytosolic NADH utilization, 185–188, 186f
  import of fatty acids into, 250–251, 251f
  phosphate exchange in, 181–182, 181f
  as reaction space, 9, 9f
  steps in fatty acid oxidation, 251–253, 252f
  superoxide formation in, 183–185
  typology, 167, 167f
Mitochondrial complexes, II, 173B
Mitochondrial electron flows, vs. photosynthetic electron flows, 194–195, 195f
Mitochondrial inner membrane
  electron and proton flow mechanisms, 172, 174–179, 174f–176f, 178f–180f
    complex I: proton pump, 172, 174–175, 174f–175f
    complex II: succinate dehydrogenase, 175
    complex III: loop mechanism, 175–177, 175f, 176f
    complex IV: pump and annihilation, 177–178, 178f
    complex V: ATP synthesis, 178–179, 179f, 180f
  in oxidative phosphorylation, 167, 167f
Mitochondrial membrane transport, 180–181, 181f, 182f
Mixed inhibition, 92
Mixed or noncompetitive inhibition, 94–96, 94b, 95f, 97f
Mobile cofactors
  pathway view and, 120–121, 121f
  in photosynthesis, 192
Mole, 362
Molecular biology, 320
Monoamine oxidase, 284, 284f
Monomers, 73
Monosaccharides. See also Sugars
  disaccharides, 48, 49f, 50–51, 50f, 51b
  nomenclature, 44t
  structure, 44–45, 45f
Morphine, 316, 316f
Motif (supersecondary structure), 71, 71b
mRNA, 350
mTOR pathway, in protein synthesis control, 356–357, 356f
Muscle
  fast or glycolytic, 222
  glycogen storage and, 54
  metabolism, regulation of, 237, 237b
Myoglobin, oxygen binding in, 74–77, 77f, 78f, 81
Myosin, 222, 355t
Myristic acid, in pizza, 34b

N
N-acetylglutamate, 291
NAD+
  in glycolysis, 138
  as mobile cofactor, 120–121, 121f
  structure, 120f
NAD-binding domain, 72–73
NADH (nicotinamide adenine dinucleotide)
  cytosolic utilization in mitochondria, 185–188, 186f
  as electron carrier, 119–120, 120f
  in glycolysis, 138
  high energy and, 7–8, 8f
  in Krebs cycle, 150, 152
  as mobile cofactor, 120–121, 121f
  in oxidative phosphorylation, 166, 168
  reactive oxygen species and, 377
NADH-Q reductase (complex I), 172, 174–175, 174f–175f
NAD+/NADH, 170
NADP+
  as mobile cofactor, 120–121
  in pentose phosphate shunt, 238, 238f
  in photosynthesis, 192
  structure, 120f
NADPH
  hydride for, 378
  as mobile cofactor, 120–121
  in photosynthesis, 192
  production, 241
  structure, 120f
NADP+/NADPH cofactor pair, in photosynthesis, 192, 194
NADP+/NADPH complex, 238
NaOH (sodium hydroxide), 22, 26
NDP (nucleoside diphosphate), 312
NDP kinase, 117–118
Near-equilibrium, 112
Negative cooperativity, 179
Negative supercoil, 325
N end (amino end), 68
Neutrons, charge, 362–363
NH3. See Ammonia
NH3 assimilation, 282–283, 282f
Niacin (vitamin B3), 373t, 374f
Nicotinamide adenine dinucleotide. See NADH
Nitric oxide synthesis, 316–317, 316f
Nitrification, 282
Nitrogen
  electronegativity, 14
  metabolism
    amino acid anabolism, 300–302, 301t, 302f, 303f, 304
    amino acid catabolism. See Amino acids, catabolism
    metabolically-irreversible nitrogen exchange reactions, 283–284, 284f
    near-equilibrium exchange reactions, 284–286, 285f
    nitrogen cycle, 281—283, 282f
    urea cycle, 286–291, 287f–290f
  pathways, other, 316–317, 316f
Nitrogenase, 282, 282f
Nitrogen cycle, 281—283, 282f
Nitrogen fixation, 281–282, 282f
Nitroglycerin, 281, 281f
NMR (nuclear magnetic resonance spectroscopy), 79
Noncompetitive (mixed) inhibition, 94b, 96b, 371–372, 371f
Noncooperative, 76
Nonenzymatic, 116b

Nonessential amino acids, 300, 301, 301t, 302f, 303f
Nonpolar lipids, 30
Nonreducing end, 52, 52f
Nonreducing sugars, 51b
Nonstandard free energy changes, 111–112, 112t
N side (phases), of thylakoid membrane, 193–194, 194f
Nuclear magnetic resonance spectroscopy (NMR), 79
Nuclear membrane, 9
Nucleic acids. See also DNA; RNA
  histones, 326–327, 327f
  replication, 327–331, 328f, 329f
  strand structures, 320–321, 320f, 321f
  supercoiling, 325–326, 326f
  transcription, 332–338, 332f–338f
Nucleophilic substitution, 99, 100f
Nucleosides, 55, 57f
Nucleotides
  metabolism, 304–312, 305f–310f, 313f–315f, 314–316
  structure, 55–56, 56f

O
Octanoate, critical micelle concentration, 38t
Oils, 33
Okazaki fragments, 328
Oleate, 253–254, 253b, 254f
Oleic acid, 31t, 34b
Oleomargarine, 253b
Olestra, 253b
Oligosaccharides, 51
Omega-6 fatty acid, 31b
Omega fatty acids, 31b
OMP (orotidine 5'-monophosphate), 307
One-carbon metabolism
  serine and, 297–298, 298f
  tetrahydrofolate and, 296, 296f, 297, 297f
On the Origin of the Species, 9
Operon, 335
Orbit, 363
Orbitals, 363
Organic, use of term, 2b
Organic chemistry, definition of, 2b
Organic compounds, in vitro synthesis of, 2
Organic foods, 2b
OriC, 328
Ornithine decarboxylase
  half-life, 355t
  in putrescine formation, 355, 355f
Orotidine 5'-monophosphate (OMP), 307
Osmolarity, 53
Osmosis, 38–40, 39f, 40b
Osmotic pressure, 39–40
Outer mitochondrial membrane, 150
Overlapping double bonds, 32, 32f
Oxaloacetate
  in citrate synthase reaction, 154
  indirect transfer from mitochondria to cytosol, 233–234, 234f
Oxidation
  coupling with phosphorylation, 182
  definition of, 119, 167
  of fatty acids, 248–254, 249f–254f
β-Oxidation, 251–253, 252f
Oxidation-reduction reactions (redox reactions), 119
Oxidative phosphorylation, 165–190
  control of mitochondria, 185
  coupled processes, 165–167, 166f
  coupling mechanism, 182
  definition of, 165